Approaches and Strategies in Next Generation Science Learning

Myint Swe Khine
University of Bahrain, Baharain

Issa M. Saleh
University of Bahrain, Baharain

Managing Director:	Lindsay Johnston
Editorial Director:	Joel Gamon
Book Production Manager:	Jennifer Yoder
Publishing Systems Analyst:	Adrienne Freeland
Development Editor:	Christine Smith
Assistant Acquisitions Editor:	Kayla Wolfe
Typesetter:	Christy Fic
Cover Design:	Jason Mull

Published in the United States of America by
Information Science Reference (an imprint of IGI Global)
701 E. Chocolate Avenue
Hershey PA 17033
Tel: 717-533-8845
Fax: 717-533-8661
E-mail: cust@igi-global.com
Web site: http://www.igi-global.com

Copyright © 2013 by IGI Global. All rights reserved. No part of this publication may be reproduced, stored or distributed in any form or by any means, electronic or mechanical, including photocopying, without written permission from the publisher. Product or company names used in this set are for identification purposes only. Inclusion of the names of the products or companies does not indicate a claim of ownership by IGI Global of the trademark or registered trademark.

   Library of Congress Cataloging-in-Publication Data

Approaches and strategies in next generation science learning / Myint Swe Khine and Issa M. Saleh, editors.
    p. cm.
  Includes bibliographical references and index.
  Summary: "This book examines the challenges involved in the development of modern curriculum models, teaching strategies, and assessments in science education in order to prepare future students in the 21st century economies"--Provided by publisher.
  ISBN 978-1-4666-2809-0 (hardcover) -- ISBN 978-1-4666-2810-6 (ebook) -- ISBN 978-1-4666-2811-3 (print & perpetual access) 1. Science--Study and teaching. 2. Classroom environment. 3. Interaction analysis in education. 4. Computer-assisted instruction. 5. Internet in education. I. Khine, Myint Swe. II. Saleh, Issa M.
  Q181.A1A696 2013
  507.1--dc23
                                                            2012032522

British Cataloguing in Publication Data
A Cataloguing in Publication record for this book is available from the British Library.

All work contributed to this book is new, previously-unpublished material. The views expressed in this book are those of the authors, but not necessarily of the publisher.

Editorial Advisory Board

Kumar Laxman, *University of Auckland, New Zealand*
Ziad Shaker, *University of North Texas, USA*
John Wilkinson, *University of Bahrain, Bahrain*
Baohui Zhang, *Nanjing University, China*

Table of Contents

Preface xiv

Section 1
Theoretical Foundation and Conceptual Frameworks

Chapter 1
Self-Regulated Learning as a Method to Develop Scientific Thinking 1
    Erin E. Peters Burton, George Mason University, USA

Chapter 2
Multiple Perspectives for the Study of Teaching: A Conceptual Framework for Characterizing and
Accessing Science Teachers' Practical-Moral Knowledge 27
    Sara Salloum, Long Island University – Brooklyn, USA

Chapter 3
Teaching a Socially Controversial Scientific Subject: Evolution 52
    Hasan Deniz, University of Nevada Las Vegas, USA

Chapter 4
A Theoretical and Methodological Approach to Examine Young Learners' Cognitive Engagement in
Science Learning 64
    Meng-Fang Tsai, Chung-Yuan Christian University, Taiwan
    Syh-Jong Jang, Chung-Yuan Christian University, Taiwan

Section 2
Modeling, Simulation, and Games

Chapter 5
Argumentation and Modeling: Integrating the Products and Practices of Science to Improve Science
Education 85
    Douglas B. Clark, Vanderbilt University, USA
    Pratim Sengupta, Vanderbilt University, USA

Chapter 6

Reification of Five Types of Modeling Pedagogies with Model-Based Inquiry (MBI) Modules for High School Science Classrooms 106

    Todd Campbell, University of Massachusetts Dartmouth, USA
    Phil Seok Oh, Gyeongin National University of Education, Korea
    Drew Neilson, Logan High School, USA

Chapter 7

Why *Immersive, Interactive* Simulation Belongs in the Pedagogical Toolkit of "Next Generation" Science: Facilitating Student Understanding of Complex Causal Dynamics 127

    M. Shane Tutwiler, Harvard University, USA
    Tina Grotzer, Harvard University, USA

Chapter 8

Teachers and Teaching in Game-Based Learning Theory and Practice 147

    Mario M. Martinez-Garza, Vanderbilt University, USA
    Douglas B. Clark, Vanderbilt University, USA

Section 3
Curriculum Innovations

Chapter 9

Opening Both Eyes: Gaining an Integrated Perspective of Geology and Biology 165

    Renee M. Clary, Mississippi State University, USA
    James H. Wandersee, Louisiana State University, USA

Chapter 10

Promoting the Physical Sciences among Middle School Urban Youth through Informal Learning Experiences 184

    Angela M. Kelly, Stony Brook University, USA

Chapter 11

Rooted in Teaching: Does Environmental Socialization Impact Teachers' Interest in Science-Related Topics? 205

    Lisa A. Gross, Appalachian State University, USA
    Joy James, Appalachian State University, USA
    Eric Frauman, Appalachian State University, USA

Chapter 12

Analysis of Discourse Practices in Elementary Science Classrooms using Argument-Based Inquiry during Whole-Class Dialogue 224

    Matthew J. Benus, Indiana University Northwest, USA
    Morgan B. Yarker, University of Iowa, USA
    Brian M. Hand, University of Iowa, USA
    Lori A. Norton-Meier, University of Louisville, USA

Section 4
Evaluation and Assessment Issues

Chapter 13
Next Generation Science Assessment: Putting Research into Classroom Practice 247
    Edward G. Lyon, Arizona State University, USA

Chapter 14
A Tool for Analyzing Science Standards and Curricula for 21st Century Science Education 265
    Danielle E. Dani, Ohio University, USA
    Sara Salloum, Long Island University, USA
    Rola Khishfe, American University of Beirut, Lebanon
    Saouma BouJaoude, American University of Beirut, Lebanon

Chapter 15
Measuring and Facilitating Highly Effective Inquiry-Based Teaching and Learning in Science Classrooms 290
    Jeff C. Marshall, Clemson University, USA

Compilation of References 307

About the Contributors 350

Index 356

Detailed Table of Contents

Preface xiv

Section 1
Theoretical Foundation and Conceptual Frameworks

Chapter 1
Self-Regulated Learning as a Method to Develop Scientific Thinking 1
  Erin E. Peters Burton, George Mason University, USA

The development of skills and the rationale behind scientific thinking has been a major goal of science education. Research has shown merit in teaching the nature of science explicitly and reflectively. In this chapter, the author discusses how research in a self-regulated learning theory has furthered this finding. Self-regulation frames student learning as cycling through three phases: forethought (cognitive processes that prepare the learner for learning such as goal setting), performance (employment of strategies and self-monitoring of progress), and self-reflection (evaluation of performance with the goal). Because students have little interaction with the inherent guidelines that drive the scientific enterprise, setting goals toward more sophisticated scientific thinking is difficult for them. However, teachers can help students set goals for scientific thinking by being explicit about how scientists and science function. In this way, teachers also explicitly set a standard against which students can self-monitor their performance during the learning and self-evaluate their success after the learning. In addition to summarizing the research on learning and teaching of self-regulation and scientific thinking, this chapter offers recommendations to reform science teaching from the field of educational psychology.

Chapter 2
Multiple Perspectives for the Study of Teaching: A Conceptual Framework for Characterizing and Accessing Science Teachers' Practical-Moral Knowledge 27
  Sara Salloum, Long Island University – Brooklyn, USA

This chapter outlines a framework that characterizes science teachers' practical-moral knowledge utilizing the Aristotelian concept of phronesis/practical wisdom. The meaning of phronesis is further explicated and its relevance to science education are outlined utilizing a virtue-based view of knowledge and practical hermeneutics. First, and to give a background, assumptions about teacher knowledge from a constructivist and sociocultural perspective are outlined. Second, the Aristotelian notion of phronesis (practical wisdom) is explicated, especially in terms of how it differs from other characterizations of practical knowledge in science education and how it relates to practical-moral knowledge. Finally, the authors discuss how the very nature of such practical-moral knowledge makes it ambiguous and hard to articulate, and therefore, a hermeneutic model that explores teachers' practical-moral knowledge indirectly by investigating teachers' commitments, interpretations, actions, and dialectic interactions is outlined. Implications for research and teacher education are outlined. Empirical examples are used to demonstrate certain points. A virtue-based view of knowledge is not meant to replace others, but as a means to enrich the understandings of the complexity of teacher knowledge and to enhance the effectiveness of teacher educators.

Chapter 3
Teaching a Socially Controversial Scientific Subject: Evolution 52
Hasan Deniz, University of Nevada Las Vegas, USA

This chapter explores teachers' and students' acceptance and understanding of evolutionary theory by using conceptual ecology (Toulmin, 1972) as a theoretical lens. Demastes, Good, and Peebles (1995) describe the conceptual ecology for evolutionary theory. Acceptance of evolutionary theory is part of this conceptual ecology, and this conceptual ecology also contains the following five components: (1) prior conceptions related to evolution (understanding of evolutionary theory); (2) scientific orientation (degree to which the learner organizes his/her life around scientific activities); (3) view of the nature of science; (4) view of the biological world in competitive and causal terms as opposed to aesthetic terms; and (5) religious orientation. A complex web of connections among components of conceptual ecology for evolutionary theory influences one's acceptance and understanding of evolutionary theory. Therefore, studying the relationship between acceptance and understanding of evolutionary theory as a part of the conceptual ecology for evolutionary theory is more promising than studying acceptance of evolutionary theory in isolation. Moreover, studying acceptance of evolutionary theory as an integral part of the conceptual ecology may enable us to explain why some teachers and students show a high degree of acceptance and others show a low degree of acceptance.

Chapter 4
A Theoretical and Methodological Approach to Examine Young Learners' Cognitive Engagement in Science Learning 64
Meng-Fang Tsai, Chung-Yuan Christian University, Taiwan
Syh-Jong Jang, Chung-Yuan Christian University, Taiwan

Children start learning science in formal school settings as early as kindergarten ages. Scientific literacy is an essential component in forming science curriculum for students' knowledge and understanding of scientific concepts across grades and with different scientific disciplines. Students' cognitive engagement is displayed through different levels of cognitive processes involved during their knowledge construction. Much of the research in students' cognitive engagement of science learning focuses on older students and the use of self-report measures on their cognitive strategies. Research on young learners' cognitive engagement in science learning is missing in current literature. Part of the reason may be young learners are just beginning to develop a repertoire of cognitive strategies as well as lacking sufficient linguistic competence to accurately articulate the cognitive strategies involved in their science learning. The chapter introduces a theoretical and methodological approach, quantitative content analysis, to fulfill the gap. The chapter manifests the approach through three perspectives: (a) methodological approaches being employed in current studies of cognitive engagement in science learning, (b) the need for a different methodological approach for young learners, and (c) how this particular approach can be adopted in future research on young learners' cognitive engagement in science learning.

Section 2
Modeling, Simulation, and Games

Chapter 5

Argumentation and Modeling: Integrating the Products and Practices of Science to Improve Science Education 85

    Douglas B. Clark, Vanderbilt University, USA
    Pratim Sengupta, Vanderbilt University, USA

There is now growing consensus that K12 science education needs to focus on core epistemic and representational practices of scientific inquiry (Duschl, Schweingruber, & Shouse, 2007; Lehrer & Schauble, 2006). In this chapter, the authors focus on two such practices: argumentation and computational modeling. Novice science learners engaging in these activities often struggle without appropriate and extensive scaffolding (e.g., Klahr, Dunbar, & Fay, 1990; Schauble, Klopfer, & Raghavan, 1991; Sandoval & Millwood, 2005; Lizotte, Harris, McNeill, Marx, & Krajcik, 2003). This chapter proposes that (a) integrating argumentation and modeling can productively engage students in inquiry-based activities that support learning of complex scientific concepts as well as the core argumentation and modeling practices at the heart of scientific inquiry, and (b) each of these activities can productively scaffold the other. This in turn can lead to higher academic achievement in schools, increased self-efficacy in science, and an overall increased interest in science that is absent in most traditional classrooms. This chapter provides a theoretical framework for engaging students in argumentation and a particular genre of computer modeling (i.e., agent-based modeling), illustrates the framework with examples of the authors' own research and development, and introduces readers to freely available technologies and resources to adopt in classrooms to engage students in the practices discussed in the chapter.

Chapter 6

Reification of Five Types of Modeling Pedagogies with Model-Based Inquiry (MBI) Modules for High School Science Classrooms 106

    Todd Campbell, University of Massachusetts Dartmouth, USA
    Phil Seok Oh, Gyeongin National University of Education, Korea
    Drew Neilson, Logan High School, USA

It has been declared that practicing science is aptly described as making, using, testing, and revising models. Modeling has also emerged as an explicit practice in science education reform efforts. This is evidenced as modeling is highlighted as an instructional target in the recently released Conceptual Framework for the New K-12 Science Education Standards: it reads that students should develop more sophisticated models founded on prior knowledge and skills and refined as understanding develops. Reflecting the purpose of engaging students in modeling in science classrooms, Oh and Oh (2011) have suggested five modeling activities, the first three of which were based van Joolingen's (2004) earlier proposal: 1) exploratory modeling, 2) expressive modeling, 3) experimental modeling, 4) evaluative modeling, and 5) cyclic modeling. This chapter explores how these modeling activities are embedded in high school physics classrooms and how each is juxtaposed as concurrent instructional objectives and scaffolds a progressive learning sequence. Through the close examination of modeling in situ within the science classrooms, the authors expect to better explicate and illuminate the practices outlined and support reform in science education.

Chapter 7
Why *Immersive, Interactive* Simulation Belongs in the Pedagogical Toolkit of "Next Generation" Science: Facilitating Student Understanding of Complex Causal Dynamics 127
    M. Shane Tutwiler, Harvard University, USA
    Tina Grotzer, Harvard University, USA

Demonstration and simulation have long been integral parts of science education. These pedagogical tools are especially helpful when trying to make salient unseen or complex causal interactions, for example during a chemical titration. Understanding of complex causal mechanisms plays a critical role in science education (e.g. Grotzer & Basca, 2003; Hmelo-Silver, Marathe, & Liu, 2007; Wilensky & Resnick, 1999), but few curricula have been developed to expressly address this need (e.g. Harvard Project Zero, 2010). Innovative education technologies have allowed content designers to develop simulations that are both immersive and engaging, and which allow students to explore complex causal relationships even more deeply. In this chapter, the authors highlight various technologies that can be used to leverage complex causal understanding. Drawing upon research from both cognitive science and science education, they outline how each is designed to support student causal learning and suggest a curricular framework in which such learning technologies might optimally be used.

Chapter 8
Teachers and Teaching in Game-Based Learning Theory and Practice 147
    Mario M. Martinez-Garza, Vanderbilt University, USA
    Douglas B. Clark, Vanderbilt University, USA

Interest in game-based learning has grown dramatically over the past decade. Thus far, most of the focus has not included the role of teachers. This chapter first summarizes the theoretical research on game-based learning and the implications of that research for the role of teachers. The authors next review the game-based learning literature that has specifically articulated a role for teachers or achieved an empirical description of teacher action within a game-based learning context. They then connect these accounts with more general research on teachers and technology use, elaborating on points of contact and identifying differences that may signal special challenges. Finally, the authors articulate an expanded role for teachers in game-based learning practices in terms of game-based learning research and new scholarship on the psychology of games.

Section 3
Curriculum Innovations

Chapter 9
Opening Both Eyes: Gaining an Integrated Perspective of Geology and Biology 165
    Renee M. Clary, Mississippi State University, USA
    James H. Wandersee, Louisiana State University, USA

The focus of this chapter is an exploration of integrated geology and biology learning—from past to present. The chapter explains why active and integrated geological and biological learning became the lodestar of the authors' decade-long EarthScholars Research Group's research program. The authors argue that using an active and integrated geobiological pedagogical approach when teaching geology or biology provides natural opportunities for students to learn and do authentic scientific inquiry in a manner similar to how contemporary scientists conduct their work. The authors further review research that concerns the active, integrated geobiological science learning approach—in middle school, secondary, and college classrooms, laboratories, and field studies. The authors favor a gradual course transition to this pedagogy, while highlighting the advantages of adopting such an approach—both for teachers and students. Finally, the authors conclude the chapter with challenges and future directions in the design of active, integrated geobiological science learning environments.

Chapter 10
Promoting the Physical Sciences among Middle School Urban Youth through Informal Learning Experiences 184
    Angela M. Kelly, Stony Brook University, USA

Numerous reform efforts in STEM education have been targeted towards increasing the number of qualified STEM professionals in the U.S., which necessitates promoting science participation among secondary and post-secondary students. Some novel designs have focused on the middle school years, when students tend to lose interest in science and formulate opinions on science self-identification. This chapter describes the effectiveness of developing informal physical science experiences for middle school students in underserved urban communities. Several cohorts of students have participated in inquiry-based physics and chemistry weekend classes that incorporated authentic applications from the urban setting, field visits to scientists' laboratories and museums, advanced educational technology tools, and learning complex scientific concepts. Participants reported significant improvements in their attitudes, knowledge, and appreciation of the physical sciences, suggesting that well designed constructivist physical science programs are potentially transformative in improving students' academic self-efficacy, confidence, and persistence in science, and positional advantage. The potential of early, rigorous experiences with the physical sciences is explored as a means for improving science participation and diversifying the ranks of future scientists.

Chapter 11
Rooted in Teaching: Does Environmental Socialization Impact Teachers' Interest in Science-Related Topics? 205
    Lisa A. Gross, Appalachian State University, USA
    Joy James, Appalachian State University, USA
    Eric Frauman, Appalachian State University, USA

Research in Environmental Socialization (ES) and the impact of significant life experiences suggest that childhood play in outdoor environments shape later adult activities or career interests. Few studies have investigated how childhood experiences influence curricular interests of preservice and inservice teachers. This preliminary study examines what ES factors of teachers raised in rural and/or non-rural environments reveal about their interests in science topics and field-based learning opportunities. Results suggest that teachers growing up in rural areas were slightly less interested than non-rural teachers in field-based learning and expressed less experience with environmental education. Teachers with more ES experiences (e.g., played in the woods, built forts) expressed greater interest in science-related topics than those who had indicated fewer experiences. Rural teachers tended to have more ES experiences than non-rural teachers. The authors discuss how environmental socialization factors influence teacher preference for curricular programs specific to environmental and ecological topics and raise questions about the changing environmental socialization experiences of preservice and novice teachers.

Chapter 12
Analysis of Discourse Practices in Elementary Science Classrooms using Argument-Based Inquiry during Whole-Class Dialogue 224
> *Matthew J. Benus, Indiana University Northwest, USA*
> *Morgan B. Yarker, University of Iowa, USA*
> *Brian M. Hand, University of Iowa, USA*
> *Lori A. Norton-Meier, University of Louisville, USA*

This chapter discusses an analysis of discourse practices found in eight different elementary science classrooms that have implemented the Science Writing Heuristic (SWH) approach to argument-based inquiry. The analysis for this study involved examining a segment of whole-class talk that began after a small group presented its claim and evidence and ended when the discussion moved on to a new topic, or when a different group presented. The framework for the analysis of this whole-class dialogue developed through an iterative process that was first informed by previous analysis, review and modification of other instruments, and notable anomalies of difference from this data set. Each classroom was then rated using the Reform Teaching Observation Protocol (RTOP), which provided a score for the extent to which the teacher was engaged with reform-based science teaching practices. Our analysis shows that elements of whole-class dialogue in argument-based inquiry classrooms were different across varying levels of RTOP implementation. Overall, low level RTOP implementation (little evidence of reformed-based practice) had a question and answer format during whole class talk that rarely included discourse around scientific reasoning and justification. Higher levels of RTOP implementation were more likely to be focused on student use of scientific evidence to anchor and develop a scientific understanding of "big ideas" in science. These findings are discussed in relation to teacher professional development in argument-based inquiry, science literacy, and the teacher's and students' grasp of science practice.

Section 4
Evaluation and Assessment Issues

Chapter 13
Next Generation Science Assessment: Putting Research into Classroom Practice 247
> *Edward G. Lyon, Arizona State University, USA*

The recent release of science education documents such as A Framework for K-12 Science Education: Practices, Crosscutting Concepts, and Core Ideas (National Research Council, 2012) marks the transition into a new generation of science education. This transition necessitates a close look at how pre-college science teachers will assess a diverse group of students in ways that are consistent with science education reform. In this chapter, the authors identify current research in science assessment and employ assessment coherence, assessment use, and assessment equity as guiding principles to address the challenges of putting science assessment research into classroom practice. To exemplify these challenges, they describe a study where a research instrument designed to measure scientific reasoning skills was translated into a high school science classroom assessment. The goal of this chapter is to stimulate conversation in the science education community (researchers, assessment developers, teacher educators, administrators, and classroom teachers) about how to put science assessment research successfully into practice and to describe what next steps need to be taken, particularly around assessing diverse student populations.

Chapter 14
A Tool for Analyzing Science Standards and Curricula for 21st Century Science Education 265
> *Danielle E. Dani, Ohio University, USA*
> *Sara Salloum, Long Island University, USA*
> *Rola Khishfe, American University of Beirut, Lebanon*
> *Saouma BouJaoude, American University of Beirut, Lebanon*

Twentieth century curricula are no longer sufficient to prepare students for life and work in today's diverse, fast-paced, technologically driven, and media saturated world of the 21st century. This chapter presents a new framework for analyzing science standards and curricula to determine the extent of alignment with 21st Century essential understandings and skills. The Tool for Analyzing Science Standards and Curricula (TASSC) was developed using the conceptual frameworks proposed by the Partnership for 21st Century Skills, the Organization for Economic Co-Operation and Development, and the typology of knowledge proposed by Jurgen Habermas. Development of TASSC relied on an iterative process of refinement, testing, and discussions resulting in an instrument with three sections and related rating scales: content, skills, and additional curricular components. TASSC was piloted using middle school science standards and curricula in the context of two US states (Ohio and New York) and two Arab countries (Lebanon and Qatar). The analysis procedure and individual case study results are presented and discussed in the chapter.

Chapter 15
Measuring and Facilitating Highly Effective Inquiry-Based Teaching and Learning in Science Classrooms 290
> *Jeff C. Marshall, Clemson University, USA*

For the last decade or so there has been a huge push to incorporate best practice into the classroom. For science, this includes bringing effective inquiry-based instruction into all classrooms as a means to engage the learner. However, all inquiry instruction is not equal in terms of improving student achievement and conceptual development. This chapter explores how four critical constructs to learning (curriculum, instruction, discourse, and assessment) can be effectively measured and then used to guide more effective instructional practice. The Electronic Quality of Inquiry Protocol (EQUIP) is an instrument that can be used to measure and then to frame the discussion regarding the quality of inquiry-based instructional practice. Specifically, this chapter provides an overview of EQUIP, details the reliability and validity of EQUIP, shares a sample lesson that is analyzed using EQUIP, explores ways that EQUIP can help with teacher transformation relative to inquiry instruction, and addresses the relationship of EQUIP scores and student achievement data. There is a very high correlation between teacher performance on EQUIP and the ensuing student growth noted during an academic year.

Compilation of References 307

About the Contributors 350

Index 356

Preface

Science and technological advances continue to play a pivotal role in improving the quality of living standard throughout the world. It is eminent that the long-term prosperity of a nation is dependent upon scientifically literate persons as consumers together with talented innovators in the science and technology fields. The emphasis on learning science must accelerate in order to provide a solid foundation of scientific knowledge to its citizens. This book builds upon the international success of the publication *Fostering Scientific Habits of Mind: Pedagogical Knowledge and Best Practices in Science Education* (2009, Sense Publishers) and takes a closer look at the contemporary and innovative practices in teaching and learning science.

This book examines the opportunities, challenges, and barriers to the development of innovative curriculum models, promising teaching strategies, and authentic assessment in science education to prepare future students for the 21st-century economies. The chapters in this volume also feature cutting-edge research endeavours, seamless integration, and intelligence use of technology tools, including modeling and games, conceptual frameworks, developmental processes that are directly transferable to classroom practices, and a range of new visions and future scenarios. This book is designed to promote meaningful conversations, encourage synergistic collaborations, and exchange useable knowledge among academics, researchers, teachers, and school administrators on the next generation learning science. This book is divided into four sections. They deal with the specific aspects and approaches and strategies in next generation science learning. The theoretical foundation and conceptual frameworks are presented in section 1 and followed by the use of modeling, simulation, and games in section 2 of the book. The chapters in section 3 deal with the curriculum innovations in science courses, and section 4 covers evaluation and assessment issues.

THEORETICAL FOUNDATION AND CONCEPTUAL FRAMEWORKS

In Chapter 1, Erin Peters-Burton describes self-regulated learning as a method to develop scientific thinking. She notes that the development of skills and rationale behind scientific thinking has been the important goal in science education. In so doing, the research found that teaching the nature of science explicitly and reflectively can develop scientific thinking. The chapter presents a literature review on the methods used to teach the nature of science in the classrooms. She notes that learning the nature of science should not take the place of learning science content, rather they should be taught simultaneously. She urges secondary education must pursue instruction that intertwines both content instruction and instruction on ways of knowing in science. The chapter discusses how research in self-regulated learning theory has furthered this finding. The author presents that self-regulation frames student learning as cycling through three phases: forethought (cognitive processes that prepare the learner for learning such as goal setting), performance (employment of strategies and self-monitoring of progress), and self-reflection (evaluation of performance with the goal). She notes that it is possible for teachers to help students in setting goals for scientific thinking by making it explicit how scientists and science function. In this way, teachers also explicitly set a standard against which students can self-monitor their performance during the learning, and self-evaluate their success after the learning. The chapter concludes that self-regulation of the nature of science has great potential to support students in developing appropriate strategies to become more science-minded learners.

Sara Salloum from Long Island University presents the multiple perspectives for the study of science in Chapter 2. The purpose of the chapter is to elucidate a framework for understanding teaching practice as more than an arena for the application of theoretical knowledge and sets of practical-moral knowledge. When outlining the framework, the author uses the Aristotelian concept of phronesis/practical wisdom. The chapter begins with providing assumptions about teacher knowledge from a constructivist and sociocultural perspective. This is followed by an explanation on the Aristotelian notion of phronesis (practical wisdom), especially in terms of how it differs from other characterizations of practical knowledge in science education and how it relates to practical-moral knowledge. The author discusses how the very nature of such practical-moral knowledge deems it ambiguous and hard to articulate. The author proposes a hermeneutic model that explores teachers' practical-moral knowledge indirectly by investigating teachers' commitments, interpretations, actions, and the dialectic interactions between them. Implications for research and teacher education are also discussed. The chapter concludes that a virtue-based view of knowledge is not meant to replace others, but as a means to enrich our understandings of the complexity of teacher knowledge and to enhance our effectiveness as teacher educators.

"Teaching a Socially Controversial Scientific Subject: Evolution" by Hasan Deniz (Chapter 3) discusses what educators need to consider when teaching evolution. He points out the four domains that educators need to know when teaching evolution. These are conceptual, epistemic, socio-cultural, and religious domains. Deniz explains each domain in detail within the chapter. He then addresses students' alternative conceptions about evolution, such as all evolutionary change is adaptive, progressive, and teleological. Then Hassan explores why students' attitudes towards evolution are a direct result of students' religious perspectives, naïve nature of science views, and social-cultural backgrounds. The author argues that in order for students to change their view about evolution, they need to move from their alternative conceptions toward scientifically accepted conceptions. He adds that for the condition of conceptual change to take place, the learned need to go through four stages according to Posner, Strike, Hewson, and Gertzog (1982). These are: (1) dissatisfaction – this condition requires that the learner fails to make sense of some event with his/her existing conception; (2) intelligibility – this condition necessitates that the learner has some understanding of the new conception; (3) plausibility – this condition is satisfied when the learner accepts the new conception; and (4) fruitfulness – this condition emphasizes that the learner should be able to use the new conception to explain novel situations as well as the situations that were formerly explained by the new conception. Then, the author compares when teachers accept the evolution theory to when teaches do not accept and how that impacts the amount of time teachers' spent teaching the evolution theory. The author supports his argument by looking at the work of Berkman and Plutzer (2010), who explore the correlation of the adoption of the NAS positions and the instructional time spent on evolution. Finally, the author warns that the conceptual change model can be a problem when it comes to "assimilate students into the culture of Western science at the expense of their indigenous culture."

In "A Theoretical and Methodological Approach to Examine Young Learners' Cognitive Engagement in Science Learning" (Chapter 4), Meng-Fang Tsai and Syh-Jong Jang argue that young learners lack "sufficient linguistic competence to accurately articulate the cognitive strategies involved in their science learning." They note that current measures for cognitive engagement are scarce and usually better suited for older students. Furthermore, scientific studies of young children's cognitive development are also rare. This results in a lack of understanding about how young children learn science. The chapter is divided into five parts. First, the authors give background on the theoretical framework, definitions of cognitive engagement, similar theories, and introduce the question or the problem of study. Tsai and Jang then explore several definitions for cognitive engagement for the purposes of the study. For example, Fredricks, Blumenfeld, and Paris (2004) define cognitive engagement as "originated from school engagement that includes three types of engagement: behavioral, emotional, and cognitive." Later they argue, "behavioral engagement is defined as students' behaviors identified in relation to their engagement such as school attendance, and participation in school activities." On the other hand, Pintrich and Schrauben (1992) defined cognitive engagement "as to include cognitive and motivational component." The authors then look at similar theories, such as perspectives of children's development, information processing theory, and Vygotsky. After looking at several methodological approaches, the authors introduce a new methodological approach to study how young learners engage in science. The new method codes qualitative data and statistical analysis to quantify traditionally qualitative data.

MODELING, SIMULATION, AND GAMES

The chapters in section 2 present the use of modeling, simulation, and games to enhance science learning. "Argumentation and Modeling: Integrating the Products and Practices of Science to Improve Science Education" by Clark and Sengupta (Chapter 5) argues that science education should focus on epistemic and representational practices of scientific inquiry and support their argumentation by examining the work of Duschl (2008), as well as Lehrer and Schauble (2006). The main focus of the chapter is argumentation and computational modeling. The authors believe that engaging students in inquiry-based activities that support the learning of complex science requires having argumentation and computational modeling. This can help to improve "higher academic achievement in schools, increased self-efficacy in science, and an overall increased interest in science." The chapter provides a theoretical framework for engaging students in argumentation and uses agent-based modeling as an example to demonstrate modeling. The authors also provide free technologies that can be adopted in science classrooms when students are exploring argumentation and modeling. According to Clark and Sengupta, students should understand three core ideas in order to grasp the implication of Argumentation and Modeling in the teaching and learning of science. "First, modeling is the central enterprise, purpose, and goal of science. Second, argumentation is the practice that allows scientists to determine the fit of their models with the world. Third, communities of scientists evaluate models, methods, and evidence through argumentation using shared criteria and analytical approaches developed and agreed upon by the community." For the first core idea, the authors agree with Hestenes (1993) that modeling plays a great role in facilitating the understanding of complex concepts in science. As for the second core idea, how argumentation facilitates in determining how the module fits, the authors use the Argument-Driven Inquiry (ADI) approach to show how modeling can be applied in a real world setting. Finally, the authors believe that their proposed approach, ADI, can be evaluated by the scientific community to illustrate the third insight.

In Chapter 6, Campbell and his colleagues present reification of five types of modeling pedagogies with Model-Based Inquiry (MBI) modules for high school science classrooms. It has been documented that practicing science is aptly described as making, using, testing, and revising models. Modeling has also emerged as an explicit practice in science education reform efforts. Modeling has been highlighted as an instructional target in the recently released *Conceptual Framework for the New K-12 Science Education Standards* and it states that students should develop more sophisticated models founded on prior knowledge and skills and refined as understanding develops. Reflecting the purpose of engaging students in modeling in science classrooms, Oh and Oh (2011) have suggested five modeling activities, the first three of which are based on van Joolingen's (2004) earlier proposal: (1) exploratory modeling, (2) expressive modeling, (3) experimental modeling, (4) evaluative modeling, and (5) cyclic modeling. This chapter explores how these modeling activities are embedded in high school physics classrooms and how each is juxtaposed as concurrent instructional objectives and scaffolds in a progressive learning sequence. Through the close examination of modeling in situ within science classrooms, the authors expect to better explicate and illuminate the practices outlined and support reform in science education.

In Chapter 7, Shane Tutwiler and Tina Grotzer from Harvard Graduate School of Education discuss why immersive, interactive simulation belongs in the pedagogical toolkits of next generation science. They note that pedagogical tools such as demonstration and simulation have long been integral parts of science education, and science teachers found that these tools are helpful to make salient unseen or complex causal interactions, for example during a chemical titration. Understanding of complex causal mechanisms plays a critical role in science education (e.g. Grotzer & Basca, 2003; Hmelo-Silver, Marathe, & Liu, 2007; Wilensky & Resnick, 1999), but few curricula have been developed to address this need. It seems that in recent years, content designers developed simulations that are both immersive and engaging, and which allow students to explore complex causal relationships deeply. In their chapter, they highlight various technologies that can be used to leverage complex causal understanding. They analyze the interactive simulation program, SimCity 4, which offers a virtual simulation experience with the underlying complex causal structures. They also introduce another simulation program, EcoMUVE, which they have built and are testing. EcoMUVE was designed to teach ecosystem science concepts and to help students develop expert patterns of scientific reasoning about the causal dynamics within the ecosystem. Drawing upon research from both cognitive science and science education, the authors conclude the chapter with suggested curricular framework to use immersive and interactive simulation programs effectively.

Interest in game-based learning has grown dramatically over the past decade. There have been numerous research articles published not only in educational research journals but also in other interdisciplinary publications. Educators recognize that if the games are properly embedded and used in the curriculum, they can be an effective educational tool for teaching digital generation students. However, current application of games stresses students' involvement in the instructional process, but most of the focus has not included teachers. In Chapter 8, Martinez-Garza and Clark first summarize the theoretical research on game-based learning and the implications of that research for the role of teachers. The authors next review the game-based learning literature that has specifically articulated a role for teachers or achieved an empirical description of teacher action within a game-based learning context. The authors then connect these accounts with more general research on teachers and technology use, elaborating on points of contact and identifying differences that may signal special challenges. They point out that teachers should have more input into the forms and content of learning games, but the current situation still relies on the commercial game developers. The chapter concludes that although games hold potential to enhance education, teachers have not been sufficiently supported in using games for learning in the classroom. It also finds that teachers' expertise have not been optimally leveraged in the design of games for learning. The authors suggest that teachers have much to offer to the game design process, and if the teachers and games designers work together, instruction that focuses on engaging and deep meaningful learning for students can become a reality.

CURRICULUM INNOVATIONS

"Opening Both Eyes: Gaining an Integrated Perspective of Geology and Biology" by Renee M. Clary and James H. Wandersee (Chapter 9) focuses on the idea of integrating geology and biology in teaching science subjects. The authors start the chapter with a brief history about the integration of geology with biology. The authors then outline several benefits of this integration, such as providing natural opportunities for students to do an authentic scientific inquiry. Throughout the chapter, Clary and Wandersee explore the literature on the integrated geobiological science learning approach—in middle school, secondary, and college classrooms, laboratories, and field studies. In middle and secondary schools, the authors suggest having "Interdisciplinary scientific portals, the use of active learning and inquiry-based projects, and the incorporation of the history and philosophy of science." Teachers can easily adopt most of the activities for middle school for high school by making it more rigorous. By analyzing students' survey responses, Clary and Wandersee discover strong interest in the local landscape, large magnitude events, and unusual specimens during the high school years. Moreover, according to the authors' previous work (Clary & Wandersee, 2008a), fossils, dinosaurs, and sand can help to address several scientific constraints in high school science classrooms. The authors also suggest "*student-led* inquiry in which students generate their own questions and design subsequent investigations" as a pedagogical approach where teachers try to connect scientific concepts for high school students. Moreover, the authors argue that in order to have an effective student-led inquiry, teachers have to tap into students' prior knowledge and try to link it with their existing cognitive frameworks. In university courses, the authors mention the lack of inquiry learning by looking at the work of Drew (2011). University science courses tend to be lecture-driven and intensive in mathematics. The authors did a study of 515 students over three semesters using their Petrified Wood Survey (PWS). The results show that students gained conceptual understanding in the areas of geologic time, evolution, and fossilization processes. The content area with the smallest gains was geochemistry, according to the study. The authors recommend a gradual introduction of their approach in a curriculum. They point out that when integrating geology and biology, teachers benefit by having more motivated and interested students in classrooms. As for students, integration will help to facilitate active learning, a holistic understanding of how science actually works, and a more accurate understanding of the nature of science. Finally, students will benefit of having an "in-depth understanding of carefully selected, important scientific constructs."

"Promoting the Physical Sciences among Middle School Urban Youth through Informal Learning Experiences" by Angela M. Kelly (Chapter 10) looks at the issue of underrepresentation of minorities in Science, Technology, Engineering, and Mathematics (STEM) careers. Kelly traces the problem back to the types of courses that students are taking in middle and high school. She notes that most students do not take chemistry and physics as electives in high school and that most middle schools focus on life sciences rather than physical sciences. According to the authors, this has resulted in the under representation of minority students in STEM careers. To address this concern, Kelly introduces a study that explores the effectiveness of exposing middle school students in underserved urban communities to informal physical science experiences. The study took place over a weekend. The activities in the study included "incorporated authentic applications from the urban setting, field visits to scientists' laboratories and museums, advanced educational technology tools, and learning complex scientific concepts." The result of the study shows significant improvements in students' attitudes, knowledge, and appreciation of the physical sciences. Improved students' academic self-efficacy, confidence, and persistence in science, and positional advantage were also noted. The chapter is divided into three sections. The first section explores studies of trends in physical sciences among American high school students and the importance of chemistry and physics in STEM. The author states that there has been an increase in the number of student enrollments in high school sciences in the U.S. as whole. For example, a study done by The National Science Board (2006) that looks at transcripts between 1990 and 2000 shows that the percentage of high school graduates completing a chemistry course rose from 45% to 63% and a physics course rose from 21.5% to 31%. The second section of the chapter explores the context for the study, which is the program structure of the physics and chemistry coursework at the Bronx Institute. Finally, the author concludes the chapter by examining the effectiveness of the program and possible future replication.

In Chapter 11, "Rooted in Teaching: Does Environmental Socialization Impact Teachers' Interest in Science-Related Topics?" Lisa Gross, Joy James, and Eric Frauman describe a research study about Environmental Socialization (ES) and try to answer the question, does significant life experience affect adult activities or career interests. The authors observed that very few previous studies examine the impact of childhood experience on curricular interests of pre-service and in-service teachers. The chapter explores what "ES factors of teachers raised in rural and/or non-rural environments reveal about their interests in science-topics and field-based learning opportunities." The study investigates differences between teachers who grew up in a non-rural environment versus a rural environment and their respective interest in: 1) field-based learning and environmental education, 2) teaching science related topics, 3) environmental socialization, and 4) field-based learning and environmental education. The instrument was developed by students enrolled in an undergraduate level recreation assessment course, two representatives from environmental education facility, and three faculty members representing two college departments (education and recreation). The survey instrument had 27 questions and included demographic information, environmental socialization measures, and field-based topic interest questions. The sample size for the study was 88 teachers and student teachers, 45 with rural upbringings and 39 non-rural. The result of the study indicated that teachers who grew up in a non-rural environment have slightly higher interest in field-based learning and environmental education compared to teachers from a rural environment, which was contrary to what the authors anticipated. With respect to interest in teaching science-related topics, non-rural teachers expressed greater interest in 6 out of 10 of the science-related topics compared to rural teachers, which was also contrary to what the authors anticipated to find. The results also indicated a similarity in both rural and non-rural teachers in terms of environmental socialization experiences. Finally, the results showed similarity in both groups in terms of their interest in science-related topics as well. In other words, those teachers who are interested in environmental experiences tend to be more interested in science-related topics in general.

"Analysis of Discourse Practices in Elementary Science Classrooms during Whole-Class Discussion" by Matthew J. Benus, Morgan B. Yarker, Brian M. Hand, and Lori A. Norton-Meier (Chapter 12) explains the discourse practice in eight elementary science classrooms that applied the "Science Writing Heuristic (SWH) approach to argument-based inquiry." The chapter looks at the discourse that takes place after finishing a topic in science and before moving to a new topic. The study uses Reform

Teaching Observation Protocol (RTOP) to rate the teacher engagement with the reform-based science teaching practices. According to the authors, the outcome of the study indicates that whole-class dialogue in argument-based inquiry classrooms differed across classrooms based on the degree of implementation of RTOP. Low-level RTOP implementation was correlated with a low level of student engagement in "discourse around scientific reasoning and justification." In addition, the authors emphasize that in order to fully develop elementary-aged students with reasoning in scientific arguments takes time, courage, and ongoing professional development for teachers. Moreover, the authors argue that teachers ultimately determine the success of discourse practices in classrooms.

EVALUATION AND ASSESSMENT ISSUES

Lyon, in Chapter 13, begins by highlighting the recent release of science education documents such as *A Framework for K-12 Science Education: Practices, Crosscutting Concepts, and Core Ideas* (National Research Council, 2012). that mark the transition into a new generation of science education. This requires science educators to look at how pre-college science teachers will assess a diverse group of students in ways that are consistent with science education reform. In the chapter, the author notes that in enhancing next generation science assessment, teacher educators need to consider challenges science teachers will face when employing assessment in the diverse and dynamic classroom. The author identifies current research in science assessment and employs assessment coherence, assessment use, and assessment equity as guiding principles to address the challenges of putting science assessment research into classroom practice. To exemplify these challenges, the author describes a study where a research instrument designed to measure scientific reasoning skills was translated into a high school science classroom assessment. The study took place as part of the Assessing Scientific Inquiry and Leadership Skills (AScILS) project. This project was developed to study high school and undergraduate programs that promote entry into biomedical careers, especially for minority students. The chapter aims to stimulate conversation in the science education community (researchers, assessment developers, teacher educators, administrators, and classroom teachers) about how to put science assessment research successfully into practice and to describe what next steps need to be taken, particularly around assessing diverse student populations.

In Chapter 14, "A Tool for Analyzing Science Standards and Curricula for 21st Century Science Education," by Danielle E. Dani, Sara Salloum, Rola Khishfe, and Saouma BouJaoude argue that conceptualizations of scientific literacy in the curricula in the 20th century is not enough to prepare students for the 21st century because the set of skills that were required for the 20th century are not the same as those needed in the 21st century. The authors then discuss the essential understandings and skills that are needed for the 21st century. They support their argument by looking at work that was done by the American Association for the Advancement of Science (1993), Bybee (1997), and the National Research Council (1996). Moreover, the authors argue that the Framework for the Analysis of Education Programs (FAEP) is inadequate for analyzing science standards and curricula. Hence, the focus of this chapter is the process of developing a new framework, the Tool for Analyzing Science Standards and Curricula (TASSC), to analyze science standards and curricula. The chapter explores the application of TASSC in multiple contexts such as two US states (Ohio and New York) and two Arab countries (Lebanon and Qatar). The study used middle school students rather than high school for two reasons. First, the unified nature of science courses in middle school compared to high school and because all middle school students have to take science courses. Secondly, students in middle school are just starting to develop "attitudes and dispositions towards science" compared to high school students. The result of the study indicated that all four contexts are at different stages in meeting the essential understandings and skills for the 21st century. Finally, the authors make it clear the intention for developing TASSC is not "for rank-ordering states with respect to the degree to which they incorporate 21st century essential understandings and skills," but to help reformers and curriculum designers focus on areas for further development.

Jeff Marshall's chapter (Chapter 15) draws attention to measuring and facilitating highly effective inquiry-based teaching and learning in science classrooms. The chapter begins with the importance of high-quality inquiry-based instructional practice into science classrooms as part of the educational reform efforts. As the National Research Council's (2012) framework for K-12 science education documented, new visions of teaching and learning include inquiry forms of instruction and integrate cross-disciplinary concepts and core ideas in learning. Marshall argues that for science this includes bringing effective inquiry-based instruction into all classrooms as a means to engage the learner. However, all inquiry instruction is not equal in terms of improving student achievement and conceptual development. This chapter explores how four critical constructs to learning (curriculum, instruction, discourse, and assessment) can be effectively measured and then used to guide more effective instructional practice. The Electronic Quality of Inquiry Protocol (EQUIP) is an instrument that can be used to measure and then to frame the discussion regarding the quality of inquiry-based instructional practice. Specifically, this chapter provides an overview of EQUIP, details the reliability and validity of EQUIP, shares a sample lesson that is analyzed using EQUIP, explores ways that EQUIP can help with teacher transformation relative to inquiry instruction, and addresses the relationship of EQUIP scores and student achievement data. There is a very high correlation between teacher performance on EQUIP and the ensuing student growth noted during an academic year. The author notes that the use of EQUIP as an instrument will guide teachers in their practice to a greater quantity and quality of inquiry-based instruction.

CONCLUSION

As pointed out by Lyon (in this book), recent publications of the National Research Council (2012) recommend a framework for K-12 science education that is built around three major dimensions. These include "scientific and engineering practices, crosscutting concepts that unify the study of science and engineering through their common application across fields, and disciplinary core areas of physical sciences, life sciences, earth and space sciences, and engineering, technology, and application of science" (p. 29). Based on this framework, next generation science standards are being developed that will enrich the content and practice across disciplines and be internationally benchmarked (Achieve, 2012). Inputs from stakeholders, science teacher educators, and researchers will assist in implementation these standards. This book explores the approaches and strategies of next generation science learning from multiple perspectives. The contributors in the volume articulate theoretical foundations and conceptual frameworks, explore the use of pioneering technologies, and propose curriculum innovations. They also discuss assessment issues pertaining to science standard and curricula for 21st century science education. It is hoped that this book will be a valuable resource for science teacher educators, science teachers, researchers, and administrators who have a goal of enhancing teaching and learning in next generation science to prepare students for the future workforce.

Myint Swe Khine
University of Bahrain, Bahrain

Issa M. Saleh
University of Bahrain, Bahrain

REFERENCES

American Association for the Advancement of Science. (1993). *Benchmarks for science literacy: Project 2061*. Oxford, UK: Oxford University Press.

Archive Inc. (2012). *Next generation science standards.* Retrieved 17 June 2012 from http://www.nextgenscience.org

Berkman, M. B., & Plutzer, E. (2010). *Evolution, creationism, and the battle to control America's classrooms*. Cambridge, UK: Cambridge University Press. doi:10.1017/CBO9780511760914

Bybee, R. (1997). *Achieving scientific literacy: From purposes to practices*. Portsmouth, NH: Heinemann Educational Books.

Clary, R. M., & Wandersee, J. H. (2008a). Marquee fossils: Using local specimens to integrate geology, biology, and environmental science. *Science Teacher (Normal, Ill.)*, *75*(1), 44–50.

Drew, C. (2011, November 4). Why science majors change their minds: It's just so darn hard. *New York Times*. Retrieved from http://www.nytimes.com/2011/11/06/education/edlife/why-science-majors-change-their-mind-its-just-so-darnhard.html?pagewanted=1&_r=1&emc=eta1

Duschl, R. (2008). Science education in three-part harmony: Balancing conceptual, epistemic, and social learning goals. *Review of Research in Education*, *32*, 268–291. doi:10.3102/0091732X07309371

Fredricks, J. A., Blumenfeld, P. C., & Paris, A. H. (2004). School engagement: Potential of the concept, state of the evidence. *Review of Educational Research*, *74*, 59–109. doi:10.3102/00346543074001059

Grotzer, T. A., & Basca, B. B. (2003). Helping students to grasp the underlying causal structures when learning about ecosystems: How does it impact understanding? *Journal of Biological Education*, *38*(1), 16–29. doi:10.1080/00219266.2003.9655891

Hestenes, D. (1993). *Modelling is the name of the game*. Paper presented at the National Science Foundation Modelling Conference. Dedham, MA.

Hmelo-Silver, C. E., Marathe, S., & Liu, L. (2007). Fish swim, rocks sit, and lungs breathe: Expert-novice understanding of complex systems. *Journal of the Learning Sciences*, *16*, 307–331. doi:10.1080/10508400701413401

Lehrer, R., & Schauble, L. (2006). Cultivating model-based reasoning in science education. In *Cambridge Handbook of the Learning Sciences* (pp. 371–388). Cambridge, UK: Cambridge University Press.

National Research Council. (2012). *A framework for K-12 science education: Practices, crosscutting concepts and core idea*. Washington, DC: The National Academy Press.

National Science Board. (2006). *Science and engineering indicators 2006*. Arlington, VA: National Science Foundation.

Oh, P. S., & Oh, S. J. (2011). What teachers of science need to know about models: An overview. *International Journal of Science Education, 33*(8), 1109–1130. doi:10.1080/09500693.2010.502191

Pintrich, P. R., & Schrauben, B. (1992). Students' motivational beliefs and their cognitive engagement in classroom academic tasks. In Schunk, D. H., & Meece, J. L. (Eds.), *Student Perceptions in the Classroom* (pp. 149–183). Hillsdale, NJ: Erlbaum.

Posner, G. J., Strike, K. A., Hewson, P. W., & Gertzog, W. A. (1982). Accommodation of a scientific conception: Toward a theory of conceptual change. *Science Education, 66*(2), 211–227. doi:10.1002/sce.3730660207

Saleh, I. M., & Khine, M. S. (Eds.). (2009). *Fostering scientific of mind: Pedagogical knowledge and best practices in science education*. Rotterdam, The Netherlands: Sense Publishers.

van Joolingen, W. (2004). *Roles of modeling in inquiry learning*. Paper presented at the IEEE International Conference on Advanced Learning Technologies. Joensuu, Finland.

Wilensky, U., & Resnick, M. (1999). Thinking in levels: A dynamic systems approach to making sense of the world. *Journal of Science Education and Technology, 8*(1), 3–19. doi:10.1023/A:1009421303064

Section 1
Theoretical Foundation and Conceptual Frameworks

Chapter 1
Self-Regulated Learning as a Method to Develop Scientific Thinking

Erin E. Peters Burton
George Mason University, USA

ABSTRACT

The development of skills and the rationale behind scientific thinking has been a major goal of science education. Research has shown merit in teaching the nature of science explicitly and reflectively. In this chapter, the author discusses how research in a self-regulated learning theory has furthered this finding. Self-regulation frames student learning as cycling through three phases: forethought (cognitive processes that prepare the learner for learning such as goal setting), performance (employment of strategies and self-monitoring of progress), and self-reflection (evaluation of performance with the goal). Because students have little interaction with the inherent guidelines that drive the scientific enterprise, setting goals toward more sophisticated scientific thinking is difficult for them. However, teachers can help students set goals for scientific thinking by being explicit about how scientists and science function. In this way, teachers also explicitly set a standard against which students can self-monitor their performance during the learning and self-evaluate their success after the learning. In addition to summarizing the research on learning and teaching of self-regulation and scientific thinking, this chapter offers recommendations to reform science teaching from the field of educational psychology.

INTRODUCTION

Learning how to think scientifically is important for an informed citizenry. In this era of information exchange and connectedness, knowledge continues to grow in an exponential way and technology fortifies this progress. Students graduating from K-12 schools must have the skills and knowledge to be independent learners, which includes the ability to think scientifically. Although it is important to generate students who are interested in pursuing science as a career, we also must be mindful that all students will be making future decisions based in science for their community. Therefore it is

DOI: 10.4018/978-1-4666-2809-0.ch001

imperative that all students are scientifically literate when they leave formal schooling. However, there are still many unresolved issues regarding teaching students to understand science as a way of knowing effectively.

The purpose of this chapter is to present a literature review of the methods used to teach the nature of science in the classroom with particular emphasis on explicit and reflective approaches. Additionally, this chapter will address the parallels found between science education literature on explicit and reflective approaches and processes found in self-regulated learning theory. Because self-regulated learning theory has more articulated processes of learning that are not currently described in science education, adoption of a self-regulation oriented framework to study explicit and reflective teaching of the nature of science affords clearer methods of measuring learning. Empirical studies that have used self-regulated learning theory to teach the nature of science using explicit and reflective approaches are presented, and recommendations are made for future work in teaching the nature of science.

THE ROLE OF SCIENTIFIC THINKING IN SCIENCE EDUCATION

Scientific literacy contains two components: scientific knowledge and knowledge about the scientific discipline (Duschl, 1990). Scientific knowledge is the body of information that is factual and content-based. Knowledge about the scientific discipline is considered the methods that generate and validate scientific knowledge, which are the inherent guidelines that scientists use to ensure that the information that is generated from their scientific activities are valid and reliable (Lederman, 1992). In a standards-based and high-stakes testing environment, scientific knowledge is given priority, with little or no time left for teaching about the scientific discipline and knowledge validation strategies. A focus on static, factual knowledge results in a lack of understanding of how that knowledge comes about, and little understanding of what it is to be a scientist (Tobin & McRobbie, 1997). Without knowledge of the basic guidelines regarding the dependence of the scientific enterprise on characteristics such as rationality, precision of language, and attempts to limit bias as a standard for understanding the world around them, one must depend on other forms of knowing, such as tradition or instinct. Although ways of knowing such as instinct and tradition create a well-rounded human being, thinking scientifically is vital in making rational decisions. The famous physicist, Richard Feynman is attributed the quote, "Philosophy of science is about as useful to scientists as ornithology is to birds." Ignoring that birds do not have the cognitive capacity to understand ornithology, there are two reasons to disagree with this statement: (a) thinking is more powerful when the execution of the thinking is apparent to the thinker and (b) philosophy of science should be taught in science classes so that the learners who may not consider themselves to be "science-minded" have a grasp of the guidelines for knowledge generation. Scientific thinking skills have been advocated as an important component of science education because they provide a framework on which the students can incorporate content knowledge (Duschl, 1990; Lederman, 1992; McComas, Almazroa, & Clough, 1998; Parkinson, 2004, Peters, 2006; Turner, 2000). Learning the nature of science should not take the place of learning science content, rather they should be taught simultaneously. Students who have a deep understanding of the nature of a scientific endeavor can use this knowledge to create more scientifically valid content knowledge (Akerson & Abd-El-Khalick, 2003; Crawford, 2005; Duschl, 1990). When knowledge generation guidelines are hidden, then they may not even be agreed upon or apparent to all involved. Knowledge is most powerful when one understands both the content and the course of action in which the content came to be.

Conversely, ways of knowing in science should be taught in conjunction with content in a secondary science classroom. Scientific knowledge and knowledge about science have a relationship where one cannot be fully understood without the other; the two are inextricable. Much has been written about the danger in teaching science as a way of knowing devoid of significant science content (Brickhouse, Dagher, Letts, & Shipman, 2000; Johnson & Southerland, 2002; Olson & Clough, 2001; Ryder, Leach, & Driver, 1999; Smith & Sharmann, 1999). These authors contend that a robust understanding of the mores of the scientific discipline cannot be formed without the reality of the content to support the rationale. If students only understand the technical aspects of the nature of the scientific endeavor, as would be the case in a decontextualized educational setting, then they are unable to apply their knowledge to socio-scientific issues. The future population is dependent on a citizenry that is scientifically literate, and secondary education must pursue this fully with instruction that intertwines both content instruction and instruction on ways of knowing in science.

INCORPORATING SCIENTIFIC THINKING INTO INSTRUCTION

Secondary science educators have demonstrated their proficiency in teaching content, but scientific epistemology instruction that occurs in a meaningful way is not yet pervasive in secondary school systems. Teaching scientific epistemology well in the schools requires defining what is meant by ways of knowing in science. There have been many different orientations regarding what science is and how it is performed such as positivism, logical empiricism, critical rationalism, structuralism, semantic views, and postmodernism. Perspective of scientific epistemology must also be defined, as it could be oriented by scientists, philosophers of science, or science educators (Stenhouse, 1985; Duschl, 1988; Hodson, 1993), the scope of the view taken (Ryan & Aikenhead, 1992), and the use of qualitative or quantitative measurement (Gallagher, 1991; Lederman, Abd-El-Khalick, Bell, & Schwartz, 2002). Fortunately, the education community has been able to come to somewhat of a consensus on several key aspects that should be taught in the K-12 setting (McComas, 2005; Lederman, 2007; Osborn, Simon, & Collins, 2003). The convergent aspects of scientific epistemology recognized in the education community are commonly known as the nature of science. The nature of science can be defined as the inherent guidelines that scientists use to develop and verify scientific knowledge. The guidelines that have been explicated include: a) scientific knowledge is durable yet tentative, b) empirical evidence is used to support ideas in science, c) social and historical factors play a role in the construction of scientific knowledge, d) laws and theories play a central role in developing scientific knowledge, yet they have different functions, e) accurate record keeping, peer review, and replication of experiments help to validate scientific ideas, f) science is a creative endeavor, and g) science and technology are not the same, but they impact each other (Lederman, 1992; McComas, 2008). In the United States, all 50 states have adopted nature of science knowledge standards, to varying degrees, into their curriculum framework and map exactly to this list of aspects (McComas, Lee, & Sweeney, 2009). The standards for teaching the nature of science are in curriculum and are being assessed by high-stakes testing. Educational researchers and teacher educators must work to establish effective, meaningful ways to incorporate nature of science instruction into the already time-honored content instruction in K-12 science classrooms.

TEACHING EXPLICIT AND REFLECTIVE APPROACHES TO THE NATURE OF SCIENCE

Although it is intuitive to think that by conducting inquiry that students will understand how scientists operate, there is a body of research that demonstrates explicit instruction in the nature of science has been found to be more effective. Several researchers have advocated that novice learners should be provided with direct instructional guidance (Cronbach & Snow, 1977; Klar & Nigam, 2004; Mayer, 2004; Shulman & Keisler, 1966; Sweller, 2003). Implicit instruction refers to conducting scientific activities and expecting students to synthesize ways of knowing in science from these activities. Explicit instruction refers to teacher scaffolding in developing students' abilities to develop goals aligned with science and to be metacognitively aware of the strategies they are employing in their scientific activities. A review of the literature regarding the teaching and learning of the Nature of Science (NOS) shows a progression of methods of teaching NOS using implicit means, then a shift to explicit means, and most currently, a movement of explicit and reflective approaches to teaching NOS.

An emphasis on implicit means to teach the nature of science can be traced back to curricula that emerged in the 1960s and 1970s such as the Physical Science Study Curriculum (PSSC) and the Biological Sciences Curriculum Study (BSCS), which advocated for hands-on activities and science process skills in an inquiry-based format (Lawson, 1982; Rowe, 1974), but did not explicitly emphasize NOS ideas. Several large-scale studies on these curricula came to the same conclusion: implicit NOS instruction was not effective as compared to traditional means of teaching science. Studies of the PSSC curricula found that they did not enhance students' NOS views as compared to traditional textbook-centered curricula (Crumb, 1965; Trent, 1965) as measured by Test on Understanding Science (TOUS) (Klopfer & Cooley, 1961). Likewise, a study of the BSCS Yellow curricula demonstrated that students using these materials, which made no explicit references to NOS but utilized an inquiry-based approach, did not perform significantly better on the TOUS or the Processes of Science Test than students in a traditional high school biology program (Jungwirth, 1970). A comparison of the influence of BSCS Yellow, PSSC, and traditional science courses on students' NOS views using 3,500 students from 44 high schools yielded no significant differences among treatments (Tamir, 1972) as measured by the Science Processes Inventory (Welch & Pella, 1967). Meichtry (1992) compared a large group of students who used the BSCS curriculum (n=1004) with students who were engaged in a traditional biology course (n=603) and found the BSCS group's understandings of developmental and testable NOS significantly decreased as compared to the traditional group, as measured by the Modified Nature of Scientific Knowledge Scale (MNSKS). Other interventions that utilize an implicit NOS approach, such as the one carried out by Moss, Abrams, and Robb (1998) which engaged 11th and 12th grade students in inquiry-oriented projects with a scientist partner for one year, have also been ineffective in improving student NOS views. In light of these large-scale studies over an extended period of time, the expectation that students will develop NOS ideas as a consequence of engaging in inquiry-activities alone is unsupported and other techniques to develop student ideas about NOS were explored.

In response to the lack of effectiveness of curricula that employ implicit means, a shift in NOS instruction occurred toward more explicit means, where NOS is considered a cognitive learning outcome to be planned for and actively pointed out during instruction (e.g., Akindehin, 1988). At this time it was suggested that nature of science should no longer thought of as an affective element of instruction, rather it is to be considered a cognitive objective. Orientation of planned for NOS ideas came from various NOS aspects (Billeh &

Hasan, 1975) and history and philosophy of science (Ogunniyi, 1983). Because it is understood that teachers cannot possibly teach something they do not understand themselves, emphasis on teacher understanding of NOS is prominent, in additional to an added element of reflective instruction (e.g., Abd-El-Khalick, 2001; Abd-El-Khalick, Bell, & Lederman, 1998; Akerson, Abd-El-Khalick, & Lederman, 2000). In these studies, an explicit and reflective approach included overtly instructing NOS aspects and then providing structured opportunities to reflect on the aspects in a context of the activity or of science content. In this way the various NOS aspects were identified by the instructor and when reasoned to be intelligible by the learner, the learner was given opportunities to articulate their understanding and apply it in other domains of knowledge. Various studies have provided evidence that teacher NOS views can be improved using an explicit and reflective approach (Abd-El-Khalick, et al., 1998; Akerson, et al., 2000; Shapiro, 1996) and that an explicit approach is more effective than an implicit approach in shifting teacher NOS views to be more sophisticated (Abd-El-Khalick & Lederman, 2000; Akerson & Hanuscin, 2007; Hanuscin, Akerson, & Phillipson-Mower, 2006; Khishfe, 2008; Peters, 2009; Peters & Kitsantas, 2010a; Scharmann, Smith, James, & Jensen, 2005; Schwartz, Lederman, & Crawford, 2004). A succession of studies with pre-service elementary teachers has shown evidence supporting the effectiveness of an explicit and reflective approach to teaching the nature of science (Abd-El-Khalick, et al., 1998; Abd-El-Khalick & Akerson, 2004; Abd-El-Khalick, 2005). Continued studies on implicit approaches for NOS instruction remained consistent that this approach is generally not successful in improving learners' NOS views (Sandoval & Morrison, 2003; Schwartz, Lederman, & Thompson, 2001). There is consistent, clear, and convincing evidence that explicit and reflective method of instruction is productive and continued research on how and why these methods work should be explored.

Although generally successful, explicit and reflective means of NOS instruction is not without difficulties for learners. Many studies have found that explicit and reflective means of instruction do not result in improved NOS views for *all* learners involved in the treatments (Carey, Evans, Hona, Jay, & Unger, 1989; Khishfe & Abd-El-Khalick, 2002; Leach, Hind, & Ryder, 2003; Liu & Lederman, 2002; Morrison, Raab, & Ingram, 2009; Tao, 2003), and some do not retain their improved views, challenging the idea that the learner fully conceptualized the new concepts (Akerson, Morrison, & Roth McDuffie, 2006). Additionally, even teachers with sophisticated views of the nature of science have difficulty translating their understanding into classroom instruction (Abd-El-Khalick, et al., 1998). The mechanisms behind the usefulness of explicit and reflective instruction for teachers should continue to be examined and explicated.

STUDENTS' NOS KNOWLEDGE

Fewer studies have been conducted that examine student learning of NOS, but evidence from these studies are still consistent with the teacher learning of NOS studies; explicit and reflective methods are more effective than implicit method of instruction. Smith, Maclin, Houghton, and Hennessey (2000) found that sixth-grade students who had been in an inquiry classroom since first-grade with the same teacher held more informed views of the NOS in the areas of empirical testing, collaborating with colleagues, making explanations, and developing scientific ideas. Akerson and Abd-El-Khalick (2003) found that fourth-grade students in a classroom that emphasized implicit NOS instruction through inquiry did not have informed views of NOS. In studies have compared implicit and explicit/reflective NOS instruction with students, student who experienced explicit/reflective NOS instruction improved their NOS views significantly over students in a comparison group (Khishfe & Abd-El-Khalick, 2002; Peters & Kitsantis, 2010b). In both the teacher and the student populations, explicit and reflective instruction has resulted in improved NOS views over implicit methods of instruction.

MECHANISMS OF EXPLICIT AND REFLECTIVE INSTRUCTION

Evidence points to the effectiveness of an explicit and reflective approach, so future investigation into these instructional components has merit. A review of the description of "explicit" and "reflective" in the above-mentioned studies is helpful in understanding the cognitive processes behind an explicit and reflective approach to NOS instruction. Foremost, explicit teaching of NOS should not be confused with an instructional technique called "explicit teaching" (Rosenshine & Stephens, 1986) where there is a statement of goals at the beginning of instruction and a step-by-step procedure to achieve those goals is provided to the learner. Rather, explicit methods of teaching NOS involve "cognitive instructional outcomes that should be intentionally targeted and planned for in the same manner that abstract understandings associated with high-level scientific theories are intentionally targeted" (Khifshe & Abd-El-Khalick, 2002, p. 555). Several NOS researchers describe an explicit instructional approach as one that deliberately focuses learners' attention on various aspects of NOS during classroom instruction, discussion, and questioning (Akerson & Volrich, 2005; Bell, Matkins, & Gansneder, 2010; McDonald, 2010). Not to be confused with didactic teaching, explicit approaches are based on the assumption that NOS instruction "should be planned for instead of being anticipated as a side effect or secondary product" (Akindehin, 1988, p. 73).

Reflective instructional elements have been defined in the NOS literature as "providing students with opportunities to analyze the activities in which they are engaged from various perspectives (e.g., a NOS framework), to map connections between their activities and ones undertaken by others (e.g., scientists), and to draw generalizations about a domain of knowledge (e.g., epistemology of science)" (Khifshe & Abd-El-Khalick, 2002, p. 555). In a more holistic description of reflective approaches to NOS, Abd-El-Khalick and Lederman (2000) explain that "reflective elements are designed to emulate the sort of activities that historians, philosophers, and sociologists engage in their efforts to understand the workings of science" (p. 691). Likewise, Scharmann, Smith, James, and Jensen (2005) believe reflective instruction should challenge learners to think about how their work illustrates NOS, and how their inquiries are similar to or different from the work of scientists. Although the terms "explicit" and "reflective" have been described in the literature, they are very general and difficult to operationalize. Other models of learning that are more descriptive and have precision of language are available to further explicate the concepts of "explicit" and "reflective," and research along these lines may help to explain some of the limits to the success of explicit and reflective approaches to NOS instruction.

SELF-REGULATED LEARNING AND THE NATURE OF SCIENCE

Self-regulated learning is a field that has potential to elucidate the mechanisms behind the successes and barriers of explicit and reflective NOS teaching and learning. Self-regulation refers to self-generated thoughts, feelings and actions that are planned and cyclically adapted to the attainment of goals (Zimmerman, 1990) and is a process that learners adopt when they are actively metacognitively, motivationally, and behaviorally participants in their own learning (Zimmerman, 1986). Self-regulated learners are self-aware of what they know and what they do not know are strategic in their approach to learning, and attribute their successes and failures in learning to their own processes, rather than any uncontrollable factor or innate talent. The features of self-regulated learning can be categorized even further into measurable processes that map to the features of explicit and reflective NOS instruction. Mapping the ideas provided in explicit and reflective NOS instruction to self-regulated learning will help to provide a more detailed description of explicit and reflective NOS instruction and will also provide devices from which to measure each factor of a learner's process given explicit and reflective NOS instruction. Self-regulatory methods of learning have been shown to be effective in aiding learners to explicitly analyze the skills or knowledge that is needed to achieve a particular goal. To accomplish the goal of learning about the nature of science and science content, students can perform inquiry-based activities, be prompted to think about why they are conducting certain processes, and, in turn, evaluate their thinking in terms of the agreed-upon processes of the scientific community. Research has shown improvement using self-regulated learning strategies in the diverse areas related to academic learning: intrinsic motivation (Ryan, Connell, & Deci, 1984), academic studying (Thomas & Rohwer, 1986), classroom interaction (Rohrkember, 1989; Wang & Peverly, 1986), use of instructional media (Henderson, 1986), metacognitive engagement (Corno & Mandinach, 1983), self-monitoring learning (Ghatala, 1986; Paris, Cross, & Lipson, 1984), dart throwing (Kitsantas, Zimmerman, & Cleary, 1999), and writing revision (Zimmerman & Kitsantas, 2002). Using the course of action in the iterative self-regulation model as a lens for NOS learning, the research can move forward and become more explicit itself.

Self-regulated learners enter three phases of a learning cycle: forethought, performance, and self-reflection as illustrated in Figure 1 (Zimmerman, 2000). The forethought phase refers to influential processes that precede efforts to act and set the stage for action such as analyzing tasks and setting process-oriented goals (e.g., asking students to organize the content they already knew about the inquiry problem). The performance phase includes processes that occur during the action, such as implementation of the task and self-monitoring (e.g., asking students to conduct hands-on inquiries and to monitor their progress). The self-reflection phase refers to the processes that occur after the performance efforts which

Figure 1. Self-regulated learning in science inquiry

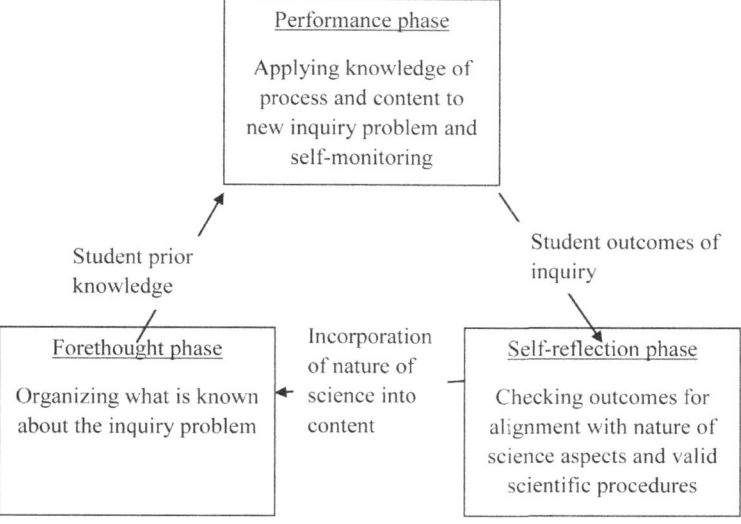

influence a person's response to the action such as the use of standards to make self-judgments about the performance (e.g., students compare their activities in the inquiry against the nature of science). Because students continue to cycle through the self-regulation feedback loops, when students enter successive iterations of the loop, they have more sophisticated forethought, performance, and self-reflection. There is evidence that attainment of high levels of academic achievement requires a self-regulatory dimension of competence in addition to basic talent and high-quality instruction (Zimmerman & Kitsantas, 2007). One possible reason that teachers and students have difficulty understanding the nature of science is their lack of exposure to the same inherent ways of knowing as a scientist (Hogan, 2000). In order to construct accurate ideas about how science works as a discipline, a novice must have guidance to the particular processes and knowledge that are necessary to the field (Kirschner, Sweller, & Clark, 2006). Although self-regulatory processes are internally driven, they can be encouraged by mentors or an appropriately constructed learning environment (Zimmerman, 2000), thus allowing an explicit and reflective NOS approach to serve as a type of "mentor" to teach students how science is performed in a directed way. Self-regulated learning strategies provides a framework that can scaffold naïve views of the nature of science to more developed views of the nature of science, which can transform the way nature of science knowledge is taught because the explicit features of NOS instruction from the literature map directly to the forethought phase, and the reflective features of NOS instruction map directly to the self-reflective phase of the self-regulation model. The performance phase describes the student performance during the learning activity.

FORETHOUGHT

Within each of the three processes of forethought, performance, and self-reflection there are variables that further explain how the self-regulation cycle works to improve NOS learning. The forethought phase includes the subprocesses of goal setting, strategic planning, self-efficacy beliefs, task interest, and goal orientation. A list of subprocesses for forethought, performance, and self-reflection can be found in Figure 2. The means in which "explicit NOS instruction" has been used in science education literature is directly connected to each of the subprocesses within forethought, and will be discussed in this section. Goal setting refers to the decisions made regarding specific outcomes of learning (Locke & Latham, 1990) and determines the point at which self-reflection occurs at the end of one cycle. Strategic planning refers to the methods the learner chooses to attain the preferred goals (Zimmerman & Martinez-Pons, 1992). Self-efficacy refers to personal beliefs about one's capability to perform at a particular level (Bandura, 1986). Task interest refers to an orientation to continue learning efforts, even without extrinsic rewards (Deci, 1975). Goal orientation is a spectrum of the person's rationale underlying the intended goal that ranges from performance, where the orientation is toward rewards such as good grades, to mastery, where the orientation is toward learning to be competent at a task despite any outward recognition (Ames, 1992). Goal setting and strategic planning are deliberate, cognitive choices, but they are also dependent on the variables of self-efficacy beliefs, task interest, and goal orientation, which are based in belief systems. For example, a student with high self-efficacy in science will tend to set higher goals (Zimmerman, Bandura, & Martinez-Pons, 1992) and will plan more effective strategies to reach those goals than less self-efficacious students (Zimmerman & Bandura, 1994). A student who has an intrinsic task interest and a mastery goal orientation will be more likely to pursue learning how to accomplish a task regardless of any recognition from outside entities. The end goal of learning NOS for a proficient learner is to understand how scientific knowledge came to be and why it is distinct from other ways of knowing, rather than to merely get good grades for the sake of perception.

Planning and focusing student efforts are characteristics of an explicit approach to NOS, which are identical characteristics in goal setting and strategic planning, the two variables that can be actively set during instruction during the forethought phase. Explicit NOS instruction have been described as "cognitive instructional outcomes that should be intentionally targeted" (Khifshe & Abd-El-Khalick, 2002, p. 555) and an approach that deliberately focuses learners' attention on various aspects of NOS during classroom instruction, discussion, and questioning (Akerson & Volrich, 2005; Bell, Matkins, & Gansneder, 2010; McDonald, 2010), which "should be planned for instead of being anticipated as a side effect or secondary product" (Akindehin, 1988, p. 73). Advocates of an explicit approach to NOS instruction are recommending similar processes as goal setting and strategic planning of NOS aspects within inquiry instruction. Table 1 illustrates the parallel nature of descriptions of explicit NOS approaches and their relationship to goal setting and strategic planning. A better understanding of how and why explicit NOS approaches are successful and not successful, a model for self-regulation gives guidance in unpacking processes of goal setting.

Within goal setting, there are seven advantageous properties of goals that emphasize why an explicit approach would lead to more successful attempts at learning NOS (Zimmerman, 2008): (a) specificity of goals, (b) proximity of goals, (c) hierarchical organization of short-term and long-term goals, (d) congruence or lack of conflict among one's goals, (e) level of challenge of goals, (f) self-set or assigned origins of goals, (g) conscious quality of goals, and (h) alignment of goals

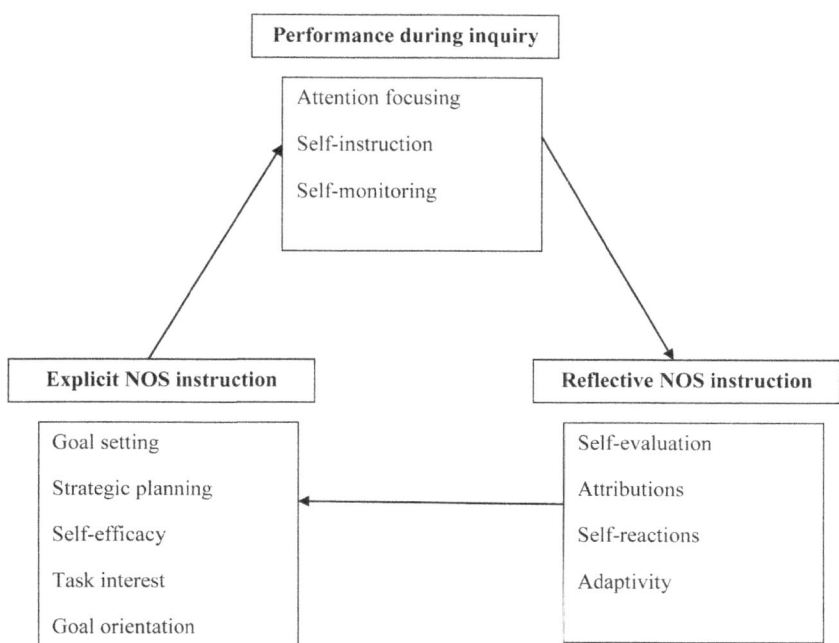

Figure 2. Subprocesses of self-regulated learning oriented to explicit and reflective approaches

Table 1. Examples of alignment of explicit NOS approaches to subprocesses of forethought phase of self-regulation

Forethought subprocess	Explicit NOS instruction
Goal setting	"cognitive instructional outcomes that should be intentionally targeted" (Khifshe & Abd-El-Khalick, 2002, p. 555) Focuses learner attention on various aspects of NOS (Akerson & Volrich, 2005; Bell, Matkins, & Gansneder, 2010; McDonald, 2010) "should be planned for instead of being anticipated as a side effect or secondary product" (Akindehin, 1988, p. 73).
Strategic planning	Learning declarative knowledge about each NOS aspect, then application of each NOS aspect, then integration of NOS aspects with a way of knowing in science through context of content
Self-efficacy	Explicit approach to NOS instruction leads to attainable goal in which student experiences success, increasing self-efficacy of understanding the scientific enterprise
Task interest	Knowing how knowledge is constructed in science leads to more interest in content
Goal orientation	Student who wants to understand how inquiry should be approached needs to understand the underpinnings of the development of scientific knowledge, otherwise the student may be oriented to merely complete the inquiry to get a grade

on learning outcomes. When goals are more specific, they are more attainable (Bandura, 1988). For example, a specific goal regarding the nature of science would be "I will be sure that each of my assertions in my conclusions are directly related to evidence I observed," whereas a general goal would be "I want to do science better." The specific goal is more distinct and measurable, and the learner is able to determine when they have achieved the goal. The general goal is too nebulous to be measured, and the learner will be unsure when the task has been successfully completed. Proximity of goals refers to the length of time in which the goal is intended to be reached and is related to the hierarchical organization of short-term and long-term goals (Zimmerman, 2000). An example of a proximal goal in the short term would be "I will understand the relationship between models and theories in biology by the end of the week," as opposed to a long-term, distal goal, "I will learn all of the major theories in biology and their role over time in developing the field of biology by the end of the year." Whereas they are both admirable goals, the proximal goal provides more prospects to remain on task and to produce a successful outcome than the distal goal, and if properly organized into hierarchy that the proximal goals must be reached first, is profitable to the learner. If only distal goals are set, the learner will have more difficulty being motivated to reach the goal than if the learner set both proximal and distal goals. An example of how NOS instruction parallels this principle is the breakdown of NOS into aspects, such as the seven aspects discussed above. Learning the whole of NOS would be overwhelming to a learner if it were not described in more reduced terms. Learners who have congruence of goals regarding NOS would aspire to understand how knowledge is acquired and validated in the scientific enterprise, and would not be deterred by conflicting goals set from external sources, such as a classroom oriented only toward memorization of scientific facts. For example, a learner who wants to understand about the inherent guidelines from which science as a discipline operates and is in a learning environment that explicitly points toward NOS examples within the context of the classroom learning has congruence of goals. Goals that are set higher than can be achieved are not productive, so learners must be able to set goals that are challenging, yet within their reach (Locke & Lathan, 1990). The nature of science, being an epistemology, is difficult to for a novice learner to explicate given only implicit instruction. Explicit instruction is necessary to help learners set goals that are reasonable in a field where they do not have much experience. Another advantageous property of one's goals is the origin of the goal. Learners who set goals for themselves are more committed to attaining those goals because of their perception of self-determination and autonomy (Deci & Ryan, 1991; Schunk, 1985). Unfortunately, students rarely have access to the world of scientists or their epistemology in everyday life (Hogan, 2000). In this case, NOS instruction should begin with assigned origin of goals, where the instructor explains the learning objective explicitly for the nature of science, and then shifts instruction to encourage a self-set goal for students to become more aware of the inherent guidelines of the scientific enterprise (Peters, 2009; Peters & Kitsantas, 2010a; Peters & Kitsantas, 2010b). Conscious quality of goals refers to the level of intentionality in setting the goals. Learners are influenced by intentional as well as low conscious goals, and high conscious goals are more reliably reached than low conscious goals (Howard & Bray, 1988; Locke & Latham, 2002). In NOS instruction, high conscious goals are set when the instructor makes the features of NOS explicit to students because they are emphasized as important and prominent in the learning. Lastly, alignment of goals with learning outcomes refers to three circumstances: the focus of learning product, learning process, or general goal. A goal that is "do your best" is a general goal; "make three empirical observations" is a product goal; and "consider that observations should be based on standards such as the metric system" is a process goal. General goals tend to not result in productive learners at any point (Kanfer & Ackerman, 1989). However, novice learners have a higher performance when they begin learning with a process goal, but as that skill becomes automatic to the learner, switching to a product goal produces higher performance (Zimmerman & Kitsantas, 1997). In the case of NOS instruction, prompting students to behave in a way of scientists explicitly (such as making empirical observations and accurate record keeping) helps them to understand why there are particular traditions in doing science (such as multiple trials). Once the behaviors that yield valid scientific information are automatic to learners, then product goals can be introduced such as "be sure to generate valid conclusions." Both goals must be initiated explicitly. Unpacking elements of goal setting is helpful in explicating what explicit NOS instruction means beyond planning and focusing learners' attention deliberately on NOS aspects.

Strategic planning is closely related to goal setting and is distinguished by proactive or reactive learners. Proactive learners design their own strategies to help with cognition, control interruption due to affective factors, and direct performance into productive lines (Corno, 1993; Paris, Byrnes, & Paris, 2001). Proactive learners are committed to their goals, exert greater efforts in reaching the goals, and demonstrate persistence in their learning (Wigfield, Tonks, & Eccles, 2004). Conversely, reactive learners take little responsibility for their own learning and rely on outside indicators or extrinsic rewards to determine success. Proactive learners find value in the learning, and this finding can be used to enhance NOS instruction, because the very cognitive outcomes of NOS are the rationale of *why* scientific processes are done in a specific way. For example, a student may design an experiment to include multiple trials of each level of a variable. However, if NOS instruction changes the student's understanding from "I record three trials of all levels because that is what we always do" to "multiple trials are necessary in ensuring valid information and it depends on the content as to what is the minimum multiple trials necessary," then the student has placed value on the task and has transformed their understanding of NOS.

Goal setting and strategic planning can be addressed in the classroom as overt processes, but they are also dependent on affective elements such as self-efficacy of NOS understanding, task interest, and goal orientation. Including self-efficacy of NOS understanding, task interest, and goal orientation revisits the idea that NOS can be thought of in affective terms into understanding factors that encourage NOS understanding, but does not exclude cognitive outcomes such as the ones that come from goal setting and strategic planning. Considering self-regulation as a lens to view productive NOS instruction requires both a cognitive orientation and an affective orientation that are interactive.

Self-efficacy is a powerful component in the learning process because it is the key to changing one's actions (Bandura, 2006). If a learner does not believe they can change their actions, then they will not persevere in the face of adversity, which is common for learning new knowledge. Learners with a strong self-efficacy approach problems with confidence and welcome challenge. Self-efficacy, if low, can be debilitating and limiting to a learner. Improving explicit and reflective NOS instruction in a more articulate way requires a consideration of self-efficacy. One of the most influential sources of self-efficacy is the interpretation of the result of one's own previous attainments, otherwise known as mastery experience (Bandura, 1997). One's judgment of competence in NOS would depend on prior tasks that involved NOS and the interpretation of the results. When learners believe they are successful based on their interpretation, their confidence of their ability to learn NOS increases and they seek more challenging tasks in the field. Self-efficacy is a key for developing a more sophisticated NOS understanding, as it is well known that NOS must be addressed iteratively over time and in many contexts. Another influential source of self-efficacy is modeling, where learners learn by watching a proficient mentor complete the task, and coping models have been shown to be more helpful in raising confidence than mastery models (Schunk, 1987; Schunk & Hanson, 1988). Coping models are mentors who struggle to solve the problem, and mastery models are those who treat mistakes as though they did not occur. Models who overtly struggle through problems provide more information about the processes involved in getting to the solution of the problem than mastery models. Explicitly planning NOS events in class has an influence on self-efficacy because it reveals the processes involved in attaining NOS knowledge, educating the learner in how to do the learning, not merely what content to learn.

Task interest is an additional affective factor that contributes to forethought in learning. Individuals who have more interest in a task tend to focus their attention more, have more persistence in learning the task, and are more emotionally invested than an individual with less task interest. Here interest means interest in the content, not in extraneous features such as the "fun" environment in which the content is taught (Hidi, 1990; Schiefele, 1991). Value is thought of as the learner's perceptions of utility and importance of the subject matter (Garcia & Pintrich, 1994). If learners are unaware of the underpinning guidelines that scientists use in their discipline, then it is difficult to have value and interest in the content. Often, learner goals will be oriented toward recognition or merely passing the course by completing all of the assignments. Science teaching that only addresses the factual content does not foster interest or the belief that learning science is valuable beyond passing the test. Teaching learners how the factual knowledge is generated and accepted in addition to the factual knowledge has potential to increase task interest in learning science, and thus support rich forethought for the learner, preparing the learner to perform well in the task at hand.

Goal orientation ranges from mastery goal orientation, which is a state where a learner seeks to improve intellectually, to performance goal orientation, which is a state where a learner seeks to gain recognition relative to others by performing the learning task. Goals orientation is related to self-efficacy and task interest in that mastery goal orientation is positively correlated with a preference for challenging activities (Ames & Archer, 1988), high levels of interest, task involvement, and persistence (Elliot & Dweck, 1988; Harackiewicz, Barron, Pintrich, Elliot, & Thrash, 2002). Conversely, performance goal orientation is correlated positively with surface level learning strategies (Graham & Golan, 1991, Nolen, 1988) and self-handicapping strategies (Urdan, Midgley, & Anderman, 1998). Productive learners possess a mastery goal orientation where they learn to develop their competencies and deal with their deficiencies. Teaching NOS explicitly provides a means for learners to be competent in acquiring science knowledge outside of class, because with a developed understanding of NOS, a learner knows why particular methods in science are used and can adapt their performance to be more "science-minded." Consider that a science class, which only teaches science content, devoid of NOS instruction, cannot equip a student to develop their own scientific knowledge from investigations. A learner who is competent in a content-only science class can master memorization of theories, laws, and other factual information, but little else if it is not explicit. However, a class that incorporates explicit NOS instruction provides ways to support claims with empirical evidence, to be creative, to connect the development of scientific knowledge with social and historical factors, and to develop habits of mind of scientists, among other competencies related to a scientific way of knowing.

Explicit NOS instruction is described in science education literature as intentionally targeted cognitive outcomes. The forethought phase of self-regulatory theory (Zimmerman, 2000) provides a mechanism that directly maps to explicit NOS instruction, as it is the introductory processes that a learner must engage to be successful in the performance of a task, as shown in Table 1. Teachers can help students adapt active cognitive processes such as goal setting and strategic planning, while enhancing self-efficacy of science learning, task interest and goal orientation in the design of the learning environment. Within the forethought phase, proficient learners should set challenging yet attainable goals, choose productive strategies to achieve those goals, foster high self-efficacy about learning science, find tasks meaningful, and be oriented to achieving the goal in order master the content and task. Explicit instruction in NOS has the potential to address all of these factors that enhance learning. Pointing at learning outcomes for NOS, as suggested in science education literature, is a beginning to this process, but could be more clearly articulated with the help of the self-regulatory model. Future research into the interactions of the self-regulatory features of forethought and of explicit NOS instruction is needed.

PERFORMANCE

The processes that occur during the learning effort are categorized as the performance phase. Although the performance phase does not map directly to explicit and reflective NOS instruction, there is merit in describing the performance phase processes in terms of NOS instruction because of the cyclical nature of the self-regulatory model. Three types of control processes have been identified during the performance phase of self-regulated learning: attention focusing, self-instruction, and self-monitoring (Zimmerman, 2000). Attention focusing occurs when learners can shift attention away from competing factors and focus on a task that will help them achieve their goal (Corno, 1993). Self-instruction refers to the ability of a learner to tell themselves how to proceed in a task, and has been shown to improve students' learning (Schunk, 1982). Self-monitoring occurs when learners check on their progress during the task. By examining the roles of these three subprocesses, one can see the importance of an explicit and reflective approach in learning NOS. For example, without explicit instruction in NOS, a novice science learner may not pay attention to appropriate knowledge and skills in class because they do not know what is important. Without explicit instruction in NOS, a novice science learner may not know what steps are needed to self-instruct to complete the task of inquiry in a way that is scientifically valid. Finally, if a learner has no standard to compare against, in this case the guidelines of the scientific enterprise, they cannot self-monitor during their performance. Explicit NOS instruction helps to articulate the standards in science and consequently students overtly know how to do science in a way that is compatible to the scientific enterprise.

SELF-REFLECTION

It is not surprising that "reflective NOS instruction" corresponds directly to the self-reflection phase of self-regulated learning theory, as the names might suggest. Reflective instructional elements in NOS instruction are described in science education literature as direct opportunities to compare student science work against the principles defined in NOS and to make generalizations about domains of knowledge. Again, the self-reflection phase of self-regulated learning offers a more articulate model of what NOS instruction strives to attain in the reflective approach as seen in Table 2. There are four subprocesses within self-reflection: self-evaluation, attributions, self-reactions, and adaptivity. After the performance of a task, learners often engage in self-evaluation first and compare self-monitored information against a standard or goal. Khifshe and Abd-El-Khalick (2002) called directly for students learning NOS to reflect by mapping "connections between their activities and ones undertaken by scientists" (p. 555). Here the standard or goal is the activity of scientists, and they are in effect asking students to self-evaluate their performance in science against the standard of the field.

Once learners self-evaluate their performance and judge it to be aligned or misaligned with a standard, they ascribe attributions to their successes or failures. Attributing success or failure of a performance is key in promoting learning because if the learner attributes errors to their ability (they were not born with the skills of being a scientist) or an outside factor (the belief that the student made errors because teacher just does not like him/her) rather than effort or process, then the learner may give up trying to improve. That is, productive learners attribute their successes or failures in a task to strategies. Given the chance to reflect on their understanding of NOS, students

Table 2. Examples of alignment of reflective NOS approaches to subprocesses of self-reflection phase of self-regulation

Self-reflection subprocess	Reflective NOS instruction
Self-evaluation	"providing students with opportunities to analyze the activities in which they are engaged from various perspectives (e.g., a NOS framework), to map connections between their activities and ones undertaken by others (e.g., scientists), and to draw generalizations about a domain of knowledge (e.g., epistemology of science)" (Khifshe & Abd-El-Khalick, 2002, p. 555). reflective instruction should challenge learners to think about how their work illustrates NOS, and how their inquiries are similar to or different from the work of scientists (Scharmann, Smith, James, & Jensen, 2005)
Attributions	Reflection what the student has done with what is done in the discipline of science gives students the connection to attribute their successes or failures to the strategies, not a hidden "talent" for science
Self-reactions	Reflection activities in NOS instruction gives students clear messages when they are making choices that lead to valid information and when they are not making choices that lead to valid information
Adaptivity	"reflective elements are designed to emulate the sort of activities that historians, philosophers, and sociologists engage in their efforts to understand the workings of science" (Abd-El-Khalick & Lederman, 2000, p. 691).

have the opportunities to see their strategies retroactively and could be encouraged to attribute their successes or errors to effort, rather than innate ability. Additionally, students who have an understanding of NOS can comprehend that there is a rationale for the strategies utilized in science, rather than harbor an understanding that broadly a person can think like a scientists or not. An explicit and reflective approach to NOS instruction affords the learner the steps to fully participate in the scientific enterprise.

Positive attribution leads to positive self-reactions, which are the methods a person chooses to further their understanding of the learning situation or the way they react to their self-evaluation and attribution. Personal attributions of successes and failures to strategy (learning activities) have resulted in positive self-reactions, and attributions to ability (being born with a skill) have been shown to result in negative self-reactions (Zimmerman & Kitsantas, 1997). Consider when a student conducting scientific inquiry in the classroom with a group and the data analysis they are conducting does not yield a consistent trend. The learner could give up and resign themselves to not being science-minded (a negative attribution and self-reaction) or the learner could reconsider her strategies and reflect upon the empirical NOS aspect and realize that more trials are needed for a valid and reliable investigation of the scientific phenomena. An explicit and reflective NOS approach is required for the positive attribution and self-reaction to transpire.

Constructive attributions also help learners identify the source of learning errors and adapting their future performance (Zimmerman & Martinez-Pons, 1992). The more experience a learner has, the more adaptable their reactions to self-evaluations become. In this process, learners try a variety of learning strategies until they find the ones the work best for them in the situation. An example connection between NOS learning and adaptivity can be illustrated in the learning that occurs during "hands-on" but not "minds-on" labs. These "hands-on" (but not "minds-on") labs require only variable manipulation to solve a problem, with no consideration of the phenomena behind the variables, such as a superficially executed bridge building activity. Bridge building is a traditional activity in classes where students work together to build a bridge which will hold a particular amount of weight given specifications on variables such as materials and size. Rarely does the process of building a bridge require students to investigate phenomena such as static equilibrium forces or loads. Instead, this activity often requires students to try out different arrangements of a bridge to see which one holds more weight through trial and error without regard to the phenomena governing the bearing of weight. If the students were required to understand the basics of static equilibrium forces and center of gravity, then they would be more adaptive thinkers when their bridge did not hold the required amount of weight. That is, they would have more information to draw from in considering alternate arrangements for the bridge. In the same way, knowledge of NOS allows students to think more flexibly about *why* they undertake the processes they do in scientific inquiry and can adapt their process if an error occurs. Reflecting on the aspects of NOS when considering findings of an inquiry activity, learners have more information to draw from regarding the validity of their scientific work.

The forethought, performance and self-reflection phases exist in an iterative cycle in self-regulation theory, and the connections between self-reflection and forethought complete this cycle. Accurate self-evaluations and attributions of strategy lead to favorable self-reactions. Positive self-reactions augment constructive forethought about one's ability to be a productive learner, which leads to productive goal setting and strategic planning, higher self-efficacy, and higher task interest (Zimmerman & Kitsantas, 1997). Forethought prepares the learner for the optimum performance possible, and then accurate self-reflection on the performance leads to a more powerful forethought phase for the next iteration of the cycle. In terms of NOS knowledge, an explicit approach to NOS learning in science instruction prepares the learner for high performance in learning science, especially via inquiry, and provides a structure for students to compare their performance against the customs of the scientific field, which develops into more sophisticated knowledge about scientific endeavors in each iterative cycle.

COMPARING NAÏVE AND SKILLFUL SELF-REGULATED NOS LEARNERS

The interactions of subprocesses in the three phases of Zimmerman's (2000) self-regulated learning model are complex, and a description of a naïve and a skillful self-regulated learner within the context of NOS learning aids in illustrating how to self-regulated learning articulates an explicit and reflective approach. A naïve learner, we will call her Natalie, begins by setting non-specific, distant goals such as "I will get an A in science." Natalie has set a goal, but it is not a productive or informed goal and will be difficult to monitor and assess during her performance in science class. Natalie has decided that she wants to be a higher performer in science, no matter what, and is less interested in the subject matter than she is in getting on the honors list. Outwardly, this may sound like a goal that will drive Natalie toward success, but that success is not focused on learning and can possibly be attained by memorizing material for assessments. She wants to do well this year in science, but she has not done well in science in the past and she cannot coherently express the plans that she has to change her learning strategies. She knows that science operates differently than other subjects, but cannot describe the "rules' of scientific inquiry, which makes her less interested in pursuing her studies of science outside of class. When Natalie goes to science class and performs her tasks in class, she does not have a focus beyond getting the assignments completed. Often, she will enter into a lab by saying to herself, "I don't really know how to be a scientist, so I won't be surprised if I fail this lab" which sets up her attribution for failure to be an unchangeable, innate entity. She only knows if she is doing well in science when her teacher gives her a grade or some verbal feedback, which Natalie avoids as much as possible because she just wants to complete the assignments. When she fails a scientific inquiry assignment, she resigns herself that she just is not scientifically minded, so she does not try anything differently to learn how scientists think, and continues to fail at learning science. In this example of a naïve science learner, there are many opportunities to redirect poorly chosen learning strategies by explicit and reflective NOS instruction. Natalie has little direction, and explicit NOS instruction would provide direction, while reflective NOS instruction would give her opportunities to evaluate her own learning.

A skillful learner in science, Sabrina, begins by setting a goal for the year that is focused on learning rather than on recognition ("I will learn the role of evidence in science") and sets benchmark goals to reach this distant goal ("I will first learn why many trials are needed in labs, then I will learn what is meant by valid results"), and she understands the hierarchy of her set of goals. This process of setting goals has worked for her in other classes, so she believes that this strategy will work for her in science class this year. She feels ownership of this type of learning because the teacher has explicitly described how scientific epistemologies relate to the content knowledge and to the labs they do in class, so she in intrinsically interested in learning how to think scientifically. She feels comfortable being challenged in her scientific thinking because the teacher supports her synthesis of domains of knowledge by pointing out how different fields of science discover and validate knowledge. During class, Sabrina often asks why particular processes in science are used, and the teacher elaborates on the scientific enterprise in response. She is aware of what she needs to do to solve problems in science because instead of following a prescribed set of steps, Sabrina has developed a way of knowing in science that allows her to compare how professional scientists act and think to her own thinking. She actively seeks out help when her ways of knowing in science do not correspond to the nature of science. She does not worry too much about "messing up" during science labs, because she knows that the teacher provides students opportunities to reflect on the logic of their own thinking. When Sabrina fails, she knows that she needs to try some other learning strategy, and she draws from the explicit and reflective NOS instruction to adapt the ways that she is thinking about her scientific inquiry in class. The compilation of successes and failures along with their corresponding learning strategies help Sabrina to make a library of academic strategies that work for her and she draws from them to create new goals, with the help of the teacher's explicit and reflective instruction into other NOS aspects.

EVIDENCE OF SELF-REGULATION OF NATURE OF SCIENCE KNOWLEDGE

Several studies have investigated the effect of a self-regulatory intervention and explored the role of subprocesses of goal setting, self-monitoring, and self-evaluation in learning NOS. Quasi-experimental studies comparing explicit NOS instruction with implicit NOS instruction using an intervention of metacognitive prompts designed to help students with goal-setting, strategic planning, self-monitoring, self-instruction, and adaptivity demonstrated success in improving NOS knowledge and content knowledge (Peters, 2009; Peters & Kitsantas, 2010a, 2010b; Peters Burton, 2010). The implicit group in these studies was not given overt opportunities to engage in the self-reflection phase of a self-regulatory loop of forethought, performance, and self-reflection, while the explicit group was given prompts to encourage self-regulatory processes. Evidence from these studies indicates that goal setting, self-monitoring, and self-evaluation are useful in developing both NOS knowledge and content knowledge. The students in the explicit treatment reported having a higher value placed on empirical evidence in making conclusions, and the implicit treatment depended more on didactic learning (Peters, 2009). As students are often unaware of the traditions that guide scientific knowledge development (Hogan, 2000), the prompts acted to set goals for learning about the scientific enterprise that were proximal and attainable, which the students then adopted for their own. Similarly, these results emphasize that the pedagogic knowledge gained from student engagement in science activities are not necessarily aligned with the science as a way of knowing, and that the expert domain of learning science is supported when taught explicitly (Kirschner, Sweller, & Clark, 2006). Self-monitoring was completed by the explicit group when they used the prompting intervention to check their work against the NOS aspects that were identified in the prompts. Self-evaluation in the studies were operationalized in three ways: 1) questions regarding student work alignment with scientific way of knowing, 2) social feedback of student work through reaching consensus on conclusions, and 3) comparison of student work with the examples provided in the intervention. The explicit group used the prompts as scaffolds to accurately self-evaluate, while the implicit group tended to check answers with the teacher, which illustrates that they did not understand the NOS knowledge deeply enough to know if they performed well. Attempts at utilizing self-regulation to increase NOS knowledge have shown to be promising, but the body of evidence is not yet convincing without more investigation.

FUTURE OF NOS RESEARCH

Explicit and reflective approaches to NOS instruction shows potential to inform NOS learners more effectively. However, more precision of language is needed about why explicit and reflective methods are effective. Self-regulatory theory as applied to NOS instruction helps to expand and explain many of the processes of learning that occur by the use of explicit and reflective methods. Examining the variables within each phase of self-regulation in the context of the nature of science provides a finer-grained examination of explicit and reflective by mapping directly to forethought and self-reflection respectively. A key element of self-regulated learning is goal setting, and because students are often not engaged with the scientific community, NOS goals must be set for students because naïve NOS learners are not yet capable of setting goals for themselves. Only when students have clear goals for learning about science can they understand how to strategically plan their learning activities, self-monitor their progress, and self-evaluate their work against the standards of the field of science. Much work has been done in other fields regarding self-regulation, and it is time to put this knowledge to use in moving the field of science education forward. The study of NOS knowledge is especially fruitful, given the close match of self-regulation to explicit and reflective approaches, and the resulting expansion of learning processes behind explicit and reflective yields many opportunities for future research. Keys to developing scientific literacy for all students must be approached in a systematic way if it is to be successful. Self-regulation of the nature of science has great potential to support reticent students to develop strategies that lead them to become more science-minded.

REFERENCES

Abd-El-Khalick, F. (2001). Embedding nature of science instruction in preservice elementary science courses: Abandoning scientism, but..... *Journal of Science Teacher Education, 12*, 215–233. doi:10.1023/A:1016720417219

Abd-El-Khalick, F. (2005). Developing deeper understandings of nature of science: The impact of a philosophy of science course on preservice science teachers' views and instructional planning. *International Journal of Science Education, 27*, 15–42. doi:10.1080/09500690410001673810

Abd-El-Khalick, F., Bell, R. L., & Lederman, N. G. (1998). The nature of science and instructional practice: Making the unnatural natural. *Science Education, 82*, 417–436. doi:10.1002/(SICI)1098-237X(199807)82:4<417::AID-SCE1>3.0.CO;2-E

Abd-El-Khalick, F., & Lederman, N. G. (2000). Improving science teachers' conceptions of the nature of science: A critical review of the literature. *International Journal of Science Education, 22*, 665–701. doi:10.1080/09500690050044044

Abd-El-Khalick, F. S., & Akerson, V. L. (2004). Learning about nature of science as conceptual change: Factors that mediate the development of preservice elementary teachers' views of nature of science. *Science Education, 88*, 785–810. doi:10.1002/sce.10143

Akerson, V. L., & Abd-El-Khalick, F. S. (2003). Teaching elements of nature of science: A year long case study of a fourth grade teacher. *Journal of Research in Science Teaching, 40*, 1025–1049. doi:10.1002/tea.10119

Akerson, V. L., Abd-El-Khalick, F., & Lederman, N. G. (2000). Influence of a reflective explicit activity-based approach on elementary teachers' conceptions of nature of science. *Journal of Research in Science Teaching, 37*, 295–317. doi:10.1002/(SICI)1098-2736(200004)37:4<295::AID-TEA2>3.0.CO;2-2

Akerson, V. L., & Hanuscin, D. L. (2007). Teaching nature of science through inquiry: Results of a 3-year professional development program. *Journal of Research in Science Teaching, 44,* 653–680. doi:10.1002/tea.20159

Akerson, V. L., Morrison, J. A., & McDuffie, A. R. (2006). One course is not enough: Preservice elementary teachers' retention of improved views of nature of science. *Journal of Research in Science Teaching, 43,* 194–213. doi:10.1002/tea.20099

Akerson, V. L., & Volrich, M. L. (2006). Teaching nature of science explicitly in a first grade internship setting. *Journal of Research in Science Teaching, 43,* 377–394. doi:10.1002/tea.20132

Akindehin, F. (1988). Effect of an instructional package on preservice science teachers' understanding of the nature of science and acquisition of science-related attitudes. *Science Education, 72,* 73–82. doi:10.1002/sce.3730720107

Ames, C. (1992). Achievement goals and the classroom motivational climate. In Schunk, D. H., & Meece, J. L. (Eds.), *Student Perceptions in the Classroom* (pp. 327–348). Hillsdale, NJ: Lawrence Erlbaum.

Ames, C., & Archer, J. (1988). Achievement goals in the classroom: Students' learning strategies and motivational processes. *Journal of Educational Psychology, 80,* 260–267. doi:10.1037/0022-0663.80.3.260

Bandura, A. (1986). *Social foundations of thought and action: A social cognitive theory.* Englewood Cliffs, NJ: Prentice-Hall.

Bandura, A. (1988). Self-regulation of motivation and action through goal systems. In Hamilton, V., Bower, G. H., & Frijda, N. H. (Eds.), *Cognitive Perspectives on Emotion and Motivation* (pp. 37–61). Dordrecht, The Netherlands: Kluwer Academic. doi:10.1007/978-94-009-2792-6_2

Bandura, A. (1997). *Self-efficacy: The exercise of control.* New York, NY: Freeman.

Bandura, A. (2006). Adolescent development from an agentic perspective. In Pajares, F., & Urdan, T. (Eds.), *Self-Efficacy Beliefs of Adolescents* (*Vol. 5,* pp. 1–43). Greenwich, CT: Information Age Publishing.

Bell, R. L., Matkins, J. J., & Gansneder, B. M. (2011). Impacts of contextual and explicit instruction on preservice elementary teachers' understandings of the nature of science. *Journal of Research in Science Teaching, 48,* 414–436. doi:10.1002/tea.20402

Billeh, V. Y., & Hasan, O. E. (1975). Factors influencing teachers' gain in understanding the nature of science. *Journal of Research in Science Teaching, 12,* 209–219. doi:10.1002/tea.3660120303

Brickhouse, N. W., Dahger, Z. R., Letts, W. J., & Shipman, H. L. (2000). Diversity of students' views about evidence, theory, and the interface between science and religion in an astronomy course. *Journal of Research in Science Teaching, 37,* 340–362. doi:10.1002/(SICI)1098-2736(200004)37:4<340::AID-TEA4>3.0.CO;2-D

Carey, S., Evans, R., Honda, M., Jay, E., & Unger, C. (1989). An experiment is when you try it and see if it works: A study of grade 7 students' understanding of the construction of scientific knowledge. *International Journal of Science Education, 11,* 514–529. doi:10.1080/0950069890110504

Corno, L. (1993). The best-laid plans: Modern conceptions of volition and educational research. *Educational Researcher, 22*(2), 14–22.

Corno, L., & Mandinach, E. (1983). The role of cognitive engagement in classroom learning and motivation. *Educational Psychologist, 18,* 88–108. doi:10.1080/00461528309529266

Crawford, T. (2005). What counts as knowing: Constructing a communicative repertoire for student demonstration of knowledge in science. *Journal of Research in Science Teaching, 42*, 139–165. doi:10.1002/tea.20047

Cronbach, L. J., & Snow, R. E. (1977). *Aptitudes and instructional methods: A handbook for research on interactions.* New York, NY: Irvington.

Crumb, G. H. (1965). Understanding of science in high school physics. *Journal of Research in Science Teaching, 3*, 246–250. doi:10.1002/tea.3660030312

Deci, E. L. (1975). *Intrinsic motivation.* New York, NY: Plenum Press. doi:10.1007/978-1-4613-4446-9

Deci, E. L., & Ryan, R. M. (1991). A motivational approach to self: Integration in personality. In R. Diensbier (Ed.), *Nebraska Symposium on Motivation: Perspectives on Motivation,* (pp. 237-288). Lincoln, NE: University of Nebraska Press.

Duschl, R. A. (1988). Abandoning the scientistic legacy of science education. *Science Education, 72*, 51–62. doi:10.1002/sce.3730720105

Duschl, R. A. (1990). *Restructuring science education.* New York, NY: Teachers College Press.

Elliot, E., & Dweck, C. (1988). Goals: An approach to motivation and achievement. *Journal of Personality and Social Psychology, 54*, 5–12. doi:10.1037/0022-3514.54.1.5

Gallagher, J. J. (1991). Prospective and practicing secondary school science teachers' knowledge and beliefs about the philosophy of science. *Science Education, 75*, 121–134. doi:10.1002/sce.3730750111

Garcia, T., & Pintrich, P. R. (1994). Regulating motivation and cognition in the classroom: The role of self-schemas and self-regulatory strategies. In Schunk, D. H., & Zimmerman, B. J. (Eds.), *Self-Regulation of Learning and Performance: Issues and Educational Application* (pp. 127–153). Hillsdale, NJ: Erlbaum.

Ghatala, E. S. (1986). Strategy monitoring training enables young learners to select effective strategies. *Educational Psychologist, 21*, 434–454.

Graham, S., & Golan, S. (1991). Motivational influences on cognition: Task involvement, ego involvement, and depth of processing. *Journal of Educational Psychology, 83*, 187–194. doi:10.1037/0022-0663.83.2.187

Hanuscin, D., Akerson, V., & Phillipson-Mower, T. (2006). Integrating nature of science instruction into a physical science content course for preservice elementary teachers: NOS views of teaching assistants. *Science Education, 90*, 912–935. doi:10.1002/sce.20149

Harackiewicz, J. M., Barron, K. E., Pintrich, P. R., & Elliot, A. J. (2002). Revision of achievement goal theory: Necessary and illuminating. *Journal of Educational Psychology, 94*, 638–645. doi:10.1037/0022-0663.94.3.638

Henderson, R. W. (1986). Self-regulated learning: Implications for the design of instructional media. *Contemporary Educational Psychology, 11*, 405–427. doi:10.1016/0361-476X(86)90032-9

Hidi, S. (1990). Interest and its contribution as a mental resource for learning. *Review of Educational Research, 60*, 549–571.

Hodson, D. (1993). Philosophic stance of secondary school science teachers, curriculum experiences, and children's understanding of science: Some preliminary findings. *Interchange, 24*(1-2), 41–52. doi:10.1007/BF01447339

Hogan, K. (2000). Exploring a process view of students' knowledge about the nature of science. *Science Education, 84*, 51–70. doi:10.1002/(SICI)1098-237X(200001)84:1<51::AID-SCE5>3.0.CO;2-H

Howard, A., & Bray, D. (1988). *Managerial lives in transition*. New York: Guilford Press.

Johnston, A. T., & Southerland, S. A. (2002). *Conceptual ecologies and their influence on nature of science conceptions: More dazed and confused than ever*. Paper presented at the Annual Meeting of the National Association for Research in Science Teaching. New Orleans, LA.

Jungwirth, E. (1970). An evaluation of the attained development of the intellectual skills needed for understanding of the nature of scientific inquiry by BSCS pupils in Israel. *Journal of Research in Science Teaching, 7*, 141–151. doi:10.1002/tea.3660070210

Kanfer, R., & Ackerman, P. L. (1989). Motivation and cognitive abilities: An integrative aptitude treatment interaction approach to skill acquisition. *The Journal of Applied Psychology, 74*, 657–690. doi:10.1037/0021-9010.74.4.657

Khishfe, R. (2008). The development of seventh graders' views of nature of science. *Journal of Research in Science Teaching, 45*, 470–496. doi:10.1002/tea.20230

Khishfe, R., & Abd-El-Khalick, F. (2002). Influence of explicit and reflective versus implicit inquiry-oriented instruction on sixth graders' views of nature of science. *Journal of Research in Science Teaching, 39*, 551–578. doi:10.1002/tea.10036

Kirschner, P. A., Sweller, J., & Clark, R. E. (2006). Why minimal guidance during instruction does not work: An analysis of the failure of constructivist, discovery, problem-based, experiential, and inquiry-based teaching. *Educational Psychologist, 41*(2), 75–86. doi:10.1207/s15326985ep4102_1

Kitsantas, A., Zimmerman, B. J., & Cleary, T. (2000). The role of observation and emulation in the development of athletic self-regulation. *Journal of Educational Psychology, 92*(4), 811–817. doi:10.1037/0022-0663.92.4.811

Klahr, D., & Nigam, M. (2004). The equivalence of learning paths in early science instruction: Effects of direct instruction and discovery learning. *Psychological Science, 15*, 661–667. doi:10.1111/j.0956-7976.2004.00737.x

Klopfer, L., & Cooley, W. (1961). *Test on understanding science, form W*. Princeton, NJ: Educational Testing Services.

Lawson, A. E. (1982). The nature of advanced reasoning and science instruction. *Journal of Research in Science Teaching, 19*, 743–760. doi:10.1002/tea.3660190904

Leach, J. T., Hind, A. J., & Ryder, J. (2003). Designing and evaluating short teaching interventions about the epistemology of science in high school classrooms. *Science Education, 87*(6), 831–848. doi:10.1002/sce.10072

Lederman, N. G. (1992). Students' and teachers' conceptions about the nature of science: A review of the research. *Journal of Research in Science Teaching, 29*, 331–359. doi:10.1002/tea.3660290404

Lederman, N. G. (2007). The nature of science: Past, present, and future. In Abell, S. K., & Lederman, N. G. (Eds.), *Handbook of Research on Science Education*. London, UK: Lawrence Erlbaum & Associates, Publishers.

Lederman, N. G., Abd-El-Khalick, F., Bell, R. L., & Schwartz, R. (2002). Views of nature of science questionnaire (VNOS): Toward valid and meaningful assessment of learners' conceptions of nature of science. *Journal of Research in Science Teaching, 39*, 497–521. doi:10.1002/tea.10034

Liu, S., & Lederman, N. G. (2002). Taiwanese gifted students' views of nature of science. *School Science and Mathematics, 102,* 114–122. doi:10.1111/j.1949-8594.2002.tb17905.x

Locke, E. A., & Latham, G. P. (1990). *A theory of goal setting and task performance.* Englewood Cliffs, NJ: Prentice Hall.

Locke, E. A., & Latham, G. P. (2002). Building a practically useful theory of goal setting and task motivation: A 35-year odyssey. *The American Psychologist, 57,* 705–717. doi:10.1037/0003-066X.57.9.705

Mayer, R. (2004). Should there be a three-strikes rule against pure discovery learning? The case for guided methods of instruction. *The American Psychologist, 38,* 79–83.

McComas, W. F. (2005). *Seeking NOS standards: What content consensus exists in popular books on the nature of science.* Paper presented at the Meeting of National Association for Research in Science Teaching. Dallas, TX.

McComas, W. F. (2008). Seeking historical examples to illustrate key aspects of the nature of science. *Science & Education, 17*(2/3), 249–263. doi:10.1007/s11191-007-9081-y

McComas, W. F., Clough, M. P., & Almazroa, H. (1998). The role and character of the nature of science in science education. *Science & Education, 7,* 511–532. doi:10.1023/A:1008642510402

McComas, W. F., Lee, C. K., & Sweeney, S. (2009). *The comprehensiveness and completeness of nature of science content in the U.S. state science standards.* Paper presented at the National Association for Research in Science Teaching International Conference. Garden Grove, CA.

McDonald, C. V. (2010). The influence of explicit nature of science and argumentation instruction on preservice primary teachers' views of nature of science. *Journal of Research in Science Teaching, 47,* 1137–1164. doi:10.1002/tea.20377

Meichtry, Y. J. (1992). Influencing student understanding of the nature of science: Data from a case curriculum development. *Journal of Research in Science Teaching, 29,* 389–407. doi:10.1002/tea.3660290407

Morrison, J. A., Raab, F., & Ingram, D. (2009). Factors influencing elementary and secondary teachers' views on the nature of science. *Journal of Research in Science Teaching, 46,* 384–403. doi:10.1002/tea.20252

Moss, D. M., Abrams, E. D., & Kull, J. R. (1998). *Describing students' conceptions of the nature of science over an entire school year.* Paper presented at the Annual Meeting of the National Association for Research in Science Teaching. San Diego, CA.

Nolen, S. B. (1988). Reasons for studying: Motivational orientations and study strategies. *Cognition and Instruction, 5,* 269–287. doi:10.1207/s1532690xci0504_2

Ogunniyi, M. B. (1983). Relative effects of a history/philosophy of science course on student teachers' performance on two models of science. *Research in Science & Technological Education, 1,* 193–199. doi:10.1080/0263514830010207

Olson, J. K., & Clough, M. P. (2001). *Secondary science teachers' implementation practices following a course emphasizing contextualized and decontextualized nature of science instruction.* Paper presented at the 6th International History, Philosophy, and Science Teaching Conference. Denver, CO.

Osborne, J., Collins, S., Ratcliffe, M., Millar, R., & Duschl, R. (2003). What "ideas-about-science" should be taught in school? A Delphi study of the expert community. *Journal of Research in Science Teaching*, *40*, 692–720. doi:10.1002/tea.10105

Paris, S. G., Byrnes, J. P., & Paris, A. H. (2001). Constructing theories, identities, and actions of self-regulated learners. In Zimmerman, B. J., & Schunk, D. H. (Eds.), *Self-Regulated Learning and Academic Achievement: Theoretical Perspectives* (2nd ed., pp. 253–287). Mahwah, NJ: Earlbaum.

Paris, S. G., Cross, D. R., & Lipson, M. Y. (1984). Informed strategies for learning: A program to improve children's reading awareness and comprehension. *Journal of Educational Psychology*, *76*, 1239–1252. doi:10.1037/0022-0663.76.6.1239

Parkinson, J. (2004). *Improving secondary science teaching*. London, UK: Routledge Falmer. doi:10.4324/9780203464328

Peters, E. E. (2006). Connecting inquiry and the nature of science. *The Science Education Review*, *5*(2), 37–44.

Peters, E. E. (2012). Developing content knowledge in students through explicit teaching of the nature of science: Influences of goal setting and self-monitoring. *Science & Education*, *21*(6), 881–898. doi:10.1007/s11191-009-9219-1

Peters, E. E., & Kitsantas, A. (2010a). Self-regulation of student epistemic thinking in science: The role of metacognitive prompts. *Educational Psychology*, *30*(1), 27–52. doi:10.1080/01443410903353294

Peters, E. E., & Kitsantas, A. (2010b). The effect of nature of science metacognitive prompts on science students' content and nature of science knowledge, metacognition, and self-regulatory efficacy. *Journal of School Science and Math*, *110*, 382–396. doi:10.1111/j.1949-8594.2010.00050.x

Peters Burton, E. E. (2010). *Learning about the human aspect of the scientific enterprise: Gender differences in conceptions of scientific knowledge*. Advancing Women in Leadership Journal, 30(12). Retrieved from http://advancingwomen.com/awl/awl_wordpress/

Rohrkemper, M. (1989). Self-regulated learning and academic achievement: A Vygotskian view. In Zimmerman, B. J., & Schunk, D. H. (Eds.), *Self-Regulated Learning and Academic Achievement: Theory, Research and Practice* (pp. 143–167). New York, NY: Springer. doi:10.1007/978-1-4612-3618-4_6

Rosenshine, B., & Stevens, R. (1986). Teaching functions. In Wittrock, M. C. (Ed.), *Handbook of Research on Teaching* (3rd ed., pp. 376–391). New York, NY: Macmillan.

Rowe, M. B. (1974). A humanistic intent: The program of preservice elementary education at the University of Florida. *Science Education*, *58*, 369–376. doi:10.1002/sce.3730580311

Ryan, A. G., & Aikenhead, G. S. (1992). Students' preconceptions about the epistemology of science. *Science Education*, *76*, 559–580. doi:10.1002/sce.3730760602

Ryan, R. M., Connell, J. P., & Deci, E. L. (1984). A motivational analysis of self-determination and self-regulation in education. In Ames, C., & Ames, R. (Eds.), *Research on Motivation in Education* (Vol. 2, pp. 13–52). New York, NY: Academic Press.

Ryder, J., Leach, J., & Driver, R. (1999). Undergraduate science students' images of science. *Journal of Research in Science Teaching*, *36*, 201–219. doi:10.1002/(SICI)1098-2736(199902)36:2<201::AID-TEA6>3.0.CO;2-H

Sandoval, W. A., & Morrison, K. (2003). High school students' ideas about theories and theory change after a biological inquiry unit. *Journal of Research in Science Teaching, 40*(4), 369–392. doi:10.1002/tea.10081

Scharmann, L. C., Smith, M. U., James, M. C., & Jensen, M. (2005). Explicit reflective nature of science instruction: Evolution, intelligent design, and umbrellaology. *Journal of Science Teacher Education, 16*, 27–41. doi:10.1007/s10972-005-6990-y

Schiefele, U. (1991). Interest, learning, and motivation. *Educational Psychologist, 26*, 299–323.

Schunk, D. H. (1982). Verbal self-regulation as a facilitator of children's achievement and self-efficacy. *Human Learning, 1*, 265-277.

Schunk, D. H. (1985). Self-efficacy and classroom learning. *Psychology in the Schools, 22*, 208–223. doi:10.1002/1520-6807(198504)22:2<208::AID-PITS2310220215>3.0.CO;2-7

Schunk, D. H. (1987). Peer models and children's behavioral change. *Review of Educational Research, 57*, 149–174.

Schunk, D. H., & Hanson, A. R. (1988). Influence of peer-model attributes on children's beliefs and learning. *Journal of Educational Psychology, 77*, 313–322. doi:10.1037/0022-0663.77.3.313

Schwartz, R. S., Lederman, N. G., & Crawford, B. A. (2004). Developing views of nature of science in an authentic context: An explicit approach to bridging the gap between nature of science and scientific inquiry. *Science Education, 88*, 610–645. doi:10.1002/sce.10128

Schwartz, R. S., Lederman, N. G., & Thompson, T. (2001). *Grade nine students' views of nature of science and scientific inquiry: The effects of an inquiry-enthusiast's approach to teaching science as inquiry*. Paper Presented at the Annual Meeting of the National Association of Research in Science Teaching (NARST). St. Louis, MO.

Shapiro, B. L. (1996). A case study of change in elementary student teacher thinking during an independent investigation in science: Learning about the face of science that does not yet know. *Science Education, 80*, 535–560. doi:10.1002/(SICI)1098-237X(199609)80:5<535::AID-SCE3>3.0.CO;2-C

Shulman, L., & Keisler, E. (Eds.). (1966). *Learning by discovery: A critical appraisal*. Chicago, IL: Rand McNally.

Smith, C. L., Maclin, D., Houghton, C., & Hennessey, M. G. (2000). Sixth-grade students' epistemologies of science: The impact of school science experiences on epistemological development. *Cognition and Instruction, 18*, 349–422. doi:10.1207/S1532690XCI1803_3

Smith, M. U., & Scharmann, L. C. (1999). Defining versus describing the nature of science: A pragmatic analysis for classroom teachers and science educators. *Science Education, 83*, 493–509. doi:10.1002/(SICI)1098-237X(199907)83:4<493::AID-SCE6>3.0.CO;2-U

Stenhouse, D. (1985). *Active philosophy in education and science*. London, UK: George Allen & Unwin.

Sweller, J. (2004). Instructional design consequences of an analogy between evolution by natural selection and human cognitive architecture. *Instructional Science, 32*, 9–31. doi:10.1023/B:TRUC.0000021808.72598.4d

Tamir, P. (1972). Understanding the process of science by students exposed to different science curricula in Israel. *Journal of Research in Science Teaching*, 9, 239–245. doi:10.1002/tea.3660090309

Tao, P. (2003). Eliciting and developing junior secondary students' understanding of the nature of science through a peer collaboration instruction in science stories. *International Journal of Science Education*, 25, 147–171. doi:10.1080/09500690210126748

Thomas, J. W., & Rohwer, W. D. Jr. (1986). Academic studying: The role of learning strategies. *Educational Psychologist*, 21, 19–41.

Tobin, K., & McRobbie, C. J. (1997). Beliefs about the nature of science and the enacted science curriculum. *Science & Education*, 6, 355–371. doi:10.1023/A:1008600132359

Trent, J. (1965). The attainment of the concept "understanding science" using contrasting physics courses. *Journal of Research in Science Teaching*, 3, 224–229. doi:10.1002/tea.3660030309

Turner, T. (2000). The science curriculum: What is it for? In Sears, J., & Sorenson, P. (Eds.), *Issues in Science Teaching* (pp. 4–15). London, UK: Routledge Falmer. doi:10.1145/3166.3168

Urdan, T. C., Midgley, C., & Anderman, E. M. (1998). The role of classroom goal structure in students' use of self-handicapping strategies. *American Educational Research Journal*, 35, 101–122.

Wang, M. C., & Peverly, S. T. (1986). The self-instructive process in classroom learning contexts. *Contemporary Educational Psychology*, 11, 370–404. doi:10.1016/0361-476X(86)90031-7

Welch, W. W., & Pella, M. O. (1968). The development of an instrument for inventorying knowledge of the processes of science. *Journal of Research in Science Teaching*, 5, 64. doi:10.1002/tea.3660050115

Wigfield, A., Tonks, S., & Eccles, J. S. (2004). Expentancy-value theory in cross-cultural perspective. In McInerney, D. M., & Van Etten, S. (Eds.), *Big Theories Revisited* (Vol. 4, pp. 165–198). Greenwich, CT: Information Age.

Zimmerman, B. J. (1986). Development of self-regulated learning: Which are the key subprocesses? *Contemporary Educational Psychology*, 11, 307–313. doi:10.1016/0361-476X(86)90027-5

Zimmerman, B. J. (1990). Self-regulating academic learning and achievement: The emergence of a social cognitive perspective. *Educational Psychology Review*, 2, 173–201. doi:10.1007/BF01322178

Zimmerman, B. J. (2000). Attaining self-regulation: A social-cognitive perspective. In Boekaerts, M., Pintrich, P., & Zeidner, M. (Eds.), *Handbook of Self-Regulation* (pp. 13–39). San Diego, CA: Academic Press. doi:10.1016/B978-012109890-2/50031-7

Zimmerman, B. J. (2008). Goal setting: A key proactive source of academic self-regulation. In Schunk, D. H., & Zimmerman, B. J. (Eds.), *Motivation and Self-Regulated Learning: Theory, Research, and Applications* (pp. 267–295). Hillsdale, NJ: Lawrence Erlbaum Associates.

Zimmerman, B. J., & Bandura, A. (1994). Impact of self-regulatory influences on writing course attainment. *American Educational Research Journal*, 31, 845–862.

Zimmerman, B. J., Bandura, A., & Martinez-Pons, M. (1992). Self-motivation for academic attainment: The role of self-efficacy beliefs and personal goal setting. *American Educational Research Journal*, 29, 663–676.

Zimmerman, B. J., & Kitsantas, A. (1997). Developmental phases in self-regulation: Shifting from process to outcome goals. *Journal of Educational Psychology*, 89, 1–10. doi:10.1037/0022-0663.89.1.29

Zimmerman, B. J., & Kitsantas, A. (2002). Acquiring writing revision and self-regulatory skill through observation and emulation. *Journal of Educational Psychology*, 94(4), 660–668. doi:10.1037/0022-0663.94.4.660

Zimmerman, B. J., & Kitsantas, A. (2007). A writer's discipline: The development of self-regulatory skill. In Hidi, S., & Boskolo, P. (Eds.), *Motivation to Write*. New York, NY: Kluwer Publishers.

Zimmerman, B. J., & Martinez-Pons, M. (1992). Perceptions of efficacy and strategy use in the self-regulation of learning. In Schunk, D. H., & Meece, J. (Eds.), *Student Perceptions in the Classroom: Causes and Consequences* (pp. 185–207). Hillsdale, NJ: Erlbaum.

Chapter 2
Multiple Perspectives for the Study of Teaching:
A Conceptual Framework for Characterizing and Accessing Science Teachers' Practical–Moral Knowledge

Sara Salloum
Long Island University – Brooklyn, USA

ABSTRACT

This chapter outlines a framework that characterizes science teachers' practical-moral knowledge utilizing the Aristotelian concept of phronesis/practical wisdom. The meaning of phronesis is further explicated and its relevance to science education are outlined utilizing a virtue-based view of knowledge and practical hermeneutics. First, and to give a background, assumptions about teacher knowledge from a constructivist and sociocultural perspective are outlined. Second, the Aristotelian notion of phronesis (practical wisdom) is explicated, especially in terms of how it differs from other characterizations of practical knowledge in science education and how it relates to practical-moral knowledge. Finally, the authors discuss how the very nature of such practical-moral knowledge makes it ambiguous and hard to articulate, and therefore, a hermeneutic model that explores teachers' practical-moral knowledge indirectly by investigating teachers' commitments, interpretations, actions, and dialectic interactions is outlined. Implications for research and teacher education are outlined. Empirical examples are used to demonstrate certain points. A virtue-based view of knowledge is not meant to replace others, but as a means to enrich the understandings of the complexity of teacher knowledge and to enhance the effectiveness of teacher educators.

DOI: 10.4018/978-1-4666-2809-0.ch002

INTRODUCTION

Teachers (practitioners) often hold visions of 'good' teaching that differ from those of researchers, teacher educators, and reform documents (e.g., AAAS, 1990; NRC, 1996). Such disparity is a facet of the gap between theory and practice in education, and entails significant difficulties for research and reform efforts aimed at getting teachers to embrace and enact visions of 'good' teaching valued by researchers and teacher educators (Kennedy, 2006; Wildy & Wallace, 1995). Despite concerted efforts, closing this gap has been more elusive than initially imagined (Carr, 1995; Crawford, 2007), where substantial reform efforts in science education have met with limited success (Lynch, 2001; Smith & Southerland, 2007). We argue that a major reason for such limited success stems from educational research and reform that aims at changing teacher *actions* without ample understanding of underlying teacher knowledge.

Teacher actions have been the focus of research efforts since the early 1960s. Initially teacher actions were studied in an attempt to isolate 'effective' teaching techniques (process-product research) (e.g., Medley, 1979; Doyle, 1977; as cited in Issler, 1983; Woolfolk & Galloway, 1985). More recently, teacher actions (including language) have been scrutinized in the broader context of examining teacher knowledge and beliefs; and their influence on student learning, attitudes, skills, and classroom dynamics (e.g., Anderson & Mitchener, 1994; Borko & Putnam, 1996; Haney & McArthur, 2002; Moje, 1995; Mulholland & Wallace, 2008; Tsai, 2002). The latter focus on teacher knowledge, beliefs, and practices has uncovered yet further complexities of the gap between theory and practice. In science teaching, changing teachers' beliefs proved difficult (Smith & Southerland, 2007), and even when teachers held or shifted towards reform consistent-beliefs, dissonance often emerged between actual classroom practices and stated beliefs about teaching, learning, and nature of science (Bell, Lederman, Abd-El-Khalick, 2000; Simmons, et al., 1999; Southerland, Gess-Newsome, & Johnston, 2003). Evidently, translation of teacher beliefs and knowledge (specifically theoretical) into practice is more complex than initially perceived. Due to such complexity, the nature of teacher knowledge remains a rich area for exploration (Mulholland & Wallace, 2008).

PURPOSE

A broad aim of this chapter is to further elucidate a framework for understanding teaching practice as more than an arena for the application of theoretical knowledge and sets of skills (craft), but as a practice where teachers continuously engage a form of non-theoretical practical-moral knowledge. Acknowledging the role of non-theoretical knowledge in teaching has gained momentum in science education and several terms have been used to refer to it: practical knowledge (e.g., Duffee & Aikenhead, 1992; Fenstermacher, 1994; Lotter, Hardwood, & Bonner, 2007; Mulholland & Wallace, 2008; van Driel, Beijaard, & Verloop, 2001); practical-moral knowledge (Salloum & Abd-El-Khalick, 2010); and personal practical theories (Smith & Southerland, 2007). A complicating aspect of studying non-theoretical teacher knowledge though is elucidating and conceptualizing its character: Is it form of knowledge, reasoning, or an aspect of one's 'being' (e.g., Breire & Ralphs, 2009; Feldman, 2002)? Is a set of conceptions, skills, values, and beliefs that teachers develop with experience (e.g., van Driel, et al., 2001)? How can we study such knowledge? These issues have more practical importance than their esoteric nature suggests (Southerland, Sinatra, & Mathews, 2001), specifically since models promoted and utilized in educational research and teacher education are greatly influenced by conceptualizations of teacher knowledge and its nature.

In this chapter, teachers' practical-moral knowledge is characterized utilizing the Aristotelian concept of phronesis or practical wisdom (e.g., Breier & Ralphs, 2010; Clark, 2005; Flyvbjerg, 2001; Korthagen, & Kessels, 1999; Korthagen, Loughran, & Russel, 2006; Salloum & Abd-El-Khalick, 2010; Schwandt, 1996, 2005). However, as Breier and Ralphs (2010) stated, the concept of phronesis and practical wisdom is gaining popularity in education but not always clarity, especially when it comes to its role within specific disciplines such as science teaching and how to study and develop it. We argue in this a chapter that a greater understanding of phronesis will contribute to our understanding of science pedagogy and teaching practices. The aim of this chapter is to further explicate the meaning of phronesis and outline its relevance to science education and educational research. To do so, a virtue-based view of knowledge (Zagzebski, 1996) and practical hermeneutics (Gadamer, 1989) are utilized. First, and to give a background, assumptions about teacher knowledge from a constructivist and sociocultural perspectives are outlined. Second, the Aristotelian notion of phronesis (practical wisdom) is explicated, especially in terms of how it differs from other characterizations of practical knowledge in science education and how it relates to practical-moral knowledge. Finally, we discuss how the very nature of practical-moral knowledge deems it ambiguous and hard to articulate; and therefore a hermeneutic model that explores teachers' practical-moral knowledge *indirectly* by investigating teachers' commitments, interpretations, actions, and dialectic interactions between them is outlined. Implications for research and teacher education are outlined. Even though this is a conceptual chapter, empirical examples from my research will be used to demonstrate certain points (Salloum, 2006; Salloum & Abd-El-Khalick, 2010, Salloum, et al., 2010; Collier & Salloum, 2011).

ASSUMPTIONS ABOUT TEACHER KNOWLEDGE

Constructivist and sociocultural traditions have been invaluable in informing the majority of research studies in science education (Anderson, 2007). Cognitive and social constructivism have been greatly informed by Piaget and Vygotsky's theories. Piaget's and Vygotsky's theories are developmental, where learners progress through stages to ultimately develop the 'adult abilities' related to logical and abstract thinking. Thus, a basic assumption of cognitivists is that individuals are rational problem solvers (Lave, 1988). Naturally, teachers are assumed to be at a stage where they can engage in logical thinking to solve problems, and model such reasoning to their students. Teachers need to help students develop rational thought and use this reasoning to solve problems. Learners are perceived as actively constructing and restructuring knowledge, and so teachers play valuable roles in providing experiences and social settings to aid such re-structuring. Accordingly, 'good' teachers need to know how to create settings and activities for students to restructure ideas and reach higher levels of thinking and knowledge and promote both verbal and practical (through experimentation and inquiry methods) thoughtfulness.

Constructivists utilized Piagetian theory to help teachers restructure their knowledge of teaching and to embrace constructivist pedagogies:

Just as young learners, so, too, do teachers. Teacher education programs based in constructivist view of learning need to do more than offer a constructivist perspective in a course or two. Teachers' beliefs need to be illuminated, discussed, and challenged (Fosnot, 1996, p. 216).

An embedded assumption was that knowledge for teaching science is normative/theoretical and belief-based involving formal-rational reasoning (thus the interests in teacher beliefs). For example, it is assumed that by following logical paths (challenge and confrontation), teachers will change their beliefs about teaching and consequently their actions. A purely logical/rational model has been critiqued for being too narrow to account for teacher knowledge and reasoning (Noel, 1999), and has lead constructivists to take an either/or stance about 'good' science teaching (Wildy & Wallace, 1995). This stance is often criticized for neglecting situational and contextual aspects of teaching:

For science teachers one cannot underestimate the importance of understanding the cultural context of schools, having clear and consistent view of the subject matter, building a learning community of trust, and adapting the curriculum to accommodate the knowledge, needs, and aspirations of the students. Surely, this is what good science teaching is all about (Wildy & Wallace, p. 154).

A sociocultural perspective on science education was critical of constructivism, particularly cognitive constructivism, for neglecting the historical, social, cultural, and physical contexts of learning processes (e.g., Lemke, 2001; O'Loughlin, 1992). Sociocultural perspectives were informed by the work of Lave (1988) and Wertsch (1991). A main aspect adopted from Lave's work was that individuals are embodied selves actively engaged in relational activities with their world. Sociocultural researchers argued that constructivism framed learners in a decontextualized manner with no reference to their environment, which is problematic because "it denies the essentially collaborative and social nature of meaning making; and it privileges only one form of knowledge, namely, the technical rational" (O'Loughlin, 1992, p. 791). Thus, socioculturalists questioned normative theoretical/rational views of knowledge including the assumption of dualism and separation between mind and body (e.g. Brickhouse, 2001; Lave, 1988; Lemke, 2001). For example, Lemke (2001) critiqued the conceptual change model claiming, "there is more at stake than rational choice among competing theories.

Changing our mind is not simply a matter of rational decision making. It is a social process with social consequences" (p. 301).

Even though learners' knowledge has been problematized within sociocultural perspectives, both constructivists and socioculturalists still seem to hold normative views of teacher knowledge, assuming a "direct linkage between cognition and action" (Feldman, 2002, p. 1038), thus deeming knowledge for teaching as primarily theoretical-rational (Feldman, 2002; van Manen, 1995). Teachers are perceived as cognizing agents, who make and manage decisions by referring to generalized rules and codes, cohesive sets of beliefs, or a knowledge base (Feldman, 2002; van Manen, 1995). As mentioned above, this lead to extensive research on teacher beliefs to understand teacher knowledge and address difficulties in putting constructivist and sociocultural pedagogies into action (e.g., Anderson & Mitchener, 1994; Borko & Putnam, 1996; Haney & McArthur, 2002; Lederman, 1992; Meyer, 2004; Tsai, 2002; Windschitl, 1999). Difficulty in changing teachers' beliefs and dissonance between stated beliefs and actual practices showed that focus on a belief-based form of knowledge is not adequate to understand teacher knowledge. Sociocultural perspectives did extend research on beliefs by exploring the influence of external and cultural factors on translation of beliefs. Nonetheless, over-emphasis on outside cultural factors to explain adherence to uniformity can be problematic as it fails to take into account the personal and moral strife of individuals who actually transform cultures (Lave, 1988; Vianna & Stensenko, 2006). The aforementioned perspectives may have (unintentionally) undervalued the non-theoretical and non-cognitive dimensions of teachers' knowledge. In the next section, Aristotelian virtue ethics and a virtue-based view of knowledge (Zagzebski, 1996) are utilized to extend views of teacher knowledge. A virtue-based view is not meant to replace other views but to enrich our understandings of teacher knowledge, its study, and development.

TEACHING AND TEACHER KNOWLEDGE: FURTHER CONCEPTUALIZATIONS

Teaching seems to be the sort of occupation in which professional effectiveness is greatly enhanced by the possession and exercise of personal qualities and practical dispositions that are not entirely (if at all) reducible to academic knowledge or technical skills (Carr, 2007, p. 369).

Calls for looking outside theoretical and technical models to characterize teacher knowledge are becoming more and more popular in education and science education (e.g. Feldman, 2002; Carr, 2005, 2007; Eisner, 2002; Mulholland & Wallace, 2008; Noel, 1999; Shulman, 2007; van Driel, et al., 2001; van Manen, 1995, 1999). Several features of such knowledge are its *tacit* and *context-dependent* nature (non-propositional); its intimate relation to one's *character* and who one is as a person; and its role in moral *action*, or simply said in doing what is 'right' in a certain situation. In Salloum and Abd-El-Khalick (2010), we referred to such knowledge as practical-moral knowledge: practical since it is associated with everyday actions and ways teachers respond to situations; and moral since actions are reflective of who we are as individuals and our values. A fundamental assumption here, as Carr (2007) stated above, is that theoretical knowledge of content and pedagogy and technical skills are not enough to account for what constitutes 'good' practice. Rather, how and when practitioners make use of theoretical knowledge is influenced by their practical-moral knowledge (Kessels & Korthagen, 1996; Pendlebury, 1995). To better understand practical-moral knowledge, it is characterized by the Aristotelian concept of phronesis. There is a need first, however, to elucidate phronesis/ practical wisdom itself.

Phronesis

Aristotle proposed phronesis as one of the intellectual virtues meant to define 'truth disclosing' active conditions (Aristotle, 2002). Aristotle distinguished between two types of intellectual virtues: "one with which we contemplate those things whose first principles are *invariable*" (Aristotle, 1976, p. 204) and ones with which we contemplate things that can be *otherwise*. Phronesis and techne were designated as ones by which we contemplate what can be otherwise. Techne (craft or art) is on things that are *made* and phronesis (practical wisdom) involves *moral actions*. With phronesis, Aristotle was concerned with outlining the "right estimation of the role that reason has to play in moral action" (Gadamer, 1989, p. 310). The other truth-seeking condition of interest is episteme (scientific knowledge), which Aristotle defined as "a kind of judgment that concerns things that are universal" (Aristotle, 2002, p. 107); episteme is context-independent knowledge with invariable first principles. Episteme can be seen as the basis of theoretical and scientific knowledge. Compared to episteme or scientific knowledge, which is universal, conceptual, and decontextualized, phronesis is perceptual, situated, and involves concerns that are variable by nature (Eisner, 2002; Kessles & Korthahgen, 1996). As for differences with techne (craft or art), Aristotle stated that the ends for *making* are different than those for *acting*. In 'making' the ends are in producing something tangible separate from the 'making' process, whereas the end of acting well is acting well.

Aristotle identified practical wisdom as the ability to deliberate correctly or 'beautifully' about what is good and advantageous for oneself and what is conducive to living well in general (Aristotle, 1976, 2002). Practical wisdom is defined as "a truth-disclosing active condition involving reason about human goods that governs action" (Aristotle, 2002, p. 107). According to Aristotle, phronesis is the intellectual virtue necessary to realize *praxis* or morally committed and morally informed action concerned with how humans conduct their life as members of society (Kemmis, 2009, 2010; Schwandt, 2005).

Phronesis or practical wisdom is concerned with (a) what can be deliberated about and can be otherwise, (b) realizing a desired end, (c) reaching the end is by *voluntary action* based on deliberation. Practical wisdom is perceptual rather than conceptual (to discern what is to be deliberated on) and entails knowledge of particulars: practical wisdom "is not only about what is universal, but needs to discern particulars as well, since it has to do with action and action is concerned with particulars" (Aristotle, 2002, p. 109). Knowledge of particulars can be more important than knowledge of universals to be effective in action. An example to that effect is a teacher who knows the general principle stating that engagement through connecting science concepts to prior knowledge is essential for deep conceptual learning, but does not know what constitutes relevant prior knowledge for a particular group of students and/or for that particular topic. Such teacher would be less effective in action. Consequently, development of practical wisdom requires experience and thoughtful reflection on experience.

Functions of Phronesis

Aristotle distinguished phronesis from techne (craft and art) and episteme (scientific knowledge) and gave it prominence as the intellectual virtue that governs action and one that cultivates and mediates other virtues (such as temperance, honesty, courage, etc.). The relation between practical wisdom and virtuous character is interdependent: It is not possible to be good in the governing sense without practical wisdom, nor to have practical wisdom without virtue of character (Aristotle, 2002). Virtues are demarcated as acquired excellences that take time and effort to develop, and are distinguished from natural capacities and skills.

Zagzebski (1996) utilized Aristotle's virtue ethics to outline a virtue-based theory of knowledge that binds knowledge intimately with moral concepts. Within such a view, virtues can be moral or intellectual: The difference between moral and intellectual virtues is that all intellectual virtues act as motivation to develop new knowledge and skills (Zagzebski, 1996). For example, the virtues of intellectual sobriety, including careful inquiry, and intellectual fairness would motivate a teacher to continuously gain new knowledge about students and ways to represent content in multiple ways to provide access to all of them (Collier & Salloum, 2011).

Phronesis mediates the different intellectual and moral virtues by serving several functions (Zagzebski, 1996). One is to determine the virtuous course required or called for in a particular situation. For example, how empathic should a teacher be in different situations and with different students? The question is not to be empathic or not as two extremes, but what degree of empathy is appropriate considering the situation at hand or students' particular needs. Another function of phronesis is to balance choices among virtues that may lead to conflicting lines of action. For example, in certain situations both empathy for students and responsibility for them learning abstract and complex content are called for; however, engaging in one may conflict with the other. Teachers need to engage phronesis or practical wisdom to resolve such tension. A teacher would need to know how to push students as they learn abstract science concepts and master them, without reaching a point where their frustration can turn them off the content. A third function for phronesis is to coordinate different virtues (and understandings) into a course of action or a line of thought that would lead to action. A course of action resulting from exercising practical wisdom will always involve indeterminacy and actions cannot be entirely accounted for using purely rational arguments (Gadamer, 1989; Zagzebski, 1996).

Practical Wisdom and Teaching

Phronesis and practical wisdom as an intellectual virtue encompasses features and functions that seem to characterize several aspects of knowledge engaged in good teaching. Pendlebury (1995) stated that:

Practical wisdom is the sovereign virtue of a good teacher—a virtue whose realization in teaching requires a subtle interplay between several binary oppositions: reason and imagination, experience and innocence, cleanness of argument and richness of story, respect for principles and attunement to particulars (p. 50).

Situations in the classroom are way too complex to involve one form of knowledge or to entail a clear set of discrete goals and desired ends. Teachers always find themselves having to balance several ones (Kennedy, 2006): inquiry-based teaching for meaningful and conceptual understanding in science *and* test-prep teaching for high achievement on standardized test; serving students with diverse abilities and needs *and* moving the classroom forward within a mandated and tested curriculum; simultaneously the teacher strives to build a respectful and trusting classroom community (Kennedy, 2006; Salloum & Abd-El-Khalick, 2010; Salloum, et al., 2010). The desired ends above do not just involve theoretical or cognitive aspects, but at their heart are moral and intellectual virtues inherent to good teaching (Carr, 2007): *fairness* to students with diverse needs and abilities and to science as a discipline, *responsibility* for students' learning, their attitudes towards science, and their achievement in gatekeeper tests. For example, in interviews, teachers always rationalized their practices and actions drawing on notions such as being responsible towards students and leading them to a "safe haven" when it came to passing high stakes tests (Salloum & Abd-El-Khalick, 2010); or as being *fair* to diverse students by providing access either to difficult content or to opportunities for a better future (e.g. college admission) (Salloum, et al., 2010; Collier & Salloum, 2011).

Moreover, and specifically for science and science teaching, virtues such as the following seem especially relevant and necessary for promoting inquiry in science and deep understandings of nature of science: intellectual sobriety (careful inquiry and accepting what is warranted by evidence); honesty; open mindedness and impartiality; intellectual humility; intellectual courage; perseverance, determination, and thoroughness; and fairness in evaluating arguments of others (Zagzebski, 1996). Probably the most famous intellectual virtues in teaching are the attitudes outlined by Dewey (1933) for reflective practice: wholeheartedness, responsibility, and open-mindedness. Other relevant virtues for teaching are thorough inquiry and reflection on inquiry, the social virtues of being communicative, intellectually honest, and empathic (Carr, 2007; Zagbeski, 1996). What qualifies that above virtues as 'intellectual' is the fact that their enactment entails motivating teachers to develop new understandings, knowledge, and skills.

The enactment of virtues as ones outlined above and knowledge development motivated by them cannot be prescriptive nor is it an easy undertaking. Such enactment will involve a whole range of tensions and dilemmas among virtues themselves, requirements of specific content, and external factors such as mandated curricula and testing. Practical wisdom is essential in resolving tensions and in balancing desired ends: How do we go about managing situations that entail different and sometimes competing desired ends and virtues? How do we go about finding the appropriate mean of a virtue between excess and deficiency? It is such questions that practical wisdom is involved in.

Practical Wisdom and Science Teaching

In science education, non-theoretical practical knowledge has also emerged as an important aspect of science teachers' practice and change (e.g., Duffee & Aikenhead, 1992; Lotter, Hardwood, & Bonner, 2007; Mulholland & Wallace, 2008; van Driel, Beijaarrd, & Verloop, 2001). However, upon examining literature on practical knowledge in science education, it is noted that practical knowledge is characterized by craft knowledge rather than practical wisdom and prudence towards moral action (e.g., Mulholland & Wallace, 2008; van Driel, et al., 2001). Craft is extended to include reflective practice, which entails intellectual virtues such as openmindedness, wholeheartedness, and responsibility. Nonetheless, "reflective" craft knowledge still does not adequately capture the enactment of virtues nor the engagement in strong judgments to serve embedded 'goods' in teaching with respect to students' interests, integrity towards content, and one's own development as a practitioner.

Knowledge as craft is based on techne, one of Aristotle's intellectual virtues. Both phronesis and techne are contextual and pragmatic, but techne is exhibited in production, whereas phronesis is essential for moral practice (Carr, 2007). Gadamer's (1989) elaboration on distinctions between phronesis/practical wisdom and techne/craft is relevant to show limitations of techne/craft in characterizing teacher knowledge. Techne can be encapsulated in sets of skills that can be directly learned to produce something tangible. These are at our disposal to use or not to use (Gadamer, 1989). A craftsperson may choose to acquire them and later they can be forgotten. Whereas phronesis or practical wisdom is involved in situations where we have to act towards desired ends pertaining to teaching science well. Knowledge informing our actions to do 'well' is bound with images of who we are as individuals/science educators and our ideas of right and wrong (e.g., intellectual honesty, open-mindedness, responsibility, empathy, etc.). For example, a teacher can acquire skills in creating efficient lab set-up, clear lab directions and worksheets, and multi-level and diverse assessments. The teacher may choose to use or not use these skills based on the group of students or the topic area. However, a teacher cannot choose not to be intellectually honest without this affecting her character and how she sees herself. Rather her practical wisdom would inform her on what level of intellectual honesty is appropriate in a specific situation. An empirical example will be used to demonstrate this point (Salloum, 2006). Ms. Yasmine is a middle school physical science teacher. She attended a professional development workshop, where they were told that the connotation $HCl_{(aq)}$ is not chemically acceptable, since HCl does not exist in water but rather it is $H^+_{(aq)}$ and $Cl^-_{(aq)}$ ions that are found in the solution. Ms. Yasmine was not satisfied with the proposed concept and actually checked the grade 8 book and found $HCl_{(aq)}$ used several times. When asked about her personal opinion of what is a more correct connotation, she said it is easier for the students to see $HCl_{(aq)}$ as one entity, especially for solving displacement reactions. I asked her to elaborate on how she would deal with this issue in her teaching:

as a chemist I want what is found in the solution (here referring to having the ions and not HCl in an aqueous state), But as a teacher, I will never tell the students something not scientifically accurate, but I start with the correct (and easier) and then give the more correct, I never tell them something that is 'wrong!'

In the preceding excerpt, Ms. Yasmine's was trying to resolve the 'extent' to which she needs be intellectually honest and at the same time help students understand the material. Ms. Yasmine deliberated on a tension between herself as a 'good' teacher and as a 'good' chemist and her resolution was to give students the *less correct* but *more understandable*, at least as a starting point. She felt she had to maintain being intellectually honest (because this is how she sees herself), but the extent was determined by her assessment of students' needs. In the case above, Ms. Yasmine engaged her practical wisdom to determine the appropriate extent of intellectual honesty needed in a particular situation. As mentioned above on functions of practical wisdom, the question is not whether she needs to be intellectually honest or not as two extremes, but rather the question entailed determining the extent appropriate in light of the situation at hand.

Another distinction between phronesis and techne is that techne can have well defined ends, and means to achieving them are separate from the ends (Aristotle, 2002; Gadamer, 1989). Whereas with phronesis, virtues makes the end 'right,' and practical wisdom makes the means and things related to it 'right' (Aristotle, 1976). With phronesis, the ends are boarder and somewhat uncertain since they pertain to several 'goods' that science teaching is to serve. Moreover, "the considerations of the means is itself a moral consideration and it is this that concretizes the moral rightness of end" (Gadamer, 1989, p. 319); therefore, phronesis is about choosing the 'right' means to get to the 'right' end. An example from science education can be the phenomena of 'teaching to the test.' A teacher can maintain 'students passing the test' as a particular end and can become very skillful in teaching students techniques to tackle test problems, with or without students understanding and retaining the underlying science concepts[1]. Even when students only learn science through rote or the 'plug and chug' method, the desired end of doing well on certain tests is still achieved with techne. With phronesis or practical wisdom the desired end cannot just be about passing the test, rather the ends would be broader and pertain to several 'goods' that science teaching needs to serve: e.g., students passing gatekeeper exams *and* promoting understandings of science concepts and science as discipline. Practical wisdom would be about finding the means to balance the sometimes competing desired ends, that is, teaching to the test and enhancing students' conceptual understandings and reasoning simultaneously. In Salloum and Abd-El-Kahlick (2010), we see Ms. Yasmine doing so by associating deep conceptual understandings of science concepts with better concept retention for doing well in tests.

Finally, teacher knowledge as characterized by phronesis is seen as simultaneously a way of being and thinking/acting without a clear separation between the two (Schwandt, 2008, personal communication) or as Gadamer points out, "… knowledge, not detached from being that is becoming, but determined and is determinative of it" (Gadamer, 1989, p. 310). Accordingly, several assumptions about knowledge emerge: knowledge in this sense is not governed by rules or encapsulated in a theoretical or formal knowledge base or propositional beliefs. Rather, practitioners' knowledge actualizes by being immersed in situations, and by "interpreting formal knowledge[2], beliefs, past experiences, and commitments according to perceived situations" (Salloum & Abd-El-Khalick, 2010).

Practical Wisdom and Practical-Moral Knowledge

In Salloum and Abd-El-Khalick (2010), the authors utilized the term Practical-Moral Knowledge (P-MK) to emphasize its intimate relation with moral concepts within a practice and the cultivation of intellectual and moral virtues such as honesty, fairness, responsibility, open-mindedness etc. The main reason for using this term rather than practical wisdom is that I see teacher's practical-moral knowledge as potentially developing into practical wisdom, but not as equivalent to it. Practical wisdom requires experience and advanced kinds of thoughtful reflection and understanding: the kind of wisdom that a master science teacher would have. Practical wisdom may not characterize knowledge of all teachers (either due to lack of experience or more seriously the lack of passion or an essential virtue). However, it can be suggested that *all* teaching involves engaging practical-moral knowledge even when such knowledge is misguided and does not lead to wise choices. We believe that such practical-moral knowledge needs to be further investigated for better teacher education and that we need to especially explore insights on how to cultivate practical-moral knowledge into practical wisdom.

Different forms of knowledge are all important and essential to teaching (and good teaching). The view suggested here is meant to enrich our understandings of teacher knowledge rather than replace other views. Knowledge for and of teaching will always involve different levels and domains of knowledge: theoretical knowledge of subject matter and pedagogy (episteme), practical-moral knowledge, and craft knowledge (techne). However, what is suggested here is that practical-moral knowledge with its essentially moral nature is how teachers manage different knowledge domains and make value-judgments in action; and those judgments are aligned with virtues they see as defining their practice. In the next section, we discuss general implications for characterizing teacher knowledge as practical-moral knowledge and practical wisdom.

GENERAL IMPLICATIONS

Difficulties in Exploring Practical-Moral Knowledge

Practical-moral knowledge, similar to other forms of practical and craft knowedge, is tacit and cannot be captured by propostions, therefore is illusive and hard to examine and assess; nor can it be directly demonstrated and 'taught' to others (Aristotle, 2002). Practitioners may not be able to articulate what such knowledge is about or how they engage it (Breier & Ralphs, 2009; Lotter, Hardwood, & Bonner, 2007; van Driel, et al., 2001; van Manen, 1999). van Driel et al. explained that both narrative research approaches and research on teacher beliefs have been utilized to elucidate practical forms of knowledge (e.g., Craig, 2006; Lotter, et al., 2007; Xu & Connelly, 2009). They added that narrative research, though valuable, needs to be theoretically interpreted lest teachers' narratives become a set of stories that lack theoretical grounding. On the other hand and as mentioned above, research on beliefs does not always capture teachers' strong judgments in enacting conflicting beliefs.

Another concern in research on practical-moral knowledge is its perceived content-less nature[3]. For instance, notions such as pedagogical tact (van Manen, 1991, 1995) capture the embodied, perceptive, and moral aspects of non-theoretical knowledge, but what does it means to be "tactful" towards physics or chemistry. The role of content knowledge is especially important in science education and consequently interconnections between content and virtues. Therefore, it is especially important in science education to construct a conceptual framework that allows for empirical exploration and analyses in both academic-theoretical and practical-moral domains. In Salloum and Abd-El-Khalick (2010), the authors proposed a hermeneutic framework that empirically explores and illuminates practical-moral knowledge by examining teachers' interpretations, commitments, actions, and interactions among them (see Figure 1). This framework is further developed below to show connections between virtues and teacher interpretations and commitments. But first we explain briefly why interpretations (from a hermeneutic perspective) and commitments are central to understanding practical-moral knowledge

Figure 1. A framework for investigating science teachers' practical-moral knowledge (Salloum & Abd-El-Khalick, 2010)

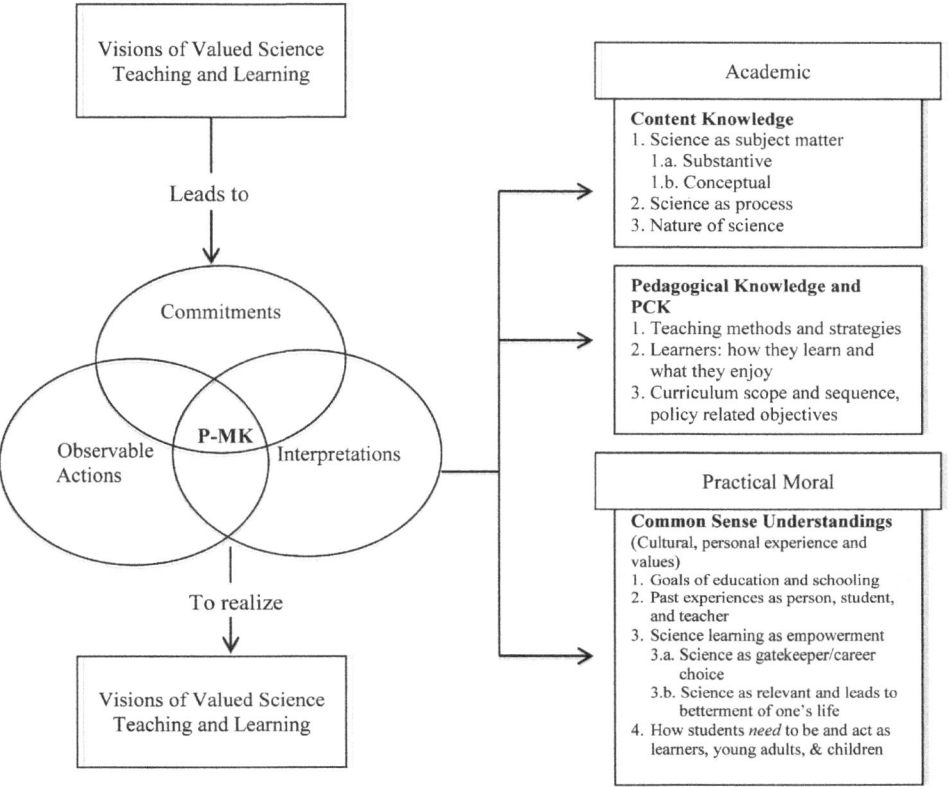

Centrality of Interpretations and Commitments to Practical-Moral Knowledge

From a hermeneutic perspective, interpretation is closely bound to understanding and therefore knowledge development, "Interpretation is not an occasional, post facto supplement to understanding; rather understanding is always interpretation, and hence interpretation is the explicit form of understanding" (Gadamer, 1989, p. 306). Moreover, Schwandt (2000) stated that "Philosophical hermeneutics argues that understanding is not, in the first instance, a procedure- or rule-governed undertaking; rather it is a very condition of being human" (p. 194). Interpretation as a human condition is a process we always find ourselves in: we cannot totally control and it is always constrained by a point of view (Gadamer, 1989). Interpretation as a way of being entails insight, discernment of salient features and prior knowledge, and emotions. It accrues in new understandings and knowledge. From a hermeneutic perspective, interpretation goes beyond the concept of 'beliefs,' which are defined as cognitive imperatives expressed in propositions (Pajares, 1992). Interpretation, on the other hand, is not necessarily propositional but "is carried out primordially not in a theoretical statement but an *action*" (Heidegger, 1962; as cited by Gallagher, 1992, p. 43, emphasis added).

Like practical wisdom, practical-moral knowledge is perceptual and thus involves interpretation of situations in light of prior knowledge and emotions about issues and individuals. From interpretation, understandings will emerge that lead to certain lines of action. Based on how a teacher interprets and frames the situation, the line of action may be constructive or counterproductive. For example, a student or group of students in a classroom may act up; depending on who the students are, the material being presented, and/or context of day and time, the teacher may interpret the situation as one requiring either punishing and reprimanding students or as one requiring reengaging students by slowing and revising instruction. The two lines of actions will have implications on how the lesson proceeds and what goals are achieved.

Teacher interpretations involve coordinating not just formal knowledge of subject matter and students (cognitive aspects), but also managing emotions, commitments, and virtues, we see definitive of our practice; accordingly, Carr (2007) maintained:

In fact, the compelling appeal of virtue ethics is that it shows precisely how the often distinguished and separated cognitive, affective, social, and motivational aspects of moral life may be coherently re-connected (p. 373).

Events of interpretation and understanding are central to the development and cultivation of teachers' practical-moral knowledge. Alternatively, practical-moral knowledge itself influences further events and acts of interpretations (the hermeneutic circle).

Story of a Teacher

An empirical example will be used below to demonstrate how perceptions, interpretation, and handling cognitive, affective, and social aspects interacted in a science classroom and the positive and negative implications that ensued (Salloum, 2006). Ms. Dunia[4] taught middle school physical science at a medium sized village public school in Mount Lebanon. Village settings tend to be more conservative and attached to tradition; people are involved in each other's lives and know details and intricacies of each other's lives and conditions. On one hand, this enhances the sense of community in schools, and on the other, it blurs boundaries between personal and professional spaces. The school served students of lower middle class socioeconomic status, with some families being more economically challenged than others. The author spent five months observing Ms. Dunia's science class and a total of nine months interacting with her and collecting data about her teaching and how she makes sense of her practice (I still maintain a personal relation with Ms. Dunia). Ms. Dunia's practice showed a strong commitment to 'challenging' students. This 'challenge,' however, took two forms: academic and social. The academic form of 'challenging' took a positive form evidenced by Ms. Dunia's use of inductive methods and careful questioning to have students develop science understandings independently. For example, Ms. Dunia was asked to go through one day's field notes and elaborate on what she saw as an important event. During that lesson, Ms. Dunia had used careful questions to have students come up with formula for pressure ($P=F/A$) from their qualitative understandings of the relation between pressure, force and area. Ms. Dunia stressed the value of students deriving the formulae themselves and her building on their qualitative understanding of the concepts to reach the formulaic quantitative expression ($P=F/A$):

The statement (qualitative) was produced by them (students); but they have to change to formula because physics is formulae and they have to work with formulae ...first they understand and then they change it into a formula.

Multiple Perspectives for the Study of Teaching

In a different interview, Ms. Dunia elaborated about the importance of challenging students to derive new knowledge themselves and thus promoting their academic independence:

Students can memorize anything, but they need to discover things in science, especially if related to their daily life.

Unfortunately, this academic and positive form of Ms. Dunia's commitment to 'challenging' students did not always translate into higher student achievement and positive attitudes towards science. Even though Ms. Dunia sought conceptual understandings of science concepts, her strife for conceptual understanding did not always actualize and many students did not do as well as she hoped, nor did they have positive attitudes towards science. Of course, many factors contribute to the issue, including low second language proficiency (science in Lebanon in taught in English or French) and lower aspirations and support for students' of lower socioeconomic status. Yet, classroom observations, interviews with Ms. Dunia, and conversations with students and the administration uncovered an additional factor that may have contributed to the situation. Ms. Dunia, who is a dedicated teacher, demonstrated a different form of 'challenging' students. This second form of challenge took a social shape that seemed to lead to negative effects. The excerpt below from grade 8 field notes displays the two forms of 'challenge' Ms. Dunia held on to:

Students reach the question where they need to indicate the reaction type. Ms. D asks them not to look at the books. She says the reactants and has the student at the board find the products and write symbols properly, then balance it. After that, students have to indicate the type of reaction. Ms. D does not accept when they say the type of reaction without proper justification. She uses questions to guide them through the different reactions, and for the most part, she wants the student on the board themselves to reach the acceptable answer:

Ziad is at the board Ms. D asks him to underline the ions of the reaction after he wrote it:

$$\underline{Na}\ \underline{OH} + \underline{H}\ \underline{Cl} \rightarrow \underline{Na}\ \underline{Cl} + \underline{H_2O}$$
$$\ \ A\ \ \ \ \ B\ \ \ \ \ \ \ \ C\ \ D\ \ \ \ \ \ \ \ A\ \ \ D\ \ \ \ \ \ \ C\ \ \ B$$

Ms. D: *What part does the displacement: H or H_2? (no response)*
Ms. D: *Na replaces what?*

Ziad looks at the products and says Na and Na ...

Then he realizes that Na replaces H and says it. Ms. D keeps on asking till he says confidently that Na replaced H.

Ms. D: *what do we call the reaction?*
Ziad: *double displacement.*
Rami: *now the idea is clearer! (I think he refers to underlining the ions on the board)*
Ms. D: *isn't this how we explained it before?! (With exasperation, in a challenging tone)*

Above we see that Ms. Dunia 'challenged' Ziad (student on the board) to independently come up with answers and identify the reaction type, which is in line with academic 'challenge.' Some students had asked earlier for further explanation of reaction types and that is why Rami remarked: *Now the idea is clearer!* Ms. Dunia challenged him: *Isn't this how we explained it?!* Here the challenge took a social form of pointing out to students' lack of attention the first time the concept was explained. Ms. Dunia confronted Rami (who is usually the class joker) with her statement. Ms. Dunia's social 'challenge' of students was evidenced by her harshness in dealing with students, especially ones she perceived as not giving enough effort to do well. Some students had expressed annoyance to me at what they saw as harsh and sometimes demeaning language towards them. I suggest that

Ms. Dunia's harshness and rigidity are grounded in a commitment to 'challenge' students in a local social and disciplinarian view. Such (common) social view maintains that when individuals are faced with their 'failure' and 'inadequacy' by an authority figure, they will be motivated to better themselves[5].

From a virtue perspective, Ms. Dunia seemed to uphold the importance of students' *intellectual independence,* which is related to valuing intellectual sobriety and inquiry; at the same time, she upheld to a moral commitment of being a strict disciplinarian responsible for shaping students' behavior. A commitment as a disciplinarian championed teacher authority and devalued students' *social independence,* which is perceived as disrespect. Managing those two virtues (intellectual independence and responsibility as a disciplinarian) proved somewhat difficult for Ms. Dunia and this was demonstrated in a very tense episode that occurred towards the end of the year when grade 9 students took mock national examinations. In Lebanon, Grade 9 students sit for high stakes national examinations to determine if they continue in an academic track or switch to a vocational track. In preparation for the national exams, schools conduct several mock exams towards the end of the year. It is important to note that students' performance in the national exams is used as a source of prestige for schools.

Students did poorly in the mock exams and this resulted in high tension between Ms. Dunia and students, who she was especially harsh to; and between her and administration, who saw her as being too harsh on students. When discussing the episode with the author, Ms. Dunia framed the situation using interpretations of students lacking care and hard work. For example, she recounted that a lower performing student did well on the physics exam because he studied with his cousin Nasma (one of Ms. Dunia's best students). She added that Nasma is not much smarter than the others, but her hard work and her parent's keenness on her getting a good education make her an excellent student. She added by drawing on personal and cultural interpretations to support her arguments:

They (meaning Ms. Dunia's parents) told us this is your future (good education) and your inheritance; this is what we heard growing up...

I asked about her harsh words to students the day before, specifically telling them that she will be happy for the ones who pass and will also be happy for the ones who do not:

Sara: *Don't you think that students may interpret this as you not caring for them?*
Ms. D: *They should not think so, because when we come every day and give them all the lessons and we try that they miss nothing. We come even when there are strikes, days that we do not get paid for, we come to school and teach; why would they think and feel like that?*

During that particular interview, Ms. Dunia used a combination of cultural, personal, subject matter, and pedagogical interpretations to make sense of students' performance and her strictness and harshness. Her frustration materialized in feelings of indignation and disappointment at both students and administration and actual physical discomfort (stomach ache). More importantly, it seemed that her indignation also rested in her being challenged back by students, and to a certain extent by the administration, who she saw as not being appreciative. As a researcher, I cannot interpret Ms. Dunia's practice as 'uncaring,' as her care for the students and the school community was evident throughout the fieldwork, but there is evidence that her management of the situation in light of the conflicting aspects in her practice, that is valuing academic independence but resenting social independence, was not fruitful to her or to others.

Ms. Dunia's practical-moral knowledge was engaged in trying to make sense of the situation. Unfortunately, outcomes did not serve students nor did they serve her, as she was not able to exercise her commitment to 'challenge' students in ways she valued. It is in situations such as the one described above where the engagement and cultivation of practical wisdom becomes essential to manage the complexity of a situation for the benefit of oneself and others (Schwartz & Sharpe, 2010; Zagzebski, 1996). Yet, cultivation of practical wisdom requires modeling and mentorship and open dialogues. Ms. Dunia needed support to question and revise her interpretations and commitments, especially in light of the cognitive, social, and emotional needs of students; and this was not always available, especially as educational setting become more and more test-driven.

Virtues, Interpretation, and Commitments

Teacher interpretations can be within academic domains such as understandings of content, its nature, pedagogical theories, etc. (see Figure 1). Interpretation can also be within moral domains where teachers interpret academic knowledge according to situations at hand and attempt to exercise discretion and 'wise' judgment in daily teaching actions based on commitments seen as inherent to good teaching. As we saw in the example above, those two domains overlap. Interpretation within moral domains are closely connected to teachers' interpretations of the 'goods' science education ought to serve (e.g., promoting scientific inquiry, curiosity, and literacy; science as gatekeeper to career choices, etc.). Accordingly, teachers construct visions and images of valued science education and of themselves as worthy practitioners. Images of what a person ought to be are integral in actions resulting from practical wisdom and practical-moral knowledge: for example images of fairness, responsibility, openness to the ideas of others, empathy, intellectual honesty, and perseverance, etc. (Gadamer, 1989). Teachers maintain standing commitments to such 'goods[6].' Commitments teachers verbalize and act upon reflect virtues they see as important to them as teachers. For example, a teacher who sees her major role as having students achieve good grades in national or state exams to get into certain colleges and majors, will become committed as such and work diligently towards such end as a 'responsible' and 'fair' teacher (as in one of the case studies in Salloum and Abd-El-Khalick, 2010). Also as we saw above, Ms. Dunia valued herself as a disciplinarian as well as a science teacher who challenges students as they 'discover' science concepts; she saw these two aspects of her practice as 'goods' that needed to be served. The dialectical interactions of interpretations and commitments are expressed in actions and classroom practices.

Implications for Educational Research

In order to empirically explore practical-moral knowledge and virtues embodied in it, it is asserted that: (a) teachers need to be engaged in dialogues and inquiries about their interpretations and commitments as practitioners; and (b) investigations of teachers' interpretations and commitments, verbalized as narrative and enacted, along with actions will illuminate practical-moral knowledge and embodied virtues to both practitioners and researchers. Such inquiry, though, needs to be done *with* teachers rather than on them (Schwandt, 1996). Schwandt proposed that the aim of such an inquiry would not be to produce theoretical/scientific knowledge to be prescribed to practitioners or to replace their practical-moral knowledge, rather the inquiry would be to engage practitioners in critical reflections and revision of their own practical-moral knowledge through scrutinizing and questioning interpretations and commitments.

Teacher-generated data and narratives resulting from such inquiries would illuminate interpretations, commitments, and ultimately virtues. However, teachers' narratives themselves may not lead to pedagogical growth and the cultivation of practical wisdom. What is important are critical and collaborative reflections (by teachers and teacher educators) on interpretations, commitments, and virtues embodied in practices; especially whether these were the most appropriate and conducive to serving students and the content area. Practical-moral knowledge develops into practical wisdom when practitioners become better capable at regulating and mediating multiple interpretations, beliefs, emotions and prior knowledge, and utilizing ones that help them interpret situations in ways conducive to practices that serve oneself and others (Schwartz & Sharpe, 2010; Zagzebski, 1996). The last section discusses how clinically rich teacher education and practitioner inquiry can help promote such critical reflections, growth in different forms of knowledge including practical-moral knowledge, and cultivation of practical-moral knowledge into practical wisdom.

Implications for Teacher Education

An assumption of the conceptual framework suggested here is that even mundane teaching actions involve certain practical-moral concerns. For example, in Salloum and Abd-El-Khalick (2010), teachers' practical arguments always invoked ideas such as being fair and responsible for student achievement (even if only for the deserving group). Therefore, it can be argued that teacher practical-moral knowledge develops in ways aligned with what is 'good' to be as a teacher and what is 'right' to do in a certain situation. The aim is not to mystify teaching, but to emphasize how teaching involves sense making that allows teachers to represent themselves as good teachers/individuals. Teacher educators need to explore ways of making good teaching from research and reform perspectives parallel to teachers' notions of 'good' and to connect reform with virtues teachers see as defining their practice. Teacher education can help teachers develop new interpretations and balance new commitments, and hence cultivate virtues conducive to good and equitable teaching.

Korthagen, Loughran, and Russel (2006) stated, "the development of practice in light of competing demands requires an approach that revolves around the need to create meaningful collaboration in learning and teaching, collaboration of peers and collaboration of teacher educators and student teachers" (p. 1027). Teachers' (in-service and pre-service) interpretations need to be explored and understood by themselves and by teacher educators. As advocated by Korthagen, Loughran, and Russel (2006), and Schwandt (1996) above, teachers and teaching candidates need to be given adequate opportunities for collaborative inquiries and critical reflections on how their practical-moral knowledge is being engaged and developed, and to further understand, value, and question their interpretations and commitments.

A virtue-based view of teacher knowledge supports teacher education models that meaningfully integrate clinically rich teacher apprenticeships/residency (co-teaching and mentoring), collaborative and disciplined practitioner inquiry/action research, along with a strong base in theoretical and skills-based education (Darling-Hammond, 2008; Korthagen, Loughran, & Russel, 2006; Tobin & Roth, 2005, van Driel, et al. 2001). Virtues are acquired excellences that take time and effort; therefore, the expectation in teacher education is for teachers and teaching candidates to understand and cultivate virtues that potentially will enable them and further their knowledge as good teachers. Developing phronesis/practical wisdom and intellectual and moral virtues relevant to teaching involves a social dimension, where good practice needs to be modeled and teacher experiences and narratives shared and reflected upon (Breier & Ralphs, 2009; Kemmis, 2010).

It is important for new teachers, from outset of their teacher preparation programs, to observe, experience, and systematically and critically reflect on how master teachers draw on various knowledge, interpretations and commitments. Furthermore, they need to be mentored and supported for sustaining such critical reflections on their own practices into induction years through collaborative practitioner inquiry.

It is important to keep in mind that handling and reflecting on practical experiences is qualitatively different when undertaken from within a phronetic perceptual model than an epistemic rational model (Kessels & Korthagen, 1996). Within a phronetic model, preservice and in-service teachers verbalize and document their interpretations of the intricacies and particularities of their teaching and learning experiences rather than elicit the appeal to abstract rules and theoretical concepts; or as Kessels and Korthagen (1996) stated:

To be able to develop this wider perception-based type of knowledge... what we need is not so much theories, articles, books, and other conceptual matters, but first and for most, concrete situations to be perceived, experiences to be had, persons to be met, plans to be exerted, and their consequences to be reflected upon (p. 21).

This is not to say that theoretical concepts are not important. Considerable time needs to be allotted to addressing theoretical concepts in science content, pedagogy and Pedagogical Content Knowledge (PCK), because it is in trying to apply concepts to specific situations that teachers become better aware of their interpretations, commitments, and how their practical-moral sense of being good teachers is inherent in practice. Teaching candidates need to build solid theoretical knowledge (episteme) and skills (techne) that will allow them to be effective in action as they enact relevant virtues. However, when immersed in extensive field experiences and systematic inquiries from the outset of their teacher preparation, they would be developing the different forms of understandings, knowledge, and skills (phronesis, episteme, and techne) simultaneously and dialectically. Simultaneous and dialectic development stands opposed to a teacher education model that starts with emphasis on theoretical knowledge and concepts that are to be later 'applied.'

Being immersed in experiences may not be enough though for developing their practical-moral knowledge, and has to be augmented with disciplined and collaborative teacher inquiry in order for new teachers, master teachers, and teacher educators to critically reflect on experiences. It is important for new and veteran educators to engage in a systematic and step by step inquiry process, where they plan for and aim some kind of change; act and observe consequences of change; reflect on the process and consequences, and then re-plan; act, observe, reflect, and so on (Kemmis & McTaggart, 2001). Disciplined and collaborative teacher inquiry allows new and experienced practitioners to: (a) create documentation (data) and narratives of teaching, (b) share documentation and narratives with perceptions and interpretations, and (c) develop new understandings, knowledge, and competencies. Through sharing, teachers can critically reflect on interactions among academic (content and pedagogy) interpretations and moral situated ones. They also become cognizant of the interaction of commitments, interpretations, and actions, thus discerning virtues underlying them. As importantly, they will be able to question the appropriateness of their interpretation, commitments, and actions in serving students and the subject matter. Kemmis (2010) has argued that reflection on individual and collective consequences, such as what occurs in practitioner inquiry, is how we cultivate practical wisdom (phronesis), "action research can help us learn phronesis, the disposition to live wisely and well, by facilitating our reflection on our individual and collective praxis" (p. 422).

Through systematic inquiry, pre-service and novice teachers can cultivate virtues that will guide their practice and act as motivation for new knowledge, skill development, and continual pedagogical growth. Indeed, responsibility, open-mindedness, and wholeheartedness are inherent to reflective practice (Dewey, 1939) and, as mentioned above, these are intellectual virtues that act as motivation for developing new knowledge for teaching. For example, by cultivating virtues such as intellectual honesty, responsibility, fairness, and perseverance, teachers would see the need to learn about multiple representations of abstract and difficult science concepts and how to establish their relevance to diverse groups of students. Similarly, the aforementioned virtues would act as motivation to learning about and utilizing diverse forms of assessment to ensure that students' knowledge is properly represented. Becoming knowledgeable about multiple representations, functions, and assessments of science concepts are important components of Pedagogical Content Knowledge (PCK) and so systematic inquiry not just cultivates virtues but enhances teacher competency and effectiveness in practice. Two examples will be presented below from my work with novice urban teachers in a practitioner inquiry graduate course to demonstrate types of growth through teacher inquiry (Collier & Salloum, 2011; Salloum, et al., 2010).

Multiple Representations and Access

In Collier and Salloum (2011), Traci (co-author) a science teacher in an urban middle school sought to investigate ways to enhance conceptual understandings in science for her diverse students. Schools in urban settings are distinguished by high levels of students' diversity in terms of culture and sociolinguistics, and so it is important to devise instruction that acknowledges this diversity and perceives it as a strength. Traci conducted a teacher inquiry on "how to use discrepant events with higher order and scaffolding questioning to develop students' conceptual understanding?" She also explored "how can students' comprehension and conceptual knowledge be measured through teacher-created materials using academic language of standardized testing?"

What Traci did was to broaden her conception of discrepant events as not just the use intriguing science demonstrations and hand-on activities, but as also utilizing forms of popular culture that students identify with, such as popular songs (e.g. Cha Cha slides to teach the cross-cross method in chemistry) and comic strips (e.g. Dr. Birdley Teaches Science series). She used discrepant events, as an expanded concept, and different levels of questions to enhance students' science understandings and problem-solving competencies. In her final inquiry paper, Traci wrote:

Overall, the clarification of expectations through type of questioning and visual presentation [pop-culture based discrepant events] allowed for the majority of the students to understand each topic's aim ...

The ability for them (students) to see science as academic, transforming, and creative would allow them to see science as less foreign and unattainable.

She later added:

My inquiry provided a step-by-step process where I began to understand how students learn and remember.

Traci came to realize that "Students are no longer passive recipients of information therefore must be engaged in learning," and that "the inquiry's focus on discrepant events provided a springboard for discussion and discovery." As she conducted her inquiry with two different groups, she also noted that certain discrepant events and activities were more effective for one group more than the other. Inherent to Traci's inquiry were the notions of responsibility and fairness towards students by providing access to students and thus best serving them. From a virtue perspective, Traci's inquiry helped her cultivate and mediate virtues such as intellectual fairness and responsibility, in addition to careful inquiry about students' strengths and challenges to better serve them. Concurrently she was building her pedagogical content knowledge on properties of matter.

Teacher Inquiry and Empathy

Gizelle[7] was a high school biology and special education teacher, her teacher inquiry was about developing a sense of "personal and scientific inquiry" in her students, especially students with special needs (Salloum, et al., 2010). Working with her science co-teachers, she aimed to make science learning more inquiry-based and relevant to students' lives, thus encouraging their natural curiosity. The following excerpt is from an interview about her teacher inquiry and how it affected her practice:

Sara: Do you feel your definition of what inquiry-based biology—or how it would look like in the classroom—do you feel this may have changed when you were actually trying to implement it in your classroom and collecting data about.

Gizelle: I think it changed. I knew my students were very involved with art. And they loved the process of incorporating art. But I didn't think it was that much involved. And this was something that would really connect all students, regardless of their grade levels or their reading levels. The one thing they all shared was the concept of art. And they love sharing their opinions. I think they were great debaters. And they also loved the process of the kinetic activity of doing something with their hands. And I didn't think that was going to be such a big deal in my inquiry…. And when we were looking at past activities, we began to see a pattern that the students were more engaged. The less we talked, the more they worked. So the more kinetic activities that there were, they were able to obtain the information and present the information and research information and question their own—instead of us questioning their work, their presentation, their classmates were the ones asking them questions about their work….

As we see from the excerpt above, Gizelle's inquiry lead to new interpretations and understandings about students and their needs and strengths. Generating data (e.g. observations and narrative, student interviews, analyzing student work, etc.) and reflecting on data expanded her concept of 'scientific inquiry' and how to actualize an ideal of inquiry-based learning for *all* students, including special populations. From a virtue perspective, Gizelle's inquiry was successful as it helped her cultivate and mediate several virtues inherent to good teaching: open mindedness, careful inquiry, and empathy; and enact them in her teaching.

CONCLUDING THOUGHTS AND CHALLENGES AHEAD

Creating meaningful collaboration among practitioners, teacher educators, and new teacher poses many challenges and cannot adopt a 'one size fits all.' For example as collaborative inquiry, practitioner inquiry, and reflective practice are being pushed in teacher education programs, they themselves may become prescriptive and hegemonic, and so lose some of their meaningfulness for new teachers. They need to be carried out in ways that address particular needs of teaching candidates. The author is currently generating data in that respect and even though generally pre-service and novice teachers find the inquiry experience rewarding, certain issues are surfacing (Salloum, 2009; Salloum, et al., 2010). Some issues are ideological such as how teachers see the validity and quality of their inquiries. Others are practical, such as the added burden teachers see in engaging in practitioner inquiry, which is a not a *choice* but rather a requirement for credit.

Collaborative inquiry needs to be dialectical and involve questioning our own perceptions, interpretations, and commitments as practitioners in educational research and teacher education, or as Loughran (2002) suggested of practicing what we preach. Gadamer (1989) has indicated that Aristotle extended phronesis/practical wisdom with 'sympathetic understanding' or 'being understanding.' Sympathetic understanding is meant to be exercised by the one who does need to act, but by one who is trying to understand the actions and practical wisdom of another (or understanding practical-moral knowledge as in the case of novice teachers). A person with 'sympathetic understanding' would judge another by transposing him/herself fully into the concrete situation of the person who has to act, "the person who is understanding does not know and judge as one who stands apart and unaffected but rather he thinks along with the other from the perspective of a specific bond of belonging, as if too he was affected" (Gadamer, 1989, p. 320). Sympathetic understanding can help us as teacher educators better facilitate the teacher inquiry experiences. I work with teaching candidates in urban settings, where candidates are either novice teachers going through an alternative certification route or are pre-service teacher. Each group requires a different approach and framing of the inquiry process itself. For novice teachers struggling with demands of a new practice, it is better framed as an extension to formative assessment and as a way to make assessment more authentic and fair. This way they can build on students' strengths and address their challenges, planning instruction that provides multiple access points for diverse students. For pre-service teachers, the process can be more structured to entail careful inquiry for enhancing candidates' pedagogical content knowledge, empathy for student, and their understandings of students as learners and thinkers.

Continuously developing our sympathetic understanding can help us as researchers and teacher educators better understand the complex nature of teacher knowledge in general, and specifically practical-moral knowledge and practical wisdom (including our own). Along with pre-service, novice, and experienced teachers, we can all develop our practical wisdom by engaging in inquiries that examines how a teaching practice involves the complex interaction of three modalities: the actual, the necessary, and the possible (Heidegger, 1993; as cited in Shulman, 2007). With careful and reflective investigation of how practical-moral knowledge and practical wisdom becomes concretized we will hopefully become better informed on how to bridge the gap between educational theory and practical and how to better manage our interpretations and commitments and those of classroom practitioners.

REFERENCES

AAAS. (1990). *Science for all Americans*. Oxford, UK: Oxford University Press.

Anderson, C. (2007). Perspectives on science learning. In Abell, S., & Lederman, N. (Eds.), *Handbook of Research on Science Education* (pp. 3–30). Mahwah, NJ: Lawrence Erlbaum.

Anderson, R. D., & Mitchener, C. P. (1994). Research on science teacher education. In Gabel, D. L. (Ed.), *Handbook of Research on Science Teaching and Learning* (2nd ed., pp. 3–44). New York, NY: McMillan.

Aristotle. (1976). *The Nicomachean ethics* (Thomson, J. A. K., Trans.). London, UK: Penguin.

Aristotle. (2002). *The Nicomachean ethics* (Sachs, J., Trans.). New York, NY: Riverhead Books.

Bell, R. L., Lederman, N. G., & Abd-El-Khalick, F. (2000). Developing and acting upon one's conception of the nature of science: A follow-up study. *Journal of Research in Science Teaching, 37*, 563–581. doi:10.1002/1098-2736(200008)37:6<563::AID-TEA4>3.0.CO;2-N

Borko, H., & Putnam, R. T. (1996). Learning to teach. In Berliner, D. C., & Calfee, R. C. (Eds.), *Handbook of Educational Psychology* (pp. 673–708). New York, NY: McMillan.

BouJaoude, S., Salloum, S., & Abd-El Khalick, F. (2004). Relationships between selective cognitive variables and students' ability to solve chemistry problems. *International Journal of Science Education, 26*, 63–84. doi:10.1080/09500690320000703l5

Breier, M., & Ralphs, A. (2009). In search of phronesis: Recognizing practical wisdom in the recognition (assessment) of prior learning. *British Journal of Sociology of Education, 30*, 479–493. doi:10.1080/01425690902954646

Brickhouse, N. W. (2001). Embodying science: A feminist perspective on learning. *Journal of Research in Science Teaching, 38*, 282–295. doi:10.1002/1098-2736(200103)38:3<282::AID-TEA1006>3.0.CO;2-0

Carr, D. (2005). Personal and interpersonal relationships in education and teaching: A virtue ethical perspective. *British Journal of Educational Studies, 53*, 255–271. doi:10.1111/j.1467-8527.2005.00294.x

Carr, D. (2007). Character in teaching. *British Journal of Educational Studies, 55*(4), 369–389. doi:10.1111/j.1467-8527.2007.00386.x

Carr, W. (1995). *For education*. Berkshire, UK: Open University Press.

Clark, C. (2005). The structure of educational research. *British Educational Research Journal, 31*, 289–308. doi:10.1080/01411920500082128

Collier, T., & Salloum, S. (2011). *The use of discrepant events and higher order/scaffolding questions for deeper science learning*. Paper presented at the Annual Ethnography in Education Research Forum. Philadelphia, PA.

Crawford, B. A. (2007). Learning to teach science inquiry in the rough and tumble of practice. *Journal of Research in Science Teaching, 44*, 613–642. doi:10.1002/tea.20157

Darling-Hammond, L. (2008). A future worthy of teaching for America. *Phi Delta Kappan, 89*, 730–735.

Dewey, J. (1933). *How we think: A restatement of the relation of reflective thinking to the educative process*. Boston, MA: D. C. Heath.

Doyle, W. (1977). Paradigms for research on teacher effectiveness. In Shulman, L. S. (Ed.), *Review of Research in Education* (*Vol. 5*, pp. 163–179). Washington, DC: American Educational Research Association.

Duffee, L., & Aikenhead, G. (1992). Curriculum change, student evaluation, and teacher practical knowledge. *Science Education, 76*, 493–506. doi:10.1002/sce.3730760504

Eisner, E. W. (2002). From episteme to phronesis to artistry in the study and improvement of teaching. *Teaching and Teacher Education, 18*, 375–385. doi:10.1016/S0742-051X(02)00004-5

Feldman, A. (2002). Multiple perspectives for the study of teaching: Knowledge, reason, understanding, and being. *Journal of Research in Science Teaching, 39*, 1032–1055. doi:10.1002/tea.10051

Fenstermacher, G. D. (1994). The knower and the known in teacher knowledge research. In Darling-Hammond, L. (Ed.), *Review of Research in Education* (Vol. 20, pp. 3–56). Washington, DC: American Educational Research Association.

Flyvbjerg, B. (2001). *Making social sciences matter*. Cambridge, UK: Cambridge University Press.

Fosnot, C. T. (1996). Teachers construct constructivism: The center for constructivist teaching/teacher preparation project. In Fosnot, C. T. (Ed.), *Constructivism: Theory, Perspectives, and Practices* (pp. 205–216). New York, NY: Teachers College Press.

Gadamer, H.-G. (1989). *Truth and method* (2nd ed.). (Weinsheimer, J., & Marshall, D. G., Trans.). New York, NY: Crossroad.

Gallagher, S. (1992). *Hermeneutics and education*. New York, NY: SUNY.

Haney, J. J., & McArthur, J. (2002). Four case studies of prospective science teachers' beliefs concerning constructivist teaching practices. *Science Education*, *86*, 783–802. doi:10.1002/sce.10038

Heidegger, M. (1962). *Being and time* (Macquarrie, J., & Robinson, E., Trans.). New York, NY: Harper and Row.

Issler, K. (1983). A conception of excellence in teaching. *Education*, *103*, 338–344.

Kemmis, S. (2009). Action research as a practice-based practice. *Educational Action Research*, *17*, 463–474. doi:10.1080/09650790903093284

Kemmis, S. (2010). What is to be done? The place of action research. *Educational Action Research*, *18*, 417–427. doi:10.1080/09650792.2010.524745

Kemmis, S., & McTaggart, R. (2000). Participatory action research. In Denzin, N., & Lincoln, Y. (Eds.), *Handbook of Qualitative Research* (2nd ed., pp. 567–605). Thousand Oaks, CA: Sage.

Kennedy, M. M. (2006). Knowledge and vision in teaching. *Journal of Teacher Education*, *57*, 205–211. doi:10.1177/0022487105285639

Kessels, J., & Korthagen, F. (1996). The relationship between theory and practice: Back to the classics. *Educational Researcher*, *25*, 17–22.

Kortahgen, F., & Kessels, J. (1999). Linking theory and practice: Changing the pedagogy of teacher education. *Educational Researcher*, *28*, 4–17.

Korthagen, F., Loughran, J., & Russell, T. (2006). Developing fundamental principles for teacher education programs and practices. *Teaching and Teacher Education*, *22*, 1020–1041. doi:10.1016/j.tate.2006.04.022

Lave, J. (1988). *Cognition in practice*. Cambridge, UK: Cambridge University Press. doi:10.1017/CBO9780511609268

Lederman, N. G. (1992). Students' and teachers' conceptions of the nature of science: A review of the research. *Journal of Research in Science Teaching*, *29*, 331–359. doi:10.1002/tea.3660290404

Lemke, J. L. (2001). Articulating communities: Sociocultural perspectives on science education. *Journal of Research in Science Teaching*, *38*, 296–316. doi:10.1002/1098-2736(200103)38:3<296::AID-TEA1007>3.0.CO;2-R

Lotter, C., Harwood, W. S., & Bonner, J. J. (2007). The influence of core teaching conceptions on teachers' use of inquiry teaching practices. *Journal of Research in Science Teaching*, *44*, 1318–1347. doi:10.1002/tea.20191

Loughran, J. (2002). Effective reflective practice: In search of meaning in learning about teaching. *Journal of Teacher Education, 53*(1), 33–43. doi:10.1177/0022487102053001004

Lynch, S. (2001). Science for all is not equal to one size fits all: Linguistic and cultural diversity and science education reform. *Journal of Research in Science Teaching, 38*, 622–627. doi:10.1002/tea.1021

Medley, D. M. (1979). Effectiveness of teachers. In Peterson, P. L., & Walberg, H. J. (Eds.), *Research on Teaching: Concepts, Findings, and Implications*. Berkley, CA: Mc-Cutchan.

Meyer, H. (2004). Novice and expert teachers' conceptions of learners' prior knowledge. *Science Education, 88*, 970–983. doi:10.1002/sce.20006

Moje, M. (1995). Talking about science: An interpretation of the effects of teacher talk in a high school science classroom. *Journal of Research in Science Teaching, 32*, 349–371. doi:10.1002/tea.3660320405

Mulholland, J., & Wallace, J. (2008). Computer, craft, complexity, change: Exploration into science teacher knowledge. *Studies in Science Education, 44*, 41–62. doi:10.1080/03057260701828135

National Research Council. (1996). *National science education standards*. Washington, DC: National Academic Press.

Noel, J. (1999). On the varieties of phronesis. *Educational Philosophy and Theory, 31*, 273–289. doi:10.1111/j.1469-5812.1999.tb00466.x

O'Loughlin, M. (1992). Rethinking science education: Beyond Piagetian constructivism toward a sociocultural model of teaching and learning. *Journal of Research in Science Teaching, 29*, 791–820. doi:10.1002/tea.3660290805

Pajares, M. F. (1992). Teachers' beliefs and education research: Cleaning up a messy construct. *Review of Educational Research, 62*, 307–332.

Pendlebury, S. (1995). Reason and story in wise practice. In McEwan, H., & Egan, K. (Eds.), *Narrative in Teaching, Learning and Research* (pp. 50–65). New York, NY: Teachers College Press.

Salloum, S. (2006). *Teaching as practice: Blending intellectual and the moral in pursuit of science teachers' practical knowledge*. (Unpublished Doctoral Dissertation). University of Illinois at Urbana-Champaign. Champaign, IL.

Salloum, S. (2009). *Pedagogical growth and practitioner inquiry: ESL urban teachers' perceptions of their growth while conducting inquiries into their practice*. Paper presented at the Annual Meeting of the American Educational Research Association. San Diego, CA.

Salloum, S., & Abd-El-Khalick, F. (2010). Practical knowledge in teaching: Case studies from physical science classrooms. *Journal of Research in Science Teaching, 47*, 929–951.

Salloum, S., Jennings, M., Arrabito, N., Schmidt, M., McCall, C., & Frederick, T. ... Benn-Scantlebury, A. (2010). *Novice urban teachers engaging in practitioner inquiry: Lessons learned, rewards, and challenges*. Paper presented at Annual Ethnography in Education Research Forum. Philadelphia, PA.

Schwandt, T. A. (1996). Farewell to criteriology. *Qualitative Inquiry, 2*, 58–72. doi:10.1177/107780049600200109

Schwandt, T. A. (2000). Three epistemological stances for qualitative inquiry. In Denzin, N. K., & Lincoln, Y. S. (Eds.), *Handbook of Qualitative Inquiry* (2nd ed., pp. 189–213). Thousand Oaks, CA: Sage.

Schwandt, T. A. (2005). On modeling our understanding of the practice fields. *Pedagogy, Culture & Society, 13*, 313–332. doi:10.1080/14681360500200231

Schwartz, B., & Sharpe, K. (2010). *Practical wisdom*. New York, NY: Riverhead Books.

Shulman, L. (2007). Practical wisdom in the service of professional practice. *Educational Researcher, 36*, 560–563. doi:10.3102/0013189X07313150

Simmons, P. E., Emory, A., Carter, T., Coker, T., Finnegan, B., & Crockett, D. (1999). Beginning teachers: Beliefs and classroom actions. *Journal of Research in Science Teaching, 36*, 930–954. doi:10.1002/(SICI)1098-2736(199910)36:8<930::AID-TEA3>3.0.CO;2-N

Smith, L., & Southerland, S. A. (2007). Reforming practice or modifying reforms? Elementary teachers' response to the tools of reform. *Journal of Research in Science Teaching, 43*, 396–423. doi:10.1002/tea.20165

Southerland, S., Sinatra, G. M., & Mathews, M. (2001). Beliefs, knowledge, and science education. *Educational Psychology Review, 13*, 325–351. doi:10.1023/A:1011913813847

Southerland, S. A., Gess-Newsome, J., & Johnson, A. (2003). Portraying science in the classroom: The manifestation of scientists' beliefs in classroom practice. *Journal of Research in Science Teaching, 40*, 669–691. doi:10.1002/tea.10104

Tobin, K., & Roth, W.-M. (2005). Implementing coteaching and cogenerative dialoguing in urban science education. *School Science and Mathematics, 105*, 313–322. doi:10.1111/j.1949-8594.2005.tb18132.x

Tsai, C.-C. (2002). Nested epistemologies: Science teachers' beliefs of teaching, learning, and science. *International Journal of Science Education, 24*, 771–783. doi:10.1080/09500690110049132

van Driel, J. H., Beijaard, D., & Verloop, N. (2001). Professional development and reform in science education: The role of teachers' practical knowledge. *Journal of Research in Science Teaching, 38*, 137–158. doi:10.1002/1098-2736(200102)38:2<137::AID-TEA1001>3.0.CO;2-U

van Manen, M. (1995). On the epistemology of reflective practice. *Teachers and Teaching: Theory and Practice, 1*, 33–50. doi:10.1080/1354060950010104

van Manen, M. (1999). The practice of practice. In M. Lange, J. Olson, & W. BŸnder, (Eds.), *Changing Schools/Changing Practices: Perspectives on Educational Reform and Teacher Professionalism*. Luvain, Belgium: Garant.

Vianna, E., & Stetsenko, A. (2006). Embracing history through transforming it. *Theory & Psychology, 16*, 16–81.

Wildy, H., & Wallace, J. (1995). Understanding teaching or teaching for understanding: Alternative frameworks for science classrooms. *Journal of Research in Science Teaching, 32*, 143–156. doi:10.1002/tea.3660320205

Windschitl, M. (1999). A vision educators can put into practice: Portraying the constructivist classroom as a cultural system. *School Science and Mathematics, 99*, 189–196. doi:10.1111/j.1949-8594.1999.tb17473.x

Woolfolk, A. E., & Galloway, C. M. (1985). Nonverbal communication and the study of teaching. *Theory into Practice, 24*, 77–85. doi:10.1080/00405848509543150

Zagzebski, L. T. (1996). *Virtues of the mind: An inquiry into the nature of virtue and the ethical foundations of knowledge*. Cambridge, UK: Cambridge University Press. doi:10.1017/CBO9781139174763

ENDNOTES

1. Teaching for tested algorithmic skills rather than conceptual understandings is well documented in science education (e.g., BouJaoude, Salloum, & Abd-El-Khalick, 2004).
2. Theoretical knowledge is an umbrella term for theoretical academic knowledge involved in science teaching such knowledge of science disciplines, psychological, and sociocultural theories of learning.
3. Carr (2007) stated that the perceived 'content-less' aspect has been put forth as a critique of virtue ethics in education.
4. All names are pseudonyms.
5. For example, a parent or teacher may tell a child: "Is this low grade worthy of you?" or "This is not what I expected from you," etc.
6. 'Goods' that teachers perceive are usually subjective and related to their experiences, and not necessarily reflecting reform notions in science education.
7. Pseudonym.

Chapter 3
Teaching a Socially Controversial Scientific Subject:
Evolution

Hasan Deniz
University of Nevada Las Vegas, USA

ABSTRACT

This chapter explores teachers' and students' acceptance and understanding of evolutionary theory by using conceptual ecology (Toulmin, 1972) as a theoretical lens. Demastes, Good, and Peebles (1995) describe the conceptual ecology for evolutionary theory. Acceptance of evolutionary theory is part of this conceptual ecology, and this conceptual ecology also contains the following five components: (1) prior conceptions related to evolution (understanding of evolutionary theory); (2) scientific orientation (degree to which the learner organizes his/her life around scientific activities); (3) view of the nature of science; (4) view of the biological world in competitive and causal terms as opposed to aesthetic terms; and (5) religious orientation. A complex web of connections among components of conceptual ecology for evolutionary theory influences one's acceptance and understanding of evolutionary theory. Therefore, studying the relationship between acceptance and understanding of evolutionary theory as a part of the conceptual ecology for evolutionary theory is more promising than studying acceptance of evolutionary theory in isolation. Moreover, studying acceptance of evolutionary theory as an integral part of the conceptual ecology may enable us to explain why some teachers and students show a high degree of acceptance and others show a low degree of acceptance.

INTRODUCTION

Evolution as a unifying theme in biology education has been supported by major science education policy documents and understanding of evolution has been considered as an important part of scientific literacy (American Association for the Advancement of Science, 1993; National Research Council, 1996). Major science education organizations such as National Association of Biology Teachers (2011) and National Science Teachers Association (2003) in the United States have also supported evolution as a unifying theme in biology. More recently evolution has been identified as one

DOI: 10.4018/978-1-4666-2809-0.ch003

of four disciplinary core ideas in life sciences in "A Framework for K-12 Science Education: Practices, Crosscutting Concepts, and Core Ideas" (NRC, 2012). According to this new framework which will serve as the basis for the next generation of science education standards biological evolution includes four components:

- Evidence of Common Ancestry and Diversity
- Natural Selection
- Adaptation
- Biodiversity and Humans

The framework takes a learning progressions approach and describes what students need to know by the end of grades 2, 5, 8, and 12 for each component. It is clear that evolution continues to be considered as a major or overarching idea in life sciences curriculum, but it is less clear to what extent evolution has been taught in actual classroom settings and how students handle learning the evolution content.

There is a difference between teaching a socially controversial scientific subject and a non-socially controversial scientific subject.

Educators need to consider these five integrated domains when teaching evolution:

- The conceptual domain.
- The epistemic domain.
- The socio-cultural domain.
- The religious domain.
- The legal domain.

The first three domains are important to consider in teaching other science content as well, but the last two domains become particularly important when it comes to teaching evolution.

The conceptual domain: The conceptual domain includes both scientifically accepted major evolutionary ideas and students' alternative conceptions about evolution. It is well known that students hold intuitive conceptions of the natural world and these conceptions are often in conflict with the scientifically accepted conceptions (Driver, 1981; NRC, 2007). After decades of research on misconceptions we now know that students' minds are not "*tabula rasa*" or "*empty vessels.*" Students do have alternative conceptions. Students' alternative conceptions need reorganization in order to accommodate the scientifically accepted contemporary views. Many researchers emphasized that students' prior conceptions might interfere with the learning (e.g., Bransford, et al., 1999; Chinn & Brever, 1993; Pintrich, et al., 1993). Pintrich et al. (1993) suggested that prior knowledge can play two contradictory roles during the learning process. They contemplated that prior knowledge can either impede the learning process through students' alternative frameworks, or it can facilitate it by providing students a conceptual basis for evaluating the validity of newly encountered ideas.

Ausubel's frequently quoted statement captures the importance of students' intuitive conceptions during the learning process in a dramatic way. This quote appears before the preface of the book that Ausubel co-authored with Novak and Hanesian (Ausubel, Novak, & Hanesian, 1978).

If had to reduce all of educational psychology to just one principle, it would say this: The most important single factor influencing learning is what the learner already knows. Ascertain this and teach him accordingly.

Common students' alternative conceptions about evolution include the following ideas (Werth, 2012):

- All evolutionary change is adaptive.
- Evolutionary change is progressive.
- Evolutionary change is teleological (goal-directed).

Educators need to be aware that such ideas are common among students when teaching about major evolutionary concepts. Apart from these common misconceptions, many students come to the classroom with a strong tendency to believe the evolutionary theory is not true. For majority of these students, their apprehensive attitude toward evolution is related to factors such as students' religious perspectives, naïve nature of science views, and social-cultural backgrounds. Therefore, evolution instruction that embraces constructivist notions of learning should consider these factors. It is true that public school science classes are not a place to teach religion but disregarding the importance of students' religious perspective while learning about evolution is against the fundamental assumptions of constructivist pedagogy.

Science education community has long been trying to find answers to the following questions. Can students have a robust conceptual understanding of evolutionary theory without accepting its validity? Is there a relationship between understanding and acceptance of evolutionary theory? Researchers reached into different conclusions about the nature of this relationship. Some researchers found no relationship between understanding and acceptance of evolutionary theory (Sinatra, Southerland, McCounaughy, & Demastes, 2003) while others reported a positive relationship (Deniz, et al., 2008, 2011; Rutledge & Warden, 2000). Deniz et al. (2009, 2012) found that students' acceptance of evolutionary theory significantly increases after students improve their understanding of evolutionary theory.

The epistemic domain: Nature of science (epistemology of science) refers to values and beliefs specific to scientific knowledge and its development (Lederman, 1992). Students' nature of science views or epistemological beliefs about science determine the rules and criteria used for differentiating what counts as good scientific knowledge. Therefore, students' epistemological beliefs play a crucial role during the learning process.

It was acknowledged that there is no agreed-upon single definition of NOS among philosophers of science, historians of science, scientists, and science educators, but certain aspects of NOS are uncontroversial and relevant to K-16 education (Abd-El-Khalick & Akerson, 2004; Schwartz, Lederman, & Crawford, 2004). These NOS aspects include but are not limited to conceptions that scientific knowledge is empirically-based, tentative, subjective, inferential, socially and culturally embedded, and depends upon human creativity and imagination. Students should have appropriate nature of science understandings before they develop a robust conceptual understanding of any science subject. This is particularly true for developing a deeper scientific understanding about evolutionary theory. Dagher and BouJaoude (1997) found that allowing students to discuss their alternative conceptions about evolution and teaching them about nature of science is more likely to increase their understanding of evolution. Martin-Hansen (2008) reported that nature of instruction reduced the percentage of students feeling apprehensive about learning evolution from 38% to 12% in a first-year college course. Martin-Hansen (2008) also reported that the nature of science instruction did not help some students (Young Earth Creationists) reduce their apprehension level toward learning evolution. Verhey (2005) reported interesting findings with regard to the impact of instruction on understanding of evolutionary theory. Verhey (2005) found that strong emphasis on evolution did not significantly increase student conceptions of evolutionary theory, but comparing "intelligent design" with evolutionary theory, with a special emphasis on nature of science led to extensive change toward scientifically accepted conceptions.

Table 1 describes the nature of science views that students need to develop to be able to place evolutionary theory in its appropriate standing in science.

Staver (1999) reported that Kansas State Board of Education considered an alternative version of nature of science provided by Creation Science Association for Mid-America while deciding whether or not the evolutionary theory should be

Table 1. Description of the nature of science aspects

NOS aspect	Description
Bounded NOS	Science is a limited way of knowing. Science cannot answer moral and ethical questions. Scientists do not invoke supernatural explanations when doing science.
Empirical NOS	Scientific knowledge is based on empirical evidence. Knowledge claims in science are made with evidence and observations of nature. However, this empirical base does not provide a secure base for science because observations are influenced by scientists' creativity, and personal and theoretical subjectivity.
Inferential NOS	There is a difference between observation and inference. Observations are descriptive statements about the nature that are available to the senses. Observes can agree upon observation statements with relative ease. Inferences are interpretations of observations. They are not immediately available to the senses.
Creative NOS	Creativity and imagination of scientists play a major role in the scientific inquiry. The role of creativity and imagination is not limited to any specific phase of the scientific inquiry. Creativity and imagination are of importance before, during, and after data collection. Creativity and imagination allow scientist to build theoretical models that inferentially explain the natural phenomena.
Subjective NOS	Scientists try to achieve objectivity, but absolute objectivity is not possible in science. Theoretical orientations of scientists make them unavoidably subjective. In addition to scientists' theoretical orientations their personal characteristics and social and cultural backgrounds contributes to subjectivity of scientists. All these factors influence scientists' choice of research questions, methods of research, observations, and interpretations of their observations.
Tentative NOS	Scientific knowledge is tentative but durable. Scientific knowledge is subject to change with the availability of new evidence and with the interpretation of the old evidence, but this change does not happen on the daily basis. Science is not concerned with finding the final truth.
"Scientific Method"	There is not a general and universal scientific method that is followed by all research scientists to solve scientifically oriented questions.
Social and cultural NOS	Science is a human activity. It is influenced by social and cultural factors. These social and cultural factors include social composition, religion, worldview, political and economic factors. Science is not only influenced by these factors but also it influences these factors.
Social NOS	Science is no longer a solitary pursuit. Scientific knowledge is constructed through social negotiation. Despite their individual differences members of a scientific community of practice share common traditions, values, and theoretical frameworks. This social dimension enhances the objectivity of scientific knowledge. The double-blind peer-review process used by scientific journals is a major component of this NOS aspect.
Theory/ Law	There is no hierarchical relationship between theories and laws. Laws are mathematical descriptions of natural phenomena. Theories do no turn into laws. Different usages of the word theory are problematic. In science, theories are extremely well-supported web of hypotheses that are constructed to explain natural phenomena. However, everyday use of the word theory refers to some sort of a wild idea which may or may not have an empirical support.

included in the revised Kansas state science standards. It is interesting to note that both proponents and opponents of evolutionary theory use their own version of nature of science to justify whether or not evolutionary theory should be included in the science education standards. The way in which the nature of science is conceptualized is inextricably connected to the curricular decision making about evolution. For example, if one accepts Creation Science Association for Mid-America version of nature of science evolution does not qualify to be a scientific theory which is worthy of inclusion in the curriculum.

Nature of science items included in a proposed draft of Kansas state standards by the Kansas State Board of Education (Staver 1999).

1. Good science is science that is verifiable, falsifiable, and repeatable.
2. Historic science, which includes the study of past events such as the origin of life and the universe, is not good science because these ideas are not testable, as the past is not verifiable, falsifiable, or repeatable.
3. Scientific law is considered to be more important than scientific theory.

4. Inductive reasoning is emphasized over and above deductive reasoning, which is downplayed.

Many religious people consider evolution as a form of atheism. They think that acceptance of evolution and belief in God are not compatible. Improving people's nature of science views can help them resolve this unnecessary conflict. Nature of instruction emphasizing the boundaries of science can help certain people come to the conclusion that their acceptance of evolution does not harm their religious beliefs. Science is a limited way of knowing. Science cannot answer questions about the supernatural. Another important nature of science understanding that can help religious people reduce their apprehensive attitude toward evolution is recognizing the difference between methodological and philosophical materialism. Scientists operate according to methodological materialism. In other words, they do not invoke supernatural explanations when doing science and explaining natural phenomena. Although scientists may or may not have personal religious beliefs they have to be silent about the existence of supernatural when doing science. Contrary to the methodological materialism, proponents of the philosophical materialism do not accept supernatural. In other words, philosophical materialists deny the existence of God and therefore they are atheists. All scientists are methodological materialists but not all scientists are philosophical materialists. In fact, some scientists accept the theistic evolution position.

The socio-cultural domain: Lerner (2000) examined how different states within the United States treated evolution in their state science education standards. Lerner (2000) reported that only 10 states have very good-excellent evolution education standards, 14 states have good standards, 7 states have satisfactory standards, 6 states have unsatisfactory standards, and 13 states have standards that are worse than unsatisfactory. Evolution education is vulnerable to social influences more than any other subject in science. The poor condition of evolution education standards in many states can be interpreted in light of attitude surveys toward evolution. According to surveys administered by Gallup in 2010 and 2012, forty and forty six percent of Americans respectively endorsed the view that "God created human beings pretty much in their present form at one time within the last 10,000 years or so." Thirty-eight and thirty-two percent agreed that "human beings have developed over millions of years from less advanced forms of life, but God guided this process." Sixteen and fifteen percent agreed that "human beings have developed over millions of years from less advanced forms of life, but God had no part in this process," and six and seven percent respectively did not report any opinion in 2010 and 2012. Percentage of people supporting Young Earth Creationist view increased from 40% to 46% and percentage of people supporting theistic evolution position decreased from 38% to 32% within the last two years. Percentage of people supporting evolution without any involvement from God remained pretty much the same within the last two years. However, examination of the poll results between 1982 and 2012 indicates no dramatic change within the last thirty years.

Berkman and Plutzer (2010) conducted a comprehensive study titled "National Survey of High School Biology Teachers." They used three positions endorsed by the National Academy of Sciences (1998, 2008) to frame their study: (a) there is no debate whether evolution occurred within the scientific community, (b) evolution is necessary to understand other topics in biology, (c) evolution should be used a unifying theme in teaching biology. Berkman and Plutzer (2010) turned these three positions into Likert-type survey questions:

1. When I teach evolution (including answering student questions) I emphasize the broad consensus that evolution is a fact, even as

scientists disagree about the specific mechanisms through which evolution occurred.
2. It is possible to offer an excellent general biology course for high school students that includes no mention of Darwin or evolutionary theory.
3. Evolution serves as the unifying theme for the content of the course.

These questions were administered to a national sample of 926 biology teachers. According to the study, 74% of teachers agreed with the first question, 21% of teachers disagreed with the question, and 5% did not respond. Twelve percent of teachers agreed with the second question, 83% of teachers disagreed with the second question, and 5% did not offer any answer. As for the third question, 62% of biology teachers agreed with the third question, 34% of biology teachers disagreed, and others did not respond. As it can be understood from these numbers a significant percentage of American biology teachers remains aloof to evolution despite the endorsements from AAAS (1993), NRC (1996), NABT (2011), and NSTA (2003).

Miller, Scott, and Okamoto (2006) compared the acceptance of evolutionary theory among adults in 34 countries. Countries such as Iceland, Denmark, Sweden, France, Japan, and United Kingdom have 75 percent or more acceptance rates of evolutionary theory. Countries such as United States and Turkey have acceptance rates of 40 percent and 25 percent respectively. If a concept has little leverage within a cultural milieu, it will not be readily acceptable and it will be difficult for that concept to be included in the school curriculum. Costa (1995) stated that successful transition of students from their own world to school science depends on the compatibility of family and school cultures.

Deniz et al. (2008) found that Turkish preservice biology teachers whose parents have more education are more likely to accept evolution as a scientifically valid theory. This makes sense considering the fact Turkish education system is historically modeled based on Western education principles especially after the declaration of Republic of Turkey in 1923.

According to Aikenhead and Jegede (1999) when the culture of school science is in sync with students' social and cultural values science instruction tend to happen smoothly. However, if there is a conflict between the culture of science and students' socio-cultural values science instruction tends to damage students' socio-cultural values by forcing them to abandon their indigenous values. For this reason, Aikenhead and Jegede (1999) called for developing culturally sensitive curricula and teaching methods to be able to avoid the clash between students' cultural values and the culture of Western science. Cobern (1996) also underscored this point by stating that affective and contextual factors are evoked more strongly when a person struggles to learn a topic which does not have much leverage within that person's worldview.

The religious domain: Researchers reported negative relationships between religious orientation and acceptance of evolutionary theory. Grose and Simpson (1982) found that students in an introductory college biology class who perceived that their church generally influenced their thought had lower acceptance of evolutionary theory. Osif (1997) reported that that religiosity of biology teachers is positively correlated with the rejection of the evolutionary theory. Miller, Scott, and Okamoto (2006) found that religious beliefs were almost twice as much predictive of attitude toward evolution in the United States compared to European countries. Obviously, in both the United States and Europe, the correlation between religious beliefs and attitude toward evolution was negative, but this correlation was much bigger in the United States. This indicates that religious beliefs are negatively correlated with attitude toward evolution but the strength of the correlation varies in different countries. Deniz et al. (2011) found that preservice Turkish biology teachers' acceptance of evolutionary theory is negatively correlated with their religiosity. They also found that preservice teachers with lower acceptance of evolutionary theory and strong religious beliefs are less likely to teach evolution in their future classrooms.

Table 2 describes how students' religious beliefs may influence their learning about evolution and their acceptance-rejection of evolutionary theory.

The legal domain: Who should determine the curriculum in public schools? Should the citizens as taxpayer decide what to teach in schools? Should educators as professionals decide what to teach? Should this decision be left to scientific community? The controversy over who should determine the curriculum in public schools was historically being tackled by the judicial system in the United States.

Table 2. Students' learning approach and their evolution rejection-acceptance position according to their religiosity

	Learning approach	Evolution rejection-acceptance position
Extremely Religious	Students prefer to stick to their religious values. They are alienated from the culture of science and they reject to learn science.	**Young Earth creationism**: Proponents of this position hold literal interpretations of Bible and they think that Earth is less than 10,000 years old. They reject scientific findings with regard to the age of Earth and biological evolution.
↑	Students prefer to stick to their religious values but at the same time, they are not fully alienated from the culture of science. Prior conceptions of students in this category preclude them from accepting evolutionary theory.	**Old Earth creationism**: Proponents of this position do not interpret six days of Biblical creation as twenty-four hours but as longer time periods. They accept other modern science but they reject biological evolution. **Intelligent design creationism**: Proponents of intelligent design holds the view that the evolution of organs such as vertebrate eye is too complex to be explained naturally and this complexity requires an intelligent designer. They do accept natural selection but they do not accept that mutation and natural selection can explain evolution of species.
	Students prefer to stick to both their religious values and the culture of science. Students develop two conceptual frameworks one for school science and one for everyday experiences. At first, students in this category consciously or unconsciously keep two conceptual frameworks separate. Students will activate one conceptual framework or another depending upon the context. Later, students in this category can make a concerted effort to reconcile these two conceptual frameworks. Prior conceptions of students in this category may preclude them from learning evolution.	**Theistic evolutionism**: Proponents of this position think that God creates through the natural processes and mechanisms. They do accept that natural selection and mutation can explain evolution of species. They are methodological materialists but not philosophical materialists.
↓	Students prefer to assimilate into the culture of science and they are silent about the existence of God. Prior conceptions of students in this category will not preclude them from learning evolution. The extent of their learning will be determined by the quality of instruction and their personal characteristics such as motivation and metacognitive awareness.	**Agnostic evolutionism**: Proponents of this position accept evolution but they neither reject nor accept the existence of God. They are methodological materialists but they avoid accepting or rejecting philosophical materialism.
Not Religious	Students prefer to assimilate into the culture of science and they reject the existence of God. Prior conceptions of students in this category will not preclude them from learning evolution. The extent of their learning will be determined by the quality of instruction and their personal characteristics such as motivation and metacognitive awareness.	**Materialistic evolutionism**: Proponents of this position accept evolution and they reject the involvement of God with evolution. They are both methodological and philosophical materialists.

Table 3 shows that The United States Supreme Court has consistently ruled in favor of evolution.

Public opinion polls consistently indicated that people support teaching alternatives such as creationism and intelligent design to evolution in the United States (Plutzer & Berkman, 2008). It seems like alternatives to teaching evolution were beaten in the courtrooms, but the public desire to introduce these alternatives into the curriculum is still strong. Similarly, such desires are also present even in Britain, which enjoys about 75% evolution acceptance rate according to the international study conducted by Miller, Scott, and Okamoto (2006). According to a survey conducted by Ipsos MORI (2006) for the British Broadcasting Corporation (BBC), when participants are allowed to pick one or more choices among creationism, intelligent design, and evolution, 44% of the participants said that creationism should be taught in science lessons in British schools, 41% indicated that they would like to see intelligent design taught in schools, and 69% wanted evolution as part of science curriculum.

CONCLUSION

Students' learning about evolution can be conceived in two different ways: *learning as conceptual change* and *learning as cultural border crossing*.

According to *learning as conceptual change* approach, students are supposed to move from their alternative conceptions toward scientifically accepted conceptions. *Learning as conceptual change* model supports replacing students' alternative conceptions with the scientifically accepted conceptions.

This type of learning can be represented as $C_1 \rightarrow C_2$.

Posner et al. (1982) described the conditions required for conceptual change: (1) dissatisfaction, this condition requires that the learner fails to make sense of some event with his/her existing conception, (2) intelligibility, this condition necessitates that the learner has some understanding of the new conception, (3) plausibility, this condition is satisfied when the learner accepts the new conception, and (4) fruitfulness, this condition

Table 3. Sample court cases about evolution in the United States

Date	Case	Detail	Final Ruling
1925	Tennessee vs. Scopes	Scopes, a science teacher, admitted using a text dealing with human evolution. This was in violation with Tennessee law.	The Tennessee Supreme Court decided that as a public employee it was teachers' responsibility to teach whatever the state determines. The case was not taken to the United States Supreme Court.
1968	Epperson vs. Arkansas	Epperson, tenth grade biology teacher, adopted a textbook including a chapter about evolution. This was in violation of a state law that made illegal to teach human evolution in a public school or university. Arkansas Charcery Court decided that school teachers should enjoy a substantial degree of academic freedom. The Arkansas State Supreme Court overturned the ruling.	The United States Supreme Court ruled in favor of Epperson.
1987	Edwards vs. Aguillard	Aguillard challenged a Lousiana law that required the teaching of creation science along with evolution.	The United States Supreme Court found the law in violation of the First Amendment.
2005	Kitzmiller vs. Dover Area School District in Pennsylvania	The school board of Dover Area School District introduced intelligent design in the curriculum through a mandated disclaimer to be read by teachers. This was challenged by a group of science teachers and some parents.	District Court for the Middle District of Pennsylvania ruled that introducing creationism and intelligent design in the curriculum is unconstitutional because creationism and intelligent design are religion not science.

emphasizes that the learner should be able to use the new conception to explain novel situations as well as the situations that were formerly explained by the new conception. As it can be understood from this description, learning process is unidirectional in this model. This conceptual change model was revised in response to criticisms (e.g., Cobern, 1996; Solomon, 1987). The revised model acknowledged the importance of intuition, emotion, motivation, and social factors (Strike & Posner, 1992). However, the model still stayed as unidirectional at its core.

Aikenhead and Jegede (1999) criticized *learning as conceptual change* model because the model aims to assimilate students into the culture of Western science at the expense of their indigenous culture. Aikenhead and Jegede (1999) conceptualized the transition between a student's everyday concepts and scientific concepts as *cultural border crossing*. According to this model, students can construct both indigenous and scientific conceptions of natural phenomena side by side and they can make transitions from indigenous conceptions to scientific conceptions and vice versa depending upon the context.

This type of learning can be represented as $C_1 \longleftrightarrow C_2$.

It is reasonable to expect that biology teachers' acceptance of evolutionary theory will influence how they treat evolution in their own classes. Biology teachers who accept evolutionary theory are more likely to use evolution as an overarching theme in their teaching. In fact, a significant number of studies indicated that acceptance of evolutionary theory is a good predictor of instructional approach taken toward evolution (Aguillard, 1999; Eve & Dunn, 1990; Shankar & Skoog, 1993).

Berkman and Plutzer (2010) reported that there is a correlation between the adoption of the NAS positions and the instructional time spent on evolution. They found that biology teachers who strongly support all three positions spent 20 hours a year on evolution, whereas biology teachers who do not support these positions spent only 11 hours on evolution.

Learning is a complex process. Controversial nature of teaching evolution adds to this complexity making the learning process even more complicated. Teaching evolution requires paying attention to all five domains described in this chapter. These domains altogether shape how students make sense of evolutionary theory in biology classrooms. Conceptual, epistemic, socio-cultural, religious, and legal factors influence the learning process in both conceptualizations of learning. It is up to biology teachers to decide what conceptualization of learning can help their students learn evolution in a meaningful way.

REFERENCES

Abd-El-Khalick, F., & Akerson, V. L. (2004). Learning as conceptual change: Factors that mediate the development of preservice elementary teachers' views of nature of science. *Science Education*, 88(5), 785–810. doi:10.1002/sce.10143

Aguillard, D. (1999). Evolution education in Louisiana public schools: A decade following Edwards v. Aguillard. *The American Biology Teacher*, 61(3), 182–188. doi:10.2307/4450650

Aikenhead, G. J., & Jegede, O. J. (1999). Cross-cultural science education: A cognitive explanation of a cultural phenomenon. *Journal of Research in Science Teaching*, 36(3), 269–287. doi:10.1002/(SICI)1098-2736(199903)36:3<269::AID-TEA3>3.0.CO;2-T

American Association for the Advancement of Science. (1993). *Benchmarks for science literacy: A project 2061 report*. Oxford, UK: Oxford University Press.

Ausubel, D. P., Novak, J. D., & Hanesian, H. (1978). *Educational psychology: A cognitive view* (2nd ed.). New York, NY: Holt, Rinehart, and Winston.

Berkman, M. B., & Plutzer, E. (2010). *Evolution, creationism, and the battle to control America's classrooms*. Cambridge, UK: Cambridge University Press. doi:10.1017/CBO9780511760914

Bransford, J., Brown, A., & Cocking, R. (Eds.). (1999). *How people learn: Brain, mind, experience, and school*. Washington, DC: National Academy Press.

Chinn, C. A., & Brewer, W. F. (1993). The role of anomalous data in knowledge acquisition: A theoretical framework and implications for science instruction. *Review of Educational Research, 63*, 1–49.

Cobern, W. (1996). Worldview theory and conceptual change in science education. *Science Education, 80*(5), 579–610. doi:10.1002/(SICI)1098-237X(199609)80:5<579::AID-SCE5>3.0.CO;2-8

Costa, V. B. (1995). When science is another world: Relationships between worlds of family, friends, school, and science. *Science Education, 79*(3), 313–333. doi:10.1002/sce.3730790306

Dagher, Z. R., & BouJaoude, S. (1997). Scientific views and religious beliefs of college students: The case of biological evolution. *Journal of Research in Science Teaching, 34*(5), 429–445. doi:10.1002/(SICI)1098-2736(199705)34:5<429::AID-TEA2>3.0.CO;2-S

Deniz, H., Cetin, F., & Yılmaz, I. (2011). Examining the relationships among Turkish preservice biology teachers' acceptance of evolution, religiosity and teaching preference for evolution. *Journal of Science Education and Technology, 31*(4), 2.1-2.8

Deniz, H., Donnelly, L., & Yilmaz, I. (2008). Exploring the factors related to acceptance of evolutionary theory among Turkish preservice biology teachers: Toward a more informative conceptual ecology for biological evolution. *Journal of Research in Science Teaching, 45*(4), 420–443. doi:10.1002/tea.20223

Deniz, H., Shrader, P. G., & Keilty, J. (2012). *Impact of evolution instruction on understanding and acceptance of evolutionary theory and the nature of relationships among understanding, acceptance, and religiosity*. Paper presented at the Annual Meeting of National Association for Research in Science Teaching. Indianapolis, IN.

Driver, R. (1981). Pupils' alternative frameworks in science. *European Journal of Science Education, 3*(1), 93–101. doi:10.1080/0140528810030109

Eve, R. A., & Dunn, D. (1990). Psychic powers, astrology & creationism in the classroom? *The American Biology Teacher, 52*(1), 10–20. doi:10.2307/4449018

Gallup (2012). *Evolution, creationism, intelligent design*. Retrieved from http://www.gallup.com/poll/21814/evolution-creationism-intelligent-design.aspx

Grose, E. C., & Simpson, D. (1982). Attitudes of introductory college biology students toward evolution. *Journal of Research in Science Teaching, 19*(1), 15–23. doi:10.1002/tea.3660190103

Ipsos, M. O. R. I. (2006). *BBC news*. Retrieved from http://news.bbc.co.uk/2/hi/science/nature/4648598.stm

Lederman, N. G. (1992). Students' and teachers' conceptions about the nature of science: A review of the research. *Journal of Research in Science Teaching, 29*(4), 331–359. doi:10.1002/tea.3660290404

Lerner, L. S. (2000). *Good science, bad science: Teaching evolution in the states*. Washington, DC: Thomas B. Fordham Foundation.

Martin-Hansen, L. M. (2008). First-year college students' conflict with religion and science. *Science & Education, 17*(4), 317–357. doi:10.1007/s11191-006-9039-5

Miller, J. D., Scott, E. C., & Okamoto, S. (2006). Public acceptance of evolution. *Science, 313*, 765–766. doi:10.1126/science.1126746

National Academy of Science. (1998). *Teaching about evolution and the nature of science*. Washington, DC: National Academy Press.

National Academy of Science. (2008). *Science, evolution, and creationism*. Washington, DC: National Academy Press.

National Association of Biology Teachers. (2011). *NABT's statement on teaching evolution*. Retrieved from http://www.nabt.org/websites/institution/index.php?p=92

National Research Council. (1996). *National science education standards*. Washington, DC: National Academy Press.

National Research Council. (2007). *Taking science to school: Learning and teaching in grades K-8*. Washington, DC: National Academy Press.

National Research Council. (2012). *A framework for K-12 science education: Practices, crosscutting concepts, and core ideas*. Washington, DC: National Academy Press.

National Science Teachers Association. (2003). *NSTA position statement: The teaching of evolution*. Retrieved from http://www.nsta.org/pdfs/positionstatement_evolution.pdf

Osif, B. A. (1997). Evolution and religious beliefs: A survey of Pennsylvania high school teachers. *The American Biology Teacher, 59*(9), 552–556. doi:10.2307/4450382

Pintrich, P. R., Marx, R. W., & Boyle, R. A. (1993). Beyond cold conceptual change: The role of motivational beliefs and classroom contextual factors in the process of conceptual change. *Review of Educational Research, 63*(2), 167–199.

Plutzer, E., & Berkman, M. B. (2008). Evolution, creationism, and the teaching of human origins in schools. *Public Opinion Quarterly, 72*(3), 540–553. doi:10.1093/poq/nfn034

Posner, G. J., Strike, K. A., Hewson, P. W., & Gertzog, W. A. (1982). Accommodation of a scientific conception: Toward a theory of conceptual change. *Science Education, 66*(2), 211–227. doi:10.1002/sce.3730660207

Rutledge, M. L., & Warden, M. A. (2000). Evolutionary theory, the nature of science & high school biology teachers: Critical relationships. *The American Biology Teacher, 62*(1), 23–31. doi:10.1662/0002-7685(2000)062[0023:ETTNOS]2.0.CO;2

Schwartz, R. S., Lederman, N. G., & Crawford, B. A. (2004). Developing views of nature of science in an authentic context: An explicit approach to bridging the gap between nature of science and scientific inquiry. *Journal of Research in Science Teaching, 88*(4), 610–645.

Shankar, G., & Skoog, G. (1993). Emphasis given evolution and creationism by Texas high school biology teachers. *Science Education, 77*(2), 221–233. doi:10.1002/sce.3730770209

Sinatra, G. M., Southerland, S. A., McConaughy, F., & Demastes, J. W. (2003). Intentions and beliefs in students' understanding and acceptance of biological evolution. *Journal of Research in Science Teaching, 40*(5), 510–528. doi:10.1002/tea.10087

Solomon, K. (1987). Social influences on the construction of pupils' understanding of science. *Studies in Science Education, 14*, 63–82. doi:10.1080/03057268708559939

Staver, J. R. (1999). When public understanding of science thwarts standards-based science education. *Electronic Journal of Science Education, 3*(4). Retrieved from http://ejse.southwestern.edu/article/viewArticle/7613/5380

Strike, K., & Posner, G. J. (1992). A revisionist theory of conceptual change. In Duschl, R. A., & Hamilton, R. J. (Eds.), *Philosophy of Science, Cognitive Psychology, and Educational Theory and Practice* (pp. 147–176). New York, NY: State University of New York.

Verhey, S. D. (2005). The effect of engaging prior learning on student attitudes toward creationism and evolution. *Bioscience, 55*(11), 991–1003. doi:10.1641/0006-3568(2005)055[0996:TEOEPL]2.0.CO;2

Werth, A. (2012). Avoiding the pitfall of progress and associated perils of evolutionary education. *Evolution: Education & Outreach, 5*(2), 249–265. doi:10.1007/s12052-012-0417-y

Chapter 4
A Theoretical and Methodological Approach to Examine Young Learners' Cognitive Engagement in Science Learning

Meng-Fang Tsai
Chung-Yuan Christian University, Taiwan

Syh-Jong Jang
Chung-Yuan Christian University, Taiwan

ABSTRACT

Children start learning science in formal school settings as early as kindergarten ages. Scientific literacy is an essential component in forming science curriculum for students' knowledge and understanding of scientific concepts across grades and with different scientific disciplines. Students' cognitive engagement is displayed through different levels of cognitive processes involved during their knowledge construction. Much of the research in students' cognitive engagement of science learning focuses on older students and the use of self-report measures on their cognitive strategies. Research on young learners' cognitive engagement in science learning is missing in current literature. Part of the reason may be young learners are just beginning to develop a repertoire of cognitive strategies as well as lacking sufficient linguistic competence to accurately articulate the cognitive strategies involved in their science learning. The chapter introduces a theoretical and methodological approach, quantitative content analysis, to fulfill the gap. The chapter manifests the approach through three perspectives: (a) methodological approaches being employed in current studies of cognitive engagement in science learning, (b) the need for a different methodological approach for young learners, and (c) how this particular approach can be adopted in future research on young learners' cognitive engagement in science learning.

DOI: 10.4018/978-1-4666-2809-0.ch004

A Theoretical and Methodological Approach

OUTLINE OF THE CHAPTER

The chapter contains five sections based on the three perspectives proposed. The first section provides a theoretical framework covering the definition of cognitive engagement, its related theories, and statement of the problem elicited on the issue of the chapter. The second section provides an overview of current methodologies used in the studies on students' cognitive engagement and instructional discourse that promotes this engagement in science learning. Thirdly, a methodological approach and the employment of the approach are then introduced followed by a discussion at the end of the chapter.

THEORETICAL FRAMEWORK

Definition of Cognitive Engagement

Academic engagement is defined as students' commitment or investment in school and is thought to influence students' task management, classroom behaviors (Greenwood, Horton, & Utley, 2002), motivation (Linnenbrink & Pintrich, 2003), science attitudes, and academic achievement (Singh, Granville, & Dika, 2002). Cognitive engagement is originated from school engagement that includes three types of engagement: behavioral, emotional, and cognitive (Fredricks, Blumenfeld, & Paris, 2004). Behavioral engagement is defined as students' behaviors identified in relation to their engagement such as school attendance, and participation in school activities. Emotional engagement is defined as students' affective reactions to classroom such as happiness, sadness, and anxiety. Thirdly, cognitive engagement is defined as the students' psychological investment in learning. Pintrich and Schrauben (1992) defined cognitive engagement as to include cognitive and motivational component. Motivational components include expectancy, value, and affect. Cognitive components include knowledge, learning strategies, and thinking strategies. Other researchers specifically view students' self-regulation of their use of cognitive and metacognitive processes during acquisition and transformation of knowledge as a form of cognitive engagement (Corno & Mandinach, 1983; Mandinach & Corno, 1985). Therefore, cognitive engagement has been studied with different constructs such as motivation (Greene & Miller, 1996; Meece, Blumenfeld, & Hoyle, 1988; Walker, Greene, & Mansell, 2006) and perceptions of classroom tasks (Greene, Miller, Crowson, Duke, & Akey, 2004). Since cognitive engagement is defined and studied differently with disparate constructs by researchers, in this chapter, we adopt the definition by Fredricks et al. (2004) and Dole and Sinatra's descriptions of levels of information processing during knowledge construction, to discuss the issue on examining young children's cognitive engagement in science learning.

Cognitive Engagement and Related Theories

Perspectives of Children's Development

Dole and Sinatra (1998) draw upon Piaget's notion that the process of knowledge construction involves the active transformation and organization of knowledge by the learner (Gallagher & Reid, 2002; Piaget, 1977) through processes of assimilation and accommodation. Knowledge acquisition involves ongoing mental constructions consisting of reorganization and reconstruction that occur when there is disequilibrium within children's current schemes. Developmental research suggests that children undergo important forms of cognitive development through the school years (Flavell, 1999). Their content knowledge becomes better organized with age (Gallagher & Reid, 2002; Siegler, 1989) and children develop increased control over their cognitive processes. Children's ability to use metacognitive knowledge and reasoning also develops with age (Brown, 1987;

Flavell, 1979). Children's strategic knowledge and their ability to select appropriate cognitive and metacognitive strategies develop with age as well (Siegler, 1989). Researchers with domain-general perspectives believe that human beings form cohesive knowledge structures. For these reasons, cognitive engagement may look very different in kindergarten and elementary classrooms as opposed to middle and high school classrooms.

Additionally, from developmental perspectives, different children of the same age show a great deal of variability in cognitive strategy use (Siegler, 1994; Siegler & Booth, 2004; Siegler & Chen, 2002). For example, one student may use multiple strategies including previously learned and new strategies in solving a multiplication word problem. Applying the concept to the chapter, young learners of same age may show various patterns of cognitive engagement through what they talk about (e.g., classroom discourse) in science learning environments. Therefore, unlike using measures to examine older students' cognitive engagement, inferences of cognitive engagement for young learners must be drawn from (videotaped) observations of classroom learning behavior, rather than using self-report measures.

Different from domain general approaches that focus on children's cognitive development and learning in general, domain specific approaches, in contrast, propose that human knowledge is organized in unique domain-specific systems such as knowledge of language, knowledge of science, and knowledge of mathematics (Carey & Spelke, 1994; Hirschfeld & Gelman, 1994). Researchers with domain-specific perspectives suggest that these domain-specific systems demonstrate functions independent and distinct from each other (Carey & Spelke, 1994), in other words, each domain holds different characteristics and may demonstrate unique and independent paths of development and learning. A key assumption of many scholars who favor a domain-specific perspective is that learning is continuous, gradual, and context dependent, rather than global, discrete, and stage-like and can be developed with sufficient practice and experience disregarding ages of learners. This is an important distinction between domain-specific approaches and domain-general, stage theory approaches such as those of Piaget and the neo-Piagetians (Case & McKeough, 1989).

From a domain-general perspective, young children cannot be expected to show higher order forms of cognitive engagement, which are typically defined in the research literature as forms of metacognition. Within a domain-general framework, metacognitive thinking (such as comprehension monitoring, planning, reflection, and evaluation), as well as declarative metacognitive knowledge (e.g., metamemory) are thought to emerge in the elementary school period and develop through adolescence and adulthood (Schneider, 2008). Therefore, self-report measures of cognitive engagement, those that codify engagement in global metacognitive terms, are unlikely to capture young children's cognitive engagement.

The need to examine young learners' cognitive engagement in scientific contexts is essential, as children's cognitive engagement is likely to be manifested contextually through participation in the domain-specific investigative activities of that learning environment. For example, in scientific learning contexts, the specific cognitive activities would be making predictions or asking questions about what might happen; making and recording observations; providing explanations and justifications for explanations. Articulating and examining more domain-specific forms of cognitive engagement is also important since different disciplines may call for different types of thinking and reasoning. Young children may employ nascent forms of higher order cognitive engagement such as monitoring, self-evaluation, and correction, contextually as they engage in specific tasks even though they cannot declaratively report on their engagement in such activities. These forms of cognitive engagement that are specific to scientific learning tasks may be missed by an overly broad, domain-general way of codifying cognitive engagement. Therefore, a different methodology approach is needed to examine young learners' cognitive engagement in a domain-specific science learning environment.

Information Processing Theory

Information processing perspectives extend on Piaget's ideas but emphasize the specific kinds of cognitive and metacognitive processes that constitute cognitive engagement during the acquisition and transformation of knowledge. Cognitive engagement refers to students' use of directed or purposeful mental processes for learning (Fredricks, et al., 2004). Dole and Sinatra (1998) define cognitive engagement in information processing terms and suggest that information processing should be viewed as a continuum "from low cognitive engagement to high metacognitive engagement" (p. 121). They further suggest that when students are low in cognitive engagement, they process low amounts of new information by using simple strategies such as rote memorizing. Students who are moderately engaged show more depth of processing, use more elaborative strategies (e.g., they rephrase information in their own words during note taking or study; they connect new information to prior knowledge), and reflect more on their learning (e.g., they identify relevant versus irrelevant information; they identify gaps in understanding) than those with low cognitive engagement. The highest engagement involves "deep processing, elaborative strategy use, and significant metacognitive reflection" (p. 121). According to Dole and Sinatra, this continuum of engagement is the most salient component of conceptual change processes during learning.

Research suggests that students with higher cognitive engagement perform better compared to students with lower cognitive engagement (Greene & Miller, 1996). Students with high cognitive engagement tend to self-regulate their learning, organize and structure knowledge, acquire, select, and integrate information during the process of knowledge construction, and use elaborative learning and thinking strategies.

Vygotsky

According to Vygotsky (1978), social interactions between experienced people and novices are essential in supporting cognitive development and learning. It is imperative to provide a learning environment to support children's development and learning through adult-child and peers interactions in context (Brown & Reeve, 1987). This supportive environment with scaffolding provided can help children reach their maturation stage from their actual developmental level into their Zone of Proximal Development (ZPD). Children develop their cognitive skills and knowledge through the internalization of higher psychological functions that they first acquire through interactions with and support or scaffolding from others such as parents, teachers, or more expert peers (Vygotsky, 1978). The knowledge gained through this interpersonal process is transformed into an intrapersonal process that results in children's cognitive development. Vygotsky's research suggests that the nature and form of social support and interaction during classroom learning is likely to influence both how and what students learn. Therefore, young children's cognitive engagement can be enhanced by an enriched learning environment where provides support for young children's cognitive development and learning through sufficient interactions.

From Piaget's perspectives on cognitive development in learning, individual learners construct knowledge on their own that involve internal cognitive processes, whereas from Vygotsky's viewpoints, individuals primarily gain and construct knowledge through their external interactions with people and/or objects such as tools used in the environments. Both concepts of Piaget and Vygotsky are important components for students' cognitive engagement in learning science and teachers' discourse produced in teaching science. Researchers have studied both constructs of students' cognitive engagement in science learning and instructional practices involved in the learning environments for this engagement.

Statement of the Problem

The chapter is triggered by a topic, students' cognitive engagement in science learning, which researchers have studied it with various methodological approaches. After reviewing current studies on the examination of this construct, the authors find the need to broaden our understanding of what involves cognitive engagement for young learners in science learning by introducing a new methodological approach.

Young children's cognitive engagement has been studied in reading. There is little research examining young children's cognitive engagement in science learning. Part of the reason may be that current measures used to examine cognitive engagement are best for older students. Measures to assess young learners' cognitive engagement are scarce. With the consideration of young learners' limitation of metacognitive capacity, observational techniques is the most constructive method to examine their nature of cognitive engagement. Furthermore, carefully designed scientific contents with the perspectives of young children's cognitive development are rare. The methodological approach, quantitative content analysis, is introduced to serve the purposes of examining young learners' cognitive engagement through coding qualitative data that quantitative methodological approach lacks, as well as conducting statistical analyses that qualitative methodological approach usually cannot do.

CURRENT METHODOLOGICAL APPROACHES TO EXAMINE COGNITIVE ENGAGEMENT

Several researchers have studied cognitive engagement through subjects' self-reports or questionnaires (Greene & Miller, 1996; Greene, et al., 2004; Meece, et al., 1988; Pintrich & De Groot, 1990). These researchers typically employed students' self-reported strategy use on learning tasks as a measure of their cognitive engagement. Researchers have also examined the relations between a variety of motivational variables (e.g., students' intrinsic and extrinsic motivation, self-efficacy, goal orientations, and values for academic achievement) and cognitive engagement (Greene & Miller, 1996; Walker, et al., 2006). Other researchers have adopted qualitative approaches, using open-ended interviews with students and classroom observations to investigate students' engagement patterns (Ainley, 1993; Chin & Brown, 2000). These two groups of studies will be discussed in more detail as it follows.

Studies with Self-Report Measures

Much of the research pertaining to cognitive engagement focuses on older students and uses self report measures such as survey instruments to determine cognitive engagement (Blumenfeld & Meece, 1988; DeBacker & Crowson, 2006; Dupeyrat & Marine, 2005; Flowerday & Schraw, 2003; Greene & Miller, 1996; Greene, et al., 2004; Meece, et al., 1988; Metallidou & Vlachou, 2007; Walker, et al., 2006). Examples of such self-report measures include: (a) Motivation and Strategy Use Survey (Greene & Miller, 1996) used to measure motivational and cognitive dimensions of college students' engagement, (b) Science Activity Questionnaire (SAQ) (Meece, et al., 1988) used to measure 5th and 6th-grade students' goal orientation and cognitive engagement, and (c) Learning Process Questionnaire (LPQ) used to measure the depth of cognitive engagement in 11th-grade

A Theoretical and Methodological Approach

students (Ainley, 1993). Table 1 provides a summary of various self-report instruments used in research on cognitive engagement with sample items used to measure cognitive engagement in different studies.

One instrument that is widely used in these types of research studies is the Motivated Strategies for Learning Questionnaire (MSLQ). Early researchers, such as Pintrich and De Groot (1990), used the MSLQ as a measure to examine the relationship between motivation, self-regulation, and performance of seventh-grade students in their English and Science classes. Students' cognitive engagement was analyzed both in terms of the students' cognitive and metacognitive strategy use as well as their persistence on difficult tasks. Participants received the MSLQ that contained 56 Likert scale items to indicate their motivation, cognitive and metacognitive strategy use, and management of effort.

Greene et al. (2004) conducted a study to examine how high school students' perceptions of classroom structures influenced their self-efficacy, perceived instrumentality, mastery goal, performance goal, strategy use, and achievement. Students' strategy use was examined to be an indicator of their cognitive engagement. The participants included sophomores, juniors, and seniors from English classes in a suburban high school. The researchers used the Survey of Classroom Goals Structures to examine students' perceptions of classroom structures (e.g., motivating tasks, autonomy support, and mastery evaluation). The Approaches to Learning instrument was used to examine students' goal orientation, perceived instrumentality, and cognitive strategies. The instrument was modified based on the survey developed by Miller, Greene, Montalvo, Ravindran, and Nichols (1996).

Several studies have also indicated the relations for cognitive engagement and motivation with different ages of learners (Greene & Miller, 1996; Meece, et al., 1988; Walker, et al., 2006). For example, Greene and Miller (1996) conducted a study to investigate the associations of goal orientation, perceived ability, cognitive engagement, and course achievement with 104 pre-service teachers from an Educational Psychology class. The Moti-

Table 1. Summary of instruments with self-report items for cognitive engagement

Author(s) of the Study	Instrument	Sample Item of Cognitive Engagement
Pintrich and De Groot (1990)	MSLQ: Cognitive Strategy scale	Organizational strategies: I outline the chapters in my book to help me study (Item 54).
	Self-regulation scale	Even when study materials are dull and uninteresting, I keep working until I finish (Item 41).
Greene et al. (2004)	Approaches to Learning instrument: Cognitive strategies	I plan my study time for this class.
Walker et al. (2006)	Motivation and Strategy Use Survey: Meaningful cognitive engagement	When I study I take note of the material I have or have not mastered.
	Shallow cognitive engagement	I try to memorize exactly what my instructors say in class.
Greene and Miller (1996)	Motivation and Strategy Use Survey: Meaningful cognitive engagement	I made a plan for achieving the grade I wanted on this exam.
	Shallow cognitive engagement	In order for me to understand what technical terms meant, I memorized the textbook definitions.
Meece et al. (1988)	Science Activity Questionnaire (SAQ): Active engagement	I went back over things I didn't understand.
	Superficial engagement	I copied down someone else's answer.

vation and Strategy Use Survey was administered. Students' strategy use was examined with two levels of cognitive engagement: meaningful and shallow cognitive engagement. Similarly, Walker et al. (2006) examined college students' learning strategies to indicate their cognitive engagement. Students' perceptions of belonging and valuing for academic contexts, intrinsic/extrinsic motivation, and self-efficacy were examined to predict their cognitive engagement. The participants included freshmen through seniors from Educational Psychology classes and a career exploration class. Motivation and Strategy Use Survey was used to examine students' meaningful and shallow cognitive engagement.

Moreover, Meece et al. (1988) conducted a study for fifth and sixth-grade students' active and superficial engagement and their mastery goals, self-efficacy, perceived instrumentality, and perceptions of autonomy. The participants completed surveys/questionnaires to access goal orientation and their use of high-level or effort-minimizing learning strategies. Some items of Science Activity Questionnaire (SAQ) were used to assess students' cognitive engagement (see Table 1 for sample item). Eleventh-grade students' cognitive engagement was also examined with their learning strategy (Ainley, 1993).

The researchers of previously discussed studies (Greene & Miller, 1996; Greene, et al., 2004; Walker, et al., 2006) have analyzed students' learning strategies as their cognitive engagement through self-report measures. The designations of learning strategies to measure students' cognitive engagement are set in self-report measures so this kind of methodology may not capture the natural occurrences of strategies students use while they are learning in different learning contexts. Therefore, researchers have used observational measures to examine students' cognitive engagement naturally occurred in different learning contexts.

Studies with Direct Observation

Some researchers have employed direct observation to conduct studies for cognitive engagement with older students. The following section provides a discussion of these studies.

Chin and Brown (2000) identified the associations of questions students asked and the levels of cognitive engagement. Questions associated with a surface level of cognitive engagement tended to be factual (e.g., What color is that?) or procedural questions (e.g., Could we put this into the beaker now?). Questions associated with a deep level tended to be wonderment questions such as comprehension, prediction, anomaly detection, application, and planning or strategy questions. They also investigated middle school students' cognitive structures involving deep and surface approaches to cognitive engagement. There were six eighth-grade students purposively selected based on the results of the Learning Approach Questionnaire, observations of group laboratory activities, and their teacher's evaluation of students' learning approaches. The students showed different degrees of cognitive engagement (from deep to surface) in science learning. The data sources of the study included field notes, classroom transcriptions of audiotapes and videotapes, students' interviews, and students' written work. The researchers used qualitative interpretive analysis techniques to identify five emergent categories of cognitive engagement: generative thinking, nature of explanations, asking questions, metacognitive activity, and approach to tasks. Within each category, different levels of engagement were explored. The findings were presented by providing excerpts from the qualitative data.

In a different study, students' cognitive engagement was examined by analyzing four, Year 8, mathematics video lessons and teacher and student interviews (Helme & Clarke, 2001). The authors investigated four classroom formats (a) individuals working in parallel, (b) collaborative small group activity, (c) small group interactions

A Theoretical and Methodological Approach

with teacher, and (d) whole class interactions with teacher, to measure cognitive engagement. The results indicated that classroom formats were associated with students' cognitive engagement. Different patterns of students' cognitive engagement were analyzed with conversations and behaviors presented in the four classroom formats. Table 2 provides detailed information of the indicators of students' cognitive engagement (Helme & Clarke, 2001, Table 1, p. 141).

Lee and Anderson (1993) examined the quality of students' task engagement during science learning, using classroom observations and informal interviews as data sources. In the study, task engagement was defined as students' goal choices and strategies in engaging in tasks. The classroom observations contained four kinds of class activities: (a) students read and discuss science book, (b) students conduct experiments, (c) students write their ideas and record their experiments in their activity book, and (d) students engage discussions about their ideas and conclusions. The participants were twelve sixth-grade students with low, medium, and high achievement levels selected by their classroom teachers to examine their quality of task engagement. The three achievement levels were identified based on students' science performance records when they entered their middle school and their first few months of performance before the study began.

The researchers gave clinical interviews to examine students' knowledge about the scientific concepts that they learned in the class. Students received a semi-structured interview and a self-report questionnaire to indicate their goals (e.g., understanding goal, fact acquisition goal, ego-social goal, and work-avoidant goal) in science class prior to and after the unit implementation. Students' behavior was coded for the quality of task engagement as follows: (a) self-initiated cognitive engagement (e.g., initiating science activities for better understanding, engaging in classroom tasks beyond the requirements), (b) cognitive engagement (e.g., integrating personal knowledge with scientific knowledge, applying scientific knowledge to understand the world around them), and (c) behavioral engagement (e.g., students show attention and involvement in class activities). Each type of task engagement described above was further coded to seven levels of engagement within type from the highest level (1) to the lowest level (7).

Four patterns of task engagement emerged that indicated students' goals differ and their level of task engagement varies (Lee & Anderson, 1993). The emergent patterns were intrinsically motivated to learn science, motivated to learn science, task avoidance, and active task resistance. The authors provided four case students' discourse of classroom observations and informal interviews, to discuss the four emergent patterns.

The researchers in the studies previously described have typically used self-report measures to examine students' cognitive engagement. The

Table 2. Indicators of cognitive engagement in four classroom situations

Classroom situation	Behavior
Individuals working in parallel	Verbalising thinking, self-monitoring, concentration (resisting distractions or interruptions), gestures (interpreted as externalizing thought processes), seeking information and feedback
Collaborative small group activity	Questioning, completing peer utterances, exchanging ideas, giving directions, explanations, or information, justifying an argument, gestures
Small group interactions with teacher	Answering teacher's questions, giving information, explaining procedures and reasoning, questions addressed to teacher, reflective self-questioning
Whole class interactions with teacher	Asking and answering questions, making evaluative comments, contributing ideas, completing teacher utterances

self-report measures are likely to be ineffective measures of young children's cognitive engagement. Unlike observational measures, self-report measures do not provide direct evidence of students' cognitive engagement during learning. First, many of these measures of cognitive engagement depend on students' reports of their past cognitive and metacognitive strategy choices. Young learners are just beginning to develop a repertoire of cognitive strategies and may not have acquired mastery of a sufficient repertoire of cognitive strategies for a "choice" measure to make sense. Additionally, metacognitive competence is just beginning to emerge in young children and continues to develop significantly through the upper elementary school years and beyond, well into adolescence and young adulthood (Brown, 1987; Flavell, Miller, & Miller, 1993; Kuhn, 1999, 2000). Therefore, it is unlikely that young children have metacognitive awareness of and can report accurately on their own cognitive processes and strategies. Young children may also lack the linguistic competence to understand and rate verbal descriptions of cognitive engagement of the kind used in the scales and questionnaires described above. This does not mean that young children are not cognitively engaged in learning. Rather, it implies that we may need more direct observational measures for different patterns of mental activity young children can naturally display during their learning, rather than relying on them to report their strategies of cognitive processing. Although direct observations were used in some studies, the participants were older students and the findings were generally reported with narratives. Additionally, the categories identified could not indicate cognitive processes displayed in the contexts. In this chapter, we introduce a methodology, quantitative content analysis, to allow researchers to have another option when choosing an analytical technique to examine students' cognitive engagement in science learning with qualitative data, as this methodology combines the analytical operations of both qualitative and quantitative approach.

As learning and teaching indicate a reciprocal relationship, we also review the studies that examine what instructional discourse influences students' cognitive engagement, as cognitive engagement is more likely to be enhanced in a supportive learning environment. The following section provides an overview on what instructional discourse and methodology researchers have studied and employed in relation to students' cognitive engagement.

Instructional Discourse and Students' Cognitive Engagement

There is little research with the emphasis on the relationships between instructional discourse and students' cognitive engagement. In reviewing prior empirical studies for this particular interest, we found four studies related to this interest. Researchers have considered different instructional constructs such as classroom tasks (Blumenfeld & Meece, 1988; Helme & Clarke, 2001), teacher-centered versus student-centered instruction (Trigwell, Prosser, & Waterhouse, 1999) and the role of teacher questioning (Taylor, Pearson, Peterson, & Rodriguez, 2003) into their research design to examining its relations to students' cognitive engagement. The instructional practices studied and research methodologies employed in these studies are discussed.

Classroom Tasks

Students' cognitive engagement is influenced by classroom tasks and teachers' behaviors. Blumenfeld and Meece (1988) conducted a mixed-method study to examine how task elements such as task content, social organization, procedural complexity, and products and teacher behavior related to students' involvement on tasks and cognitive engagement. Participants were fourth and sixth-grade students from eight science classrooms in four middle-class schools. Students' cognitive engagement was measured by a self-report checklist including items that measured superficial (e.g., "I just made my best guess") and high-level strategy ("I tried to remember how this was like something we did before") forms of cognitive engagement. Some students were selected for interviews to supplement the results of the questionnaires on involvement and cognitive engagement.

The activity-based lessons in the study were selected from two units. The researchers observed the lessons consisting of electricity, food system, physical and chemical changes, and the human body. The patterns of teacher behavior that promoted cognitive engagement were examined through the observer's detailed narratives of teacher behavior in the lessons and transcripts of audiotaped lessons. The researchers found two clusters of teacher behaviors related to students' engagement: (a) maintaining student participation that helped students develop on-task behaviors (e.g., giving prompts to help students' attention, checking students' progress) and (b) focusing on mastery learning that enhanced students' cognitive engagement (e.g., requiring students' demonstration of understanding for lesson content, communicating with students that they were expected to be active in learning, providing feedback on the procedures of an experiment). The authors also concluded that teachers' behaviors to support students' involvement and cognitive engagement varied by different social organization of class tasks (e.g., whole class, small group and individual).

In another study, Helme and Clarke (2001) analyzed video lessons and teacher and student interviews to find that classroom discourse promoting students' cognitive engagement varied in different classroom task settings (e.g., student-student interactions versus teacher-student interactions) in learning mathematics. Students' cognitive engagement was higher through student-student interactions than through teacher-student interactions. Classroom tasks are influenced by different domains and content areas, as in a mathematics class, teachers may encourage students to try multiple ways of solving a mathematical problem, whereas in a science class, teacher may encourage students to develop and evaluate scientific explanations.

Teacher-Centered vs. Student-Centered Instruction

Researchers have defined two instructional approaches: teacher-centered (e.g., asks students to follow exact procedures of an experiment from a textbook) and student-centered (e.g., encourages students to conduct an experiment in a group), characterized by different patterns of teacher-student interaction (Trigwell, et al., 1999). These differences in instruction are thought to influence students' learning outcomes.

Trigwell et al. (1999) studied college students and their teachers. Teachers' instructional approaches were examined by the Approaches to Teaching Inventory that included two scales (a) Information Transmission/Teacher-Focused approach and (b) Conceptual Change/Student-focused approach. Table 3 provides information of the scales administered and sample items of each scale.

Teachers who reported their teaching as student-centered (e.g., encouraging self-directed learning, allowing students to encounter their difficulties and make conceptual change through classroom interactions between teacher and students) had students with deeper approaches to learning compared to the students whose teachers reported their teaching as a teacher-centered of transmitting information. Students in the teacher-centered approach tended to show surface learning in class. Teacher instructional discourse patterns, such as challenging students' thinking, providing scaffolding for students' difficulties, coaching students' learning strategies, and asking higher-level questions, influences students' cognitive engagement in reading and writing (Taylor, et al., 2003).

Table 3. Name of scales and sample item of the scales

Scale	Sample Item
Approaches to Teaching Inventory: Information Transmission/Teacher-Focused approach (intention)	I feel it is important to present a lot of facts in the classes so that students know what they have to learn for this subject.
Approaches to Teaching Inventory: Information Transmission/Teacher-Focused approach (strategy)	I design my teaching in this subject with the assumption that most of the students have very little useful knowledge of the topics to be covered.
Approaches to Teaching Inventory: Conceptual Change/Student-focused approach (intention)	I feel a lot of teaching time in this subject should be used to question students' ideas.
Approaches to Teaching Inventory: Conceptual Chang/Student-focused approach (strategy)	We take time out in classes so that students can discuss among themselves the difficulties that they encounter studying this subject.
Study Process Questionnaire	In reading new material for this topic, I find that I'm continually reminded of material I already know, and see the latter in a new light (Item 8).

The Role of Teacher Questioning

Teachers who implement instruction using higher-level questioning with high poverty students in grades 1-5 facilitate their writing and reading and thus reinforce their cognitive engagement (Taylor, et al., 2003). The types of questions teachers ask influence students' psychological investment and their development in reading and writing. The researchers implemented a reading instruction that maximized students' cognitive engagement in both reading and writing. The reading instruction had four dimensions in the framework which included: (a) support of higher-level thinking, (b) encouragement of independent strategies for word-recognition and comprehension, (c) involve student support perspectives, and (d) foster active involvement. Data included results of scores on academic assessments in reading and writing, classroom observations, and teacher interviews.

To make classroom observations, the researchers coded certain categories to examine which instructional practices facilitated students' achievement in reading and writing and to identify how effective the proposed framework was in fostering students' cognitive engagement. Examples of the instructional categories coded were as follows: coaching in word-recognition strategies, vocabulary instruction, lower-level questioning or writing about text, and higher-level questioning or writing about text. The higher-level questioning approach helped develop students' cognitive engagement as manifest in higher-level thinking (e.g., makes connections between prior knowledge and new knowledge; involves monitoring in the process of solving reading problems) in reading and writing. Thus, the types of questions teachers ask influence students' cognitive engagement.

Studies on the role of instructional discourse in promoting students' cognitive engagement suggest that teachers can use different instructional practices to enhance students' cognitive engagement in learning such as providing different classroom tasks, asking higher-level questions, challenging students' ideas, and encouraging students' participations. Similar to the studies on students' cognitive engagement, these studies on instructional practices focus on older students. As we discussed previously, direct observations will be the best data sources for researchers to examine young learners' cognitive engagement and what instructional discourse that promotes this engagement naturally displayed in a designed learning environment through classroom discourse. To analyze verbal data, we then introduce a methodology, quantitative content analysis, to serve an option for researchers who have the research interests to apply it in their own research.

Methodological Approach to Examine Young Learners' Cognitive Engagement

Quantitative content analysis is originated from research of media data such as newspapers (Riffe, Lacy, & Fico, 2005) and has been used to analyze various forms of data sources that are messy but contain rich information such as webpages, transcripts of audios and videos, discussions, interviews, observations. Recently, this methodology not only has been used in the research on science communication through newspapers (Crawley, 2007), it has also been used in different research areas such as business (Barringer, Jones, & Neubaum, 2005), computer mediated communications (Iseke-Barnes, 1996; Mowrer, 1996), and development between use of technology and pedagogical and content knowledge (Koehler, Mishra, & Yahya, 2007). Other researchers in science education have also employed this methodology to examine students' science learning (Samarapungavan, Westby, & Bodner, 2006; Samarapungavan & Wiers, 1997).

Quantitative content analysis is defined as "the systematic assignment of communication content to categories according to rules, and the analysis of relationships involving those categories using statistical methods" (Riffe, et al., p. 3). This methodological approach focuses on the textual content of verbal communication or discourse in context and is designed to analyze, compare, and contrast patterns and categories identified from data as well as relationships among them. It follows an objective and systematic way to analyze data by employing bootstrapping procedures combining top-down or theoretically derived and bottom-up or data-derived coding categories in the analysis of verbal data (Chi, 1997; Riffe, et al., 2005). Top-down codes are derived from theory and prior empirical research; bottom-up codes are derived from the new empirical data in the study at hand, based on the researchers' theoretical judgments about the emergence of new or previously unidentified patterns. Once a set of initial coding categories is defined and applied to the data, the process of testing the fit of categories to data is repeated in cycles in order to refine and modify the definitions of categories based on the theoretical and empirical findings relevant for a designated study (Chi, 1997). The defined categories then analyzed with statistical methods. Some researchers may argue this methodological process to bring another level of subjectivity, however, this methodological process is obviously superior to apprehend any possible phenomenon occurred for the construct (i.e., young children's cognitive engagement in science learning) studied and make more thorough explorations with fundamental conclusions than simply using self-report measures or presenting results with few selected narratives.

To introduce the methodology in an integral way, we selected two references related to quantitative content analysis, to serve as the main sources in discussing each procedure involved in the methodology (Chi, 1997, Riffe, et al., 2005). Riffe et al. (2005) proposed a general model of linear progression for this particular methodology (see Table 3). Since most researchers should be familiar with the first part of the model (i.e., conceptualization and purpose), we will only discuss the procedures that involve in the parts of design and analysis according to the eight procedures proposed by Chi (1997). These eight concrete procedures of coding and data analyses are: "(1) Reducing or sampling the protocols, (2) Segmenting the reduced or sampled protocols (sometimes optional), (3) Developing or choosing a coding scheme or formalism, (4) Operationalizing evidence in the coded protocols that constitutes a mapping to some chosen formalism, (5) Depicting the mapped formalism (optional), (6) Seeking pattern(s) in the mapped formalism, (7) Interpreting the pattern(s), and (8) Repeating the whole process, perhaps coding at a different grain size (optional)" (Chi, 1997, p. 283). These eight procedures exclude the steps of collecting and transcribing data. The procedures can be adopted by researchers based on the design of each research study. We use an example study to explain how to employ each procedure of the methodology in detail in the next section (see Table 4).

EMPLOYMENT OF THE METHODOLOGY WITH AN EXAMPLE STUDY

In order for researchers to better understand how to employ this methodology, we came up with an example study to model each procedure of this methodology with explanations provided.

The Example Study

The example study is designed to examine 1st-grade students' cognitive engagement in a problem-based scientific learning environment with a unit named "Internal movements of plants." There are four 1st-grade elementary classes from the same elementary school participating in the study. Two teachers implement the unit designed with problem-based lessons for science learning, and their classes serve as Intervention Group (IG). The other two teachers use direct instruction with the use of textbooks for the same science topic, and their classes serve as Traditional Group (TG). The problem-based unit is designed with three phases that students (1) discover a problem as a group that relates to the topic "internal movements of plants," (2) design and apply experiment procedures to solve the problem, and (3) reflect on and synthesize what they have learned through students' own explanations. Each phase lasts one week, and the whole unit is 3-week long. The IG teachers have to implement the problem-based lessons designed for the unit two times each week and mainly serve as facilitators to guide students' learning and provide information when necessary. Each lesson lasts 30-45 minutes. There are 91 first graders participating in the study – IG has

Table 4. General linear progression model of quantitative content analysis

Conceptualization and purpose
Identify the problem
Review theory and research
Pose specific research questions and hypotheses
Design
Define relevant content
Specify formal design
Create dummy tables
Operationalize (coding protocol and sheets)
Specify population and sampling plans
Pretest and establish reliability procedures
Analysis
Process data (establish reliability and code content)
Apply statistical procedures
Interpret and report results

Note. Adopted from Riffe et al. (2005, p. 55).

48 students (one class has 23, and the other class has 25) and TG has 43 students (one class has 21, and the other class has 22). All lessons are videotaped, and the primary data source of the study is videotaped lessons.

The main purposes of the study are to examine patterns of 1st-grade students' cognitive engagement in the IG and TG as well as teachers' instructional discourse that involves in promoting their students' cognitive engagement in the two learning contexts. After the patterns are examined, differences of cognitive engagement and instructional discourse between the two groups then can be compared. We assume that there are four research questions proposed for the example study: (1) Does 1st graders' cognitive engagement vary by the two intervention classes? (2) How does teacher instructional discourse relate to 1st graders' cognitive engagement in the intervene group? (3) Does 1st graders' cognitive engagement vary in the two groups (i.e., IG and TG)? (4) Does teachers' instructional discourse vary in the two groups (i.e., IG and TG)? For all the research questions, we propose no significant difference for research questions 1, 3, and 4 and no significant associations for research question 2.

Description of Each Procedure of the Methodology with the Example Study

Chi (1997) proposed eight functional steps for quantitative content analysis in coding and analyzing verbal data. By using the example study, we will model the procedures with explanations of how to employ each step in the example study followed by discussions for the reliability and validity of quantitative content analysis.

Step One is to Reduce the Samples

Researchers can decide to reduce the sample of videotaped lessons, as there are 24 video lessons as a total in the study. Researchers can choose sampling method depending on their concerns in the research design including content and time. Riffe et al. (2005) introduced non-probability and probability sampling. Probability sampling includes simple random, systematic, and stratified sampling. Non-probability sampling contains convenience sampling and purposive sampling. There are two other sampling methods introduced: cluster and multistage sampling. In the current example study, the main purpose is to explore patterns of cognitive engagement for each phase, therefore, purposive sampling is used to select representative data for each phase. Through purposive sampling, six separate videos are selected from IG and TG, totaling 12 videos for analyses.

Step Two is to Segment the Reduced or Sampled Protocols

This is to segment the unit of analysis. Each classroom discourse of the 12 selected videos can be broken down into segments called Turn Constructional Units (TCUs) and further coded each segment into defined categories. A discourse can be perceived as a segment being presented with a structure of utterances grouped together (Grosz, Pollack, & Sidner, 1989). A segment broken down from the classroom discourse or TCU is the unit of the analysis. A unit of the analysis may involve more than one sentence in order for a participant to generate an idea clearly (Chi, 1997; Riffe, et al., 2005).

TCUs are used to identify meaningful segments of classroom discourse (e.g., completion of an idea or a scientific explanation) from the transcripts for each participant (Sacks, Schegloff, & Jefferson, 1974). TCU indicates a complete unit of conversation in a semantic form (e.g., making a counterargument to a previous students' claim, elaborating on a previous students' contribution by adding descriptive detail). Each TCU needs to be coded into preliminary coding categories for the construct of the study. Each TCU can be coded by lesson and class. In an Excel file of TCUs, there are columns indicating classes, phases of problem-based lessons, and length of the lesson over which TCU is coded.

Step Three is to Develop or Choose a Coding Scheme or Formalism

According to bootstrapping procedures, the initial coding scheme needs to be developed based on the prior literature and empirical data itself. For example, asking conceptual question could be identified as one of the categories as it has been found from previous empirical studies in examining students' cognitive engagement in science learning (Chin & Brown, 2000). If researchers identify a category that has not been stated from prior studies on cognitive engagement in science learning, then this category could be identified as data-derived. The initial coding scheme of the categories and definitions of the categories can be revised when analyzing more data during the coding process (Riffe, et al., 2005). New categories can be added during the process of coding all empirical data.

Step Four is to Operationalize Evidence in the Coded Protocols that Constitutes a Mapping to Some Chosen Formalism

Each unit of analysis is coded into the defined categories and the utterance needs to indicate the evidence for the category that the utterance is coded to. The context of classroom discourse needs to be considered when coding verbal data as the previous classroom discourse could be connected to latter classroom discourse of students and peers. The focus of the analysis is not to code the correctness or incorrectness of the scientific knowledge representation; it is to code the content of the verbal utterance based on what that utterance indicates about how learners engage their cognition in science learning. Researchers can employ this method to examine 1st graders' cognitive engagement for different levels of mental processes engaged in their knowledge reorganization and reconstruction. By coding the 12 selected videotaped lessons, the final coding scheme is mapped and formalized for first graders' cognitive engagement in science learning. We assume that there are 13 categories identified for 1st graders' cognitive engagement and 11 categories identified for teacher instructional discourse that promotes this engagement.

Step Five is Depicting the Mapped Formalism (Optional)

This step is to use a way to depict coded data, depending on what formalism is chosen. Researchers can provide a simple table presenting the proportions (i.e., frequency/time length of the lesson) of identified categories to depict the coded data.

Step Six is to Seek Pattern(s) in the Mapped Formalism

This step is to seek patterns in the depicted data in other types of dependent measures for coherence. For example, researchers can draw a model based on the patterns sought by the depicted data. According to the results of prior empirical studies, cognitive engagement can be grouped into deep and surface levels (Ainley, 1993; Greene & Miller, 1996; Meece, et al., 1998; Walker, et al., 2006). Therefore, researchers can try to seek whether there is a pattern displayed in the final coding scheme. The identified categories can be grouped into deep and surface levels of cognitive engagement. Researchers can provide a bar graph presenting frequencies, proportions for the two levels of cognitive engagement and instructional discourse for both students and teachers by classes to see the patterns involved in it.

Step Seven is Interpreting the Pattern(s)

In this step, hypotheses are being tested. Frequency distributions of different coding categories can be analyzed using non-parametric methods to explore relationships among categories of responses within samples or across samples. Researchers need to select its types of statistical analysis based on their proposed research questions. To confirm or disconfirm the statistical hypotheses for the example study, researchers can conduct statistical analyses. For the purposes of the example study, MANOVA can be used to address research questions 1, 3, and 4, and Chi Square can be employed to answer research question 2.

Step Eight is Repeating the Whole Process, Perhaps Coding at a Different Grain Size (Optional)

Researchers can choose to recode the data at a different grain size or if they want to address different research questions. Since verbal data contain rich sources, new research questions may occur during the process of recoding the data. In the example study, we assume that there is no other research question to be addressed so recoding the data is unnecessary.

We want to note here that although each procedure is introduced separately, the procedures may be employed in an integral way. For example, the procedures of segmenting unit of analysis, developing the coding scheme, and seeking patterns can be employed together during the coding process.

Reliability

There are two ways introduced for the reliability of quantitative content analysis. Reliability can be established by the percentages of agreement among two or more coders, and the acceptable percentage is more than 80%. The other way of conducting reliability is to conduct Cohen's Kappa (Cohen, 1960; Ary, Jacobs, & Sorensen, 2006; Riffe, et al., 2005). Cohen's Kappa is used to assess the inter-rater reliability when coding categorical variables and refers to the proportion of observed units beyond that expected by chance alone. It is the measure of agreement between two individuals in coding qualitative data into categories. Numbers of agreements and disagreements between the raters can be entered into statistical software to gain the value of reliability. A minimum value of 80% should be expected for an adequate reliability.

Validity

According to Chi (1997), to establish the validity of the pattern coded from data, researchers can code the data twice for an identified pattern so that the second process of validity check can avoid any subjective interpretation that may occur in the first process to identify the pattern. For example, in the example study, researchers can first list an initial pattern with preliminarily identified categories of cognitive engagement by reviewing the verbal data. The pattern with the preliminarily set categories can be validated by presenting at least one verbal example as the evidence to support the pre-identified categories. Another way to increase the validity is to couple qualitative analysis with quantitative measures, the unique feature of this methodology. For example, frequencies and statistical analysis (i.e., Chi Square) can confirm or disconfirm the interpretations and test the hypotheses of the example study.

DISCUSSION

Our purpose in this chapter is not to manifest quantitative content analysis as the best way of conducting studies on young children's cognitive engagement in science learning, rather, we propose this methodology that current researchers in this field may not be familiar with but can serve as an effective way to examine this particular construct. The benefits of employing the methodology on examining young learners' cognitive engagement in science learning are discussed with four aspects (1) broadening our views on analysis, (2) dealing with disorganized data, (3) conducting research with domain-specific perspectives, and (4) directing future research.

As quantitative content analysis has been used in different areas of research (Barringer, et al., 2005; Crawley, 2007; Iseke-Barnes, 1996; Mowrer, 1996; Koehler, et al., 2007; Samarapungavan, et al., 2006; Samarapungavan & Wiers, 1997), there are still very few researchers using this methodology to examine students' cognitive engagement in science learning. The lack of studies on young learners' science learning may be due to the reason that it is challenging to find an appropriate methodology to examine young learners' cognitive engagement. Quantitative content analysis is unique to solve the difficulties and problems as it creates broader views on analysis to overcome the constraints of qualitative and quantitative methodology by combining the distinguishing features of both qualitative and quantitative approach. Simply using measures can result in having a surface understanding of the construct, whereas plainly using selected narratives from qualitative data can result in researchers being subjective to data itself. Another benefit is that this methodology shares some flexibility that researchers can use this methodology along with other methodological analyses.

Researchers can use this methodology to analyze messy data into a systematic and organized pattern for any possible levels in the pattern and present findings with statistical results. Quantitative content analysis can be used for aims of examining data with rich information such as media data for young learners' cognitive engagement. The use of measures is almost impossible to understand the complex cognitive processes involved in cognitive engagement in science learning, not to mention the intention to examine this engagement with young learners. Therefore, to examine this construct, data sources with direct observations are needed. Through quantitative content analysis, messy data such as videotaped lessons can be analyzed into a systematic and organized way to investigate young children's cognitive engagement and teacher's instructional discourse that enhances this engagement as well as the cognitive levels. Young learners' metacognitive skills are also more likely to be explored through interactive conversations naturally produced in a designed scientific learning environment, which data collected with measures cannot explore.

This methodology is useful for a domain-specific context as specific categories in a context can be identified by applying the procedures. Researchers have considered domain-specific as a critical component for children's cognitive development that mental structures are organized separately based on different areas of knowledge (Carey & Spelke, 1994; Gelman & Brenneman, 2004; Hirschfeld & Gelman, 1994). When conducting studies, participants' states of cognitive development need to be considered as young learners may not have the required capability in applying the desired learning outcomes already set for them. In line with these perspectives, quantitative content analysis can be an effective methodological approach to examine the natural display of cognitive engagement with different domains and ages of learners.

Researchers interested in cognitive engagement need to direct more attention to explore the variation in patterns of cognitive engagement with younger ages of learners and in different scientific learning contexts, to gain a better understanding of what is involved in cognitive engagement in science learning. The methodology shares some flexibility in allowing researchers to purposively apply the steps to meet the needs and design of each research study. Researchers can also combine this methodology with the use of collecting quantitative data as well.

REFERENCES

Ainley, M. D. (1993). Styles of engagement with learning: Multidimensional assessment of their relationship with strategy use and school achievement. *Journal of Educational Psychology*, 85, 395–405. doi:10.1037/0022-0663.85.3.395

Ary, D., Jacobs, L. C., & Sorensen, C. (2006). *Introduction to research in education* (8th ed.). Wadsworth, UK: Cengage Learning.

Blumenfeld, P. C., & Meece, J. L. (1988). Task factors, teacher behavior, and students' involvement and use of learning strategies in science. *The Elementary School Journal*, 88, 235–250. doi:10.1086/461536

Brown, A. (1987). Metacognition, executive, control, self-regulation, and other more mysterious mechanisms. In Weinert, F. E., & Kluwe, R. H. (Eds.), *Metacognition, Motivation, and Understanding* (pp. 65–116). Hillsdale, NJ: Lawrence Erlbaum Associates.

Brown, A. L., & Reeve, R. A. (1987). Bandwidths of competence: The role of supportive contexts in learning and development. In Liben, L. S. (Ed.), *Development and Learning: Conflict or Congruence?* (pp. 173–223). Hillsdale, NJ: Lawrence Erlbaum Associates.

Carey, S., & Spelke, E. (1994). Domain-specific knowledge and conceptual change. In Hirschfeld, L. A., & Gelman, S. A. (Eds.), *Mapping the Mind: Domain Specificity in Cognition and Culture* (pp. 169–200). Cambridge, UK: Cambridge University Press. doi:10.1017/CBO9780511752902.008

Case, R., & McKeough, A. (1989). Schooling and the development of central conceptual structures: An example from the domain of children's narrative. *International Journal of Educational Research*, *13*, 835–855. doi:10.1016/0883-0355(89)90068-2

Chi, M. T. H. (1997). Quantifying qualitative analyses of verbal data: A practical guide. *Journal of the Learning Sciences*, *6*, 271–315. doi:10.1207/s15327809jls0603_1

Chin, C., & Brown, D. (2000). Learning in science: A comparison of deep and surface approaches. *Journal of Research in Science Teaching*, *37*, 109–138. doi:10.1002/(SICI)1098-2736(200002)37:2<109::AID-TEA3>3.0.CO;2-7

Cohen, J. A. (1960). A coefficient of agreement for nominal scales. *Educational and Psychological Measurement*, *20*, 37–46. doi:10.1177/001316446002000104

Corno, L., & Mandinach, E. B. (1983). The role of cognitive engagement in classroom learning and motivation. *Educational Psychologist*, *18*, 88–108. doi:10.1080/00461528309529266

Crawley, C. E. (2007). Localized debates of agricultural biotechnology in community newspapers: A quantitative content analysis of media frames and sources. *Science Communication*, *28*, 314–346. doi:10.1177/1075547006298253

DeBacker, T. K., & Crowson, H. M. (2006). Influences on cognitive engagement: Epistemological beliefs and need for closure. *The British Psychological Society*, *76*, 535–551.

Dole, J. A., & Sinatra, G. M. (1998). Reconceptualizing change in the cognitive construction of knowledge. *Educational Psychologist*, *33*, 109–128.

Dupeyrat, C., & Marine, C. (2005). Implicit theories of intelligence, goal orientation, cognitive engagement, and achievement: A test of Dweck's model with returning to school adults. *Contemporary Educational Psychology*, *30*, 43–59. doi:10.1016/j.cedpsych.2004.01.007

Flavell, J. H. (1979). Metacognition and cognitive monitoring: A new area of cognitive-developmental inquiry. *The American Psychologist*, *34*, 906–911. doi:10.1037/0003-066X.34.10.906

Flavell, J. H. (1999). Cognitive development: Children's knowledge about the mind. *Annual Review of Psychology*, *50*, 21–45. doi:10.1146/annurev.psych.50.1.21

Flavell, J. H., Miller, P. H., & Miller, S. A. (1993). *Cognitive development* (3rd ed.). Upper Saddle River, NJ: Prentice Hall.

Flowerday, T., & Schraw, G. (2003). Effect of choice on cognitive and affective engagement. *The Journal of Educational Research*, *96*, 207–215. doi:10.1080/00220670309598810

Fredricks, J. A., Blumenfeld, P. C., & Paris, A. H. (2004). School engagement: Potential of the concept, state of the evidence. *Review of Educational Research*, *74*, 59–109. doi:10.3102/00346543074001059

Gallagher, J. M., & Reid, D. K. (2002). *The learning theory of Piaget & Inhelder* (2nd ed.). New York, NY: Authors Choice Press.

Gelman, R., & Brenneman, K. (2004). Science learning pathways for young children. *Early Childhood Research Quarterly*, *19*, 150–158. doi:10.1016/j.ecresq.2004.01.009

Greene, B. A., & Miller, R. B. (1996). Influences on achievement: Goals, perceived ability, and cognitive engagement. *Contemporary Educational Psychology*, *21*, 181–192. doi:10.1006/ceps.1996.0015

Greene, B. A., Miller, R. B., Crowson, H. M., Duke, B. L., & Akey, K. L. (2004). Predicting high school students' cognitive engagement and achievement: Contributions of classroom perceptions and motivation. *Contemporary Educational Psychology*, *29*, 462–482. doi:10.1016/j.cedpsych.2004.01.006

Greenwood, C. R., Horton, B. T., & Utley, C. A. (2002). Academic engagement: Current perspectives on research and practice. *School Psychology Review*, *31*, 328–349.

Helme, S., & Clarke, D. (2001). Identifying cognitive engagement in the mathematics classroom. *Mathematics Education Research Journal*, *13*, 133–153. doi:10.1007/BF03217103

Hirschfeld, L. A., & Gelman, S. A. (1994). Toward a topography of mind: An introduction to domain specificity. In Hirschfeld, L. A., & Gelman, S. A. (Eds.), *Mapping the Mind: Domain Specificity in Cognition and Culture* (pp. 3–35). Cambridge, UK: Cambridge University Press. doi:10.1017/CBO9780511752902.002

Iseke-Barnes, J. M. (1996). Issues of educational uses of the Internet: power and criticism in communications and searching. *Journal of Educational Computing Research*, *15*(1), 1–23. doi:10.2190/FLYP-YNQC-9T55-MKB5

Koehler, M. J., Mishra, P., & Yahya, K. (2007). Tracing the development of teacher knowledge in a design seminar: Integrating content, pedagogy and technology. *Computers & Education*, *49*, 740–762. doi:10.1016/j.compedu.2005.11.012

Kuhn, D. (1999). Metacognitive development. In Balter, L., & Tamis-LeMonda, C. S. (Eds.), *Child Psychology: A Handbook of Contemporary Issues*. Philadelphia, PA: Psychology Press.

Kuhn, D. (2000). Metacognitive development. *Current Directions in Psychological Science*, *9*, 178–181. doi:10.1111/1467-8721.00088

Lee, O., & Anderson, C. W. (1993). Task engagement and conceptual change in middle school science classrooms. *American Educational Research Journal*, *30*, 585–610.

Linnenbrink, E. A., & Pintrich, P. R. (2003). The role of self-efficacy beliefs in student engagement and learning in the classroom. *Reading & Writing Quarterly*, *19*, 119–137. doi:10.1080/10573560308223

Mandinach, E. B., & Corno, L. (1985). Cognitive engagement variations among students of different ability level and sex in a computer problem solving game. *Sex Roles*, *13*, 241–251. doi:10.1007/BF00287914

Meece, J. L., Blumenfeld, P. C., & Hoyle, R. H. (1988). Students' goal orientations and cognitive engagement in classroom activities. *Journal of Educational Psychology*, *80*, 514–523. doi:10.1037/0022-0663.80.4.514

Metallidou, P., & Vlachou, A. (2007). Motivational beliefs, cognitive engagement, and achievement in language and mathematics in elementary school children. *International Journal of Psychology*, *42*, 2–15. doi:10.1080/00207590500411179

Miller, R. B., Greene, B. A., Montalvo, G. P., Ravindran, B., & Nichols, J. D. (1996). Engagement in academic work: The role of learning goals, future consequences, pleasing others, and perceived ability. *Contemporary Educational Psychology*, *21*, 388–422. doi:10.1006/ceps.1996.0028

Mowrer, D. E. (1996). A content analysis of student/instructor communication via computer conferencing. *Higher Education*, *32*, 217–241. doi:10.1007/BF00138397

Piaget, J. (1977). *The development of thought: Equilibration of cognitive structures* (Rosin, A., Trans.). New York, NY: The Viking Press.

Pintrich, P. R., & De Groot, V. (1990). Motivation and self-regulated learning components of classroom academic performance. *Journal of Educational Psychology*, *82*, 33–40. doi:10.1037/0022-0663.82.1.33

Pintrich, P. R., & Schrauben, B. (1992). Students' motivational beliefs and their cognitive engagement in classroom academic tasks. In Schunk, D. H., & Meece, J. L. (Eds.), *Student Perceptions in the Classroom* (pp. 149–183). Hillsdale, NJ: Erlbaum.

Riffe, D., Lacy, S., & Fico, F. (2005). *Analyzing media messages: Using quantitative content analysis in research* (2nd ed.). New York, NY: Routledge.

Sacks, H., Schegloff, E. A., & Jefferson, G. (1974). A simplest systematics for the organization of turn-taking for conversation. *Linguistic Society of America*, *50*, 696–735. doi:10.2307/412243

Samarapungavan, A., Westby, E. L., & Bodner, G. M. (2006). Contextual epistemic development in science: A comparison of chemistry students and research chemists. *Science Education*, *90*, 468–495. doi:10.1002/sce.20111

Samarapungavan, A., & Wiers, R. (1997). Children's thoughts on the origin of species: A study of explanatory coherence. *Cognitive Science*, *21*, 147–177. doi:10.1207/s15516709cog2102_2

Schneider, W. (2008). The development of metacognitive knowledge in children and adolescents: Major trends and implication for education. *Mind, Brain, and Education*, *2*, 114–121. doi:10.1111/j.1751-228X.2008.00041.x

Siegler, R. S. (1989). Mechanisms of cognitive development. *Annual Reviews*, *40*, 353–379. doi:10.1146/annurev.ps.40.020189.002033

Siegler, R. S. (1994). Cognitive variability: A key to understanding cognitive development. *Current Directions in Psychological Science*, *3*, 1–5. doi:10.1111/1467-8721.ep10769817

Siegler, R. S., & Booth, J. L. (2004). Development of numerical estimation in young children. *Child Development*, *75*, 428–444. doi:10.1111/j.1467-8624.2004.00684.x

Siegler, R. S., & Chen, Z. (2002). Development of rules and strategies: Balancing the old and the new. *Journal of Experimental Child Psychology*, *81*, 446–457. doi:10.1006/jecp.2002.2666

Singh, K., Granville, M., & Dika, S. (2002). Mathematics and science achievement: Effects of motivation, interest, and academic engagement. *The Journal of Educational Research*, *95*, 323–332. doi:10.1080/00220670209596607

Taylor, B. M., Pearson, P. D., Peterson, D. S., & Rodriguez, M. C. (2003). Reading growth in high-poverty classrooms: The influence of teacher practices that encourage cognitive engagement in literacy learning. *The Elementary School Journal*, *104*, 3–28. doi:10.1086/499740

Trigwell, K., Prosser, M., & Waterhouse, F. (1999). Relation between teachers' approaches to teaching and students' approaches to learning. *Higher Education*, *37*, 57–70. doi:10.1023/A:1003548313194

Vygotsky, L. S. (1978). *Mind in society: The development of higher psychological processes*. Cambridge, MA: Harvard University Press.

Walker, C. O., Greene, B. A., & Mansell, R. A. (2006). Identification with academics, intrinsic/extrinsic motivation, and self-efficacy as predictors of cognitive engagement. *Learning and Individual Differences*, *16*, 1–12. doi:10.1016/j.lindif.2005.06.004

Section 2
Modeling, Simulation, and Games

Chapter 5
Argumentation and Modeling:
Integrating the Products and Practices of Science to Improve Science Education

Douglas B. Clark
Vanderbilt University, USA

Pratim Sengupta
Vanderbilt University, USA

ABSTRACT

There is now growing consensus that K12 science education needs to focus on core epistemic and representational practices of scientific inquiry (Duschl, Schweingruber, & Shouse, 2007; Lehrer & Schauble, 2006). In this chapter, the authors focus on two such practices: argumentation and computational modeling. Novice science learners engaging in these activities often struggle without appropriate and extensive scaffolding (e.g., Klahr, Dunbar, & Fay, 1990; Schauble, Klopfer, & Raghavan, 1991; Sandoval & Millwood, 2005; Lizotte, Harris, McNeill, Marx, & Krajcik, 2003). This chapter proposes that (a) integrating argumentation and modeling can productively engage students in inquiry-based activities that support learning of complex scientific concepts as well as the core argumentation and modeling practices at the heart of scientific inquiry, and (b) each of these activities can productively scaffold the other. This in turn can lead to higher academic achievement in schools, increased self-efficacy in science, and an overall increased interest in science that is absent in most traditional classrooms. This chapter provides a theoretical framework for engaging students in argumentation and a particular genre of computer modeling (i.e., agent-based modeling), illustrates the framework with examples of the authors' own research and development, and introduces readers to freely available technologies and resources to adopt in classrooms to engage students in the practices discussed in the chapter.

DOI: 10.4018/978-1-4666-2809-0.ch005

INTRODUCTION

Science education has historically attempted "to cultivate students' scientific habits of mind, develop their capability to engage in scientific inquiry, and teach them how to reason in a scientific context" (NRC, 2011). These three foci have often been treated separately in traditional approaches to science education; however, with the result that science is often treated as isolated rote facts or artificial and arbitrary five-step methods (Driver, Leach, Miller, & Scott, 1996; Lemke, 1990). There is now growing agreement that students need to understand science and the processes of science as functions of argumentation and modeling (Duschl, 2008; Kelly, 2005; Lehrer & Schauble, 2006). The framework for the new science standards in the United States therefore "stresses the importance of developing students' knowledge of how science and engineering achieve their ends while also strengthening their competency with related practices" (NRC, 2011, p. 3.1). The new standards use the term "practices" rather than "skills" to "stress that engaging in scientific inquiry requires coordination both of knowledge and skill simultaneously" (NRC, 2011, p. 3.1). This chapter discusses the practices of argumentation and modeling in terms of their roles in the scientific disciplines and in terms of practices appropriate for students in the classroom.

WHAT ARE ARGUMENTATION AND MODELING?

True scientific literacy involves understanding how knowledge is generated, analyzed, justified, and evaluated by scientists and how to use such knowledge to engage in inquiry in ways that reflect the practices of the scientific community (Driver, Newton, & Osborne, 2000; Duschl & Osborne, 2002). Scientific inquiry is often described as a knowledge building process in which explanations are developed to make sense of data and then presented to a community of peers so they can be critiqued, debated, and revised (Driver, et al., 2000; Duschl, 2000; Sandoval & Reiser, 2004; Vellom & Anderson, 1999). Argumentation and modeling are at the heart of the scientific enterprise. As Lehrer and Scahuable (2012) point out, in the world of science, inquiry may take on various forms. Inquiry may be observational, theoretical, or computational. Inquiry may be carried out on a theorist's desk, in a physics lab, or a biological field station. However, despite these variations, all scientists engage in constructing, revising, applying, and defending models of the natural world (Giere, 1999; Hesse, 1966). Modeling has been described as the signature of research in the sciences (Nersessian, 2009), and argumentation is the process through which communities of scientists test, refine, and tentatively accept or reject models as a community. The ability to engage in scientific argumentation (i.e., the ability to examine and then either accept or reject the relationships or connections between and among the evidence and the theoretical ideas invoked in an explanation or the ability to make connections between and among evidence and theory in an argument) is, therefore, viewed by many as an important aspect of scientific literacy (Driver, et al., 2000; Duschl & Osborne, 2002; Kuhn, 1993; Siegel, 1989). Thus scientific theories, modeling, and argumentation are not separate decontextualized entities. Scientific theories, modeling, and argumentation are dynamically interwoven and interdependent.

Learning to engage in scientific modeling and argumentation is challenging for students. Furthermore, opportunities for students to learn how to engage in scientific argumentation in a productive manner as part of the teaching and learning of science are rare (Newton, Driver, & Osborne, 1999; Simon, Erduran, & Osborne, 2006) as are opportunities to engage in authentic modeling. Traditional science curricula portray scientific theories as fixed and immutable facts to be memorized and accepted. Argumentation, when included at all, tends to either be a de-contextualized game of creating rebuttals or an unreflective statement of "evidence" for theories that are treated as foregone conclusions. Similarly, models and modeling tend not to be integrated in school science in authentic forms. To the extent that they do appear in school, models usually play an illustrative, rather than scientific theory building role (Windschitl & Thompson, 2006).

This lack of integration in traditional curriculum between the products and processes of science is evidenced by research on students engaging in inquiry. Research suggests, for example, that students often do not seek out or generate data that can be used to help test their ideas or discriminate between competing hypotheses (e.g., Klahr, Dunbar, & Fay, 1990; Schauble, Klopfer, & Raghavan, 1991). In addition, students often rely on their personal views rather than use the data at hand to generate and verify hypotheses (Hogan & Maglienti, 2001). In other situations, students may use inappropriate data from an investigation to draw conclusions, or they may fail to attend to important patterns in the data (McNeill & Krajcik, 2007; Sandoval & Millwood, 2005; Kuhn, 1993; Schauble, Glaser, Duschl, Schulze, & John, 1995; Chinn & Brewer, 1993; Driver, et al., 1994). When reasoning about scientific phenomena which involve multiple "levels" (e.g., both macroscopic and microscopic), students often tend to confuse the attributes and behaviors present in one level with that of the other (Sengupta & Wilensky, 2009, 2010; Wilensky & Resnick, 1999; Resnick, 1994). Students also have difficulty generating explanations that are scientifically rigorous (Carey, Evans, Honda, Jay, & Unger, 1989; Lawson, 2003; Sandoval & Reiser, 2004). They may face similar challenges justifying and warranting their explanations (Clark & Sampson, 2008; Sandoval & Millwood, 2005; Sadler, 2004; McNeill & Krajcik, 2007; Kuhn, 1991; Brem & Rips, 2000; Kuhn & Reiser, 2005; Bell & Linn, 2000; Jimenez-Aleixandre, et al., 2000; Lizotte, McNeill, & Krajcik, 2004; Aikenhead, 2004; Linn, Eylon, & Davis, 2004), and establishing and evaluating their validity or acceptability in the context of a given phenomenon during scientific argumentation (Hogan & Maglienti, 2001; Linn & Eylon, 2006; Kuhn & Reiser, 2005; Zeidler, 1997; Clark & Sampson, 2006a; Kuhn, 1989). Finally, novice learners often underestimate the time and effort that will be required to learn successfully—their self-judgment abilities are not well developed, and they may not be motivated enough to learn with understanding (Schunk & Zimmerman, 1998).

WHAT SHOULD STUDENTS UNDERSTAND?

So what should students understand? First, modeling is the central enterprise, purpose, and goal of science. Second, argumentation is the practice that allows scientists to determine the fit of their models with the world. Third, communities of scientists evaluate models, methods, and evidence through argumentation using shared criteria and analytical approaches developed and agreed upon by the community.

Modeling is the Central Enterprise, Purpose, and Goal of Science

Students should understand modeling as the *language* of science. As Rapp and Sengupta (2012) pointed out, models are physical, computational, or mental representations that are intended to stand in for some other thing, set of things, or phenomena. Scientific models are tools for expressing scientific explanations or theories in a form that can be directly manipulated, allowing for description, prediction, and explanation. As Lehrer, Schauble, and Lucas (2008) pointed out, the "big ideas" in science derive their power from the models that instantiate them, so to fulfill the promise of the "big ideas" outlined in national science standards, students must realize these ideas as models.

Modeling is the core epistemic action through which scientists generate new knowledge, and modeling is inherently tied to constructivism (Hestenes, 1993). From the constructivist perspective, meaning is constructed and matched with experience in a manner that makes that experience meaningful and the meaning experiential. Similarly, modeling, which is the process of development and refinement of a model, is a dialectical process between model construction and model matching. Therefore, as students engage in modeling-based curricula over an extended period of time, students should understand that modeling, by its nature, involves repeated cycles of developing, representing, and testing knowledge (Rapp & Sengupta, 2012; Duschl, et al., 2008; Lehrer, Schauble, & Lucas, 2008).

Argumentation is the Practice that Allows Scientists to Determine the Fit of their Models with the World

As Lehrer, Schuable, and Lucas (2008) discuss, scientific models are also forms of argument. In the scientific world, models are regularly mobilized to support socially grounded claims and counterclaims about the nature of physical reality (Bazerman, 1988; Latour, 1999; Lynch & Woolgar, 1990; Watson & Crick, 1953). Students should understand that argumentation is a central foundation upon which scientists make decisions. This decision includes what data to collect, how to collect it, which data to select, how to represent that data, and how to determine the implications of that data as they test and refine their models in terms of the fit of those models with the data and phenomena they are modeling in the world. Students need to understand that argumentation can act as the framework that can guide their exploration of causal mechanisms of a phenomenon using a model and their exploration of the fit of a model with the world. This parallels the ideas of "getting nature to speak" (i.e., the methods and tools used to collect and select data) and "portraying nature's voice" (the interpretation and representation of the implications of that data) as outlined by Ford and Forman (2006). Thus, a focus on argumentation can guide evaluations of the appropriateness of scientific methods, data selection, data representation, data interpretation, warrants, and claims and the fit and implications of models in terms of the underlying causal mechanisms in the phenomena being modeled. In order to engage in authentic argumentation, students need to understand the role of claims, data, and warrants in scientific disciplines. They also need to understand that acceptable and appropriate criteria, methods, and representational forms are to some degree specific to individual scientific disciplines depending on the nature of the phenomena investigated by that discipline.

Communities of Scientists evaluate Models, Methods, and Evidence through Argumentation using Shared Criteria

Students should also understand that argumentation is the mechanism through which communities of scientists evaluate the models proposed by members of those communities in terms of the claims, evidence, and warrants involved in a proposed model as well as in the methods used to generate the evidence itself. As discussed above, scientific disciplines come to agree on a shared interpretation of acceptable and appropriate questions, methods, criteria, and representational forms for investigating the phenomena of interest to those disciplines. Dialogic argumentation amongst scientific community members is the primary process through which this shared interpretation evolves. Dialogic argumentation focuses on the interaction of individuals or groups attempting to convince one another of the acceptability and validity of alternative ideas. Thus, students should also come to understand that the shared interpretations of the community are not fixed or preordained in terms of acceptable and appropriate questions, methods, criteria, and representational forms. Instead, the shared interpretation of appropriate questions, methods, criteria, and representational forms continues to evolve through argumentation as the community advances in its understanding of the phenomena under investigation and as the tools and methods available to the community themselves evolve, often directly as a result of community's own explorations. In addition to the epistemological value of helping students understand the processes through which scientific communities' understandings of the world evolve, engaging students in dialogic argumentation is considered a powerful mechanism for increasing students' understanding of challenging concepts (e.g., Andriessen, Baker, & Suthers, 2003; Hogan, Nastasi, & Pressley, 2000; Leitão, 2000) as well as for increasing students' ability to engage in productive argumentation and reasoning practices (e.g., Baker, 2003; Bell, 2004; Kuhn, Shaw, & Felton, 1997).

INTEGRATING ARGUMENTATION AND MODELS

How might teachers integrate argumentation and modeling in the classroom in support of these goals? This first section discusses an excellent approach for engaging students in argumentation around pre-existing models developed by Sampson and colleagues (Sampson & Gleim, 2009; Sampson, Grooms, & Walker, 2009).

Many websites provide free access to fantastic pre-existing models that students can use to explore a wide range of scientific phenomena. Netlogo (http://ccl.northwestern.edu/netlogo/), PhET (http://phet.colorado.edu/), and Concord (http://www.concord.org), for example, have created large libraries of models that are freely available for teachers and students. Simply providing students with computational models, however, has not proven very effective, just as generic, traditional approaches to hands-on labs have not proven very effective according the National Research Council's "America's Lab Report" on the efficacy of traditional approaches to science labs in schools (NRC, 2005). The National Research Council suggests that effective hands-on lab activities and computational model lab activities: (1) focus on true inquiry to help students develop skills for grappling with the ambiguity and complexity of scientific investigations, (2) engage students in reading, writing, and critical discussions about the process and ideas, and (3) engage students in constructing and critiquing arguments about the phenomena and evidence associated with the explanations that they develop.

Sampson and colleagues (Sampson & Gleim, 2009; Sampson, Grooms, & Walker, 2009, 2011) have developed the Argument-Driven Inquiry (ADI) approach to help science teachers transform traditional laboratory activities and computational models into short, integrated instructional units that incorporate all of the features outlined by the National Research Council. The ADI approach provides "opportunities for students to design their own investigations, gather and analyze data, communicate their ideas with others during structured and interactive argumentation sessions, write investigation reports to share and document their work, and engage in peer review during a laboratory investigation" (Sampson, Grooms, & Walker, 2009). As outlined by Sampson and colleagues, the full version of the ADI instructional model consists of eight steps, which we outline in more detail in the following paragraphs.

Identification of a Task

In this stage, the students and the teacher first consider the phenomena to be investigated in light of previous experiences and other materials. The students then develop or select a question to explore.

Generation and Analysis of Data

This stage is a hands-on or virtual model-based investigation of the students' questions. While most ADI units focus on hands-on labs, several have been developed for use with computational models, such as a NetLogo model that allows students to explore the impact of camouflage on the survival of butterflies. The students design controlled comparisons with the model to collect data for their question.

Production of a Tentative Argument

In this stage, students construct an argument that includes an explanation, evidence, and their reasoning in a format that can be shared with other students. Sampson and colleagues recommend whiteboards for this purpose. The explanation is essentially an answer to the research question and may articulate a qualitative relationship or causal mechanism. The evidence includes measurements or observations to support the explanation in terms of traditional numerical data or observations. Sampson and colleagues specify that, "in order for this information to be considered evidence, it should show (a) a trend over time, (b) a difference between groups, or (c) a relationship between variables" (Sampson & Gleim, 2009, p. 467). The reasoning clarifies how the evidence supports the claim and why the evidence is justifiable and appropriate for the claim.

Argumentation Session

The students then share their arguments with one another and critique and refine one another's explanations and the connections of the data to those explanations, in small groups or as a whole class. This step serves multiple purposes: it exposes students to the ideas of other students, allows students to respond to the questions and challenges of students who have created different explanations, and engages students in the knowledge-building processes core to the scientific disciplines. The argumentation sessions also allow teachers to assess students' progress and thinking as well as to encourage students to think about overlooked issues or data. Through this process, students are exposed to the theory-laden nature of science and have the opportunity to come as a group to develop and share criteria for judging the plausibility of explanations, warrants, and reasoning.

Investigation Report

The students then write up an investigation report that explains the goals of the work, the methods employed, and their refined arguments about their findings.

Double-Blind Peer Review

The students next review reports from other students in a double-blind format. The class works together to develop criteria, which may be supplemented by the teacher in the form of guide sheets or critique sheets. The goal is to generate high-quality feedback and to help students understand how the process works in the disciplines.

Revision of the Report

Following the peer review, students have the opportunity to revise and refine their reports.

Explicit and Reflective Discussion

The class then engages in an explicit and reflective discussion about the inquiry process and the causal mechanisms underlying the phenomena under investigation.

Sampson and colleagues have developed several tools that can be used to help scaffold students as they work through each step of the ADI approach. For example, they have developed an "investigation proposal" that teachers can use to help students design better investigations during

Argumentation and Modeling

the process. Similarly, they have developed multiple peer review guides with varying degrees of scaffolding that teachers can provide for students to use during the peer-review process depending on the needs of their students. Sampson and colleagues have also developed some simpler approaches that have components of ADI but are not a full ADI, to provide a progression that can ultimately support students engaging in the full ADI process. For example, if teachers do not want to have students design an investigation and collect data but still want to do the argumentation sessions, Sampson and colleagues have an approach called "generate-an-argument" (Sampson & Grooms, 2010; Sampson & Gerbino, 2010). Similarly, if teachers want students to collect data and to do the whiteboards, but do not want students to write reports or go through the peer-review process, they can do the "evaluate-alternatives" approach (Sampson & Gerbino, 2010; Sampson & Grooms, 2009).

Thus, the ADI approach to lab instruction "fosters scientific literacy and allows students to develop scientific habits of mind, provide evidence for explanations, and think critically about suggested alternatives" (Sampson, Grooms, & Walker, 2009). Teachers can use the ADI instructional approach as a way to transform traditional computational models and hands-on experiences (where students typically follow a set procedure and answer relatively rote "analysis" questions) into powerful inquiry activities integrating inquiry with models and argumentation in a manner paralleling the actual inquiry processes within the scientific disciplines themselves. As Sampson and colleagues explain, this approach thus has "great potential and should enable more students to develop a sophisticated understanding of both the concepts under study and the process through which scientific knowledge is developed, evaluated, and refined" (Sampson & Gleim, 2009).

INTEGRATING ARGUMENTATION AND MODELING

The previous section outlines an excellent approach for authentically incorporating argumentation and inquiry with hands-on labs, which in turn can be easily extended to the use of pre-built computational models. In what follows, we present an approach for moving beyond pre-existing models to instead engage students in argumentation and inquiry that focuses on modeling itself. More specifically, how might we structure and scaffold students' modeling activities in terms of argumentation in a manner that parallels the authentic practices of the scientific disciplines? As discussed in the overview of this chapter, argumentation and modeling are the core practices at the heart of the scientific enterprise. Developing an approach for meaningfully integrating them would represent an authentic experience for students integrating the processes and products of science.

Hestenes (1993) argued that there are three kinds of epistemic modeling *games* [1] in which scientists usually engage: model building, model ramification, and model deployment. In the first type, "modeling building," the objective is to build a model to meet given specifications. These specifications are often derived from empirical data of observations. The second type, "ramification," involves analyzing the properties (that is, the ramifications) of complex systems, i.e., systems that involve interactions between multiple factors or variables. The third type, "deployment," involves the matching of models to empirical phenomena and data. In what follows, we present a general outline of how argumentation can be integrated to support such modeling in the context of computational models. Specifically, we outline an approach for the productive integration of argumentation and modeling through which students can engage in all three epistemic types of modeling.

Many model-based or modeling-based curricula typically engage in only one of these types of modeling. For example, a common use of computer models in science curricula occurs in the form of using pre-built simulations as demonstrative lecture aids that allow students to interact with simulations, primarily through controlled experimentation (variable manipulation) that helps them understand relevant aspects of the lecture. Another common form of classroom use involves students conducting guided inquiry using prebuilt models through cycles of predict-observe-explain (Sengupta & Wilensky, 2009). In these learning activities, students primarily engage in model-deployment. In other words, students are provided with the model with mathematical relationships specified by the designer, and students discover these relationships through conducting experimentation activities based on the control of variables.

Curricula that involve students developing their own computational models are significantly more challenging to implement, and often require extensive modeling expertise on the part of the teacher, as well as extensive one-on-one scaffolding that is often beyond the scope of usual K-12 classroom instruction. For example, Harel and Papert (1991) reported a study in which students constructed Logo models of fractions over an extended period of five months, during which they received support from experienced peers such as MIT graduate students. Sherin et al. (1992) reported a study in which students engaged in learning Newtonian mechanics by constructing Logo programs in the Boxer programming environment (diSessa & Abelson, 1986), but that course involved 5 weeks of programming instruction followed by 10 weeks of physics instruction. This second course was taught by expert programmers and teachers with extensive experience using and teaching with Logo. In such curricula, students do indeed engage in model construction and model deployment, and possibly even model ramification, but such curricula are challenging to implement in K-12 settings. This is due to the demands on teacher preparation (i.e., teachers need to be domain experts as well as programming experts), and due to the challenges of integrating programming with science content, which in turn has implications for class time (length of the course).

From the perspective of designing a learning environment, our approach to integrating modeling and argumentation has the following three objectives. First, the integration of argumentation into modeling should evolve as progressions in terms of both modeling and argumentation spread across multiple investigations in the curriculum. Second, aligning with constructivist perspectives, these investigations should build on one another. Third, integrated learning environments should support key practices of argumentation (e.g., critique, evaluation, explanations) as well as all three epistemic forms of modeling outlined by Hestenes (1993) in terms of building, ramification, and deployment.

Based on our review of the literature, we believe that such integration requires the development of a new kind of modeling platform that enables students to construct models without requiring programming experience, and provides software-embedded scaffolds that use argumentation as a focal activity during the process of modeling. Over the past year, we have been developing a software-based learning environment that is freely available to teachers that supports both modeling and argumentation in the context of learning science in K-12 classrooms. We describe below the key design principles guiding the design of the learning environment.

DESIGN GUIDELINES FOR INTEGRATING MODELING AND ARGUMENTATION IN AGENT-BASED COMPUTATIONAL ENVIRONMENTS

Our work is grounded in a constructivist paradigm of learning (Smith, diSessa, & Roschelle, 1994). In this paradigm, new knowledge is constructed actively by the learners by bootstrapping, rather than discarding, their prior knowledge. This idea guided our focus on agent-based modeling. The term "agent" in the context of an Agent-Based Model (or ABM) denotes individual computational objects or actors (e.g., cars), which are controlled by simple rules assigned by the user. ABMs are particularly suited for representing and understanding complex aggregate behaviors. It is the interactions between agents (based on these rules) that give rise to emergent aggregate-level behaviors of the model (e.g., formation of a traffic jam as an emergent aggregate outcome). A traffic jam can be thought of as a result of an aggregation of interactions between many individual "agents" or cars. At the individual level, the operating "rules" for each car are simple: each car accelerates if there is no car right ahead, and it slows down if it sees another car close ahead (Wilensky & Resnick, 1999). The pattern that emerges as an aggregation of many such interactions between individual cars is the traffic jam. Emergent phenomena are often counter-intuitive to understand; for example, while individual cars move forward, the overall jam propagates in the backward direction (Resnick, 1994).

When students work with ABMs, they use their intuitive knowledge at the agent level as they are asked to manipulate and reason about the behaviors of individual agents. Then, by visualizing and analyzing the aggregate-level behaviors that are dynamically displayed in ABM simulations that involve interactions between multiple agents, students can develop multi-level explanations by connecting their relevant agent-level intuitions with the emergent phenomena (Resnick, 1994; Wilensky & Resnick, 1999; Klopfer, Yoon, & Um, 2005; Sengupta & Wilensky, 2011; Blikstein & Wilensky, 2009). The scholars cited above have argued that in most science classrooms, aggregate-level formalisms are typically used to teach scientific concepts and phenomena, such as using the Lotka-Volterra differential equation to explain how populations of different species in a predator-prey ecosystem change over time (Wilensky & Reisman, 2006). While mathematically correct, these formalisms do not immediately make explicit the underlying agent-level attributes and interactions of the system, and, as such, remain inaccessible for most students. In contrast, agent-based reasoning (i.e., reasoning about the attributes and behaviors of the individual agents) has been claimed to be more accessible and to provide a bridge to aggregate reasoning (Levy & Wilensky, 2008). These claims are substantiated by experiments showing that when complex phenomena traditionally taught in high school (e.g., microscopic processes of electrical conduction) are represented in the form of multi-agent based models, much younger students (e.g., 4th and 5th graders) can access and understand those phenomena (Sengupta & Wilensky, 2011; Dickes & Sengupta, 2011).

Designing Modeling Primitives and Other Scaffolds to Support Model Development

Our goal is to engage students in learning through developing a computational model. This necessitates some form of programming (i.e., students specifying computational variables and their relationships). However, rather than introducing students to domain-general concepts in programming, our goal is to introduce them to domain-specific computational primitives for modeling particular phenomena. Over the past year, we have been developing ViMAP-Arg, an agent-based computational modeling environment to support the integration of computational modeling and argumentation, based on the ViMAP architecture (Sengupta, 2011; Sengupta & Wright, 2010). The ontology of primitives we have chosen in ViMAP-Arg is based on node-link representations. Nodes and links indicate domain-specific conceptual entities, and students design a model by selecting nodes and relevant links between them (see Figure 1). For example, the upper portion of Figure 1a shows a list of the nodes through which students can control the behavior of an ABM simulation of a wolf-sheep predation ecosystem (bottom portion of Figure 1a). These nodes represent the different types of agents in the system (e.g., wolf, sheep, and grass) and the actions pertaining to each type of agents (e.g., move, multiply, need-food). The key interactions between agents (e.g., eating) are specified through links that appear as options only when a student clicks on relevant nodes (e.g., wolf, sheep, or grass). Once the student selects two nodes, (e.g., wolf and multiply), the students can also specify the quantitative level of the interaction by choosing between options (e.g., "eats a lot" versus "eats a little") (see Figure 1b). Figure 1c shows a screenshot of a sample model developed by selecting nodes and links that specify all the relevant actions and interactions between the different types of agents in the simulation.

Because modeling is an iterative process, students can run the model during any stage of the model construction phase to test how components of their model affect the overall behavior of the simulation. Note that nodes and links represent agent-level behaviors and interactions between agents. Students can also visualize aggregate-level effects of these agent-level interactions in the form of graphs (e.g., graphs showing populations of different species over time).

Designing Software Supports to Leverage Argumentation in Order to Support Model Development, Ramification, and Deployment

In order to engage students in model ramification and deployment, students are also provided with multiple simulations displayed side-by-side in the learning environment (see Figure 2). The scaffolding here is provided in terms of the different levels of control that the student has over the models underlying each of these simulations. Besides the students' own model, which they construct from scratch, the other two models include a target "world"[2] simulation and a partially built faulty model. The target "world" simulation that the students are working to model provides target outcomes for the students' own models. The faulty or partially correct simulation is pre-built by the teacher or curriculum authors in order to create

Figure 1. Stages in creating a model using ViMAP-ARG

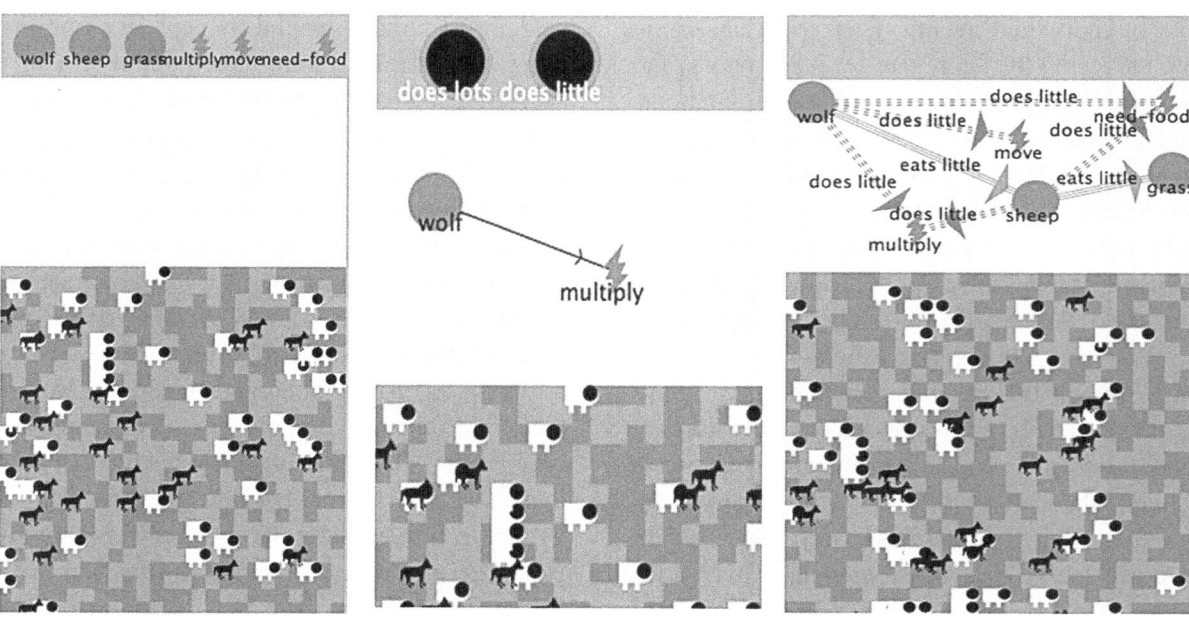

Argumentation and Modeling

Figure 2. Screenshot of the user interface of ViMAP-ARG

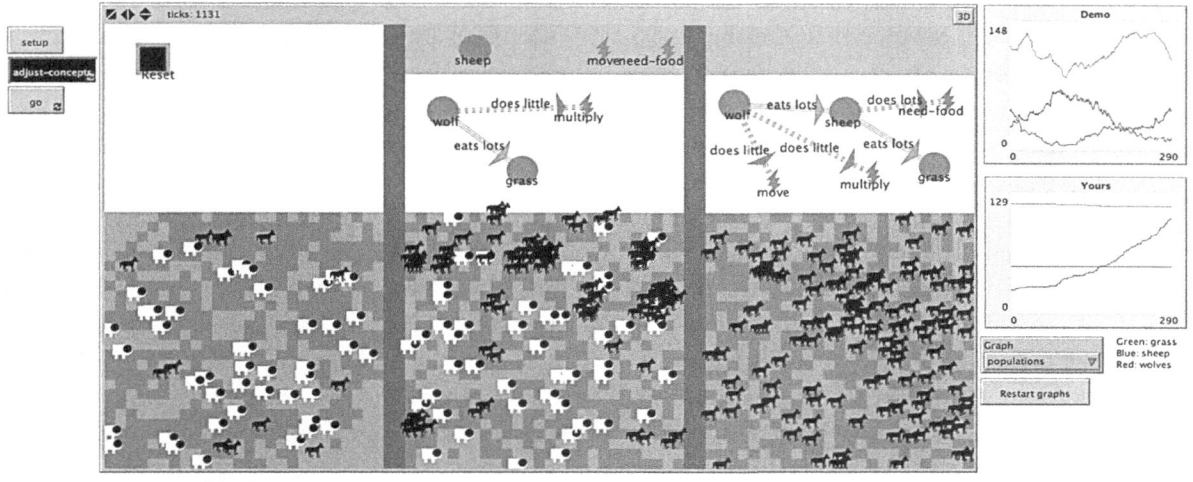

cognitive conflict in the students' minds about important contextually relevant variables. The model (i.e., the node-link relationships) underlying the world simulation is completely hidden from the students, while the model underlying the faulty simulation can be modified by the students.

In terms of the models they develop, students begin with a template of a simple model in their modeling environment that is largely functional but has one or two nodes or parameters that need to be adjusted or changed to make the model reflect the target "world" data more closely. This might be framed for the students as a proposed model that needs to be critiqued and refined. As students progress in their modeling sophistication, the students can work to modify more sophisticated template models involving more parameters and nodes, and can also begin to develop new models using the same environment without the template of an initial model. Subsequent target data sets and representations involve increasing numbers of agents and explore elaborations on the preceding target models. The sequence of models therefore allows students to systematically explore the underlying science phenomena in a systematic sequence.

Providing students with multiple simulations can facilitate the data collection and analysis process by allowing students to simultaneously compare the target world data with output data from two versions of their own models. Students can then simultaneously compare the similarities and differences between the target world data in the world simulation and the two versions of the simulations over which they have control of the underlying models. In doing so, students engage in model deployment by scientifically assessing the aggregate-level effects of their model through conducting experiments. This comparison between the multiple simulations makes this a meaningful endeavor for the students. Students also engage in model ramification, as they explore how different assumptions and specification about the behavior of agents impact the overall aggregate-level outcomes.

A highly scaffolded version of this phase could then ask students to collect data in two variants of the constructed model and identify the salient data as evidence in a comparison about the validity of their claims. This would facilitate the students' ability to connect differences in the data outputs with structural differences within their own models. This approach thus supports "debugging," which is a core epistemic and presentational practice central to model development that involves identification of bugs (i.e., dissonances between the intended behavior of the model and the actual behavior of the model). However, in this context, the process of debugging indicates refining of models in a data and theory driven manner that supports practices central to argumentation, as opposed to a less productive process of model refinement through random trial and error.

Over the course of a given modeling project and also across the span of projects across the curriculum, the specific prompts for specific comparisons would fade or be reduced as students begin to understand the role and process of critique, allowing the students to take more responsibility for determining which aspects of the models are salient to compare, what data would support such a comparison, and what might constitute random sampling variation versus fundamental differences in both the target and student models.

A CLOSER LOOK AT THE DISCURSIVE PRACTICES RELATED TO ARGUMENTATION IN THE INTEGRATED MODELING ENVIRONMENTS

In this section, we take a closer look at how discursive practices related to argumentation are supported in the integrated learning environment shown in Figure 2. We focus on three central practices: critique, explanation, and argumentation. We discuss how each of these practices can be supported in our learning environment, how these practices help develop students' multi-level understandings of the target phenomena, and what a sample progression of the activities might look like.

Critique

In terms of argumentation, our proposed approach focuses on critique, explanation, and argumentation regarding the fit between the student's model and the target "world" data in Figure 2, or similar comparisons using other modeling tools. Each cycle of modeling begins with the student critiquing the fit between the current version of the student model and the target world data in terms of both aggregate behaviors and agent level behaviors.

Aggregate behaviors involve the overall populations of each agent in the models as recorded in graphs created when running the student model or as provided in the target world data sets and representations. Examples in an ecology model would include the average population levels of each agent, patterns of change of the populations compared to one another (e.g., the squirrel population starts lower than the rabbit population but ends up higher), and the overall shapes of each population graph (e.g., discussing the amplitude and frequency of the population sine curves in a predator-prey interaction). Critique of aggregate behaviors involves comparing the graphs created when running the student model and the target world data and critiquing how closely the two sets of graphs match one another. Because agent-based models inherently incorporate variability from run to run, this activity includes a high level of authenticity because students need to run each model multiple times to determine which differences are likely due to sampling variation, and which differences might represent fundamental differences in the structures of the underlying models. It is also authentic in terms of students needing to determine which differences might be more or less salient for the phenomena under investigation.

Agent level behaviors involve the characteristics and behaviors of individual agent types in the model. Examples in an ecology model might include how fast an agent of a given type moves, what it eats, whether it reproduces, how much energy it needs, and whether and how it changes its behavior depending on the proximity of other agents. In some ways, one might think about the difference between aggregate behaviors and agent level behaviors as paralleling the ideas of "intrinsic" and "extrinsic" properties from chemistry, with aggregate behaviors being extrinsic in the sense that they are a population function of how many agents are involved at any given moment and the properties emerging from groups of agents, while agent level behaviors are "intrinsic" in the sense that they are the behaviors of individual agents. Comparing agent behaviors could involve tracking and watching individual agents in each model and comparing how they act and interact with the other agents in the model. Comparing agent behaviors might also involve tracking graphical output of mean behaviors or histograms of behaviors for individual agent behaviors (e.g., the speed of an agent type or the rate of reproduction for an agent type in an ecology model). As with the critique of aggregate behaviors, this involves a high degree of authenticity in the sense that students need to account for sampling variation as well as make decisions about what types of agent behaviors would be salient to track in terms of the phenomena under investigation. Thus, this is not a cookie-cutter activity, but one in which students need to think deeply about their goals, the processes of measurement and sampling, and the underlying phenomena.

While all modeling activities implicitly involve critique, traditional modeling activities often do not explicitly articulate this critique process or provide explicit scaffolds to support students engaging in the critique. Our approach proposes that students should have an explicit critique phase prior to each successive iteration in their model construction and revision. In terms of progressions across the curriculum and within a modeling project, students should initially be provided with specific prompts for the aspects of aggregate and agent level outputs they should compare, and directions as to how they should do so. Students should be prompted to explicitly address each of these comparisons through written description or other interfaces. Simply including fields in an electronic word processing document or paper handout can be sufficient. More sophisticated computer-based interfaces integrated into a modeling environment, however, could provide more flexible approaches. In an integrated computer-based modeling environment, for example, this could involve specifying the degree of matching point-by-point. A computer-based integrated environment could provide feedback regarding the critique point-by-point and support students in adjusting their critique in terms of any of those points.

Explanation

If discontinuities between the target world data and the student's model are identified through the critique phase, the student should then be scaffolded in creating an explanation/claim about the source/mechanism in the constructed model that leads to the discontinuity. This is essentially the creation of a claim about the underlying causal mechanism that simultaneously identifies and proposes corrective action. As with critique, traditional modeling activities implicitly include explanation of the sources of difference between the target world data and the student's model, but students in traditional modeling activities are also free to attempt random trial and error in revising the underlying structure of their model rather than investigating reasoned explanations for observed differences. Students should also be scaffolded in thinking about the implications of their proposed structural explanation in terms of whether the proposed changes fit all aspects of the world data and how the proposed change in the model might change what we think about the phenomena under investigation. For example, a student might conceivably improve fit between world data population graphs and the student's model by allowing deer to "fly" in the model, but that student should be encouraged to consider whether or not that fits with what else is known about deer. Thus, students should be encouraged to consider explanations in terms of causal mechanisms that increase model fit with all aspects of the world data, rather than treating the modeling process solely as a programming challenge.

Argumentation

After creating an explanation about the source of differences between the agent and aggregate level outputs in terms of changes that need to be made to the student's model, the student should then be scaffolded in identifying and collecting further evidence for and against the proposed causal mechanism. This can include evidence collected during the Critique phase or during subsequent experimentation. Of specific importance, the teacher and the scaffolding should help students search for evidence that might contradict their claims, as well as evidence that might supports their claims, because research has shown that students initially tend to focus only on evidence that supports their claims and ignore evidence that contradicts their claims in a manner very similar to the ways in which scientists have historically dealt with anomalous data (Chinn & Brewer, 1993).

As with the Critique and Explanation phases, this scaffolding might simply comprise explicit prompts in a paper handout or electronic word processing document that provide instructions and fields for students to complete at the beginning of each successive iteration of their models, or these scaffolds might be integrated into the modeling environments themselves. Electronic word processing documents can be superior to paper handouts because they can allow students to "copy" aggregate and agent level output graphs and then "paste" them into the document. Students can then mark and annotate the graphs in their document to illustrate the specific comparisons or data points of interest. Paper handouts rely on students to draw and describe their data, which can focus students more explicitly on salient points, but can also become cumbersome. Ultimately, however, either approach can prove productive.

In modeling environments developed specifically to support this process, more flexible approaches for students to identify data might include the ability for students to "click" on specific points in the target world data and the data from their own models. A sophisticated environment might then provide feedback about how well the highlighted data supports the claim that has been made and prompt further reflection. Similarly, a sophisticated environment might allow students to create playable "movie" clips tracking the behaviors of individual agents in the model world to support claims about the agent characteristics in the model in comparison to the target world data (which might also include feedback from the environment about the fit between the data and claims).

A very important component of this phase involves engaging the student not only in identifying evidence that supports and conflicts with a claim but also in providing reasoning for why that data represents valid evidence regarding that claim.

Progressions

Many aspects of suggested progressions are discussed in the critique, explanation, and argumentation sections above in terms of progressions within a given modeling project as well as over time in successive modeling projects within the curriculum. Essentially, students will initially require a great deal of explicit scaffolding and support for each of the three phases. Furthermore, projects early in the curriculum might only include the critique phase, then gradually add the explanation and argumentation phases in later projects in the curriculum. Each phase should include explicit and detailed scaffolding when first introduced to make the underlying processes, goals, and criteria explicit. Discussions between the students and teacher should also make these processes, goals, and criteria explicit.

Scaffolding should subsequently be faded and reduced over time as students became more proficient. This reduction and fading is critical. The purpose of these scaffolds is not simply to help students explore the underlying science phenomena through inquiry-based modeling but also to support students in developing proficiency and habits of mind for engaging in authentic inquiry. As outlined earlier in this chapter, students should engage in inquiry that integrates products and processes of science. The explicitness and specificity of the scaffolding should be reduced in terms of the prompts and directions provided to the students. The goal in instruction involves fading aspects of the direct scaffolding/prompting within phases and actually gradually removing the phases themselves.

In terms of larger progressions within modeling projects, the arguments developed in the argumentation phase might then be plugged into the argumentation sessions and potentially even the blind peer review components of the ADI model. This would allow students to understand the role of argumentation within the scientific community as well as provide excellent feedback and practice highlighting the core ideas of modeling and argumentation, including the opportunity to develop shared understanding and criteria as a community for engaging in inquiry.

SUMMARY AND DISCUSSION

Science educators and historians of science have shown that scientific practices like argumentation and modeling develop only over the long term, both historically within the sciences and individually within the lifetime of individuals (Lehrer & Schauble, 2010). Early in this chapter, we claimed that students should understand three core ideas. First, modeling is the central enterprise, purpose, and goal of science. Second, argumentation is the practice that allows scientists to determine the fit of their models with the world. Third, communities of scientists evaluate models, methods, and evidence through argumentation using shared criteria and analytical approaches developed and agreed upon by the community. To address these goals, this chapter has outlined both the ADI approach developed by Sampson and colleagues for integrating argumentation with pre-built models (Sampson & Gleim, 2009; Sampson, Grooms, & Walker, 2009) and our proposed approach for integrating argumentation with modeling in agent-based computational environments.

From the perspective of the first goal of helping students develop a modeling-based epistemology of science, we argued that it is important that students should understand that modeling is the central enterprise, purpose, and goal of science, while they are engaged in longer term, authentic scientific inquiry. Based on Hestenes' (1993) categorization, our pedagogical approach involves fostering such an epistemology by engaging students in three kinds of modeling activities, including model development, model ramification, and model deployment in a computational learning environment that integrates modeling and argumentation. We have proposed design guidelines, as well as presented ViMAP-Arg, a freely available learning environment designed specifically to support both argumentation and modeling in the context of leaning about population dynamics in a predator-prey ecosystem.

In terms of the second goal of helping students understand that argumentation is the practice that allows scientists to determine the fit of their models with the world, we have presented the ADI approach, based on which teachers can design instruction that will enable students to develop argumentation practices using pre-designed models. In addition, we have also outlined our own approach for integration with modeling in terms of how key discursive practices central to argumentation (such as critique and explanation) can be integrated with modeling. We propose that a productive integration of argumentation and modeling can result in a curriculum in which students can engage in all of the three kinds of modeling discussed by Hestenes (1993) in terms of development, ramification, and deployment. This stands in contrast to many typical modeling curricula in which students engage in one of these kinds of modeling. A key feature of our designed learning environment, ViMAP-Arg, is that students are provided with multiple simulations of the same phenomena that are displayed side-by-side. These simulations vary in terms of degree of accuracy, as well as in terms of the degree of students' access to and control over the underlying mathematical relationships that govern each simulation. It is through scaffolded learning activities that involve critique and explanation, based on comparison between the mathematical behavior of these simulations, that students engage in all the three kinds of modeling activities. Although we have contextualized much of our discussion in the particular context of ViMAP-Arg, we believe that the key design features we discussed here are generalizable, in that they can guide the design and development of computational learning environments to support the integration of modeling and argumentation in the context of learning other scientific domains (such as physics and chemistry).

In terms of the third overarching pedagogical goal, our proposed approach can be integrated into an overarching ADI framework to more closely integrate modeling and argumentation in service of engaging students as communities of scientists. In such activities, guided by the ADI approach, students' models developed in ViMAP-Arg can be used so that they can engage in evaluating models, methods, and evidence through argumentation using shared criteria and analytical approaches developed and agreed upon by the community.

While science education has historically attempted "to cultivate students' scientific habits of mind, develop their capability to engage in scientific inquiry, and teach them how to reason in a scientific context" (NRC, 2011), these three foci have traditionally often been treated separately (Driver, Leach, Miller, & Scott, 1996; Lemke, 1990). There is now growing agreement that students need to understand science and the processes of science as functions of argumentation and modeling (Duschl, 2008; Kelly, 2008; Lehrer & Schauble, 2006). The framework for the new science standards in the United States therefore "stresses the importance of developing students' knowledge of how science and engineering achieve their ends while also strengthening their competency with related practices" (NRC, 2011, p. 3.1). Toward this end, this chapter has outlined how approaches for integrating argumentation and modeling can productively engage students in inquiry-based activities that support learning of complex scientific concepts as well as the core argumentation and modeling practices that are at the heart of scientific inquiry.

REFERENCES

Aikenhead, G. (2004). Science-based occupations and the science curriculum: Concepts of evidence. *Science Education*, 89(2), 242–275. doi:10.1002/sce.20046

Andriessen, J., Baker, M., & Suthers, D. (2003). *Arguing to learn: Confronting cognitions in computer-supported collaborative learning environments.* Berlin, Germany: Springer.

Bazerman, C. (1988). *Shaping written knowledge : The genre and activity of the experimental article in science*. Madison, WI: University of Wisconsin Press.

Bell, P., & Linn, M. C. (2000). Scientific arguments as learning artifacts: Designing for learning from the web with KIE. *International Journal of Science Education*, 22(8), 797–818. doi:10.1080/095006900412284

Blikstein, P., & Wilensky, U. (2009). An atom is known by the company it keeps: Constructing multi-agent models in engineering education. *International Journal of Computers for Mathematical Learning*, *14*(2), 81–119. doi:10.1007/s10758-009-9148-8

Brem, S. K., & Rips, L. J. (2000). Explanation and evidence in informal argument. *Cognitive Science*, *24*(4), 573–604. doi:10.1207/s15516709cog2404_2

Carey, S., Evans, R., Honda, M., Jay, E., & Unger, C. (1989). An experiment is when you try it and see if it works: A study of grade 7 students' understanding of the construction of scientific knowledge. *International Journal of Science Education*, *11*, 514–529. doi:10.1080/0950069890110504

Chinn, C. A., & Brewer, W. F. (1993). The role of anomalous data in knowledge acquisition: A theoretical framework and implications for science instruction. *Review of Educational Research*, *63*, 1–49.

Clark, D., & Sampson, V. (2006). *Characteristics of students' argumentation practices when supported by personally-seeded discussions*. Paper presented at the Annual Meeting of the National Association for Research in Science Teaching. San Francisco, CA.

Clark, D. B., & Sampson, V. (2008). Assessing dialogic argumentation in online environments to relate structure, grounds, and conceptual quality. *Journal of Research in Science Teaching*, *45*(3), 293–321. doi:10.1002/tea.20216

Dickes, A., & Sengupta, P. (2011). Learning natural selection in 4th grade with multi-agent-based computational models. In Sengupta, P., & Hall, R. (Eds.), *Models, Modeling, and Naïve Intuitive Knowledge in Science Learning*. Berkeley, CA: Jean Piaget Society. doi:10.1007/s11165-012-9293-2

diSessa, A., & Abelson, H. (1986). Boxer: A reconstructible computational medium. *Communications of the ACM*, *29*(9), 859–868. doi:10.1145/6592.6595

Driver, R., Asoko, H., Leach, J., Mortimer, E., & Scott, P. (1994). Constructing scientific knowledge in the classroom. *Educational Researcher*, *23*, 5–12.

Driver, R., Leach, J., Millar, R., & Scott, P. (1996). *Young people's images of science*. Philadelphia, PA: Open University Press.

Driver, R., Newton, P., & Osborne, J. (2000). Establishing the norms of scientific argumentation in classrooms. *Science Education*, *84*(3), 287–313. doi:10.1002/(SICI)1098-237X(200005)84:3<287::AID-SCE1>3.0.CO;2-A

Duschl, R. (2000). *Making the nature of science explicit. Improving science education: The contribution of research*. Philadelphia, PA: Open University Press.

Duschl, R. (2008). Science education in three-part harmony: Balancing conceptual, epistemic, and social learning goals. *Review of Research in Education*, *32*, 268–291. doi:10.3102/0091732X07309371

Duschl, R. A., & Osborne, J. (2002). Supporting and promoting argumentation discourse in science education. *Studies in Science Education*, *38*, 39–72. doi:10.1080/03057260208560187

Ford, M. J., & Forman, E. A. (2006). Redefining disciplinary learning in classroom contexts. *Review of Research in Education*, *30*, 1–32. doi:10.3102/0091732X030001001

Giere, R. (1988). Laws, theories, and generalizations. In *The Limits of Deductivism* (pp. 37–46). Berkeley, CA: University of California Press.

Giere, R. N. (1999). Using models to represent reality. In Magnani, L., Nersessian, N. J., & Thagard, P. (Eds.), *Model-Based Reasoning in Scientific Discovery* (pp. 41–57). New York, NY: Kluwer Academic/Plenum. doi:10.1007/978-1-4615-4813-3_3

Harel, I., & Papert, S. (1991). Software design as a learning environment. In Harel, I., & Papert, S. (Eds.), *Constructionism*. Norwood, NJ: Ablex.

Hesse, M. (1966). *Models and analogies in science*. Notre Dame, IN: University of Notre Dame Press.

Hestenes, D. (1992). Modelling games in the Newtonian world. *American Journal of Physics*, *60*, 732–748. doi:10.1119/1.17080

Hestenes, D. (1993). *Modelling is the name of the game*. Paper presented at the National Science Foundation Modelling Conference. Dedham, MA.

Hogan, K., & Maglienti, M. (2001). Comparing the epistemological underpinnings of students' and scientists' reasoning about conclusions. *Journal of Research in Science Teaching*, *38*(6), 663–687. doi:10.1002/tea.1025

Hogan, K., Nastasi, B. K., & Pressley, M. (2000). Discourse patterns and collaborative scientific reasoning in peer and teacher-guided discussions. *Cognition and Instruction*, *17*(4), 379–432. doi:10.1207/S1532690XCI1704_2

Jimenez-Aleixandre, M., Rodriguez, M., & Duschl, R. A. (2000). Doing the lesson or doing science: Argument in high school genetics. *Science Education*, *84*(6), 757–792. doi:10.1002/1098-237X(200011)84:6<757::AID-SCE5>3.0.CO;2-F

Kelly, G. J. (2005). Inquiry, activity, and epistemic practice. *Rutgers University*. Retrieved from http://www.ruf.rice.edu/rgrandy/NSFConSched.html

Klahr, D., Dunbar, K., & Fay, A. L. (1990). Designing good experiments to test bad hypotheses. In *Computational Models of Scientific Discovery and Theory Formation* (pp. 355–401). San Mateo, CA: Morgan Kaufman.

Klopfer, E., Yoon, S., & Um, T. (2005). Teaching complex dynamic systems to young students with StarLogo. *Journal of Computers in Mathematics and Science Teaching*, *24*(2), 157–178.

Kuhn, D. (1989). Children and adults as intuitive scientists. *Psychological Review*, *96*(4), 674–689. doi:10.1037/0033-295X.96.4.674

Kuhn, D. (1991). *The skills of argument*. Cambridge, UK: Cambridge University Press. doi:10.1017/CBO9780511571350

Kuhn, D. (1993). Science as argument: Implications for teaching and learning scientific thinking. *Science Education*, *77*(3), 319–337. doi:10.1002/sce.3730770306

Kuhn, D., Shaw, V., & Felton, M. (1997). Effects of dyadic interaction on argumentative reasoning. *Cognition and Instruction*, *15*(3), 287–315. doi:10.1207/s1532690xci1503_1

Kuhn, L., & Reiser, B. (2005). *Students constructing and defending evidence-based scientific explanations*. Paper presented at the Annual Meeting of the National Association for Research in Science Teaching. Dallas, TX.

Kuhn, L., & Reiser, B. (2006). *Structuring activities to foster argumentative discourse*. Retrieved from http://hi-ce.org/iqwst/Papers/KuhnReiserAERA2006.pdf

Latour, B. (1999). *Pandora's hope: Essays on the reality of science studies*. Boston, MA: Harvard University Press.

Lawson, A. (2003). The nature and development of hypothetico-predictive argumentation with implications for science teaching. *International Journal of Science Education*, *25*(11), 1387–1408. doi:10.1080/0950069032000052117

Lehrer, R., & Schauble, L. (2006). Cultivating model-based reasoning in science education. In *Cambridge Handbook of the Learning Sciences* (pp. 371–388). Cambridge, UK: Cambridge University Press.

Lehrer, R., & Schauble, L. (2010). What kind of explanation is a model? In Stein, M. K. (Ed.), *Instructional Explanations in the Disciplines* (pp. 9–22). New York, NY: Springer. doi:10.1007/978-1-4419-0594-9_2

Lehrer, R., Schauble, L., & Lucas, D. (2008). Supporting development of the epistemology of inquiry. *Cognitive Development, 23*(4), 512–529. doi:10.1016/j.cogdev.2008.09.001

Leitão, S. (2000). The potential of argument in knowledge building. *Human Development, 43*(6), 332–360. doi:10.1159/000022695

Lemke, J. L. (1990). *Talking science: Language, learning, and values* (Vol. 1). New York, NY: Ablex Publishing Corporation.

Levy, S. T., & Wilensky, U. (2008). Inventing a "mid-level" to make ends meet: Reasoning through the levels of complexity. *Cognition and Instruction, 26*(1), 1–47. doi:10.1080/07370000701798479

Linn, M. C., Eylon, B., & Davis, E. A. (2004). The knowledge integration perspective on learning. In *Internet Environments for Science Education* (pp. 29–46). Mahwah, NJ: Lawrence Erlbaum Associates.

Linn, M. C., & Eylon, B. S. (2006). Science education: Integrating views of learning and instruction. In *Handbook of Educational Psychology* (pp. 511–544). New York, NY: Macmillan.

Lizotte, D. J., Harris, C. J., McNeill, K. L., Marx, R. W., & Krajcik, J. (2003). *Usable assessments aligned with curriculum materials: Measuring explanation as scientific way of knowing*. Paper presented at the Annual Meeting of the American Educational Research Association. Chicago, IL.

Lizotte, D. J., McNeill, K. L., & Krajcik, J. (2004). Teacher practices that support students' construction of scientific explanations in middle school classrooms. In *Proceedings of the 6th International Conference of the Learning Sciences*, (pp. 310-317). Mahwah, NJ: Lawrence Erlbaum Associates, Inc.

Lynch, M., & Woolgar, S. (Eds.). (1990). *Representation in scientific practice*. Cambridge, MA: MIT Press.

McNeill, K. L., & Krajcik, J. (2007). Middle school students' use of appropriate and inappropriate evidence in writing scientific explanations. In *Thinking with Data: The Proceedings of 33rd Carnegie Symposium on Cognition*. Mahwah, NJ: Lawrence Erlbaum Associates, Inc.

National Research Council. (2005). *America's lab report: Investigations in high school science*. Washington, DC: National Academy Press.

National Research Council. (2011). *Conceptual framework for new science education standards*. Washington, DC: National Academy of Sciences Board on Science Education.

Nersessian, N. J., & Patton, C. (2009). Model-based reasoning in interdisciplinary engineering. In *Handbook of the Philosophy of Technology and Engineering Sciences* (pp. 687–718). Amsterdam, The Netherlands: North Holland. doi:10.1016/B978-0-444-51667-1.50031-8

Newton, P., Driver, R., & Osborne, J. (1999). The place of argumentation in the pedagogy of school science. *International Journal of Science Education, 21*(5), 553–576. doi:10.1080/095006999290570

Rapp, D. N., & Sengupta, P. (2012). Models and modeling in science learning. In *Encyclopedia of the Sciences of Learning*. New York, NY: Springer.

Resnick, M. (1994). *Turtles, termites, and traffic jams: Explorations in massively parallel microworlds*. Cambridge, MA: MIT Press.

Sadler, T. D. (2004). Informal reasoning regarding socioscientific issues: A critical review of research. *Journal of Research in Science Teaching*, *41*(5), 513–536. doi:10.1002/tea.20009

Sampson, V., & Clark, D. (2008). Assessment of the ways students generate arguments in science education: Current perspectives and recommendations for future directions. *Science Education*, *92*(3), 447–472. doi:10.1002/sce.20276

Sampson, V., & Gerbino, F. (2010). Two instructional models that teachers can use to promote and support scientific argumentation in the biology classroom. *The American Biology Teacher*, *72*(7), 427–431. doi:10.1525/abt.2010.72.7.7

Sampson, V., & Gleim, L. (2009). Argument-driven inquiry to promote the understanding of important concepts & practices in biology. *The American Biology Teacher*, *71*(8), 465–472.

Sampson, V., & Grooms, J. (2009). Promoting and supporting scientific argumentation in the classroom: The evaluate alternatives instructional model. *Science Scope*, *32*(10), 67–73.

Sampson, V., & Grooms, J. (2010). Generate an argument: An instructional model. *Science Teacher (Normal, Ill.)*, *77*(5), 33–37.

Sampson, V., Grooms, J., & Walker, J. (2009). Argument-driven inquiry: A way to promote learning during laboratory activities. *Science Teacher (Normal, Ill.)*, *76*(7), 42–47.

Sampson, V., Grooms, J., & Walker, J. (2011). Argument-driven inquiry as a way to help students learn how to participate in scientific argumentation and craft written arguments: An exploratory study. *Science Education*, *95*(2), 217–257. doi:10.1002/sce.20421

Sandoval, W. A., & Millwood, K. A. (2005). The quality of students' use of evidence in written scientific explanations. *Cognition and Instruction*, *23*(1), 23–55. doi:10.1207/s1532690xci2301_2

Sandoval, W. A., & Reiser, B. J. (2004). Explanation driven inquiry: Integrating conceptual and epistemic scaffolds for scientific inquiry. *Science Education*, *88*(3), 345–372. doi:10.1002/sce.10130

Schauble, L., Glaser, R., Duschl, R., Schulze, S., & John, J. (1995). Students' understanding of the objectives and procedures of experimentation in the science classroom. *Journal of the Learning Sciences*, *4*(2), 131–166. doi:10.1207/s15327809jls0402_1

Schauble, L., Klopfer, L. E., & Raghavan, K. (1991). Students' transition from an engineering model to a science model of experimentation. *Journal of Research in Science Teaching*, *28*, 859–882. doi:10.1002/tea.3660280910

Schunk, D. H., & Zimmerman, B. J. (Eds.). (1998). *Self-regulated learning: From teaching to self-reflective practice*. New York, NY: Guilford Press.

Sengupta, P. (2011). Design principles for a visual programming language to integrate agent-based modeling in K-12 science. In *Proceedings of the Eighth International Conference of Complex Systems (ICCS 2011)*, (pp. 1636 – 1637). ICCS.

Sengupta, P., & Wilensky, U. (2010). Multi-agent-based modeling and learning electricity: Design and epistemological issues. In Khine, M. S., & Saleh, I. M. (Eds.), *Dynamic Modeling: Cognitive Tool for Scientific Enquiry*. New York, NY: Springer.

Sengupta, P., & Wright, M. (2010). *ViMAP*. Nashville, TN: Vanderbilt University.

Sherin, B., diSessa, A., & Hammer, D. (1993). Dynaturtle revisited: Learning physics through collaborative design of a computer model. *Interactive Learning Environments*, *3*(2), 91–118. doi:10.1080/1049482930030201

Sherin, B., diSessa, A. A., & Hammer, D. M. (1993). Dynaturtle revisited: Learning physics through collaborative design of a computer model. *Interactive Learning Environments*, *3*(2), 91–118. doi:10.1080/1049482930030201

Siegel, H. (1989). The rationality of science, critical thinking, and science education. *Synthese*, *80*(1), 9–42. doi:10.1007/BF00869946

Simon, S., Erduran, S., & Osborne, J. (2006). Learning to teach argumentation: Research and development in the science classroom. *International Journal of Science Education*, *28*(2), 235–260. doi:10.1080/09500690500336957

Smith, J. P., diSessa, A. A., & Roschelle, J. (1993). Misconceptions reconceived: A constructivist analysis of knowledge in transition. *Journal of the Learning Sciences*, *3*(2), 115–163. doi:10.1207/s15327809jls0302_1

Vellom, R. P., & Anderson, C. W. (1999). Reasoning about data in middle school science. *Journal of Research in Science Teaching*, *36*(2), 179–199. doi:10.1002/(SICI)1098-2736(199902)36:2<179::AID-TEA5>3.0.CO;2-T

Watson, J. D., & Crick, F. H. C. (1953). A structure for deoxyribose nucleic acid. *Nature*, *171*, 737–738. doi:10.1038/171737a0

Wilensky, U., & Reisman, K. (2006). Thinking like a wolf, a sheep, or a firefly: Learning biology through constructing and testing computational theories—An embodied modeling approach. *Cognition and Instruction*, *24*(2), 171–209. doi:10.1207/s1532690xci2402_1

Wilensky, U., & Resnick, M. (1999). Thinking in levels: A dynamic systems perspective to making sense of the world. *Journal of Science Education and Technology*, *8*(1). doi:10.1023/A:1009421303064

Windschitl, M., Thompson, J., & Braaten, M. (2008). Beyond the scientific method: Model-based inquiry as a new paradigm of preference for school science investigations. *Science Education*, *92*(5), 941–967. doi:10.1002/sce.20259

Zeidler, D. (1997). The central role of fallacious thinking in science education. *Science Education*, *81*, 483–496. doi:10.1002/(SICI)1098-237X(199707)81:4<483::AID-SCE7>3.0.CO;2-8

ENDNOTES

[1] We will interchangeably use the terms *activity* and *game* in this chapter.

[2] Our use of the term "world" here is different than the usual use of the term in the literature on computational microworlds, where a microworld usually indicates a simulation (e.g., Resnick, 1994). Technically, ViMAP-Arg is a collection of three microworlds (or simulations). However, in our usage, "world" simulation indicates a target set of behaviors that are canonically correct and expert-like, and can act as a scaffold for modeling activities.

Chapter 6
Reification of Five Types of Modeling Pedagogies with Model-Based Inquiry (MBI) Modules for High School Science Classrooms

Todd Campbell
University of Massachusetts Dartmouth, USA

Phil Seok Oh
Gyeongin National University of Education, Korea

Drew Neilson
Logan High School, USA

ABSTRACT

It has been declared that practicing science is aptly described as making, using, testing, and revising models. Modeling has also emerged as an explicit practice in science education reform efforts. This is evidenced as modeling is highlighted as an instructional target in the recently released Conceptual Framework for the New K-12 Science Education Standards: it reads that students should develop more sophisticated models founded on prior knowledge and skills and refined as understanding develops. Reflecting the purpose of engaging students in modeling in science classrooms, Oh and Oh (2011) have suggested five modeling activities, the first three of which were based van Joolingen's (2004) earlier proposal: 1) exploratory modeling, 2) expressive modeling, 3) experimental modeling, 4) evaluative modeling, and 5) cyclic modeling. This chapter explores how these modeling activities are embedded in high school physics classrooms and how each is juxtaposed as concurrent instructional objectives and scaffolds a progressive learning sequence. Through the close examination of modeling in situ within the science classrooms, the authors expect to better explicate and illuminate the practices outlined and support reform in science education.

DOI: 10.4018/978-1-4666-2809-0.ch006

INTRODUCTION

It has been well documented that doing science is aptly described as making, using, testing, and revising models (Clement, 2008; Giere, Bickle, & Mauldin, 2006; Halloun, 2004; Nersessian, 2008). Nersessian (2008), for example, stated that "model construction, manipulation, evaluation, and adaptation are a primary means through which scientists create new conceptual representations" (p. 10) and indicated models as the basic units for scientists to work with theories. Modeling has also emerged as an explicit pedagogical practice in science education reform efforts. This is evidenced as modeling is highlighted as an instructional anchor in the recently released *A Framework for K-12 Science Education: Practices, Crosscutting Concepts, and Core Ideas* (National Research Council, 2011), whereby students are envisioned developing more sophisticated models founded on prior knowledge and skills and refined as understanding develops. Modeling is conceived as a central practice for science learning that can 1) allow "students to be themselves within a culture of scientific inquiry" (Johnston, 2008, p. 12), 2) support the development of explanations extracted from evidence (Khan, 2007; Windschitl, Thompson, & Braaten, 2008a, 2008b), and 3) engage students in scientific argumentation through sharing, comparing, and deciding between competing models (Böttcher & Meisert, 2011; Passmore & Svoboda, 2012). While these are but a few of the possible important benefits of modeling, these and other benefits are dependent on the intentional educational applications of scientific modeling practices, some of which are described next.

Reflecting the purpose of engaging students in modeling practices in science classrooms, Oh and Oh (2011) have suggested five pedagogical conceptualizations for modeling, the first three of which were based on van Joolingen's (2004) earlier proposal: 1) *exploratory modeling*, 2) *expressive modeling*, 3) *experimental modeling*, 4) *evaluative modeling*, and 5) *cyclic modeling*. These five ways of modeling are referred to as modeling pedagogies to highlight how they can assist in framing pedagogical transformations of scientific practices that teachers perceive as helpful in meeting desired student learning outcomes (e.g., scientific discourse, scientific understanding). To get a sense of how modeling is currently being leveraged in science classrooms, this chapter explores how these modeling pedagogies are embedded in high school physics classrooms. Through the close examination of the modeling pedagogies in situ within a high school physics course, we expect to better illuminate classroom inquiry outlined and supportive of reform in science education, which can in turn reveal possible ways of enacting model-based science instruction.

MODELING AS SCIENCE AND SCIENCE LEARNING

Situating modeling in science education begins to make sense by considering the roles modeling plays in the work of scientists and in the context of specific scientific fields (e.g., astronomy, chemistry, evolutionary biology, geology). Although there is no single definition of a model, models are broadly recognized as representations or systems of objects, events, processes, and ideas (Gilbert & Boulter, 2000). In modeling, extra-linguistic entities like pictures and diagrams assume fundamental roles in the functions of models when they serve to describe, explain, and predict natural phenomena and communicate scientific ideas with others (Buckley & Boulter, 2000; Oh & Oh, 2011; Shen & Confrey, 2007).

Passmore and Stewart (2002) articulated modeling as a central cognitive goal of evolutionary biology, as one example, as they explained how this field works to understand how life on Earth has changed and to develop models that can provide explanatory power in this pursuit. Therefore, the cognitive tasks evolutionary biologists are concerned with are developing chronologies of past changes in life on Earth and models that can explain these reconstructed chronologies. This is especially important because evidence for these changes are distributed over millions of years and because direct observations and replicable experimentations are less accessible compared with other disciplines. Therefore, Darwin's mechanistic model of natural selection combined with other models, such as speciation models and population genetics models, converges to provide predictive explanations that can be measured against available indirect evidence (e.g., fossil record, homologous structures).

More historical examples supporting the importance of models and modeling can be found from other scientific disciplines. For example, Newton used a model of white light composed heterogeneously of colors to enable a full range of explanations surrounding the behavior of light (Gilbert, Boulter, & Rutherford, 1998). In the history of geology, a number of visual models, such as those of subterranean convection current, sea floor spreading, and magnetic profiles near ocean ridges, played central roles in the 20th century revolution in the theory of earth dynamics (Giere, 1999). In the context of chemistry, Justi and Gilbert (1999) provided evidence of how and when scientists used models and modeling as ideas about chemical kinetics evolved and became more sophisticated. Finally, but not exhaustively, the historical role and iterative nature of models can be seen from an early Copernicus model of concentric planetary spheres to Kepler's model of planetary motion (Taylor, Barker, & Jones, 2003). Collectively from these examples, both currently and historically and across a range of scientific disciplines, it is obvious that models have and continue to contribute to advances in science and in some cases serve as the central artifacts of these advances.

Consistent with the description thus far, it is agreed that models served a central aim of scientists: an aim of research that seeks to refine explanatory models so that they can guide future research, among other things (Passmore & Stewart, 2002). Moreover, Shen and Confrey (2007) characterized a model as a hybrid of mentality and nature when they explicated how models support the development of scientific understanding:

The purpose of modeling is to describe, explain, predict, and communicate with others a natural phenomenon, an event, or an entity. There exists a mapping between a base and a target, which we denote as a construct and its referent. There are rules and structures in the operation of the construct and those in the operation of the referent (natural law). Modeling is a process of coordinating the rules and structures of the construct to those of the referent and distancing the construct from the referent by ways of simplifying, quantifying, and representing (p. 950).

Therefore, it can be seen that models help bridge the gap between observed phenomena and theoretical ideas about why those phenomena occur (Morrison & Morgan, 1999; Oh & Oh, 2011). The same principle applies to science learning: using models in science classrooms is beneficial because models support constructing and reasoning with students' mental models (Buckley & Boulter, 2000; Gilbert & Ireton, 2003; Nersessian, 1999). In fact, a number of studies have provided evidence for the effectiveness of model-based science instruction. Gobert and colleagues (Gobert, 2005; Gobert & Clement, 1999; Gobert & Pallant, 2004) showed, for example, that the process of modeling the interior of the earth and its dynamic movements was helpful both for enhancing students' understanding of the spatial and causal aspects of plate tectonics and for fostering their perceptions of the nature of models. Penner, Lehrer, and Schauble (1998) engaged third-grade children in building, testing, and revising models of the human elbows and found that with modeling even young students better understood the mechanics of the human body. In addition, models and modeling have shown their promises in science teacher education programs as well (Akerson, et al., 2009; Schwarz & Gwekwerere, 2007; Windschitl & Thompson, 2006; Windschitl, Thompson, & Braaten, 2008a). For instance, Windschitl and Thompson (2006) demonstrated how preservice science teachers developed more sophisticated knowledge of scientific models and increased the incorporation of model-based lessons in their classrooms after engaging in course activities focused on fostering deeper understanding of the epistemic roles of models as they are coordinated with theories to support arguments in scientific inquiry. Schwarz and Gwekwerere's (2007) study also revealed how using a guided inquiry approach combined with a modeling instructional framework in the context of a preservice science teaching methods course led to an increased focus on scientific inquiry and reliance on several types of models in lesson planning.

Recent science education reform documents parallel these studies as emphasizing the important roles of models in science teaching and learning (Duschl, Schweingruber, & Shouse, 2007; NRC, 2011). Especially, *A Framework for K-12 Science Education: Practices, Crosscutting Concepts, and Core Ideas* (NRC, 2011) suggests modeling as suitable for all grades by saying, "Modeling can begin in the earliest grades, with students' models progressing from concrete 'pictures' and/or physical scale models (e.g., a toy car) to more abstract representations of relevant relationships in later grades, such as a diagram representing focus on a particular object in a system" (p. 58). However, it has been reported that model-based teaching is not widely implemented in school classrooms and that when implemented, it is likely missing some important aspects of scientific modeling (Khan, 2011). Teachers are often believed to be responsible for this lack of model-based science instruction (Crawford & Cullin, 2004; Justi & Gilbert, 2002a, 2002b, 2003; Smit & Finegold, 1995; van Driel & Verloop, 1999, 2002). For instance, based on their survey with a range of participants, Justi and Gilbert (2003) asserted that teachers did not possess coherent ontological and epistemological views on the nature of models. van Driel and Verloop (1999, 2002) also indicated that teachers' knowledge of models and modeling in science were often limited and that they were not fully aware of students' views and abilities concerning the same topics. Such limited teacher perceptions would likely influence the manners in which the teachers use models in their science classrooms. That is, as Louca, Zacharia, and Constantinou (2011) argued, "[F]ew teachers know how modeling looks and what it entails in authentic learning situations and even fewer know how to support productive student modeling in science" (p. 920). This also implicates science teacher educators and in-service professional developers, as well as the science education programs the teachers have gone through (Crawford & Cullin, 2004; Smit & Finegold, 1995). But, as mentioned earlier, promising research has emerged in science teacher education (e.g., Akerson, et al., 2009; Schwarz & Gwekwerere, 2007; Windschitl & Thompson, 2006; Windschitl, Thompson, & Braaten, 2008a), which will be even more important into the future as modeling is elevated in science instruction in national standards documents.

OUR COLLABORATION PROJECT FOR MODEL-BASED INQUIRY (MBI)

In agreement with Louca et al. (2011), we recognized the need of a project to provide teachers with conceptual, as well as practical guidance that helps them apply scientific practices of modeling successfully in their classrooms. Such a project was actually realized thanks to 1) the recent proposal of five modeling pedagogies (Oh & Oh, 2011) and 2) the ongoing collaboration for improving school science between us, as university-based researchers (i.e., TC and PSO), and a high school science teacher (i.e., DN).

First, the five modeling pedagogies are based on the principle that scientific practices, including modeling, ought to be translated at the level of classroom learning in order for students to exercise the same type of intellectual activities as those of scientists (NRC, 2011). From this perspective, Oh and Oh (2011) proposed five modeling pedagogies as an extension of van Joolingen's (2004) earlier conceptualization of three kinds of modeling for inquiry learning. These modeling pedagogies reflect diverse ways modeling is practiced in science and are suggested as aids for applying the scientific practices of modeling into science classrooms. The five modeling pedagogies are defined as follows and examples of each of the modeling pedagogies are presented as well:

- *Exploratory modeling*, where students investigate the property of a pre-existing model by engaging with the model (e.g., changing parameters) and observing the effects. In Urhahne et al.'s (2010) study, for example, students manipulated parameters of computer simulations about mechanics and observed the outcomes to learn how motions can be described with graphs.
- *Expressive modeling*, where students express their ideas to describe or explain scientific phenomena by creating new models or using existing models. For instance, Gobert and Clement (1999) asked a group of students to draw diagrammatic models to demonstrate their understanding after they read an expository text about plate tectonics.
- *Experimental modeling* (called *inquiry modeling* originally in van Joolingen, 2004), where students form hypotheses and predictions from models and test them through experimenting with phenomena. As an example, in Maia and Justi's (2009) study, students were given a series of learning opportunities to study scientific models to explain chemical equilibrium, produce their own mental models, and test and modify them in light of experimental results.
- *Evaluative modeling*, where students compare alternative models addressing the same phenomenon or problem, assess their merits and limitations, and select the most appropriate one(s) to explain the phenomenon or solve the problem. In Passmore and Stewart's (2002) study, for instance, students examined three models to account for the diversity of species (i.e., Paley's model of intelligent design, Lamarck's model of use inheritance, and Darwin's model of natural selection) and compared them by assessing the explanatory power of each model as well as the underlying assumptions or beliefs.
- *Cyclic modeling*, where students are engaged in ongoing processes of developing, evaluating, and improving models to complete rather long science projects. Khan's (2007) GEM (Generation-Evaluation-Modification), Schwarz and Gwekwerere's (2007) EIMA (Engage-Investigate-Model-Apply), and Halloun's (2004) modeling learning cycle, as a few examples among many, all have a similar, cyclic structure of scientific modeling.

It should be emphasized that the five modeling pedagogies are not exclusive to each other. In fact, two or more modeling pedagogies can be combined to address a single science topic. As an example, students may learn both geocentric and heliocentric models of celestial motions by exploratory modeling (e.g., they can change planet positions in computer models and see how the planets are observed from the earth) and then participate in evaluative modeling to select an adequate model explaining a certain astronomical phenomenon (e.g., phase change of Venus). Furthermore, one type of modeling pedagogy can integrate others in its own scheme. For instance, experimental modeling may start with students presenting their ideas by constructing new models or manipulating existing models differently (expressive modeling) and proceed to engaging them in evaluating different models in light of experimental results (evaluative modeling). Also, cyclic modeling is able to include the other four modeling pedagogies in order that students can progressively exercise different types of scientific modeling (An example of this will be discussed later in the chapter).

Second, our collaboration is described as continuous effort to explore and build up Model-Based Inquiry (MBI) in high school science classrooms. Since 2008, Mr. Neilson and Dr. Campbell have worked closely as they first sought to find mechanism for making inquiry more palatable for classroom teaching. This earlier work led to identifying modeling as a promising way to organize science instruction as inquiry-based and afford students opportunities to experience 'epistemic practices' of science (Sandoval & Reiser, 2004). More extensive collaboration then followed to develop individual modeling activities or modules, which are now included strategically in Mr. Neilson's yearlong physics curriculum (see Table 1).

Our initial modeling modules were framed by the road map for MBI shown in Figure 1. The road map outlined student activities with three components in a multidirectional cycle—modeling, focused inquiry, and iterations—which are similar to those in Windschitl's scheme consisting of generation, testing, and revision of models (Windschitl & Thompson, 2006; Windschitl, Thompson, & Braaten, 2008a, 2008b). That is, our initial frame for MBI focused on engaging students in creating models and relying on the models to shape their scientific inquiry as the models serve as the anchor. In 2010, Dr. Oh joined to work with both Mr. Neilson and Dr. Campbell and provided his conceptualization of five modeling pedagogies (Oh & Oh, 2011). This helped connect Mr. Neilson's modeling instruction to theoretical foundations and restructure it to incorporate more various modeling practices into his science lessons. Since then, our collaborative work has been established as an ongoing project to create a better niche for implementing MBI and conducting classroom-based science education research.

While we have provided the effectiveness of Mr. Neilson's MBI instruction in other research reports (Campbell, Zhang, & Neilson, 2010), more data has recently been collected from Mr. Neilson's classrooms in the form of video-recordings. This data contains four science lessons from two different classes in which the Electrostatic Energy module was applied (see Campbell & Neilson, 2012, for additional details about the Electro-

Table 1. Mr. Neilson's yearlong physics curriculum

Unit Focus	Model-Based Inquiry (MBI) Module
Liner Motion Projection Motion	Motion of Objects
Newton's Law of Motion	Friction Forces and Rockets
Momentum	
Energy Circular Motion Center of Gravity Rotational Mechanics Universal Gravitation and Gravitational Interactions Satellite Motion	Work and Energy
Fluid Dynamics Temperature, Heat, and Expansion Heat Transfer Change of Phase Thermodynamics Vibration and Waves Sound Light Color Reflection and Refraction Lenses Diffraction and Interference	Buoyant Forces
Electrostatics and Electric Fields and Potential Electric Current and Circuits Magnetism and Electromagnetic Induction Atomic and Nuclear Physics	Electrostatic Energy

static Energy module). In this chapter, these video-recordings, as well as documentation of the other modeling modules were analyzed to reveal how Mr. Neilson has facilitated scientific modeling for his students. Especially, we will describe the classroom practices using the five modeling pedagogies as observational lenses, for we believe that this will help reify the five modeling pedagogies so that teachers of science can be offered informed practical guidance for better modeling instruction. Generally speaking, Mr. Neilson's physics lessons are structured in the cyclic modeling frame. That is, in his high school science classrooms, students are given opportunities to develop models to explain scientific phenomena, design investigations to test their models, and revisit their models for improvement. This instructional cycle involves central facets of the five modeling pedagogies, even if some could be emphasized more explicitly than others and some aspects of the modeling practices might be missing in a certain module. In the following, we first share the particular features of the modeling pedagogies found in analyzing the video data, before other MBI modules are discussed.

MODELING PEDAGOGIES IN PRACTICE: ELECTROSTATIC ENERGY MODULE

From the Electrostatic Energy module, it was revealed that Mr. Neilson's students were engaged in *expressive modeling* for a fairly long period of time. The task assigned to the students was to create models with which they could explain scientific phenomena about static electricity. To trigger students' modeling practices, Mr. Neilson provided a set of demonstrations related to static electricity. In the first demonstration, the teacher approached a glass rod rubbed with a piece of silk to a latex balloon and then did the same thing with a rubber rod rubbed in fur. After students confirmed their observation of the latex balloon

Figure 1. A road map for model-based inquiry (MBI)

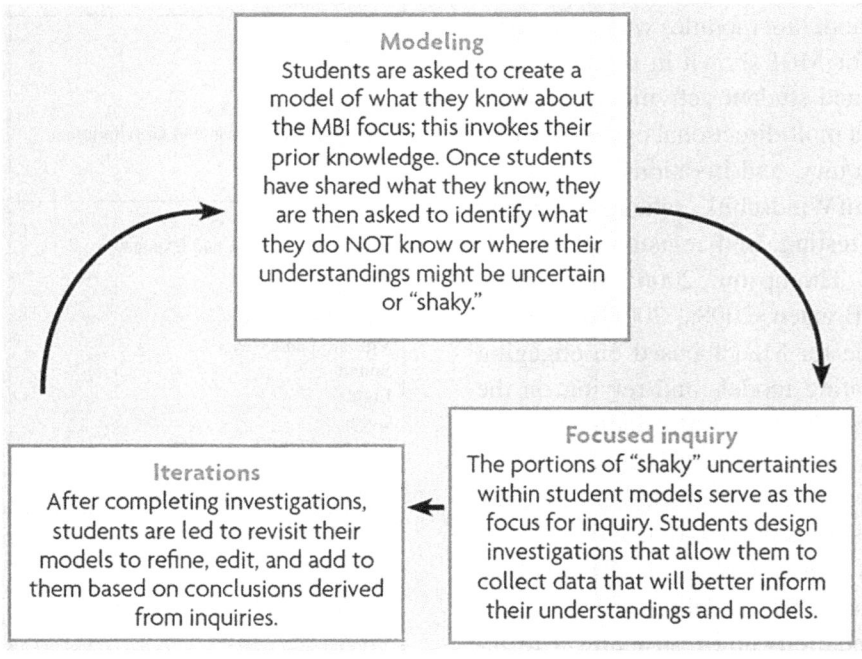

reacting differently to the rods (i.e., moved toward to the rubber rod and away from the glass rod), the teacher began his second demonstration which was about the reaction of a Mylar balloon when the rubber rod and glass rod approached. Then, as the third and final teacher demonstration, Mr. Neilson showed how the metal leaves of an electroscope reacted when touched with both rods.

The electrostatic phenomena demonstrated by the teacher became the subjects to be explained through expressive modeling by students. However, Mr. Neilson did not merely ask students to come up with models, but instead guided the students through several pedagogical actions. First, he emphasized that one purpose of scientific modeling is to explain phenomena. On several occasions during his demonstration, Mr. Neilson stated, for example, "You're going to be creating your model. Remember, your model should *explain* why you're seeing what's happening, as well as what's really happening" or simply, "Your model should *explain* these phenomena." Such utterances by the teacher were believed to be effective in reminding students that the basic functions of scientific models include explaining evidence from observations to enhance our understanding of the natural world (Buckley & Boulter, 2000; Oh & Oh, 2011; Shen & Confrey, 2007).

Second, Mr. Neilson activated students' knowledge of science so that they could base their models on the canonical or normative scientific knowledge. The static electricity phenomena studied in Mr. Neilson's classroom were those that are fundamentally explained by scientific ideas of electrons and their interactions with other electrons and subatomic particles and materials. Therefore, the teacher consistently asked students to connect their models to what scientists know about the atomic structure and the movement of electrons. The following conversation shows such an effort in which Mr. Neilson and his students shared what they knew about atoms:

Mr. Neilson: *We talked yesterday about the atom, that in the nucleus the charges that are there are what?*
Students: *Positive.*
Mr. Neilson: *What, positive charges? What else is in the nucleus?*
Students: *Neutrons.*
Mr. Neilson: *Electrons are on the outside. ... Would you say they have more protons than electrons, more electrons than protons or equal numbers generally,*
Students: *Equal.*
Mr. Neilson: *Equal numbers. What do we call that situation?*
Students: *Neutral.*
Mr. Neilson: *Neutral, right. Is that what you said?*
Student: *Yeah, I said stable.*
Mr. Neilson: *Yeah, stable. ... That's what atoms are. To really explain what's happening here, you might have to look at this model of atoms. That's what I mean by looking at small. You might actually have to talk about these things.*

In the excerpt above, Mr. Neilson's last utterance demonstrates how he reveals his pedagogical intention for students to stay close to the scientific ideas about electrons and use them in generating their own models.

Providing analogies was another way Mr. Neilson guided students in creating their models. For instance, the dialogue below occurred when a student asked Mr. Neilson a question about the properties of insulators and conductors. As evident from the excerpt, Mr. Neilson provided an analogy in a way that scaffolded the student to better understand the scientific concept of insulators and conductors and adapt it in her own modeling:

Mr. Neilson: *What if I have a candy bar and I'm just barely holding onto it? You could grab it, right?*
Student: *Yeah.*

Mr. Neilson: If I'm not holding on tight and you can grab tightly, you could steal that candy bar from me.
Student: Okay, so an insulator can get electrons easier than the conductor?
Mr. Neilson: Definitely.

Besides this, several more analogies were created by the teacher or students in Mr. Neilson's MBI classrooms, including magnetic forces as a comparison to different electric charges and exchanging money as an analogy of electrons' movement between conductors and insulators.

Finally, it should be noted that Mr. Neilson encouraged students to express their models in alternative forms of representation, rather than writing out lengthy explanations. He frequently referred to student models as "mental picture" or "your vision," and asked them repeatedly "[to] draw your model" or "illustrate that." He also remarked that "picture and diagrams are much better than a bunch of words" and that "the purpose of this model [is to] ... visualize." One of the characteristics of models in science is that they are expressed with various semiotic resources, such as diagrams, graphs, and three-dimensional figures, and these non-linguistic representations play a key role in describing complicated phenomena and communicating abstract ideas to others (Giere, 1999; Nersessian, 1999; Oh & Oh, 2011). Mr. Neilson consistently emphasized this multi-modal nature of scientific models as he helped students make their ideas observable and tangible through modeling. As a result, all the student models in his classrooms were expressed in pictures accompanied with explanatory texts.

Mr. Neilson's expressive modeling was followed by *experimental modeling* in which students were to "try and test" their models. By "try and test," Mr. Neilson meant various ways to "see if we can recreate" target phenomena using models and find "evidence" to adjust the models. In a class, he explained:

The cool thing about your model is, if it makes sense to you right now, then that's what ought to go down. As long as you can tell me why ... that's the starting point. Then, what we'll do is, we'll do some tests and see if we can recreate that. If we recreate it, then we've given some evidence to support your contention. ... We found evidence, and then we adjusted our model accordingly.

For example, as a student suggested that there might be different charges involved when a rod was rubbed with silk or fur, Mr. Neilson asked reflectively, "Is it conclusive that there are two different charges?" He then engaged the whole class in an experiment with an electroscope to further investigate the student's idea. In addition, when students came up with different models to explain why two leaves of an electroscope pushed apart with charged rods touching the top of the electroscope, he accepted all the ideas regardless of their accuracy and asked the students to suggest new demonstrations or experiments to test the ideas. Consequently, much of the observed class time was spent with the teacher or students conducting new experiments using different materials and different pairs (e.g., metal rod rubbed with fur or silk, rubber rod with silk, glass rod with fur or plastic bag).

According to Khan's (2011) recent report, 'making predictions based on models' is one of the missing practices in model-based teaching in high school science classrooms. Considering this, Mr. Neilson's classrooms are considered exceptional in that the teacher tried to clarify students' thoughts before they gathered evidence through experimenting. As an example, the following conversation shows Mr. Neilson exploring students' hypotheses and predictions regarding the electroscope's reactions to charged rods before they collected evidence to test their models (In the excerpt, all student names are pseudonyms).

Mr. Neilson: Anybody have a test you want to do right now? ... Andy?

Andy: Maybe, when you rub it [rubber rod] on the rabbit skin, that positively charges it. Then, that charge goes to both pieces of the tin foil [of the electroscope].... When two things are positively charged, they repel each other.

Mr. Neilson: Is that true, two things that are positively charged...?

Brittney: I think when you are rubbing the rubber and the fur, it's stripping the fur of its electrons.

Mr. Neilson: Why do you say that? What would you lead you to say that? ... Andy thought maybe this was positive. You're disagreeing with Andy?

Brittney: Well, because if that is positive, it would attract the electrons from that. ...

Mr. Neilson: What if it's not positive to start with? Is this positively charged right now?

Craig: Or neutrally charged....

Mr. Neilson: Okay. Why would this tend to be negatively charged? Do you think that it's negatively charged? Are you with Andy now? ... What Andy said that this is transferring positive charge to this, and that makes the leads or these strips of foil repel.

Brittney: Well, wouldn't it work the same way if it was transferring negative energy, because negative and negative will repel each other?

Mr. Neilson: Okay. Do you agree, Andy? ... (indicating Doug) Yeah.

Doug: It's definitely not positively charged, because you can't strip something of protons ... It's going to be stripping electrons if anything, just logic. ...

Mr. Neilson: Does anybody think from what we talked about this material right here that would give us some insight as to why this might be doing the stripping of electrons?

Fred: Not stripping. I think like you're creating heat and energy inside when you rub them together.

Mr. Neilson: Okay.

Fred: It positively charges it ... from the heat and the energy in there.

Mr. Neilson: May be heat? ... Okay, these are all good ideas....

Doug: I don't understand why you had to rub that one with the fur instead of the silk. What would happen if you rubbed it—that one with the silk?

Mr. Neilson: We could do any test you want.

Although conversations like the above abounded in Mr. Neilson's MBI classrooms to explore student ideas, it was not possible to listen to all students' explanations within limited class time. So, the students were guided to illustrate their models in worksheets and state their hypotheses or predictions by the models. In particular, Mr. Neilson referred to "if, then" statement as a hypothesis or prediction which the students were to form out of their models.

I want to test what happens with this balloon. I want to know why it's doing what it's doing. ... Can I do some test on that? Someone talked about distance. Does distance matter? How does it matter? All those are good. Good questions. Then, some if, then statement.... If I do this, then this will happen. Then, that will naturally go into test. If, you say, if this occurs, then this will happen. Now you know exactly what you're going to test, right?

As an example, when we looked into student worksheets, it was found that a student wrote, "Possible causes of static [electricity include] friction (heat)" and moved on to explain, "If heat is the factor, then color is the next factor: the black rubber rod absorbs more energy in the form of electrons creating a greater charge in the glass." His initial explanation, of course, was not scientifically correct, and it changed to a scientifically valid one (e.g., "Electrons flow to [a] finger [when] touching [on the top of the electroscope].") thanks to the experimental modeling and other activities in the class.

The lack of students stating and discussing hypotheses and predictions is not uncommon in school science classrooms. Also, we believe that if more students had been involved publicly in formulating hypotheses and predicting experimental results based on their models, additional benefits could have been identified in Mr. Neilson's classrooms. However, an important implication can be drawn from experimental modeling in Mr. Neilson's MBI lessons. That is, having student ideas explored and articulated through modeling can result in discourse-rich classroom environments, as well as the learners' deeper understanding of scientific concepts.

Despite this positive feature, there was also a 'missing practice' (Khan, 2011) in Mr. Neilson's modeling instruction. The explicit purpose of Mr. Neilson's experimental modeling was to validate individual models of students and generate evidence to be used for improving the models. That is to say, his modeling focused on assessing a single model rather than evaluating multiple alternative models or selecting which competing models may be best suited for explaining the observed electrostatic phenomena. We will discuss this matter later in the chapter.

In Mr. Neilson's physics classrooms, expressive and experimental modeling developed further into *cyclic modeling*. The purpose of the cyclic modeling was to provide students with continuous opportunities to test their models, collect more evidence, and improve models by pondering the evidence. Remarkably, before students started to test and revisit their models, Mr. Neilson explained the rationale of the cyclic modeling to his students:

What are we gonna be doing with your models as you learn more? Yeah, changing them. I don't like the word fixing em'. That implies you guys made a mistake. As you get more evidence, you modify it. You make changes to it. There's no right answer in science. We arrive at an answer, and then maybe new evidence shows up, and we don't like that answer anymore, and we change it.

It is obvious in the teacher's utterance above that Mr. Neilson intentionally avoids using the term "fix" and instead utilizes the terms "change" and "modify." We see this reflecting Mr. Neilson's understanding of an essential aspect of scientific models: models in science are subject to empirical and theoretical tests and revisable as a consequence of those tests (Oh & Oh, 2011; Windschitl, Thompson, & Braaten, 2008a, 2008b). It is also important for students to understand this tentative nature of scientific models, if they are to learn science by exercising scientific practices. Keeping this norm in mind, Mr. Neilson even asked students to use a pencil when they were working on models:

Remember ... do it in pencil. ... In the last hour, ... I said, "Why do it in pencil?" They said, "Because we want to fix our model." I don't think that's a very good word. We don't fix things. ... A model is something where you change it as you get new evidence. That's what we're doing.

In addition to convincing students of the rational for cyclic modeling, Mr. Neilson helped the students modify their models by providing falsifying evidence. For instance, when a student proposed that a rubber rod rubbed with fur and a glass rod with silk might have the same charge, Mr. Neilson showed a new demonstration in which the foil ribbons in an electroscope reacted differently to the rubber rod and glass rod, so the student could look back on her model and change it.

Mr. Neilson's cycling modeling resulted in progressions of student understanding of static electricity and their models about it. Figure 2 presents a piece of evidence for this conclusion. Part A of Figure 1 is a student's initial model, where he explains an electrostatic phenomenon with the difference in size of atoms between an insulator and conductor. In his worksheet, he wrote:

Because rubber [atoms] are smaller, when rubbed with fur, it will take the fur's electrons since the distance to the center is less, so the pull is greater. Rubbing the rod causes more collisions, so there are more opportunities to steal electrons. The opposite effect happens to the glass because the glass' atoms are larger than the silk's (since the rubber is a better conductor than glass, its atoms are smaller than the glass). The smaller the atom, the better the insulator.

In his modified model, Part B of Figure 2, however, the same student constructed his explanations with the idea of the movement of electrons, saying "When you rub the rubber with fur, the rubber, which is a good insulator with tightly bound electrons, will take electrons from the fur whose electrons are held less tightly, giving the rod a negatively charge and the fur a positive charge." Notably, as evident in Figure 2, his new model is not only scientifically valid, but also able to explain more phenomena related to static electricity.

MODELING PEDAGOGIES IN PRACTICE: ADDITIONAL MBI MODULES

Our collaborations over a number of years have resulted in the development of more cohesive modeling experiences for students. As already presented, Table 1 shows Mr. Neilson's yearlong physics curriculum that includes six strategically placed MBI modules. The following is a brief description of each of these modules beyond the Electrostatic Energy modeling activity already discussed:

- **Motion of Objects:** Students are introduced to modeling in this first unit of the year. In the module, students create models that illustrate their understanding of terms used to describe motion (e.g., speed, velocity, acceleration, instantaneous vs. average, and relevant equations). Students are asked to use their models to make predictions about an object in motion in a performance assessment after iterative constructions of models are complete.

- **Friction:** Students are asked to model how friction works as a type of force. They create models depicting how they understand friction and subsequently revise these models with investigations using a force probe and objects pulled across a horizontal plane (see Campbell & Neilson, 2009, for detail).

- **Forces and Rockets:** At the completion of the unit on Newton's Laws of Motion, students are asked to model their understanding of forces acting on rockets made out of two-liter bottles and propelled with pressure. Students first consider design features common on rockets (i.e., nosecones and fins) and use modeling to consider their functions. They then decide how to construct two rockets to test one portion of their models. Student groups perform an initial launch and revisit models based on their findings as well as those of their peers. Students complete a subsequent launch after revisions to their rockets and submit their models.

- **Work and Energy:** Students create models of how they understand work and its relationship to energy. Additionally, their models focus on energy transformation and what factors influences changes in energy. Students design experiments using where they drop steel balls into clay so that they have a measurable way to detect energy (i.e., depressions in the clay). Through experimentation, energy change models are iteratively revised.

- **Buoyant Forces:** Students are asked to create models to explain how they understand buoyant forces. They are introduced to a method of testing factors influencing buoyant forces whereby a string holds a mass connected to a force probe and over a container of water. Student models of buoyant force are revised iteratively with various tests (see Neilson, Campbell, & Allred, 2010, for detail).

Figure 2. Examples of student models in the electrostatic energy module

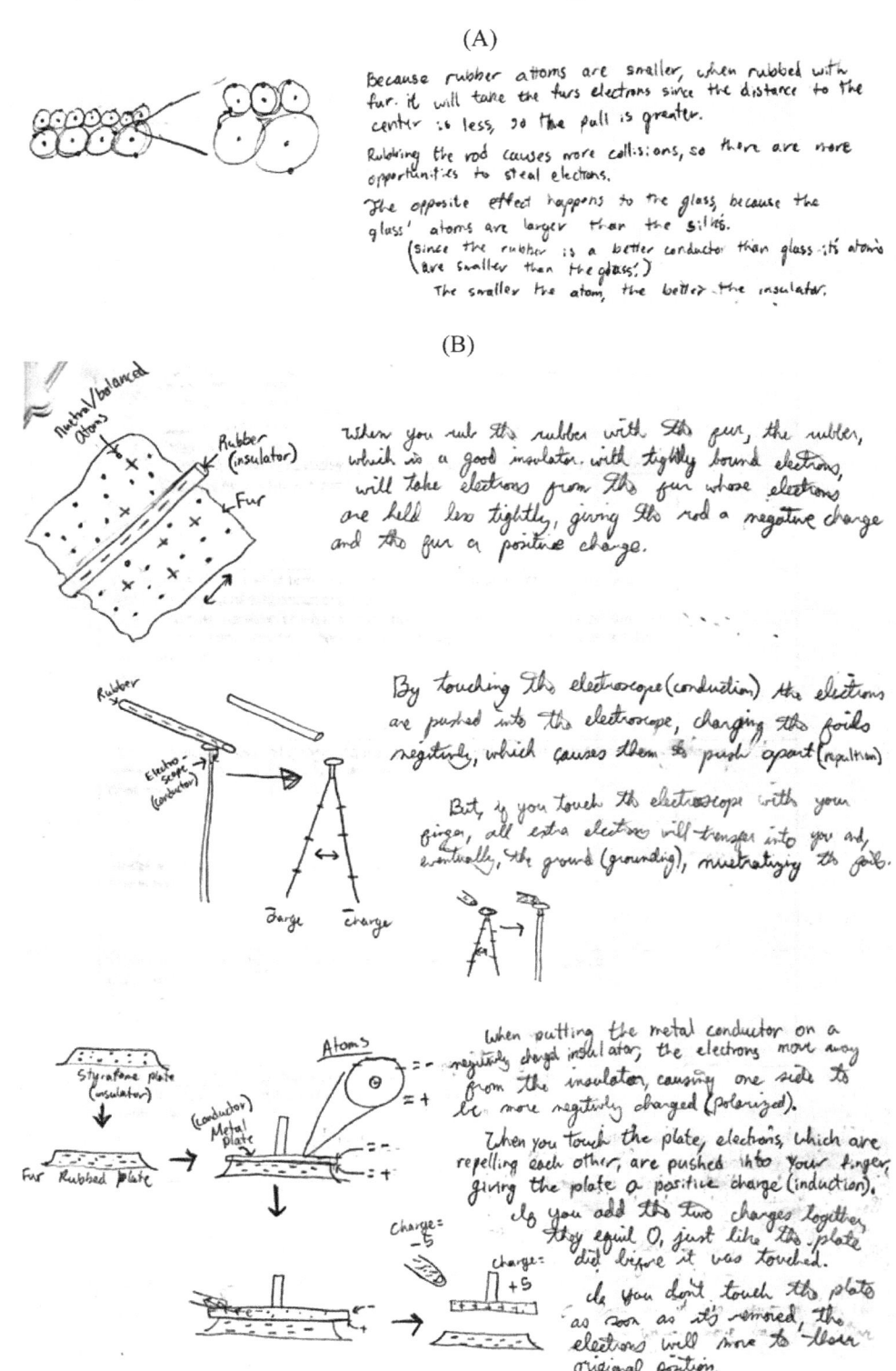

As seen previously, the video data examined closely in Mr. Neilson's Electrostatic Energy module did not include evidence of the use of exploratory and evaluative modeling. When we considered the other modules and our additional collaborating experiences throughout the year, however, it was revealed that *exploratory modeling* was applied as well in Mr. Neilson's physics classrooms. For example, while working with the Forces and Rockets module, students used mini-rocket models introduced by Mr. Neilson for testing the impact of changes in variables, such as fin placement or the number of fins on a rocket. In addition, in teaching about centripetal force, Mr. Neilson introduced a model airplane tied to a string and connected to a force probe to allow students to explore several properties of the teacher-created model and see how changes to the model influenced these properties. Among many possible benefits resulting from this exploratory modeling, we could identify three salient ones: 1) focusing student attention on phenomena that can enhance their conceptual understanding (as another example of exploratory modeling, students' manipulation of surface area exposure on an item pulled across a plane allowed them to collect data informing the idea that surface area does not change the horizontal friction acting on an object), 2) providing an opportunity for students to realize the reliance of science on empirical data for making conclusions and developing explanations, and 3) helping students better understand how to shape investigations so that they can gather evidence supportive of conclusions and explanations.

When we look further into whether *evaluative modeling* was used in the other modules implemented throughout the year, a similar pattern as in the Electrostatic Energy module was found: evaluating models was generally connected to experimental modeling that played a more central role in Mr. Neilson's classrooms. In other words, the model evaluation did occur evidently as students were engaged in investigations to determine how data fit with their current models. Nevertheless, little time was devoted to students assessing multiple alternative models or selecting between competing models either presented by the teacher or developed by their peers.

Positively, however, when we discussed future ideas for improving the MBI modules, Mr. Neilson was able to envision how evaluating modeling could enhance student experiences of scientific practices by having them evaluate others' models and provide feedback in a form of peer review. In fact, we identified a number of ways that evaluative modeling could manifest itself in Mr. Neilson's MBI classrooms. For example, evaluative modeling can be applied in isolation, so that it can be included as a part of a regular science lesson without distorting the pre-established instructional structure. It can also be integrated with other modeling pedagogies in a frame that scaffolds a progressive sequence of learning science with modeling. Figure 3 is an instance of such instructional frames that we found useful when discussing reorganizations of Mr. Neilson's MBI modules.

The instructional frame in Figure 3 involves four modeling pedagogies as basic components with cyclic modeling integrating them into its own scheme. A MBI lesson may proceed by following the sequentially organized modeling pedagogies. First, in *expressive modeling*, students express their ideas about a topic by creating models. Their expressive modeling should be preceded by enough opportunity for exploring scientific phenomena and finding some aspects to be explained. Second, through *experimental modeling*, students gather empirical data to validate and improve their models. If a laboratory experiment can hardly be performed, as it is often the case in earth scientific and evolutionary biological inquiry, evidence may be collected from extensive observations of natural objects and phenomena. Third, *evaluative modeling* provides students with an opportunity to compare their models with those of peers and exchange constructive critique with one another. In doing this, they may select models, which they think most appropriate or modify their own models to better explain the target phenomena. Fourth, in *exploratory modeling*, the teacher introduces a scientific model and operates it to demonstrate sophisticated understanding of the topic. Students also have a chance to manipulate the model so that they can grasp scientific ideas represented by the model. Fifth and finally, *cyclic modeling* is realized by students being allowed to modify their models on several occasions throughout the instructional process. For example, if students encounter evidence conflicting with their ideas while performing an experiment or communicating with peers, they will probably want to change their models. In addition, after the class explores the scientific model introduced by the teacher, additional chances can be offered for students to revise their models so as to align them with canonical scientific understanding.

The instructional frame explained is just one example, so many different types of combinations of the five modeling pedagogies can emerge and be found effective as a result of ongoing efforts of science teachers and science teacher educators. We thus consider our collaboration with Mr. Neilson as evolving and iterative, and therefore expect a gradual but significant improvement in teaching and learning in Mr. Neilson's science classrooms by incorporating all the five modeling pedagogies in various ways into his MBI modules.

CONCLUSION AND FUTURE DIRECTIONS WITH THE MODELING PEDAGOGIES

It is commonly recognized in the science education community that modeling is a significant part of science and should also be applied to students learning of science in schools. This sheds light on the importance of understanding ways scientific modeling can be translated into classroom practices. In line with this notion, we have reified five modeling pedagogies using MBI modules developed and implemented through collaborations between science education researchers and a high school physics teacher. The five modeling pedagogies explicated in this chapter can be used as frameworks for science teachers to select and organize student activities in ways that are consistent with epistemic practices and cognitive aims of scientists and consequently, recent reform in science education.

Considering that many teachers have limited perceptions regarding the nature of models and modeling and have difficulties in implementing model-based science instruction, Mr. Neilson can be thought of as an exception. As acknowledged previously, however, his teaching practices are not described as 'ideal,' but rather as 'lenses' through which teachers of science are able to look closely at their own teaching practices. Thus, it is hoped that the ideas shared in this chapter can serve to stimulate new insights and possibilities to improve science instruction and make it better reflect authentic practices of science.

As committed science educators and science education researchers striving to create meaningful inquiry learning experiences for students, we are eager to continue to learn from other teachers and researchers about scientific models and

Figure 3. An example instructional frame with the five modeling pedagogies

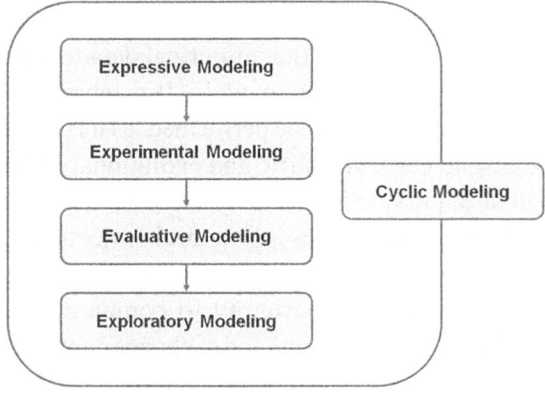

modeling. It is expected that additional theoretical and practical conceptualizations of models and modeling will emerge and help continually refine and further extend our modeling pedagogies. With respect to considering future directions with the modeling pedagogies, it is already recognized that modeling is strongly associated with explanation and argumentation—two important discursive practices of science (Böttcher & Meisert, 2011; Passmore & Svoboda, 2012). Explanation and argumentation, while inextricably linked, are distinguished for important reasons that warrant consideration of how modeling pedagogies can enhance teachers' abilities to foster both of these discursive practices within the science classroom.

Osborne and Patterson (2011) addressed the indistinguishable or conflated uses of explanation and argumentation in science education literature. Through this review, they showed lack of attention to the needed distinction between explanation and argumentation left open the possibility that neither an epistemic understanding of, nor ability in creating explanations or arguments, would be achieved. They argued that before students were able to understand and create explanations or arguments, teachers, curriculum developers, and assessment specialists, among others, should understand clearly the differences between the two and the objectives of each discursive act, as well as the central practices that can foster them. They began to make this distinction by describing explanations as attempts to account for a given phenomenon and arguments as examinations of the validity of explanations based on the question of whether they succeed in explaining a phenomenon in comparison to competing accounts. Along with a rational consistent with that of Osborne and Patterson (2011), Braaten and Windschitl (2011) also expressed a similar concern about the distinction between explanation and argumentation, by saying, "[A]ttention and guidance for scientific argumentation and for scientific explanation can be developed by analyzing the features of both and identifying what features of thinking and discourse need to be scaffolded for each" (p. 658). Therefore, to better explore future directions of the modeling pedagogies in the context of the two important discursive practices, it is prudent to clarify how they differ from each other and how each can be leveraged by different modeling pedagogies.

Braaten and Windschitl (2011), grounded in ideas about scientific explanations derived from the literature of the philosophy of science, provided the following conceptualization of explanation:

We explicitly defined for ourselves and for our novice teachers what attributes characterized "good" scientific explanations ... delineating a simplified continuum of scientific explanations that (1) employ major scientific theories, (2) seek underlying theoretical causes for observable events in nature, and (3) when appropriate, utilize mathematical models to describe patterns in data (p. 661).

This condensed description of the attributes of "good" scientific explanations offers useful insight to set the stage for the promise of modeling pedagogies in this discursive practice. As an example highlighted in the analysis of discourse from Mr. Neilson's classrooms during the Electrostatic Energy module, the teacher used expressive modeling as a pedagogical means to help students develop scientific explanations using their models. Additionally, as an important part of the expressive modeling pedagogy, Mr. Neilson ensured that students based their models on the canonical or normative knowledge of science by guiding them to consider what they shared about atoms and electrons as they tried to explain their observations of static electricity phenomena. That is, in the expressive modeling enacted in Mr. Neilson's science classrooms, the teacher was already found cultivating students' 1) use of major scientific theories and 2) reliance on underlying theoretical causes for observable events in nature—the first two characteristics of "good" scientific explanations articulated by Braaten and

Windschitl (2011). We thus see our modeling pedagogies, such as expressive modeling, as current and significant future tools that can support "the features of thinking and discourse that needs to be scaffolded" (p. 658) for scientific explanation.

On the other hand, argumentations are distinguished from explanations as presented in the following:

[T]here are two discursive entities: the explanation that attempts to account for the given phenomenon, and an argument that examines the question of whether the explanation is valid—that is, whether it succeeds in generating understanding, and whether it is better than competing accounts. The focus of any argument around an explanation, therefore, is on the claim that Explanation A is in a satisfactory/unsatisfactory explanation, or that Explanation A is a better explanation than Explanation B – and not the explanation itself (Osborne & Patterson, 2011, p. 629).

When the modeling pedagogies are considered in the context of argumentation as an essential scientific practice and student learning target (NRC, 2011), evaluative modeling seems best suited for supporting teachers in cultivating students' capabilities with scientific argumentation. Again, evaluative modeling is defined as a learning practice in which students compare alternative models addressing the same phenomenon or problem, assess their merits and limitations, and select the most appropriate one(s) to explain the phenomenon or solve the problem. It affords a context where multiple alternative explanations, in the form of either historical models or student expressed models, are compared based on the extent to which they provide satisfactory explanations of certain phenomena. Therefore, if evaluative modeling is carefully sequenced with other types of modeling pedagogies, key features of scientific argumentation can be developed and understood. For this reason, we place one important future focus of our collaborative project on investigating science classroom discourse inherent in the modeling pedagogies.

Mark Windschitl (2012), in a recent lecture, proposed that modeling has the potential to induce other prioritized practices in science, such as asking questions, planning and carrying out investigations, analyzing and interpreting data, and obtaining, evaluating, and communicating information (NRC, 2011). Likewise, beyond facilitating discursive practices of science, the modeling pedagogies are seen as capable of helping science teachers and science education researchers achieve other meaningful outcomes of science teaching and learning. For example, studies already exist exploring teachers' and students' criteria for evaluating models as an epistemic goal of science education (e.g., Nelson & Davis, 2012; Pluta, Chinn, & Duncan, 2011). In these studies, modeling pedagogies similar to the ones identified in this chapter were used and provided a context in which the researchers were enabled to reveal their participants' understanding of the nature of models in situ. Therefore, the five modeling pedagogies reified in this chapter are believed to further support wide avenues of next generation science education by continually serving teachers as heuristic tools, as well as students as learning anchors.

REFERENCES

Akerson, V. L., Townsend, J. S., Donnelly, L. A., Hanson, D. L., Tira, P., & White, O. (2009). Scientific modeling for inquiring teachers network (SMIT'N): The influence on elementary teachers' views of nature of science, inquiry, and modeling. *Journal of Science Teacher Education, 20,* 21–40. doi:10.1007/s10972-008-9116-5

Böttcher, F., & Meisert, A. (2011). Argumentation in science education: A model-based framework. *Science & Education, 20,* 103–140. doi:10.1007/s11191-010-9304-5

Braaten, M., & Windschitl, M. (2011). Working toward a stronger conceptualization of scientific explanation for science education. *Science Education*, 95, 639–669. doi:10.1002/sce.20449

Buckley, C. J., & Boulter, C. J. (2000). Investigating the role of representations and expressed models in building mental models. In Gilbert, J. K., & Boulter, C. J. (Eds.), *Developing Models in Science Education* (pp. 119–135). Dordrecht, The Netherlands: Kluwer Academic Publishers. doi:10.1007/978-94-010-0876-1_6

Campbell, T., & Neilson, D. (2009). Student ideas and inquiries: Investigating friction in the physics classroom. *Science Activities*, 46(1), 13–16. doi:10.3200/SATS.46.1.13-16

Campbell, T., & Neilson, D. (2012). Modeling electricity: Model-based inquiry with demonstrations and investigations. *The Physics Teacher*, 50(6). doi:10.1119/1.4745686

Campbell, T., Zhang, D., & Neilson, D. (2010). Model based inquiry in the high school physics classroom: An exploratory study of implementation and outcomes. *Journal of Science Education and Technology*, 20(3), 258–269. doi:10.1007/s10956-010-9251-6

Clement, J. J. (2008). *Creative model construction in scientists and students: The role of imagery, analogy, and mental simulation*. Dordrecht, The Netherlands: Springer. doi:10.1007/978-1-4020-6712-9

Crawford, B. A., & Cullin, M. J. (2004). Supporting prospective teachers' conceptions of modelling in science. *International Journal of Science Education*, 26(11), 1379–1401. doi:10.1080/0950069410001673775

Duschl, R., Schweingruber, H., & Shouse, A. (Eds.). (2007). *Taking science to school: Learning and teaching science in grades K-8*. Washington, DC: National Academy Press.

Giere, R. N. (1999). *Science without laws*. Chicago, IL: The University of Chicago Press.

Giere, R. N., Bickle, J., & Mauldin, R. F. (Eds.). (2006). *Understanding scientific reasoning* (5th ed.). Toronto, Canada: Thomson Wadsworth.

Gilbert, J. K., Boulter, C., & Rutherford, M. (1998). Models in explanations, part 1: Horses for courses? *International Journal of Science Education*, 20(1), 83–97. doi:10.1080/0950069980200106

Gilbert, J. K., & Boulter, C. J. (Eds.). (2000). *Developing models in science education*. Dordrecht, The Netherlands: Kluwer Academic Publishers. doi:10.1007/978-94-010-0876-1

Gilbert, S. W., & Ireton, S. W. (2003). *Understanding models in earth and space science*. Arlington, VA: NSTA Press.

Gobert, J. D. (2005). The effects of different learning tasks on model-building in plate tectonics: Diagramming versus explaining. *Journal of Geoscience Education*, 53(4), 444–455.

Gobert, J. D., & Clement, J. J. (1999). Effect of student-generated diagram versus student-generated summaries on conceptual understanding of causal and dynamic knowledge in plate tectonics. *Journal of Research in Science Teaching*, 26(1), 39–53. doi:10.1002/(SICI)1098-2736(199901)36:1<39::AID-TEA4>3.0.CO;2-I

Gobert, J. D., & Pallant, A. (2004). Fostering students' epistemologies of models via authentic model-based tasks. *Journal of Science Education and Technology*, 13(1), 7–22. doi:10.1023/B:JOST.0000019635.70068.6f

Halloun, I. A. (2004). *Modeling theory in science education*. Dordrecht, The Netherlands: Kluwer Academic Publishers.

Johnston, A. (2008). Demythologizing or dehumanizing? A response to Settlage and the ideals of open inquiry. *Journal of Science Teacher Education, 19,* 11–13. doi:10.1007/s10972-007-9079-y

Justi, R., & Gilbert, J. K. (1999). History and philosophy of science through models: The case of chemical kinetics. *Science & Education, 8,* 287–307. doi:10.1023/A:1008645714002

Justi, R., & Gilbert, J. K. (2002a). Modelling, teachers' views on the nature of modelling, and implications for the education of modellers. *International Journal of Science Education, 24*(4), 369–387. doi:10.1080/09500690110110142

Justi, R., & Gilbert, J. K. (2002b). Science teachers' knowledge about and attitudes towards the use of models and modelling in learning science. *International Journal of Science Education, 24*(12), 1273–1292. doi:10.1080/09500690210163198

Justi, R., & Gilbert, J. K. (2003). Teachers' views on the nature of models. *International Journal of Science Education, 25*(11), 1369–1386. doi:10.1080/0950069032000070324

Khan, S. (2007). Model-based inquiries in chemistry. *Science Education, 91,* 877–905. doi:10.1002/sce.20226

Khan, S. (2011). What's missing in model-based teaching. *Journal of Science Teacher Education, 22,* 535–560. doi:10.1007/s10972-011-9248-x

Louca, L. T., Zacharia, Z. T., & Constantinou, C. P. (2011). In quest of productive modeling-based learning discourse in elementary school science. *Journal of Research in Science Teaching, 48*(8), 919–951. doi:10.1002/tea.20435

Maia, P. F., & Justi, R. (2009). Learning of chemical equilibrium through modelling-based teaching. *International Journal of Science Education, 31*(5), 603–630. doi:10.1080/09500690802538045

Morrison, M., & Morgan, M. S. (1999). Models as mediating instruments. In Morgan, M. S., & Morrison, M. (Eds.), *Models as Mediators: Perspectives on Natural and Social Science* (pp. 10–37). Cambridge, UK: Cambridge University Press. doi:10.1017/CBO9780511660108.003

National Research Council. (2011). *A framework for K-12 science education: Practices, crosscutting concepts, and core ideas.* Washington, DC: The National Academies Press.

Neilson, D., Campbell, T., & Allred, B. (2010). Model-based inquiry: A buoyant force module for high school physics classes. *Science Teacher (Normal, Ill.), 77*(8), 38–43.

Nelson, M. M., & Davis, E. A. (2012). Preservice elementary teachers' evaluations of elementary students' scientific models: An aspect of pedagogical content knowledge for scientific modeling. *International Journal of Science Education. 34*(12), 1931-1959. doi:10.1080/09500693.2011.594103

Nersessian, N. J. (1999). Model-based reasoning in conceptual change. In Magnani, L., Nersessian, N. J., & Thagard, P. (Eds.), *Model-Based Reasoning in Scientific Discovery* (pp. 5–22). New York, NY: Kluwer Academic/Plenum Publishers. doi:10.1007/978-1-4615-4813-3_1

Nersessian, N. J. (2008). *Creating scientific concepts.* Cambridge, MA: The MIT Press.

Oh, P. S., & Oh, S. J. (2011). What teachers of science need to know about models: An overview. *International Journal of Science Education, 33*(8), 1109–1130. doi:10.1080/09500693.2010.502191

Osborne, J. F., & Patterson, A. (2011). Scientific argument and explanation: A necessary distinction? *Science Education*, *95*, 627–638. doi:10.1002/sce.20438

Passmore, C., & Stewart, J. (2002). A modeling approach to teaching evolutionary biology in high schools. *Journal of Research in Science Teaching*, *39*(3), 185–204. doi:10.1002/tea.10020

Passmore, C. M., & Svoboda, J. (2012). Exploring opportunities for argumentation in modeling classrooms. *International Journal of Science Education*. Retrieved from http://www.academia.edu/233497/Exploring_Opportunities_for_Argumentation_in_Modelling_Classrooms

Penner, D. E., Lehrer, R., & Schauble, L. (1998). From physical models to biomechanics: A design-based modeling approach. *Journal of the Learning Sciences*, *7*(3&4), 429–449.

Pluta, W. J., Chinn, C. A., & Duncan, R. G. (2011). Learners' epistemic criteria for good scientific models. *Journal of Research in Science Teaching*, *48*(5), 486–511. doi:10.1002/tea.20415

Sandoval, W. A., & Reiser, B. J. (2004). Explanation-driven inquiry: Integrating conceptual and epistemic scaffolds for scientific inquiry. *Science Education*, *88*, 345–372. doi:10.1002/sce.10130

Schwarz, C. V., & Gwekwerere, Y. N. (2007). Using a guided inquiry and modeling instructional framework (EIMA) to support preservice K-8 science teaching. *Science Education*, *91*, 158–186. doi:10.1002/sce.20177

Shen, J., & Confrey, J. (2007). From conceptual change to transformative modeling: A case study of an elementary teacher in learning astronomy. *Science Education*, *91*(6), 948–966. doi:10.1002/sce.20224

Smit, J. J. A., & Finegold, M. (1995). Models in physics: Perceptions held by final-year prospective physical science teachers studying at South African universities. *International Journal of Science Education*, *17*(5), 621–634. doi:10.1080/0950069950170506

Taylor, I., Barker, M., & Jones, A. (2003). Promoting mental model building in astronomy education. *International Journal of Science Education*, *25*(10), 1205–1225. doi:10.1080/0950069022000017270a

Urhahne, D., Schanze, S., Bell, T., Mansfield, A., & Homes, J. (2010). Role of the teacher in computer-supported collaborative inquiry learning. *International Journal of Science Education*, *32*(2), 221–243. doi:10.1080/09500690802516967

van Driel, J. H., & Verloop, N. (1999). Teachers' knowledge of models and modelling in science. *International Journal of Science Education*, *21*(11), 1141–1153. doi:10.1080/095006999290110

van Driel, J. H., & Verloop, N. (2002). Experienced teachers' knowledge of teaching and learning of models and modelling in science education. *International Journal of Science Education*, *24*(2), 1255–1272. doi:10.1080/09500690210126711

van Joolingen, W. (2004). *Roles of modeling in inquiry learning*. Paper presented at the IEEE International Conference on Advanced Learning Technologies. Joensuu, Finland.

Windschitl, M. (2012). *Ambitious teaching as the "new normal" in American science classrooms: How will we prepare the next generation of professional educators?* Paper presented at Penn State University. State College, PA.

Windschitl, M., & Thompson, J. (2006). Transcending simple forms of school science investigation: The impact of preservice instruction on teachers' understanding of model-based inquiry. *American Educational Research Journal, 43*(4), 783–835. doi:10.3102/00028312043004783

Windschitl, M., Thompson, J., & Braaten, M. (2008a). How novice science teachers appropriate epistemic discourses around model-based inquiry for use in classrooms. *Cognition and Instruction, 26*, 310–378. doi:10.1080/07370000802177193

Windschitl, M., Thompson, J., & Braaten, M. (2008b). Beyond the scientific method: Model-based inquiry as a new paradigm of preference for school science investigations. *Science Education, 92*, 941–967. doi:10.1002/sce.20259

Chapter 7
Why *Immersive, Interactive* Simulation Belongs in the Pedagogical Toolkit of "Next Generation" Science:
Facilitating Student Understanding of Complex Causal Dynamics

M. Shane Tutwiler
Harvard University, USA

Tina Grotzer
Harvard University, USA

ABSTRACT

Demonstration and simulation have long been integral parts of science education. These pedagogical tools are especially helpful when trying to make salient unseen or complex causal interactions, for example during a chemical titration. Understanding of complex causal mechanisms plays a critical role in science education (e.g. Grotzer & Basca, 2003; Hmelo-Silver, Marathe, & Liu, 2007; Wilensky & Resnick, 1999), but few curricula have been developed to expressly address this need (e.g. Harvard Project Zero, 2010). Innovative education technologies have allowed content designers to develop simulations that are both immersive and engaging, and which allow students to explore complex causal relationships even more deeply. In this chapter, the authors highlight various technologies that can be used to leverage complex causal understanding. Drawing upon research from both cognitive science and science education, they outline how each is designed to support student causal learning and suggest a curricular framework in which such learning technologies might optimally be used.

DOI: 10.4018/978-1-4666-2809-0.ch007

INTRODUCTION

"I just don't think you're thinking about all of the possible factors" the young engineer sighed as she opened a new word processing documents to take note of what her colleague was saying. Inez and Omar, partners on a city-planning project for their local municipality, were carefully reviewing their proposed development plans. Using advanced simulation software, each had independently modeled the effect of increased industrialization on pollution levels over time, and then predicted how these events might lead to further changes in their town's economy and demographics. They didn't exactly agree on the projected outcomes.

"It's clear as day!" Omar replied. "There was a link between number of factories and crime in my ten-year model. You can't deny that! I can make predictions about future trends based on that." Shaking her head, Inez calmly replied "I just don't think it's that simple. You're saying it's a direct connection, when I don't think it is. Look, my model looked at one hundred years-worth of change, and crime values went up and down, even though the number of factories in the city was the same. I don't know...it's just not simple." In time, the two decided on a complex relationship relating industrialization to crime and population over time, and prepared their findings in a report to the city zoning board.

Later, the two were tasked with helping to gauge the impact of proposed construction in a small suburban community near the city. Residents wanted to build new houses and recreational areas such as a golf course near a popular pond. Using a sophisticated multi-user virtual environment to simulate the area, Inez and Omar, along with two other teammates, discovered a scenario in which many large fish in the pond all died in a short period of time due to an unknown cause. Dividing the work equally among the team, each member set out to collect data over time within the virtual environment. Comparing their notes and constantly revising their causal models, the team finally settled on a plausible scenario and advised the zoning board of possible hazards due to their planned construction.

When asked later to reflect upon these two project, Inez and Omar thought that they learned a lot, and that they helped them to think about problems as being more complex. They also said it was more fun than just sitting in class and listening to their teacher. Inez and Omar, you see, were eighth-grade students and had been using these simulations as part of problem-based lessons in their science class throughout the year.

The vignette above is an example of how science students can interact with virtual simulations in meaningful and contextual ways. In these examples, students worked both independently and together to collect, synthesize, and analyze data, as well as to develop and test hypotheses. Moreover, they took part in these endeavors in a curricular setting that embedded the simulations within meaningful and motivating contexts relevant to the subject matter and students. In this chapter, we outline research supporting this framing and give examples of commercial and research-driven technologies that can be used as such.

Demonstration and simulation have long been integral parts of science education. These pedagogical tools are especially helpful when trying to make salient unseen or complex interactions, for example during a chemical titration. Demonstrations allow the teacher to draw students' attention to variables that otherwise might be missed and to guide students towards the explanatory narratives of a concept. Simulations, for instance, Molecular Workbench[1] (Concord Consortium, 2004) or the PhET Simulations[2] (e.g. Wieman, Adams, & Perkins, 2008), offer models of processes that cannot be directly observed—only inferred. These thoughtfully illustrated computer simulations are designed to illuminate particular processes such as conduction, osmosis, Ohm's Law, and so forth. The combination of simulation and demonstration enables students to envision these processes and gain from the expertise of an instructor in understanding them.

For example, virtual simulations are fast becoming a staple of medical education, due to the rapid rate of technological advances and growing patient anxiety over experimental procedures. Okuda, Bryson, DeMaria, Jacobson, Ingram-Goble, Zuiker, and Warren (2009) found that studies of the difference between simulation-based and traditional medical education showed gains in procedural performance and adherence to protocols, increased medical knowledge, and increased re-test performance for subjects in the simulation groups. In a similar review of studies of such simulations in medical classrooms, Issenberg, Mcgaghie, Petrusa, Gordon, and Scalese (2005) determined that the best simulations provide feedback and the chance for repetition, were well integrated into the curriculum, offered a range of difficulties, and were valid representations of the scenario being simulated. Taken together, these reviews point toward a strong positive impact of embedding simulations in medical education.

Should we expect these results to transfer to general science learners, however? After all, medical school students are pre-screened based on their content knowledge and other traits such as flexible thinking and ability to handle stress. They are, in effect, much closer to being experts than novices in their domain than the general science learner. Do these simulations have equally beneficial effects on these types of learners? Research across domains and age groups seems to suggest so.

Jimoyiannis and Komis (2001), for example, found that students using physics-based simulations showed better understanding of tasks related to velocity and acceleration when compared to a control group in a traditional lab setting.

Furthermore, Zacharia (2006) demonstrated that students in virtual physics experiment groups showed higher overall learning gains than those in a traditional experiment group. Rutten, van Joalingen, and van der Veen (2012) also found that simulations could have an impact on learning, especially in lab-based scenario training, as long as the work was properly embedded in the curriculum and valid. Looking at learning in earth science, Barab, Scott, Siyahhan, Goldstone, Ingram-Goble, Zuiker, and Warren (2009) found that students in a contextualized virtual environment based simulation showed higher gains than students using textbook-based lessons; while Koray (2011) found that students using problem-based-learning and virtual simulations fared better on post-intervention achievement measures than peers in a didactic instruction group.

These findings indicate that there are identifiable benefits to supplementing instruction with virtual simulations in various science domains. However, the framing of these simulations is critically important. From the above examples, Barab and colleagues (2009) found that the contextualized simulations were most impactful and Koray (2011) found gains when the simulations were specifically part of a problem-based-learning curriculum. In addition, Blake and Scanlon (2007) indicate that simulations such as these should contain student support, give multiple representations of concepts, and should be tailored to the learning needs of the students using them. Similarly, within the domain of chemistry, Kahn (2010) found that simulations were most effective when used in conjunction with a sound curricular development process.

Structuring of the problem space is also an important factor to consider when using virtual simulations. Shin, Jonassen, and McGee (2003) explored the types of knowledge, skills, and attitudes at play when students were presented with well-structured and ill-structured astronomy simulation tasks. When the problem was well structured, student domain knowledge and justification skills were significantly related to learning outcomes. However, students using simulations in an ill-structured problem space showed significant relationships between domain knowledge, justification skills, science attitudes, and cognitive regulation. In effect, students in messy data-spaces had to use more cognitive resources to learn and they had to call upon varied cognitive resources than students working in routine or highly structured problem spaces.

Another factor to consider when implementing virtual simulations into science classrooms is how the students themselves will interact with the technology, and whether or not that interaction might favor one type of student over another. Lin, Tutwiler, and Chang (2011) found that females in their study reported using virtual environments in the past differently than their male peers. The interaction effect between these two factors, gender and prior use, had a positive and significant relationship with post-intervention learning when controlling for pre-test scores; in other words, girls who reported playing more video games in the past outscored their female peers who reported playing fewer video games, as well as their male peers in the frequent past-play group (Lin, et al., 2011). These results were mirrored in another study, in which boys were able to complete a simulated escape from a burning building faster than girls. The time of that trial was then shown to be a significant predictor of performance on a knowledge test related to lessons learned in an unrelated virtual environment (Lin, Tutwiler, & Chang, 2012). In effect, student comfort of navigation within the simulation was gender-biased, and that ability to navigate was related to learning gains in other virtual environments. It is important to note that these studies were conducted in Taiwan, where student use of virtual environments outside of the classroom (i.e. video games) may differ from students of other cultures, as may the class environment and student motivation to perform well on assessments (such as those given during the course of the research). That being said, the trend points towards the need for careful planning and implementation in all settings where students may interact with virtual worlds.

In sum, these findings point toward a positive relationship between the use of virtual simulations and science learning. It is important to remember, however, that these positive impacts come with myriad qualifiers. The simulations must be appropriately integrated (Issenberg, et al., 2005; Kahn, 2010; Rutten, et al., 2012), contextualized (Barab, et al., 2009), provide maximum benefit when part of a problem-based lesson (Koray, 2011), implemented with student prior use of technology in mind (Lin, et al., 2011), and are often most strongly associated with performance-based outcomes (Rutten, et al., 2012; Okuda, et al., 2009).

Simulations of this type combined with demonstration are important tools in science education and will continue to play a role in engendering deep understanding in science. However, in this chapter, we argue that *immersive* and *interactive* simulations are critical to the pedagogical toolkit of "next generation" science education and that they can play a key role in helping students learn the causal dynamics of a complex world. Increasingly, scientists need to be able to reason well about complex systems dynamics—extending beyond direct, spatially, and locally immediate one to one correspondences. These dynamics may entail emergence, time delays and spatial gaps, non-linearities, extended indirect patterns, and so forth (Grotzer, 2004). In the paragraphs to follow, we make the case that simulations that have *interactive* and *immersive* qualities invite consideration of complex causal dynamics and the opportunity to reveal and revise our causal assumptions. We offer an illustration from a commercially available computer game of how simulations become more than computer games with the right set of accompanying pedagogical moves.

Research shows that without opportunities to learn the embedded causalities in scientific phenomena, students do not progress towards more expert conceptions (Hmelo-Silver, Marathe, & Liu, 2007; Perkins & Grotzer, 2005). They are likely to default to simpler patterns (Driver, Guesne, & Tiberghien, 1985; Grotzer, 2004; Raia, 2008) and to miss the inherent dynamics driving the broader systems outcomes. Further, people of all ages are typically unaware of their causal default assumptions or the role that they play in distorting scientific concepts and expert grasp of dynamic systems (e.g. Dorner, 1989; Grotzer & Lincoln, 2007). Many of the greatest ecological disasters are instantiations of this phenomenon (Grotzer, 2012; Walker & Salt, 2006).

A growing body of research calls for creating opportunities and the accompanying pedagogies for students to learn the embedded causal complexities involved in understanding the world as a systematic, integrated, and global entity. This work acknowledges that teaching reductionist, clockwork approaches to complex organic and emergent phenomenon (Jacobson, 2001) is a dangerous precedent and that we need new approaches and pedagogies to make this shift (Grotzer, Dede, Metcalf, & Clarke, 2009; Hmelo-Silver & Azevedo, 2006). Research has also suggested that new pedagogies can build upon students' existing knowledge from their experiences in the world to formalize and expand these understandings. For instance, Levy and Wilensky (2008) studied sixth grade students' understanding of scattering behavior in gym class in an exploration towards its use in science class.

Innovative education technologies have the potential to enable simulations that invite complex systems reasoning. For instance, NetLogo[3] and StarLogo[4] both offer ways to experiment with organic emergent causalities, to explore outcomes on the basis of individual agents or population levels, and to contrast the two. Students program a set of turtles that interact in parallel giving rise to population outcomes. The outcomes are often very surprising and difficult to imagine based upon the simple rules programmed at the individual level. Danish, Peppler, Phelps, and Washington (2011) introduced a wearable simulation called BeeSim to help kindergarten students learn about the systematic, decentralized behavior in a bee hive connecting their own actions (as a bee) to the individual and hive outcomes. The simulation attempts to address one aspect of what students have struggled with in thinking about decentralized causality—reasoning in different levels (Wilensky & Resnick, 1999) between individuals and populations.

Virtual Environments (VEs) extend these experiences further. Dede, Grotzer, and colleagues (Grotzer, Dede, Metcalf, & Clarke, 2009; Metcalf, Kamarainen, Tutwiler, Grotzer, & Dede, 2011) have advanced the idea that VEs can invite students to grapple with complex causal patterns in powerful ways by offering real world vignettes that are problem-based. VEs, both those designed specifically for use in classrooms and those designed for commercial use but adopted into a specific curriculum, have been used for a multitude of educational purposes (Dieterle & Clarke, 2007; Squire, 2003). As in the examples above, VEs are highly interactive and enable students to use their own actions to learn about the systems dynamics in play. However, through their virtue of being both immersive and interactive (Dede, 2009), they invite students to play out their causal default assumptions as they might in the real world, to experience the consequences, and to revise their assumptions towards more effective outcomes. They have the potential to address what causal relationships students *perceive*, *attend to*, and *reason* about in the virtual world, and this as we elaborate below, is a significant step towards better complex causal reasoning *and* action in the next generation.

Specific affordances can be built into VEs, such as the ability to zoom in and out to view the system at various levels, move backwards and forwards in time, and to receive graphical feedback about system changes makes VEs an ideal platform for helping students to understand the underlying structure of complex causal systems (Grotzer, et al., 2009; Metcalf, et al., 2011). Below, we illustrate how immersive simulations can reveal, leverage, and build complex causal understanding through exploring a specific example. Through this example, we build the case for immersion and interactivity and consider how teacher guidance can facilitate students' productive engagement towards the goals of learning the complex causal dynamics.

In the following paragraphs, we describe how one commercially produced VE, SimCity™ 4, can be used to frame and support the learning of more expert complex causal patterns in middle- to high-school students. The following sections describe SimCity™ 4 and the initial preparation that enables the students to engage in the vignette as follows, and how the simulation invites students to uncover various complex causal patterns. We unpack the specific complex causal patterns underlying the observed outcomes and detail points during which the teacher can provide support. Finally, we address some shortcomings inherent to the use of SimCity™ 4, and discuss the development of a multi-user virtual environment that more fully leverages the possibility of immersion and interactivity to enable learning of sophisticated causal dynamics.

AN OVERVIEW OF SIMCITY™ 4

Imagine that you live in a town of 30,000 people, and the population has been in a steady decline over the last decade, a scenario not uncommon in some mid-western cities. One proposed plan to reverse this trend and revitalize the community is to increase the industrial profile of the municipality, in the hopes of luring families to work in the new factories, thus bringing tax revenues back to previous levels, allowing your city to sustain services such as parks and recreations. The relationship seems very straightforward: more jobs, means more people, which equals more money for the city.

Unfortunately, things do not exactly work out as the town administration had planned. Within five years, housing prices have dropped, people are leaving the city again, and satisfaction with the administration is at an all-time low. Why did the simple model not predict these trends? What causal relationships actually led to these negative outcomes?

This scenario is a common example of the over-simplification of a complex causal system that has its parallels in abundant planning and ecological examples (See for instance, Dorner, 1989; Grotzer, 2012; Tenner, 1996). Interestingly, it also has parallels in the learning of many science concepts, such as air pressure, density, geology, or the motion of electrons through a circuit (Driver, Guesne, & Tiberghien, 1985; Feltovich, Spiro, & Coulson, 1993; Grotzer, 2004; Raia, 2008) given the common reasoning patterns that give rise to them. For instance, students typically ascribe a simple linear model to sinking and floating, by attributing the outcome solely to the weight of the object instead of realizing that the relationship between the density of the liquid and the density of the object account for the outcome (Perkins & Grotzer, 2005). Similarly, they reason that if all the green plants were to disappear, it would impact the primary consumers but not the secondary consumers (Grotzer, 1993; White, 1997)

However, this scenario is actually from a commercial computer game, SimCity™ 4. Depending upon how the game is used, it has the potential to reveal students' initial default assumptions and to model more complex causal dynamics—ones that students might at first think are simple.

Released in 2003 by software company Maxis, developers of other popular simulation games such as The Sims franchise, SimCity 4 (henceforth SimCity) is the most recent incarnation of the popular SimCity series that began in 1989 (Electronic Arts, 2011). In SimCity, players are able to plan and build cities by zoning areas as residential, commercial, or industrial. As their city's population grows, the player, acting as mayor, must decide how to allocate resources and upgrade existing infrastructures to induce growth, or to keep population at a given level, all while maintaining positive reviews from the virtual denizens of the city, the Sims.

The game itself was designed purely for entertainment purposes, and as such is not intended to be a true urban planning simulator or a means to teach complex causal patterns. Devisch (2008) outlined multiple ways in which the assumptions underlying the game's design do not make it a high-fidelity simulation. However, the underlying causal patterns *are* present and identifiable, and represent a scaled-down version of the complex interactions that underlie the social and environmental spheres of urban growth. Here, we suggest ways in which SimCity can be initially staged and used to offer a virtual simulation experience, that engages students with the underlying complex causal structures as well as how teachers can support student learning thereof.

THE INTERACTIVE SIMULATION: INDUSTRY OVERLOAD

Overview of a Productive Learning Path and Comparisons to Explore

Our interactive simulation in SimCity™ 4 Deluxe centers on the effects of a rise in industrial zoning on pollution levels, city population, and mayoral popularity within SimCity.[5] The Large City training scenario positions the student (or students, if playing in groups) as the newly appointed mayor of the demonstration city. In an effort to lure new people to the city, they have been tasked by the city council with building new industrial zones. The hope is that new industrial jobs will entice people to move to the city.

In order to set students up to notice the causal dynamics, teachers can engage them in some specific population predicting exercises. Students would pause the game and designate all unzoned areas on the map as industrial zones (see Figure 1). Then, they would make predictions about how this will affect the city's population in one, five, and ten year intervals, as well as explain the mechanism as to why population will change as they predict. They might, for example, draw a map of the city and show where they think population will change the most, and where air and water pollution changes might occur.

From here, the students can set the game to maximum speed ("cheetah mode") and observe the city's population (and population density), pollution levels (and location), and mayoral popularity via the embedded graphical data displays (see Figures 2-4). Over the course of ten in-game years, students should pause the game at six-month intervals and make observations of each of those values, and then assess whether these trends matched their predictions. If not, they would note points of deviation, and reflect upon why this might be the case.

Next, the student(s) can reset the scenario and re-zone the industrial areas as before, and let the game play through on cheetah mode, this time taking measurements at yearly intervals for ten years. Do they think the values of their data will be the same as the first ten-year period? Comparing these one-year intervals to those of their first trial, and reflecting upon why they might be different will engage students in thinking about the differences between their zoning decisions in change over time.

Finally, students can reset the scenario and make no changes to the number of industrial zones, and take the same set of yearly measurements over a ten-year in-game period. This offers a counter-factual condition (Rubin, 2005), and

Figure 1. Industrial zone (right) near a residential zone (left) in SimCity

Figure 2. SimCity air pollution data

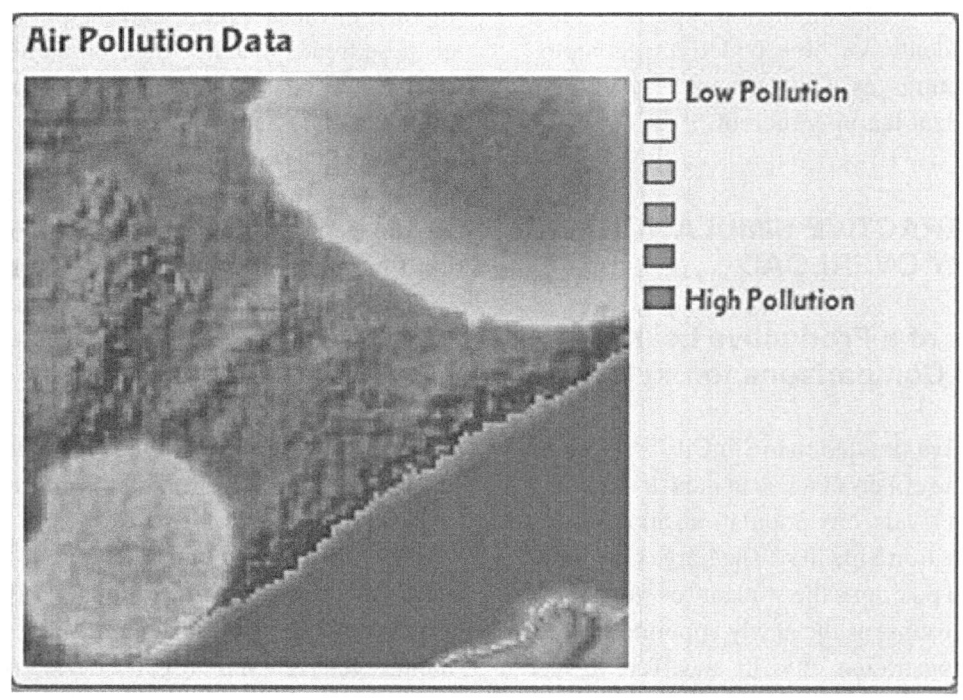

Figure 3. Graph of SimCity population, by age

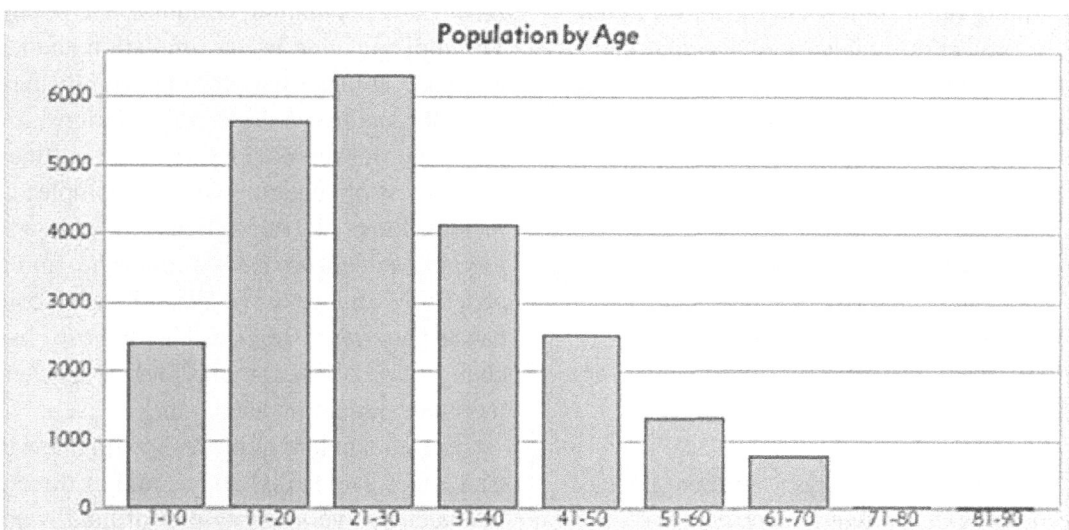

gives an example of what might have happened to the system had the mayor (student) not intervened and re-zoned. We seldom have access to counterfactual conditions in the real world so this is an important affordance that simulations can provide for learning about causal dynamics. How do these values compare to those collected from their first two trials? Why might they be different?

Timely assessment and feedback of student understanding is a critical component of the simulation. In order to facilitate this continuous assessment/feedback cycle, the teacher should ideally circulate amongst students or groups as much as possible, observing and asking students what changes they notice in the system over time. One strategy for helping students to recognize the longitudinal changes, while also assessing their understanding, is through the use of guided questions. For example, a teacher might play the role of "wise fool" and pretend that they don't fully understand the interface. They could ask students to show how population density is displayed, and in doing so, ask follow up questions about the relationship between the density and other factors. Student actions and responses can then be evaluated for understanding of underlying concepts.

In addition to the in-process changes within each trial, it is also important that student comparisons of the differences and similarities between each trial be assessed. These differences in trends form the basis for recognizing specific forms of complex causal relationships, outlined below. One method for conducting such assessment might be to embed it within a situated context (Lave & Wenger, 1991). For example, students can be assigned the role of "city planner," and act as an advisor to the mayor (the teacher). Their task, then, is to use advanced simulation software (SimCity) to evaluate the possible outcomes of the mayor's plan to increase industrialization in order to increase municipal revenue and population. The planners (students) are then tasked with

Figure 4. SimCity mayoral rating

developing lists of pros and cons for the proposal, and presenting these findings to the mayor at the end of the project through written or oral reports (or a combination of both). It is during the course of completing and reflecting upon the assignment, then, that students begin to recognize complex causal patterns, the existence of which should be noted in said reports.

Table 1 summarizes the steps detailed above.

How the Interactive, Immersive Simulation Enables Engagement with Complex Causal Dynamics

Students' simplifying assumptions about the nature of complex causal relationships are revealed in their initial simple and linear predictions about the relationship between increased industrial zoning and population, pollution, and popularity. Through engaging in the simulation and teacher guidance, students have the opportunity to reflect upon the nature of the causal relationships that led to their observed data. Here we outline what the process of engaging with the complex causal features looks like through the lens of how a prototypical student proceeds through the simulation, observing changes. We discuss what complex causal patterns are responsible for those observed changes and offer examples for how teachers can facilitate student understanding of each.

The student begins by designating new industrial zones as detailed above, and as directed by the teacher. Every empty plot of land is quickly filled, and as the student un-pauses the game, she

Table 1. Outline of conditions, steps, and suggested teacher support for using SimCity to model various causal complex patterns

Condition	Steps	Teacher support
Prediction	1. Pause game and designate free zones as "industrial"	
	2. Predict population and crime levels one, five, and ten years into the simulation	Question students about what factors led to their specific predictions
First Trial	1. Reset the scenario and designate all free zones as "industrial"	
	2. Set game speed to "cheetah mode"	
	3. Pause the game at six month intervals and take measurements	Ask students to show how each measurement is made, and what they notice
	4. Assess if values matched predictions	Ask why the value did/did not match
Second Trial	1. Reset the scenario and designate all free zones as "industrial"	
	2. Set game speed to "cheetah mode"	
	3. Pause the game at one year intervals and take measurements	Ask students to show how each measurement is made, and what they notice
	4. Assess if values matched predictions	Ask why the value did/did not match
Counterfactual	1. Reset the scenario but don't rezone any areas	
	2. Set the game speed to "cheetah mode"	
	3. Pause the game at one year intervals and take measurements	Ask students to show how each measurement is made, and what they notice

notes that activity quickly ensues in the affected areas. As time passes in-game, the density of industrial buildings rises, as the zoned areas fill. As expected, the city's population begins to rise.

However, as the student collects the requisite data at the designated annual or semi-annual points, she cycles through a few other graphs and notices that housing prices (measured in SimCity as the "desirability" of given residential plots) are beginning to fluctuate as well. Initially, she notices a spike, as the population begins to grow. However, as the population stabilizes, housing costs begin to fall, even as the density of industrial buildings continues to increase. In essence, it appears to the student that an increase in the number of industrial buildings in one area coincides with decreasing housing costs in a non-adjacent area, over time.

This is an instance of a *spatiotemporally distant* causal pattern. Many causal relationships involve a spatial delay, a temporal delay, or both, between the occurrence of the causal event and the perceived effect. For example, greenhouse gas emission in centrally located continents has contributed to the gradual warming of the arctic region over time. Research has shown, however, that people expect causes to fall in close spatiotemporal proximity to their effects, or vice versa (e.g. Grotzer, 2012).

Seeing that the student is confused, her teacher instructs her to compare the desirability data from her two intervention simulations with her control simulation. The connection is clear: increased industrial activity in one area leads to lower housing prices in another after a period of time.

Having recognized the connection between growth in the industrial sector and decreased desirability of distant homes over time, the student then begins to formulate hypotheses as to why. Her initial hypothesis is quite direct and encompasses only those two variables, degree of industrialization and desirability of homes, without taking into account any other mitigating factors. This relationship has a *simple linear* causal structure (Grotzer, 2004). However, the student quickly realizes that she cannot explain *why* the two areas, quite separate on the map, would have such a relationship.

It is at this point that the teacher can facilitate exploration of more complex dynamics by asking the student to make a list of other variables measured in SimCity, but not initially measured during the three simulations, that might also make home prices drop. Crime, for example, is one of these variables. The student could then re-run a simulation with increased industrial zones (similar to simulations 1 and 2), and note the crime map (see Figure 5) and crime graph (see Figure 6) yearly. The eventual dip in housing prices was preceded by a rise in crime spurred by the spike in industrial zones. This is, in effect, a non-linear causal relationship known as *domino causality* (Grotzer, 2004).

It is important to understand non-linear causal patterns such as *domino causality*, because very few causal relationships outside of tightly controlled laboratory studies or simple physical causal systems (such as hitting a baseball with a bat) are truly linear. In fact, it has been argued that much of the historic progress made in hard sciences, such as physics, involves making increasingly more detailed observations in order to better account for the intervening causal connections between relationships that appear to occur as action at a distance (Lange, 2002).

Later, as the student looks back over her collected data, she notices another interesting pattern. The mass industrial re-zoning took place all at once, at the beginning of the simulation. Therefore, the density of industrial buildings rose over the course of the simulation, quickly at first, and less near the end, but it generally increased at each measurement. However, some values, such as pollution, population, and mayoral popularity go through long stretches of constancy before finally changing.

This is an example of *steady state causality*, in which the system is in dynamic balance or the process occurs undetected over time until registering a perceivable change. Recognizing

Figure 5. SimCity crime map

this causal pattern is important because many causal systems are in a steady state most of the time. For example, a laptop computer sitting on a wooden desk is exerting a downward force on the table top (weight), while the table pushes back, in the opposite direction, with an equivalent force (normal force). This system is in a steady state. However, were the desk's structural integrity to become compromised (perhaps by termites), it would no longer be able to apply the appropriate normal force to the computer, and the computer would fall through. This, in turn, would trigger a shift from *steady state* to *event based* causality, which students are more likely to perceive (Chi, 1997; Grotzer, 2004).

However, more expert causal conceptions recognize steady states in addition to event-based causality—thus it is undesirable that students only recognize causal connections during or after a system collapse. In SimCity, students can be supported in their understanding of *steady state* causality by having them restart the scenario and play the game at "normal" speed. By measuring changes on a day-to-day or even week-to-week basis, and comparing the rate of change to that of the annual or semi-annual measurements taken before, students will quickly see that values tend to be stable for long stretches of time.

Finally, the student notices that the values measured in the original two simulations in which maximum industrial zones were placed are not exactly the same. Her teacher encourages her to think about why this might be, and the student decides that it might be because, even though she tried very hard to make both conditions exactly the same, she might have placed different amounts of industrial zones in each, but she's not sure how she can tell. Her teacher then asks about the other scenario, in which no extra industrial zones were added. The student decides to run that simulation once more, at cheetah speed, and take bi-annual measurements as before. Surprisingly, the numbers

Figure 6. SimCity crime graph

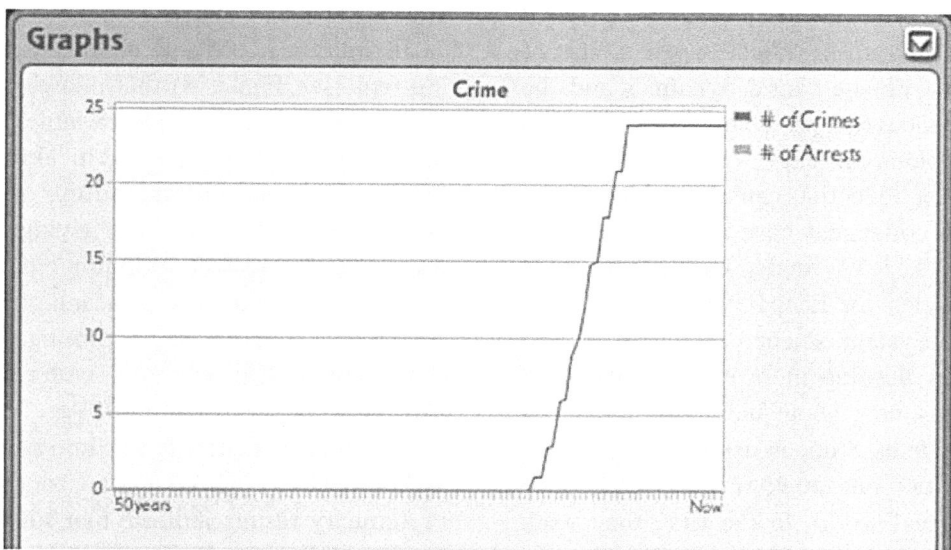

do not exactly match those from her first control trial, although they are close. The student then realizes that making changes to the game world does not directly cause the Sims to behave in a particular way, it only increases or decreases the possibility that they will.

This is an example of *probabilistic causality*, in which the causal relationship between two or more variables is not deterministic. It is important to understand this complex causal pattern because in science, effects do not always reliably follow a given cause. For example, smoking does not always cause lung cancer; it only increases the likelihood that a person will develop lung cancer over time. Students with a *deterministic* conceptualization of a causal relationship, on the other hand, might miss the existence of a causal link if the cause does not lead to the effect in every instance (Grotzer, Duhaylongsod, & Tutwiler, 2011)

In our example, the student's understanding of this pattern is leveraged by specifically drawing her attention to the differences between their two sets of data, and to reflect on what might account for them. At first, she guesses that facets of measurement error are to blame, but these can be controlled for, to a point, by having the student place the exact same number of industrial zones, and collect data on the exact same days in each trial. Ultimately, can she say with certainty exactly what the critical measurements will be at any given time, even though she was in control of placing the original causal agents (the industrial zoned areas)? If not, then the relationship, while causal, is *probabilistic*.

Our description of a SimCity-based curriculum offers a sense of what an "off the shelf" interactive, immersive program can offer students in terms of learning to reason about causal complexity. It is important to keep in mind the myriad scaffolds in place, however. Simply giving the game, which will likely be engaging and novel at first, to teachers and asking them to have their students play it is unlikely to lead to increased student causal understanding. We recommend that interested teachers first play through each scenario, so as better to anticipate problems and questions the students might have. This step, having teachers interact with the virtual simulation (or any technology being integrated), is critical but often overlooked due to time or resource constraints.

The promise of such technology can be more fully realized if the software is designed from the ground up to reflect expert modes of engagement with the causal dynamics and the many affordances that can be built in to leverage students' exploration and to reveal to them the inadequacy of reductive causal models. Dede, Grotzer, and colleagues have built and are testing EcoMUVE, a VE funded by the Institute of Education Sciences. EcoMUVE was designed to teach ecosystem science concepts and to help students develop more expert patterns of scientific reasoning about the causal dynamics within ecosystems. Students using the EcoMUVE take part in two one-week curricula embedded within a virtual world. In the first, they work together as specific types of scientists in teams to discover why all of the large fish in a pond ecosystem died. In the second, students again work in groups to explore predator-prey relationships between wolves and deer on various islands. Both modules leverage the strengths of interactive, immersive simulations, like SimCity, but they also include design elements that better allow students to detect complex causal patterns from observational data.

For example, in both modules students in EcoMUVE are able to use a Calendar Tool (see Figure 7) to move back and forth in time at will to make repeated observations. In the case of the pond module, this will allow students to measure and verify critical chemical levels and other key indicators of ecosystem health over time. In the forest module, students will be able to take more nuanced measurements of animal populations as well as explore certain aspects of the terrestrial ecosystem in more detail. In SimCity, however, the student had to reset the scenario and play through again, collecting data at designated points in time, if they wanted to re-observe a relationship. This ability to treat time more flexibly will make it more likely that students will detect changes in the VE over time, a critical part of understanding the complex nature of ecosystems.

In addition, EcoMUVE includes a data-graphing tool (see Figure 8) that allows students to compare trends in multiple variables over time. Managing data salience is a critical part of scaffolding student understanding of complex systems (Grotzer, 2012). By giving students a way to make salient individual datum, then compare sets of data in a meaningful way, the EcoMUVE supports such learning. This is in contrast to SimCity, which allows for the viewing of a single variable at a time.

Building such affordances into an interactive and immersive environment appears promising. Preliminary results indicate that students using the EcoMUVE made significant gains in their understanding of *spatiotemporally distant* causal relationships (Metcalf, et al., 2011) and *processes and steady states* (Grotzer, et al., 2011).

A final affordance of MUVEs over their commercial cousins is that they can be design with back-end data collection protocols that allow for both research as well as substantive feedback to educators using the system. For example, River City, an educational MUVE that was the technological inspiration for the EcoMUVE collected and reported student chat logs to the teacher each day that students used the system. A similar func-

Figure 7. EcoMUVE calendar tool

Figure 8. EcoMUVE data graph

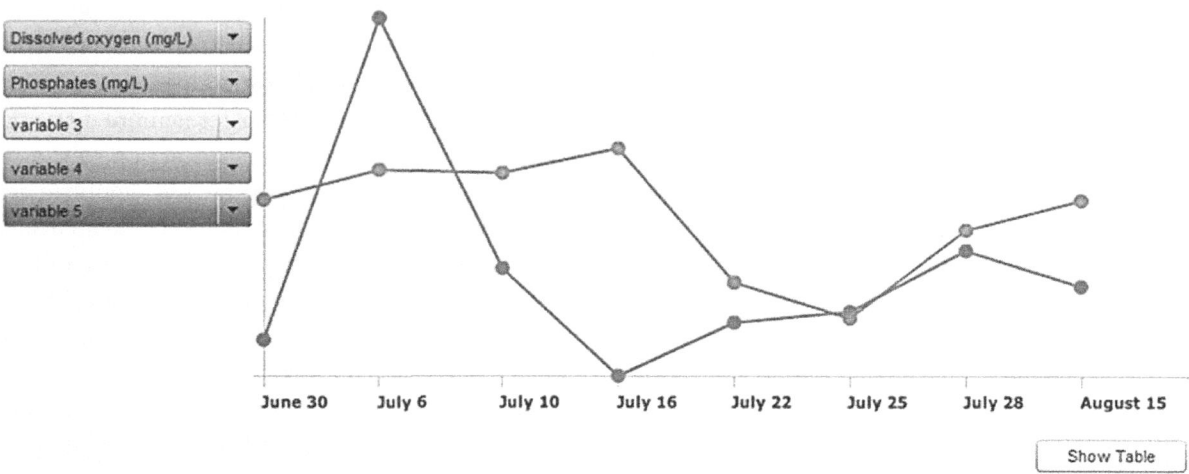

tion in the EcoMUVE could be combined with automatic data-mining algorithms that look for phrases indicative of causal misunderstandings as well as strengths. Also, student movement (Dukas, 2009) and data collection (Clarke, 2009; Ketelhut, 2007) were analyzed in River City, and similar analyses can be performed in the EcoMUVE, both for research purposes as well as feedback for teachers on student causal learning.

For example, student pathing (the patterns of areas they visit in the VE) can indicate if a student is attending to distal vice proximal causes. In addition, the order and rate at which they collect data, and what tools they use to do so, might also give us insight into their understanding of the VE as a complex ecosystem. In its final implementation, these data can be summarized and reported to help the teacher better understand student learning over the course of the curriculum.

CONCLUSION

We have argued here for the importance of including immersive and interactive simulations in the pedagogical repertoire of "next generation science" and that these simulations have a special role to play in teaching the nature of dynamic, complex causal systems. We have offered an example from "off the shelf" software of how such simulations can invite students to rethink their causal assumptions—what they perceive and attend to in an environment and how they reason about it. In framing this example, we ensured that the simulation was immersive and interactive by being appropriately integrated (Issenberg, et al., 2005; Kahn, 2010; Rutten, et al., 2012), contextualized (Barab, et al., 2009), part of a problem-based lesson (Koray, 2011), and that the student was provided the appropriate scaffolds and supports from the teacher (Blake & Scanlon, 2007).

In the example here, we included teacher support and how it might facilitate students' deepening understanding of more expert forms of reasoning about causal complexity. The level of teacher guidance has been studied in other forms of simulations. For instance, Paulson, Adams and colleagues (Adams, Paulson, & Wieman, 2009; Paulson, Perkins, & Adam, 2009) have contrasted no guidance, driving questions, gently guided, and strongly guided interactions with simulations found that no guidance or driving questions was most effective. However, we consider this an important area for study in immersive, interactive simulations given that unless driven to do so by problem-based content or guiding questions, students may be likely to fall into the same simplifying default patterns as in the real world. Further, we acknowledge that teachers are prone to similar simplifying assumptions. Supplying teachers with the necessary pedagogical content knowledge and understanding of expert epistemologies would be critical to the successful use of the simulations. In the case of the SimCity scenario proposed here, that includes having the teacher play through the scenarios and make note of trends they see before using it with their students.

Work in progress that is specifically designed to leverage the affordances of interactive and immersive technology (Dede, 2009), such as EcoMUVE, signal the promise of such simulations and argue for their place alongside other forms of simulation in science education. We caution that, as designers develop, test, and implement these immersive environments, care should be taken such that factors outside of the classroom, such as prior access and use of similar technologies, are accounted for (Lin, et al., 2011). Future work in this problem space will include detailed observation and analysis of student complex causal learning longitudinally, explorations of the relationships between prior causal knowledge and student data-gathering behavior in the causal environments, and the effect of data salience and complexity on student causal learning.

An important issue that remains to be addressed is that of transfer to real world environments. Given the rich research literature in cognitive science that shows that transfer is not straightforward and requires explicit attention and scaffolding (Perkins & Salomon, 1988), it is important to consider the challenges of mapping from a virtual world to a real one. Goldstone and Sakamoto (2003) have found that the closer the features of the virtual world fit with those of the real world, the more likely that students will be cued to the likelihood of transfer. Unfortunately, more similarity in surface features also makes it less likely that students will learn to understand the embedded deep structures—in this case, the causal complexities—deeply. The impact seems to be stronger for students who perform less well in school.

Further, the issue of transfer is more difficult when we are seeking informed action, not just better reasoning. One may be well inclined to engage in actions that are easy to take in a virtual world and yet these may have little chance of being sustainable in a real world where more effort needs to be invested to enact that action. One vivid example of this is the study by Lin et al. (2012), in which high school aged boys and girls were asked to escape from a simulated burning building in a virtual world. Though the girls and boys acted differently in the real world, with boys being more willing to take risks resulting in a lower average escape time (Lin, et al., 2012), there is no reason to believe that they would have taken the same actions in the real world, when true danger was perceived. Instead, all students would likely have chosen the route perceived to be most safe.

We expect that as powerful computing devices become ubiquitous, these devices can be leveraged to simulate complex causal dynamics. Enabling students to have an immersive and interactive experience allows them to engage with and reflect on the causal outcomes in ways that will demand their attention to the inherent complexity. Combining these technological affordances with productive pedagogies should enable powerful lessons for living more effectively in a complex world.

ACKNOWLEDGMENT

We would like to express our appreciation to our colleagues, Chris Dede, Shari Metcalf, and Amy Kamarainen as well as to the anonymous reviewers and to Dr. Myint Khine and Dr. Issa Saleh. We would also like to thank Lynneth Solis and Leslie Duhaylongsod for helping us to evaluate and think about student complex causal learning. Finally, we would like to thank the editors of this volume for their limitless patience and guidance as we worked to craft this chapter.

This work was supported in part by the following awards: Causal Learning in the Classroom Studies supported by Grant No. NSF-0845632 from the National Science Foundation, to Tina Grotzer; EcoMUVE Project supported by the Institute of Education Sciences, U.S. Department of Education, Grant No. R305A080514 to Chris Dede and Tina Grotzer; and EcoMOBILE Project supported by Grant No. NSF-1118530 to Chris Dede and Tina Grotzer. Any opinions, findings, conclusions or recommendations are those of the authors and do not necessarily reflect the views of the National Science Foundation or the Institute for Education Sciences.

REFERENCES

Adams, W. K., Paulson, A., & Wieman, C. E. (2009). What levels of guidance promote engaged exploration with interactive simulations? In *Proceedings of PERC*. PERC.

Barab, S. A., Scott, B., Siyahhan, S., Goldstone, R., Ingram-Goble, A., Zuiker, S., & Warren, S. (2009). Transformational play as a curricular scaffold: Using videogames to support science education. *Journal of Science Education*, *18*(4), 305–320.

Blake, C., & Scanlon, E. (2007). Reconsidering simulations in science education at a distance: Features of effective use. *Journal of Computer Assisted Learning*, *23*(6), 491–502. doi:10.1111/j.1365-2729.2007.00239.x

Chi, M. T. H. (1997). Creativity: Shifting across ontological categories flexibly. In Ward, T. B., Smith, S. M., & Vaid, J. (Eds.), *Creative Thought: An Investigation of Conceptual Structures and Processes* (pp. 209–234). Washington, DC: American Psychological Association. doi:10.1037/10227-009

Clarke, J. (2009). *Exploring the complexity of inquiry learning in an open-ended problem space*. (Unpublished Doctoral Dissertation). Harvard Graduate School of Education. Boston, MA.

Danish, J., Peppler, K., Phelps, D., & Washington, D. (2011). Life in the hive: Supporting inquiry into complexity within the zone of proximal development. *Journal of Science Education and Technology*, *20*(5), 454–467. doi:10.1007/s10956-011-9313-4

Dede, C. (2009). Immersive interfaces for engagement and learning. *Science*, *323*(5910), 66–69. doi:10.1126/science.1167311

Devisch, O. (2008). Should planners start playing video games? Arguments from SimCity and Second Life. *Planning Practice & Theory*, *9*(2), 209–228. doi:10.1080/14649350802042231

Dieterle, E., & Clarke, J. (2007). Multi-user virtual environments for teaching and learning. In Pagani, M. (Ed.), *Encyclopedia of Multimedia Technology and Networking* (2nd ed., pp. 1033–1044). Hershey, PA: IGI Global. doi:10.4018/978-1-60566-014-1.ch139

Dorner, D. (1989). *The logic of failure: Why things go wrong and what we can do to make go right*. New York, NY: Metropolitan Books.

Driver, R., Guesne, E., & Tiberghien, A. (Eds.). (1985). *Children's ideas in science*. Philadelphia, PA: Open University Press.

Dukas, G. (2009). *Characterizing student navigation in educational multiuser virtual environments: A case study using data from the river city project*. (Unpublished Doctoral Dissertation). Harvard Graduate School of Education. Boston, MA.

Electronic Arts. (2011). *SimCity™ 4 deluxe edition*. Retrieved on 4 December, 2011 from http://www.ea.com/simcity-4-deluxe

Feltovich, P. J., Spiro, R. J., & Coulson, R. L. (1993). Learning, teaching, and testing for complex conceptual understanding. In Frederiksen, N., & Bejar, I. (Eds.), *Test Theory for a New Generation of Tests* (pp. 181–217). Hillsdale, NJ: LEA.

Goldstone, R. L., & Sakamoto, Y. (2003). The transfer of abstract principles governing complex adaptive systems. *Cognitive Psychology, 46*, 414–466. doi:10.1016/S0010-0285(02)00519-4

Grotzer, T. A. (1993). *Children's understanding of complex causal relationships in natural systems.* (Unpublished Doctoral Dissertation). Harvard University. Cambridge, MA.

Grotzer, T. A. (2004, October). Putting science within reach: Addressing patterns of thinking that limit science learning. *Principal Leadership.*

Grotzer, T. A. (2012). *Learning causality in a complex world: Understandings of consequence.* Lanham, MD: Rowman & Littlefield.

Grotzer, T. A., & Basca, B. B. (2003). Helping students to grasp the underlying causal structures when learning about ecosystems: How does it impact understanding? *Journal of Biological Education, 38*(1), 16–29. doi:10.1080/00219266.2003.9655891

Grotzer, T. A. Dede, C., Metcalfe, S., & Clarke, J. (2009). *Addressing the challenges in understanding ecosystems: Why getting kids outside may not be enough.* Paper presented at the National Association of Research in Science Teaching (NARST) Conference. Orange Grove, CA.

Grotzer, T. A., Duhaylongsod, L., & Tutwiler, M. S. (2011). *Developing explicit understanding of probabilistic causation: Patterns and variation in young children's reasoning.* Paper presented at the American Educational Research Association (AERA) Conference. New Orleans, LA.

Grotzer, T. A., & Lincoln, R. (2007). Educating for "intelligent environmental action" in an age of global warming. In Moser, S. C., & Dilling, L. (Eds.), *Creating a Climate for Change: Communicating Climate Change and Facilitating Social Change* (pp. 266–280). Cambridge, UK: Cambridge University Press. doi:10.1017/CBO9780511535871.020

Grotzer, T. A., Tutwiler, M. S., Dede, C., Kamarainen, A., & Metcalf, S. (2011). *Helping students learn more expert framing of complex causal dynamics in ecosystems using EcoMUVE.* Paper presented at the National Association of Research in Science Teaching (NARST) Conference. Orlando, FL.

Hmelo-Silver, C. E., & Azevedo, R. (2006). Understanding complex systems: Some core challenges. *Journal of the Learning Sciences, 15*, 53–61. doi:10.1207/s15327809jls1501_7

Hmelo-Silver, C. E., Marathe, S., & Liu, L. (2007). Fish swim, rocks sit, and lungs breathe: Expert-novice understanding of complex systems. *Journal of the Learning Sciences, 16*, 307–331. doi:10.1080/10508400701413401

Issenberg, S. B., Mcgaghie, W. C., Petrusa, E. R., Gordon, D. L., & Scalese, R. (2005). Features and uses of high-fidelity medical simulations that lead to effective learning: A BEME systematic review. *Medical Teacher, 27*(1), 10–28. doi:10.1080/01421590500046924

Jacobson, M. J. (2001). Problem-solving, cognition, and complex systems: Differences between experts and novices. *Complexity, 6*(3), 41–49. doi:10.1002/cplx.1027

Jimoyiannis, A., & Vassilis, K. (2001). Computer simulations in physics teaching and learning: A case study on students' understanding of trajectory motion. *Computers & Education, 36*(2), 183–204. doi:10.1016/S0360-1315(00)00059-2

Kahn, S. (2010). New pedagogies on teaching science with computer simulations. *Journal of Science Education and Technology, 20*, 215–232. doi:10.1007/s10956-010-9247-2

Ketelhut, D. J. (2007). The impact of student self-efficacy on scientific inquiry skills: An exploratory investigation in river city, a multi-user virtual environment. *Journal of Science Education and Technology, 16*(1), 99–111. doi:10.1007/s10956-006-9038-y

Koray, O. (2011). The effectiveness of problem-based learning supported with computer simulations on academic performance about buoyancy. *Energy Education Science and Technology Part B-Social and Educational Studies, 3*(3), 293–304.

Lange, M. (2002). *An introduction to the philosophy of physics: Locality, fields, energy, and mass*. Oxford, UK: Blackwell Publishing.

Lave, J., & Wenger, E. (1991). *Situated learning: Legitimate peripheral participation*. Cambridge, UK: Cambridge University Press. doi:10.1017/CBO9780511815355

Levy, S. T., & Wilensky, U. (2008). Inventing a "mid-level" to make ends meet: Reasoning through the levels of complexity. *Cognition and Instruction, 26*, 1–47. doi:10.1080/07370000701798479

Lin, M. C., Tutwiler, M. S., & Chang, C. Y. (2011). Exploring the relationship between virtual learning environment preference, use, and learning outcomes in 10th grade earth science students. *Learning, Media and Technology, 36*(4), 399–417. doi:10.1080/17439884.2011.629660

Lin, M. C., Tutwiler, M. S., & Chang, C. Y. (2012). Gender bias in virtual learning environments: An exploratory study. *British Journal of Educational Technology, 43*(2), 59–63. doi:10.1111/j.1467-8535.2011.01265.x

Metcalf, S., Kamarainen, A., Tutwiler, M. S., Grotzer, T., & Dede, C. (2011). Ecosystem science learning via multi user virtual environments. *International Journal of Gaming and Computer-Mediated Simulations, 3*(1), 86–90. doi:10.4018/jgcms.2011010107

Okuda, Y., Bryson, E. O., DeMaria, S. Jr, Jacobson, L., Quinones, J., Shen, B., & Levine, A. I. (2009). The utility of simulation in medical education: What is the evidence? *The Mount Sinai Journal of Medicine, New York, 76*, 330–343. doi:10.1002/msj.20127

Paulson, A., Perkins, K., & Adams, W. (2009). *How does the type of guidance student use of an interactive simulation?* Retrieved from http://phet.colorado.edu/publications/Paulson_etal_2009/Paulson_etal_2009.pdf

Perkins, D., & Grotzer, T. A. (2005). Dimensions of causal understanding: The role of complex causal models in students' understanding of science. *Studies in Science Education, 41*, 117–165. doi:10.1080/03057260508560216

Perkins, D. N., & Salomon, G. (1988). Teaching for transfer. *Educational Leadership, 46*, 22–32.

Raia, F. (2008). Causality in complex dynamic systems: A challenge in earth systems science education. *Journal of Geoscience Education, 56*(1), 81–94.

Rubin, D. B. (2005). Causal inferences using potential outcomes. *Journal of the American Statistical Association, 100*(469), 322–331. doi:10.1198/016214504000001880

Rutten, N., van Joolingen, W. R., & van der Veen, J. T. (2012). The learning effects of computer simulations in science education. *Computers & Education, 58*(1), 136–153. doi:10.1016/j.compedu.2011.07.017

Shin, N., Jonassen, D. H., & McGee, S. (2003). Predictors of well-structured and ill-structured problem solving in an astronomy simulation. *Journal of Research in Science Teaching, 40*(1), 6–33. doi:10.1002/tea.10058

Squire, K. (2003). Videogames in education. *International Journal of Intelligent Simulations and Gaming, 2*(1), 49–62.

Tenner, E. (1996). *Why things bite back*. New York, NY: Vintage Books.

United States Census Bureau Website. (2011). *Factfinder*. Retrieved 4 December, 2011 from http://factfinder.census.gov/servlet/SAFFFacts?_event=Search&geo_id=&_geoContext=&_street=&_county=danville&_cityTown=danville&_state=04000US17&_zip=&_lang=en&_sse=on&pctxt=fph&pgsl=010&show_2003_tab=&redirect=Y

Walker, B., & Salt, D. (2006). *Resilience thinking: Sustaining ecosystems and people in a changing world*. Washington, DC: Island Press.

White, P. A. (1997). Naive ecology: Causal judgments about a simple ecosystem. *The British Journal of Psychology, 88*, 219–233. doi:10.1111/j.2044-8295.1997.tb02631.x

Wieman, C. E., Adams, W. K., & Perkins, K. K. (2008). PhET: Simulations that enhance learning. *Science, 322*, 682–683. doi:10.1126/science.1161948

Wilensky, U., & Resnick, M. (1999). Thinking in levels: A dynamic systems approach to making sense of the world. *Journal of Science Education and Technology, 8*(1), 3–19. doi:10.1023/A:1009421303064

Zacharia, Z. C. (2007). Comparing and combining real and virtual experimentation: An effort to enhance students' conceptual understanding of electric circuits. *Journal of Computer Assisted Learning, 23*(2), 120–132. doi:10.1111/j.1365-2729.2006.00215.x

ENDNOTES

[1] Available at: http://mw.concord.org/modeler/index.html

[2] Available at: http://phet.colorado.edu/en/simulations/category/new

[3] http://ccl.northwestern.edu/netlogo/docs/faq.html#diff

[4] StarLogo Termite Project http://education.mit.edu/starlogo/

[5] SimCity™ 4 Deluxe can be purchased online (cost, $20USD at the time of writing) through the Steam content delivery system, or through other commercial gaming vendors.

Chapter 8
Teachers and Teaching in Game-Based Learning Theory and Practice

Mario M. Martinez-Garza
Vanderbilt University, USA

Douglas B. Clark
Vanderbilt University, USA

ABSTRACT

Interest in game-based learning has grown dramatically over the past decade. Thus far, most of the focus has not included the role of teachers. This chapter first summarizes the theoretical research on game-based learning and the implications of that research for the role of teachers. The authors next review the game-based learning literature that has specifically articulated a role for teachers or achieved an empirical description of teacher action within a game-based learning context. They then connect these accounts with more general research on teachers and technology use, elaborating on points of contact and identifying differences that may signal special challenges. Finally, the authors articulate an expanded role for teachers in game-based learning practices in terms of game-based learning research and new scholarship on the psychology of games.

INTRODUCTION

Digital games are an influential and ubiquitous presence in the lives of young learners. A 2008 study by the Pew Internet and American Life Project found that 97% of teens ages 12-17 play digital games, and 50% of them report daily or nearly daily play (Lenhart, Jones, Macgill, & Pew Internet and American Life Project, 2008). With increasing access to computers, consoles, and cell phones, young people are finding that opportunities for gaming are everywhere. The emergence of video gaming as an important leisure activity among young people called into question what effects, if any, these games may have. Initially, psychologists and sociologists set out to discover whether or not video gaming is a harmful activity for children (e.g., Anderson & Bushman, 2001; Hauge & Gentile, 2003; Anderson, 2004; Skoric, Teo, & Neo, 2009); although the findings from this line of research remain controversial (e.g., Hall, Day, & Hall, 2011; Murray, Biggins, Donnerstein, Menninger, Rich, & Strasburger, 2011), there is still an enduring perception among the general public that digital games are a negative influence on young people. Not all research, however, has cast digital games in a negative light. In particular, educational researchers are increasingly interested in the positive effects of gaming. In fact, a growing body of evidence indicates that digital games can be powerful vehicles for learning. Numerous studies have linked classroom use of learning games with increased learning outcomes and improvement in students' conceptual understanding, engagement, and self-efficacy (e.g., de Freitas, 2006; Clark, Nelson, Sengupta, & D'Angelo, 2009).

The past decade has seen significant advances in the sophistication and efficacy of games designed specifically to enhance learning in a school setting. Educators who want to include digital games in their classroom activities, however, may find themselves facing significant obstacles. From the outset, game-based learning contains at its core several assumptions about teaching and learning that differ, or even run counter, to the everyday business of classrooms. Additionally, educators may feel unsupported in using games for learning, unsure about which games to use and how to use them, or they may feel they lack the knowledge or gaming experience to guide their students effectively in this activity. Taken together, these obstacles may severely constrain the degree to which games are used effectively in the classroom.

Our goal in writing this chapter is to help educators create effective strategies for implementing game-based learning activities into their classroom practices. While educational gaming presents some elements of practice that are shared with other classroom uses of technology, educational gaming presents additional unique circumstances and opportunities that research has only recently begun to address. We will present some examples from the literature that illustrate the type of personal, technological, and structural resources that teachers need to use games effectively in their classrooms, and we will outline some of the potential advantages to building partnerships between teachers and researchers and creators of educational games.

RESEARCH ON GAMES AND LEARNING

Investigation into the use of games for learning has grown from a small niche area to a major focus of research over the past decade (e.g., Clark, et al., 2009; Dieterle, 2010; Honey & Hilton, 2011). In 2006, the Federation of American Scientists issued a widely publicized report stating their belief that games offer a powerful new tool to support education and encouraging governmental and private organizational support to increase funded research into the application of complex gaming environments for learning (FAS, 2006). In 2009, a special issue of *Science* (Hines, Jasny, & Mervis, 2009) echoed and expanded this call. Mayo (2009) characterized digital games in that issue of *Science* as "capable of delivering [science, technology, engineering, and mathematics] instruction to millions" (p. 79).

As computer games became nearly ubiquitous in the 1980s, gaming became a major cultural activity among young people in the United States, Europe, and Japan. Since that time, educators are increasingly intrigued by the potential of games to empower learning. A complete overview of the development and progress of game-based learning is outside the scope of this review. By way of a historical summary, we can say that sociologists and behavioral psychologists began to conduct research on gamers as early as the late 1970s, but it was not until the early 1990s that gaming attracted significant attention from educational researchers. The earliest research efforts investigating games for learning grew out of earlier work on computer simulations, and thus inherited the theoretical focus on Piagetian notions of knowledge construction and experiential learning that has pervaded much research in that area (e.g., Papert & Harel, 1991).

In more recent times, a principal focus of educational games research has been to explore and explicate the deep links between learning and playing. Both learning and games, according to Shaffer, Squire, Halverson, and Gee (2005), "are transformative activities that are most powerful when they are personally meaningful, experiential, social, and epistemological all at the same time" (p. 105). Shaffer, Squire, Halverson, and Gee describe an effective and empowering mode of learning that, while highly sought-after in formal educational settings, is ubiquitous in digital games. Squire (2003, 2007) and Gee (2007) argue persuasively in favor of the unique affordances of digital games as learning tools and frame digital games as "the future of learning" (Shaffer, et al., 2005, p. 108) in terms of erasing barriers to learning, making knowledge accessible and personally relevant to the learner, and fostering communities of learning.

In two widely cited works, James Paul Gee (2004, 2007) grounded the discussion around games within the major theories of learning. On the subject of situated and embodied understanding, for example, Gee connects game-based learning to the works of George Lakoff, Jean Lave and Etienne Wenger, Michael Tomasello, and William Clancey. These works by Gee and work by Shaffer et al. established a theoretical frame for games for learning upon which later research largely agrees. This frame is supported around four core beliefs:

1. Games engage students in ways that normal school activities do not.
2. Games privilege experiential learning, i.e. students learn by doing.
3. Games promote identity-construction and self-efficacy.
4. Games provide opportunities for collaboration and participation in semiotic domains. (Gee, 2004).

Later scholarship expanded this theoretical frame and connected it to other works. It is now largely accepted that games make it possible for learners to develop situated understanding, that is, a form of learning that is a product of activity and the context in which the learning is used (Lave & Wenger, 1991). In this regard, games afford a broad range of contexts and variants of activity that would be otherwise inaccessible to learners, such as city planning in the game *SimCity*. Games also promote content-grounded discussion among learners (Steinkuehler & Chmiel, 2006), in which the game acts as a touchstone for inquiry and rich discussion. Learners also benefit from the creative and generative control that games allow, fostering development of their identities and self-efficacy (Ketelhut, 2007; Pelletier, 2008). These benefits work together to help students engage with complex concepts and systems, and this engagement has an epistemic character, i.e., it enables learners to observe and preserve the links between knowing and doing (Shaffer, 2006).

Compared to more established areas of educational research, inquiry into game-based learning is a new endeavor, one that has come into its own only in the last decade. For the most part, researchers in the field focus their efforts on characterizing, producing, and assessing student learning and connecting the research findings to decisions about game design as a means to bring validity and attention to the research agenda as a whole. Because of the urgency in establishing this validity, and due to certain theoretical commitments which we outline later, research in game-based learning currently has much to say about learners and the process of learning but not as much to say about teachers and the process of teaching. Since it is assumed that the efficacy of game-based learning is largely determined by the design principles that govern the way in which a particular game is built, and not by teacher action or the context of the classroom, research questions about game-based learning lend themselves to a process-product research paradigm (cf. Shulman, 1986) in which the focus is on the variables of change, i.e., how the game's design affects the student's learning. The process-product paradigm gives priority to the study of the discrete, controllable factors that, when controlled, directly produce better student learning. Since the teacher is not seen as a participant in the interaction between the learner and the game, nor as a factor that can be necessarily controlled for, research experiments into games for learning that follows the process-product paradigm generally does not attend to the role of the teacher.

While the process-product paradigm strengthens research aimed at discovering and refining principles of educational design, the lack of focus on the role of the teacher is problematic. We argue that, like any proposed classroom activity, educational games must provide a role for the teacher that, at the very least, provides him or her opportunities to assess, guide, and personalize learning. While well-designed educational games may support positive learning outcomes "in a vacuum," a teacher's influence and knowledge can greatly determine and multiply the game's beneficial effects in a classroom. Teachers are invaluable organizational resources to be leveraged in the design of games for learning; they are the principal driving forces behind the adoption of new learning tools, including games, and the key source of guidance and support for other teachers who might also want to feature digital games in their practice (Kirriemuir & McFarlane, 2004).

Several issues need to be clarified, however, if educational game designers and researchers are to ally with teachers and school administrators to increase the use and effectiveness of games for learning. Specific attention needs to be paid to: (a) the organizational and technological frameworks within which teachers locate their practice, (b) teachers' attitudes and beliefs regarding games for learning, and (c) the use of instructional technology more generally. With these prerequisites in mind, we briefly review the literature on technology use in the classroom with a focus on the resources and constraints that may affect teachers' attitudes and beliefs towards game-based learning. We also identify some tensions and limitations that operate specifically on educational gaming in the classroom in which the teacher has influence. We then review some of the game-based learning literature that has, directly or indirectly, articulated a role for teachers or achieved an empirical description of teacher action within a game-based learning context. Finally, we articulate an expanded role for teachers in game-based learning practices in terms of game-based learning research and new scholarship on the psychology of games. Our goal is not only to address the tensions, both theoretical and practical, of teaching with games, but also to propose forms of collaboration between creators of educational games and teachers. We see these collaborations as a key element in increasing the presence of games for learning in classrooms and in enhancing games for learning as more effective, fruitful, and productive parts of teaching practice.

TEACHERS' BELIEFS AND ATTITUDES TOWARD TECHNOLOGY

It is our view that, in some sense, the issues regarding game-based learning in the classroom can be better understood in the greater context of how teachers use digital technologies in their practice, which has been studied extensively (e.g. Office of Technology Assessment, 1995; Moursund & Bielefeldt, 1999; U.S. Department of Education, 2000). No particular trait of digital games suggests that our current understanding of how teachers employ computer technology should not apply to games. On the contrary, we argue that some aspects of games highlight the tensions, issues, and difficulties that teachers face when integrating any computer-based technology into their practice. Since it is likely that teachers' attitudes and beliefs about gaming will be informed to some degree by their relationship with information technology and computer-based learning activities, it is important to review the salient findings in this area in order to better place digital games in their proper context.

Current thinking on teacher preparation stresses the importance of teachers' developing knowledge, skills, and dispositions to incorporate technology effectively into their students' lives (National Council for Accreditation of Teacher Education, 1997, 2008). Teachers themselves, however, can be a nucleus of sustained resistance to the use of technology. Mumtaz (2000) found that teachers' use of computer technology can be limited by a variety of causes, including personal and psychological factors attributable to teachers themselves. Cuban (1999) asked a more fundamental question: whether it was somewhat unrealistic to expect all teachers to use technology in light of other pedagogical and achievement goals that might be more pressing. Furthermore, teacher education programs generally do not train their students in the latest developments in educational technology (Sprague, 2004). According to Moursund and Bielefeldt's (1999) report for the Milken Exchange on Education Technology, "teacher-training programs do not provide future teachers with the kinds of experiences necessary to use technology effectively in the classroom" (p. 4). Sheingold and Hadley (1990) found that teachers who effectively use technology as a part of classroom practice describe themselves as self-taught with regards to computers. Thus, while teachers' individual styles, preferences, and experience influence the amount or kinds of technology used in classrooms, there is also a significant gap in the preparation that teachers receive to help them integrate technology in support of their practice.

It may be, however, that teachers' use of technology reflects not only attitudes or lack of training but also an absence of robust convincing classroom applications of technology. For example, Russell, Bebell, O'Dwyer, and O'Connor (2003) provide a comprehensive view of the use of computer technology by teachers. In a large-scale survey, they observed that teachers generally use computers more for preparation and communication than for delivering instruction or assigning learning activities. This finding suggests that teachers largely do not have a readily accessible "toolbox" of computer-based tools to support their classroom practice.

Conversely, teachers who do have technological tools at their disposal tend to be more convinced of their value in supporting learning. Russell et al. (2003) find that by far the strongest predictor of instructional use of computers is the teacher's belief in the educational importance of technology. Among teachers who use technology for instruction delivery (the form of use that is most resonant with game-based learning), Russell and colleagues find that confidence in one's own skill with technology is an important factor in determining the quantity and quality of technology use. They also find that correlations exist between positive perceptions of access to technology, higher personal confidence in the ability to use technology, belief in the importance of teacher-directed computer use in the classroom, and the previously mentioned belief in the overall importance of technology for teaching. Their findings support previous work by Sheingold and Hadley (1990), who surveyed 600 4-12th grade teachers selected on the basis of their advanced level of classroom technology use. Sheingold and Hadley determine that a similar spectrum of traits underpin these advanced classroom practices including (a) belief in the importance of technology for teaching, (b) high personal confidence in the ability to use technology, and (c) access to computer resources. Russell and colleagues also note that there is a trend among newer teachers (i.e., those with one to five years in service) to exhibit greater confidence in technology use, more positive beliefs about student-centered teaching practices, and more openness to the value of technological tools for teaching, regardless of whether or not they have access to them in their everyday practice.

This last finding, suggesting younger teachers' uniform acceptance *in principle* of the value of technology for teaching, coupled with an almost generational divide between teachers who are more receptive to and assertive in using technology and those who are not, resonates with the distinction made in Prensky (2001) between what he terms "digital natives" and "digital immigrants." The former are defined as "native speakers" of the digital language of computers, digital games, and the Internet. The latter are those who adopted and accepted these elements of technology but have largely been educated and socialized in a pre-computerized way. Prensky makes his argument at a time when adoption of technology, while prevalent in other settings, was far from universal in schools. This distinction may be more relevant now than in 2001. Not only are most students "digital natives," but also so are most newly tenured teachers. Because of this, teachers and students increasingly share not only beliefs about the value and use of technology but also the cognitive skills (e.g., rapid information processing, multi-tasking, and collaboration), modes of communication, and media-participation habits of "digital natives." Whether or not this common ground may support game-based learning implementations is a question we will address in the following section.

These findings, taken together, support a view that if teacher preparation programs are not doing enough to encourage technology use by teachers, then such use is likely to be a result of teachers' beliefs and aptitude acquired from their personal experiences. There is, however, also an evolutionary process at work as newly trained teachers import technology-enabled forms of classroom practice from their own experiences as students and from modalities of technology use that have, in some sense, always been a part of their lives. Teachers who make productive use of technology in the classroom may not see technology as a teaching tool specifically (i.e., a tool that is defined and circumscribed by their professional activities as teachers), but see technology rather as a kind of "life tool" with affordances and benefits across contexts and activities.

While the literature may suggest a bleak outlook on the possibility of integrating games for learning into classroom practice in the short-term, it is important to note that not all teachers feel bound by lack of formal preparation, lack of effective applications, or other perceived limitations. A recent survey conducted by the Joan Ganz Cooney Center and BrainPop (Millstone & Levy, 2012) found that 50% of teachers who use educational games in their classrooms are self-taught with regards to which games to use and how to use them. Furthermore, 35% of the teachers surveyed first explored the educational possibilities of digital games through self-directed study.

SCHOOL-CENTERED AND ORGANIZATIONAL TENSIONS WITH GAMES

Moving beyond general research on teachers and technology, theoretical work on games generally agrees that the theories of learning embedded in games often run counter to the social organization of schooling (e.g., Gee, 2004, 2007). This section reviews some of these tensions between school organization and game-based learning and provides some analysis as to which of these tensions can be remediated. Generally speaking, schools have moved slowly to embrace game-based learning. Compared with established computerized learning environments such as *WISE* and *NetLogo*, games for learning hold a small share of classroom technology time. Some research has explored the sources of resistance to increasing game-based learning in schools, although this research focus is not particularly extensive. In broad terms, this research focuses on (a) teachers' activity within the school and specific relationships to game-based learning and (b) tensions between gaming and the organizational contexts of schools.

Much of our discussion up to this point has examined internal teaching factors (such as beliefs and attitudes) that constrain the integration of technology (or specifically games for learning) into classroom practice. There are also externally driven factors, however, that involve the teacher as a practitioner within the organizational setting of school. Teachers themselves appear in the literature as the most frequent reporters of these factors, and we interpret the tension produced by these external factors as part of the pragmatics of the modern school setting. Simpson and Stansberry (2009) identify several barriers perceived by teachers to efforts to bring games into their classrooms. Chief among these barriers are (a) the underdeveloped state of theory on facilitating learning through digital games and (b) the lack of familiarity most teachers have with digital games. A recurring theme in the literature we reviewed underscores the lack of articulated theories of game-based learning that resonate with teachers in terms of classroom implementation and pedagogical value. Along these lines, Kirriemuir and McFarlane's (2004) survey of teachers, curriculum, and technology experts finds that teachers:

1. Face challenges in determining which games are suitable for learning purposes,
2. Feel that they lack the time and skill to familiarize themselves with the games,
3. Feel that games largely do not provide adequate support and assessment materials, and
4. Feel unsupported in using games in classrooms in terms of available technological infrastructure.

Ertzberger (2009) reaches similar conclusions after surveying 390 in-service and pre-service teachers. Teachers in Ertzberger's study cited the lack of relevance to the curriculum, lack of time to design or adopt a game to their specific needs, and lack of available technological resources. Summarizing thirteen previous studies and surveys, Kebritchi, Hirumi, Wendy, and Kappers (2009) find that teachers and school media specialists experience several different categories of obstacles to integrating educational games in classroom settings, including time constraints, lack of sufficient technical training, scheduling issues, and concerns with keeping students on-task.

Another source of tension involves the fact that gaming is, by its open-ended and dynamic nature, an activity that runs counter to some established views of schooling, which stress orderly progress towards pedagogically significant goals (Gee, 2007). Several authors have defined this tension in terms of (a) teachers' sense of academic accountability, (b) issues of organization of time, and (c) availability of technology resources and support.

We will now examine each of these factors separately. With regard to teachers' sense of academic accountability, Kirriemuir and McFarlane explain that "it is difficult for teachers to identify quickly how a particular game was relevant to some component of the statutory curriculum, as well as the accuracy and appropriateness of the content within the game" (Kirriemuir & McFarlane, 2004, p. 18). Simpson and Stansberry (2009) build on this idea, explaining that "if teachers have not prepared the students to be able to respond to very specific knowledge based content driven assessments, the school and their jobs could be in jeopardy. Digital games are viewed by teachers as being an 'unknown' (p. 169). In these senses, teachers often perceive conflict between gaming as an activity that is productive for learning and the standards to which they and their students are accountable.

In terms of issues of organization of time, "the most frequently encountered perceived or actual obstacles were [...] the lack of time available to teachers to familiarize themselves with the game" (Kirriemuir & McFarlane, 2004, p. 3). As Sanford and colleagues explain:

Many teachers found the fixed length of lessons to be constraining in both the planning and implementation of games-based learning in schools. In part, this seemed to be a result of the novelty of the activity: teachers were unsure how much time an activity might take, and several expressed confidence that if they were to try similar activities again they would be able to manage classroom time more effectively. The fact that the available time was fixed meant that the impact of any technical issues (loading times, crashes, etc.) was more keenly felt than might have been the case had there been more flexibility in the timetable (Sanford, Ulicsak, Facer, & Rudd, 2006, p. 23).

Thus time constraints, costs, and unknowns also represent significant barriers for teachers in integrating game-based learning into their teaching.

In terms of availability of technology resources and support, "59% of all teachers would be willing to consider using such games in the future [but] 49% believed that there would be a lack of access to equipment capable of running the games" (Sanford, Ulicsak, Facer, & Rudd, 2006, p. 16). Similarly, the 390 teachers surveyed by Ertzberger "indicated the biggest deterrents to the use of video games were [...] lack of the needed technology" (Ertzberger, 2009, p. 1827). Indeed, the diffusion of computer technology into classrooms has remained problematic, even as computers have become less expensive. According to the National Center for Educational Statistics, the ratio of students-to-instructional computers in U.S. public secondary schools improved rapidly until the year 2000 but has remained largely unchanged since 2005, with 3.3 students per computer in 2005 and 2.9 students per computer in 2008 (Snyder & Dillow, 2011).

This scarcity of instructional computers is very problematic for any game-based learning practice, as most (if not all) computer games are designed around a ratio of exactly one player per computer. That said, there are at least two trends that may help mitigate this problem. First, there is the recent expansion of educational games into consoles, mobile platforms, and tablet computers, which are less expensive for schools to own and easier for teachers to operate. Second, personal ownership among older students of devices capable of running educational games is growing rapidly. According to data from the Pew Internet & American Life Project covering the same period of time, smartphone ownership by teens increased from 41% in 2004 to 71% in 2008 (Pew Internet and American Life Project, 2011). Either of these trends, or both together, could signal significant progress towards solving the problem perceived by teachers with regards to access to technology capable of supporting game-based learning practices.

In summary, game-based learning initiatives have yet to achieve wide-scale implementation, partly due to underlying tensions between the activity framework of game-based learning and the organizational constraints perceived by teachers, as well as the practical realities of everyday school activity. The literature is more articulate on limitations expressed by teachers, indicating that there may be elements of game-based learning that teachers do not fully accept. These limitations may partially originate with the nature of educational gaming as a computer-centered activity, and these limitations may therefore be understood in the more general terms of how teachers prepare, plan for, and evaluate the use of technological tools for learning. Another possible explanation, which we will explore in the next section, is that researchers and designers of games for learning have generally not articulated or communicated a role for teachers in educational gaming. Without a guiding narrative of how they "fit in" to the activity, teachers may feel particularly excluded or unprepared to scaffold their students' learning in the context of educational games.

UNDERSPECIFIED ROLE FOR TEACHERS IN GAME-BASED LEARNING PRACTICES

Thus far, we have outlined the barriers to integrating games for learning into classroom practice. A common thread in the accounts of these barriers is that it is *teachers*, not researchers or administrators, who are most keenly aware of them. The effect of these barriers is then exacerbated by the fact that the literature and research on game-based learning does not clearly articulate a position for the teacher in the classroom, so teachers lack guidance as to what their role should be, beyond simply doing their best to bring educational games into their classrooms in some form or another. For example, Kerbitchi et al. (2009) found that, out of ten educational games surveyed, only four provided a teacher's manual, and only one provided teacher aids, lesson plans, or unit plans. We must note, however, this lack of support for teachers is not a deliberate oversight on the part of educational game designers and researchers, but rather it may be a consequence of the audience for whom the game was developed or the envisioned classroom implementation from which it was designed. In a sample of reports of instances of classroom use of games for learning present in the educational research literature, Ng, Zeng, and Plass (2009) found that only 13% of the documented cases involved educational games intended for classroom use, while 68% of the documented instances featured games that were either marketed as entertainment titles or developed specifically for research purposes. Designers of the games represented in this latter 68% typically do not focus on creating teaching aids, lesson plans, and unit plans for teachers. Without a clear narrative of what teachers can do to improve the design and to facilitate the implementation of educational games, however, teachers may understandably feel unsupported in providing guidance, feedback, and scaffolding to their students in order to maximize the learning with digital games.

We believe that it is the role of educational researchers to generate, disseminate, and circulate these narratives. The current predominant absence of these narratives from game-based learning research is a complex issue, but the core reasons may stem from the same theoretical foundations underlying the perceived potential of games as tools for learning. Generally speaking, literature on games for learning features two theoretical positions that may imply, either directly or indirectly, that there is no need to articulate a role for teachers in their frameworks. First, there is the assumption that games engage learners through a channel that is completely direct. The designer and the developers of the game have created the experience and encoded the curriculum. Within that curriculum, the students' actions, attitudes, and motivation are what drive learning. This limited perspective by designers and developers does not recognize the critical intermediation of the teacher in facilitating learning in the classroom.

Second, as alluded to earlier, games reify forms and ways of learning that are largely incompatible with schooling as it is currently conceived. Teaching, in the model underlying the prevailing paradigm of schooling, is characterized as an activity likewise incompatible with games for learning (Shaffer, et al., 2005; Squire, 2005). Proponents note that students learning with educational games are placed in a central active role in the learning activity to such a degree that the activity, in fact, cannot proceed without the full participation of the learner. This complete engagement hinges on the fact that players of digital games enjoy a large degree of freedom and agency to play the game on their own terms. Gamers set their own goals and standards of performance, advance at their own pace, and add a highly individualized interpretation and meaning to the experience of play. The contrast between this description and the pervasive vision of the state of affairs in the typical classroom is constructed by some games and digital media researchers as a critique of established modes of school (Prensky, 2005; Shaffer, Squire, Halverson, & Gee, 2005).

In our view, the assumption that underpins both of these positions is that games and schools are largely inflexible cultural forms that supposedly exist in natural opposition to each other on the subject of teaching and learning. It is not uncommon for teachers, administrators, and policy-makers to hold the view that digital games are an oppositional force to learning and, vice versa, some proponents of educational gaming are very critical of formal schooling. For example, the highly influential game designer Chris Crawford wrote:

Games are... the most ancient and time-honored vehicle for education. They are the original educational technology, the natural one, having received the seal of approval of natural selection. We don't see mother lions lecturing cubs at the chalkboard; we don't see senior lions writing their memoirs for posterity. In light of this, the question, 'Can games have educational value?' becomes absurd. It is not games but schools that are the newfangled notion, the untested fad, the violator of tradition. Game-playing is a vital educational function for any creature capable of learning (Crawford, 1984, p. 16).

Crawford's position, while phrased in strong terms, is generally echoed throughout the scholarly literature on games and schools. Researchers such as Squire (1995), for example, have emphasized the limiting nature of school as a cultural form that would seek to include games. "As challenging as it is to design a good educational game," Squire writes, "it may be more challenging to design a good educational system for educational games to flourish in. [...] Our contemporary educational systems do not know how to sustain a curricular innovation built on the properties that make games compelling" (Squire, 1995, p. 6). Thus, if teachers' roles are defined as the embodied presence and spokesperson for the traditional cultural form of school, they are positioned as a natural nucleus of resistance to game-based learning initiatives.

It is clear, however, that these perspectives, whether based on received cultural forms or arguments stemming from theories of learning, are neither necessarily accurate nor constructive. We have previously stated our belief that games, like all forms of classroom activity, must articulate the role of teachers and leverage their expertise to maximize learning opportunities for students. The teacher is the ideal flexible interface between the cultures of schooling and games, capable of modulating and aligning the affordances and structures of each.

Some proposals in the literature describe such a role for teachers, while maintaining the focus of inquiry squarely on the interactions between the learner and the game. For example, Wilson (2009), writing from the perspective of the software industry and its interest in increasing the use of games and simulations in the classroom, envisions a central role for teachers that strongly emphasizes a "guide on the side" pedagogy. According to Wilson, a teacher can prepare students with the necessary background knowledge, intervene with advice during play, and guide reflective conversation after play about what students learned and how it can be applied elsewhere. Halverson (2005) casts teachers in a dual role: (a) as expert gamers who can facilitate gaming experiences for students and (b) as guides to enable reframing of game-content into forms which align with the curriculum. Becker (2007) envisions teachers as both careful critics of games and capable gamers. In Becker's format, the teacher's selection, discussion, and framing of particular games to reach curriculum goals constitute a form of instructional design. The effectiveness of each proposed role and how each role affects the teacher-perceived barriers to the inclusion of games in the classroom have not been addressed by the literature.

Teachers and Teaching in Game-Based Learning Theory and Practice

Reform perspectives on science learning stress the fact that teachers have a central role in student learning even when the tools that support that learning have their own embedded forms of guidance. For example, the influential NRC report *Taking Science to School*, states that software tools (e.g., simulations or games) "offer useful structure to student learning activities, but they cannot dictate learning. The teacher plays a critical role in realizing these designs" (Duschl, Schweingruber, & Shouse, 2007, p. 268). Although games might appear to be rigidly prescribed, with no classroom adaptation necessary or even possible, the teacher's skills and understanding of the students are critical to the success of the activity. The teacher, not the software, can orchestrate discussion, help students form and test hypotheses, guide students in forming explanations and organizing evidence, and integrate multiple strands of activity into coherent learning outcomes. As with any classroom activity, the ultimate potential of educational gaming is profoundly shaped by teachers and their beliefs, talents, and perspectives.

THE "MISSING LINK": TEACHERS' IDENTIFICATION AS GAMERS AND MAKERS OF GAMES

How might teachers and schools connect with designers and researchers of educational games in order to develop a more productive understanding of each other's affordances for enhancing student learning? Up to this point, we have based our analysis on arguments and evidence presented in the research literature on educational technology, teaching, and games for learning. We identified the divisions and places of tension. We will now explore ways in which these might be resolved. What roles or constructs might better integrate educational gaming into classroom practice?

One possible answer involves extending Prensky's distinction of "digital natives" versus "digital immigrants." This distinction touches on dual issues of identity and teaching capacity. "Digital immigrant" teachers may have fewer resources to leverage digital games for learning, whereas "digital native" teachers may have more evolved experiences and resources to connect gaming with their professional practice. Very little published research directly addresses this hypothesis. Schrader, Zheng, and Young's (2006) survey of 203 pre-service teachers found that most of the pre-service teachers had played digital games. More than half of the teachers surveyed played digital games with some frequency, but most respondents stopped short of identifying themselves as gamers. Shrader and colleagues also found that the pre-service teachers they sampled were generally positive about using digital games as learning contexts. Shrader and colleagues do not claim a causal link between the teachers' game playing experience and their generally positive attitudes toward games for learning, but it seems reasonable to assume that the pre-service teachers' own affinity for games may have contributed in some measure to their positive attitudes toward games for learning. The converse, however, may not be true (i.e., that positive attitudes towards educational gaming lead teachers to play more digital games in their personal time). In his dissertation work, James (2007) found no significant difference between teachers who are gamers and those who are not in measures of overall instructional technology usage, overall participation in innovative teaching strategies, and overall comfort in completing job-related technology tasks. This finding suggests that the "gamer" component of identity is not a necessary component of the "digital native" identity. It follows from these two findings that, while most people who view themselves as gamers are "digital natives," not all "digital natives" view themselves as gamers.

Having established this distinction, we must ask what a "gamer" identity affords teachers in terms of successful integration of games for learning that the more general "digital native" identity does not? Are there specific practices that can be traced back to teachers who identify as gamers, or particular beliefs that may or may not inform practice? No study that we could locate provides an answer to this question, although some inferences may be drawn from the reviewed literature. Some synergy exists, for example, between Halverson's (2005) vision of a teacher as a "master gamer" and the view expressed by Kafai, Franke, Ching, and Shih (1998) that teachers benefit from the design of games for learning. More specifically, it is possible that teachers who view themselves as gamers may be better prepared and motivated to successfully design games for learning or to design curricula in support of games for learning.

In terms of the vision of teachers as designers of games, Kafai et al. (1998) studied a group of 16 pre-service teachers as they participated in game-design activities where the product was a game that would help students to learn fractions. As the teachers became more familiar with the processes and constraints of game design, the games that the teachers created became more content-integrated (i.e., fraction content and game material were more closely tied together) and contained more of the teachers' knowledge of the development of children's thinking. Kafai and colleagues' study contains no information as to whether these games were more effective at helping students learn fractions, but the authors point to Loef's (1991) findings that teachers were more successful when their practice integrated both the content and their own knowledge of the development of children's mathematical thinking. Thus, game design can be seen as an activity that can drive teachers' sophisticated reasoning about student learning and ultimately empower their practice. These findings also serve to link game design with the greater framework of Pedagogical Content Knowledge (PCK) proposed by Shulman (1987), which conceives the teaching process as grounded precisely in these forms of knowledge that integrate both content and understanding of the students' learning processes.

There have been several efforts to place the tools and know-how of educational game design directly in the hands of teachers, and the results are encouraging. For example, Annetta, Mangrum, Holmes, Collazzo, and Cheng (2009) observed positive learning out comes for fifth-graders when playing *Dr. Friction*, a game designed to support a unit on simple machines. *Dr. Friction* is notable because it was created entirely by the classroom teacher. This game was created as part of the HI FIVES project, in which teachers were trained in the use of a game creation toolkit that was especially designed for users without any knowledge of programming or computer graphics design (Annetta, 2008). Another HI FIVES game, *The stolen fortune of I.M. Megabucks*, was found to help high school students develop 21st-century skills (i.e., digital literacy, inventive thinking, effective communication, and high productivity), and even helped improve the teacher's design (Annetta, Cheng, & Holmes, 2010). One important thing to note in these examples is that the teacher, as the designer and principal creator of the game, enjoyed a position within the classroom gaming activity that was central, active, and prestigious. The teacher/game designer was intimately familiar with the game, could answer students' questions about its workings, guide students when they encountered difficulty, and moderate interaction and discussion about the game's subject material.

Perhaps more feasible on a general level, teachers who view themselves as gamers may be better situated to be designers of curricula that integrated games for learning. Such game-based curricula designed by teachers who are gamers might more effectively leverages the affordances in terms of the specific knowledge that teachers possess about students' thinking and learning as well as in terms of specific knowledge about game mechanics, curriculum focus, assessment needs, and time constraints that influence their particular context. Educational game designers and researchers will find great value in partnering with teachers to help craft game-based learning initiatives. Teachers bring perspective and real-world experience that is difficult for non-teachers to access. In our own work, we have often relied on teachers not just for access to their classrooms so we can conduct our research, but also for their extensive experience and wisdom in terms of assessment, curriculum priorities, technological needs and limitations, and the current sociopolitical climate surrounding public education. Our experience with these collaborations has been uniformly positive and fruitful, and so we are constantly looking for ways to include our teacher advisors into the design and iteration process. As this form of collaboration becomes more commonplace, we believe it has strong potential to address, and perhaps successfully navigate, the issues of accountability described in Kirriemuir and McFarlane (2004) and the concern over curricular relevance expressed by Ertzberger (2009), as the assumptions and priorities that are encoded in game design would be in the hands of teachers in their classrooms.

On another level, it can be said that the capacities of a game designer are, in fact, of great value to the practice of teaching, even in systems of activity that are not computer-based. For example, an approach to instruction that is informed by game design would be more interactive, learner-centered, collaborative, and engaging; assessment, classroom management, and participation are all elements of classroom practice that could be inlaid with game elements to improve their efficacy. This process of "gamification" of school processes is advocated by McGonigal (2011) in her recent book *Reality is Broken*. McGonigal suggests that games are a unique way to structure experience and provoke positive emotion, to inspire participation, and to motivate hard work; these elements are not only highly desirable in the everyday business of classrooms but are also part of the experience that students consistently receive from games, and find nowhere else. A teacher/game designer would be able to infuse instruction with the powerful, compelling narratives of gaming, which according to McGonigal, are:

1. **Satisfying work:** Clearly defined activities that allow us to see direct impact from our efforts.
2. **Experience, or at least hope, of success:** To feel power in our own lives and show others what we are good at.
3. **Social connection:** Sharing experiences and building bonds with people we care about.
4. **Meaning:** To belong to and contribute to something that has lasting significance beyond our own individual lives (McGonigal, 2011, p. 49).

In this view, the potential of game-based learning to improve education is dwarfed by the potential of the "gamification" of learning. Students are not only engaged and motivated in the comparatively brief interludes spent playing a learning game in their classrooms, but their entire experience of school is game-like in terms of being more engaging, connected, and meaningful. Yet this transformation requires a teacher who, along with all his or her other professional capacities, is a game designer: a skillful creator of game structures and game-like experiences, challenges and rewards, what McGonigal terms a "happiness engineer."

FINAL THOUGHTS

Games hold great potential to enhance education, but generally speaking, teachers have not been sufficiently supported in using games for learning in the classroom, nor has teachers' expertise been optimally leveraged in the design of games for learning. While research has suggested that teachers should have more input into the forms and content of learning games, the development of games for learning remains firmly in the hands of educational researchers or commercial game developers. In light of the constraints teachers face when implementing game-based learning activities, this represents a missed opportunity; teachers bring specific knowledge about the limitations imposed by organizational time, curriculum, and available technology, which could be integrated into the designs of games that would be more viable and useful for classroom use.

Not only do teachers have much to offer to the game design process, but the discipline of game design also has much to offer the practice of teaching. Teachers who understand game design could integrate more and more of the compelling dynamics of gaming into a school environment that desperately needs them. As Prensky (2005) noted, "Today's kids are not ADD, they're EoE (Engage or Enrage)" (p. 1). If research were to validate the capacity of teachers as game designers and teacher preparation programs were to include it, the goal of engaging more students in deep meaningful learning may be closer at hand.

REFERENCES

Anderson, C. A. (2004). An update on the effects of playing violent video games. *Journal of Adolescence*, *27*(1), 113–122. doi:10.1016/j.adolescence.2003.10.009

Anderson, C. A., & Bushman, B. J. (2001). Effects of violent video games on aggressive behavior, aggressive cognition, aggressive affect, physiological arousal, and prosocial behavior: A meta-analytic review of the scientific literature. *Psychological Science*, *12*(5), 353–359. doi:10.1111/1467-9280.00366

Annetta, L. A. (2008). Video games in education: Why they should be used and how they are being used. *Theory into Practice*, *47*(3), 229–239. doi:10.1080/00405840802153940

Annetta, L. A., Cheng, M., & Holmes, S. (2010). Assessing twenty-first century skills through a teacher created video game for high school biology students. *Research in Science & Technological Education*, *28*(2), 101–114. doi:10.1080/02635141003748358

Annetta, L. A., Mangrum, J., Holmes, S., Collazo, K., & Cheng, M.-T. (2009). Bridging realty to virtual reality: Investigating gender effect and student engagement on learning through video game play in an elementary school classroom. *International Journal of Science Education*, 31.

Annetta, L. A., Minogue, J., Holmes, S. Y., & Cheng, M.-T. (2009). Investigating the impact of video games on high school students' engagement and learning about genetics. *Computers & Education*, *53*(1), 74–85. doi:10.1016/j.compedu.2008.12.020

Becker, K. (2007). Digital game-based learning once removed: Teaching teachers. *British Journal of Educational Technology*, *38*(3), 478–488. doi:10.1111/j.1467-8535.2007.00711.x

Clark, D., Nelson, B., Sengupta, P., & D'Angelo, C. (2009). *Rethinking science learning through digital games and simulations: Genres, examples, and evidence*. Paper presented at Learning Science: Computer Games, Simulations, and Education Workshop Sponsored by the National Academy of Sciences. Washington, DC.

Crawford, C. (1984). *The art of computer game design: Reflections of a master game designer*. New York, NY: Osborne/McGraw-Hill.

Cuban, L. (1999). The technology puzzle. *Education Week*, *18*(43), 58.

de Freitas, S. (2006). *Learning in immersive worlds: A review of game-based learning*. Bristol, UK: Joint Information Systems Committee (JISC) E-Learning Programme. Retrieved March 11, 2008, from http://www.jisc.ac.uk/media/documents/programmes/elearninginnovation/gamingreport_v3.pdf

Dieterle, E. (2010). Games for science education. In Hirumi, A. (Ed.), *Playing Games in School: Video Games and Simulations for Primary and Secondary Education* (pp. 89–112). Eugene, OR: International Society for Technology in Education.

Duschl, R. A., Schweingruber, H. A., & Shouse, A. W. (Eds.). (2007). *Taking science to school: Learning and teaching science in grades K-8*. Washington, DC: The National Academies Press.

Ertzberger, J. (2009). An exploration of factors affecting teachers' use of video games as instructional tools. In *Proceedings of the Society for Information Technology & Teacher Education International Conference 2009*, (vol. 1, pp. 1825-1831). IEEE.

Federation of American Scientists. (2006). *Summit on educational games - Harnessing the power of video games for learning*. Washington, DC: Federation of American Scientists.

Gee, J. P. (2004). *Situated language and learning: A critique of traditional schooling*. London, UK: Routledge.

Gee, J. P. (2007). *What video games have to teach us about learning and literacy* (2nd ed.). New York, NY: Palgrave Macmillan. doi:10.1145/950566.950595

Hall, R. C. W., Day, T., & Hall, R. C. W. (2011). A plea for caution: violent video games, the Supreme Court, and the role of science. *Mayo Clinic Proceedings, 86*(4), 315–321. doi:10.4065/mcp.2010.0762

Halverson, R. (2005). What can K-12 school leaders learn from video games and gaming. *Innovate: Journal of online. Education, 1*(6).

Hauge, M. R., & Gentile, D. A. (2003). Video game addiction among adolescents: Associations with academic performance and aggression. *Child Development, 40*(306), 1–3.

Hines, P. J., Jasny, B. R., & Mervis, J. (2009). Adding a T to the three R's. *Science, 323*(5910), 53. doi:10.1126/science.323.5910.53a

Honey, M. A., & Hilton, M. (Eds.). (2010). *Learning science through computer games and simulations*. Washington, DC: National Academy Press.

James, C. (2007). *Playing the game: Comparing teacher gamers to non-gamers*. (Unpublished Doctoral Dissertation). The University of Alabama. Birmingham, AL. Retrieved from http://proquest.umi.com/pqdlink?RQT=568&VInst=PROD&VName=PQD&VType=PQD&Fmt=7&did=1379569531&TS=1302994639&fromjs=1

Kafai, Y. B., Franke, M. L., Ching, C. C., & Shih, J. C. (1998). Game design as an interactive learning environment for fostering students' and teachers' mathematical inquiry. *International Journal of Computers for Mathematical Learning, 3*(2), 149–184. doi:10.1023/A:1009777905226

Kebritchi, M., Hirumi, A., Kappers, W., & Henry, R. (2008). Analysis of the supporting websites for the use of instructional games in K-12 settings. *British Journal of Educational Technology, 40*(4), 733–754. doi:10.1111/j.1467-8535.2008.00854.x

Ketelhut, D. J. (2007). The impact of student self-efficacy on scientific inquiry skills: An exploratory investigation in river city, a multi-user virtual environment. *Journal of Science Education and Technology, 16*(1), 99–111. doi:10.1007/s10956-006-9038-y

Kirriemuir, J., & McFarlane, A. (2004). *Literature review in games and learning*. Bristol, UK: Futurelab.

Lave, J., & Wenger, E. (1991). *Situated learning: Legitimate peripheral participation*. Cambridge, UK: Cambridge University Press. doi:10.1017/CBO9780511815355

Lenhart, A., Jones, S., Macgill, A. R., & Pew Internet and American Life Project. (2008). *Adults and video games*. Retrieved February 3, 2012, from http://www.pewinternet.org/~/media//Files/Reports/2008/PIP_Adult_gaming_memo.pdf.pdf

Mayo, M. J. (2009). Video games: A route to large-scale STEM education? *Science*, *323*(5910), 79–82. doi:10.1126/science.1166900

McGonigal, J. (2011). *Reality is broken: Why games make us better and how they can change the world*. New York, NY: Penguin Press.

Millstone, J., & Levy, A. (2012). *What do teachers really think about using video games in the classroom?* Paper presented at the 9th Annual Games for Change Summit. New York, NY.

Moursund, D., & Bielefeldt, T. (1999). *Will new teachers be prepared to teach in a digital age? A national survey on information technology in teacher education*. Santa Monica, CA: Milken Exchange on Education Technology. Retrieved from http://www.eric.ed.gov/ERICWebPortal/contentdelivery/servlet/ERICServlet?accno=ED428072

Mumtaz, S. (2000). Factors affecting teachers' use of information and communications technology: A review of the literature. *Journal of Information Technology for Teacher Education*, *9*(3), 319. doi:10.1080/14759390000200096

Murray, J. P., Biggins, B., Donnerstein, E., Menninger, R. W., Rich, M., & Strasburger, V. (2011). A plea for concern regarding violent video games. *Mayo Clinic Proceedings*, *86*(8), 818–820. doi:10.4065/mcp.2011.0321

National Council for Accreditation of Teacher Education. (1997). *Technology and the new professional teacher: Preparing for the 21st century classroom*. Retrieved from http://www.eric.ed.gov/ERICWebPortal/contentdelivery/servlet/ERICServlet?accno=ED412201

National Council for Accreditation of Teacher Education. (2008). *Professional standards for the accreditation of teacher preparation institutions*. Retrieved from http://www.ncate.org/Portals/0/documents/Standards/NCATE%20Standards%202008.pdf

Ng, F., Zeng, H., & Plass, J. (2009). *Research on educational impact of games: A literature review*. Report No. 02/2009. New York, NY: Institute for Games for Learning, NYU/CUNY.

Office of Technology Assessment. (1995). *Teachers and technology: Making the connection*. Report No. OTA-EHR-616. Washington, DC: U.S. Government Printing Office. Retrieved from http://www.eric.ed.gov/ERICWebPortal/contentdelivery/servlet/ERICServlet?accno=ED386155

Papert, S., & Harel, I. (1991). Situating constructionism. In *Constructionism*. New York, NY: Ablex.

Pelletier, C. (2008). Gaming in context: How young people construct their gendered identities in playing and making games. In Kafai, Y. B., Heeter, C., Denner, J., & Sun, J. Y. (Eds.), *Beyond Barbie and Mortal Kombat: New Perspectives on Gender and Gaming*. Cambridge, MA: The MIT Press.

Pew Internet and American Life Project. (2011). *Teen internet usage over time*. Retrieved from http://www.pewinternet.org/Static-Pages/Trend-Data-for-Teens/~/media/Infographics/TrendData/Teens/September2009/TeenInternetUsageOverTime–Sep2009.zip

Prensky, M. (2001). Digital natives, digital immigrants part 1. *Horizon*, *9*(5), 1–6. doi:10.1108/10748120110424816

Prensky, M. (2005). Engage me or enrage me: What today's learners demand. *EDUCAUSE Review*, *40*(5), 60.

Russell, M., Bebell, D., O'Dwyer, L., & O'Connor, K. (2003). Examining teacher technology use: Implications for preservice and inservice teacher preparation. *Journal of Teacher Education*, *54*(4), 297–310. doi:10.1177/0022487103255985

Sanford, R., Ulicsak, M., Facer, K., & Rudd, T. (2006). *Teaching with games: Using commercial off-the-shelf games in formal education*. Bristol, UK: Futurelab.

Schrader, P. G., Zheng, D., & Young, M. (2006). Teachers' perceptions of video games: MMOGs and the future of preservice teacher education. *Journal of Online Education*, *2*(3).

Shaffer, D. W. (2006). *How computer games help children learn* (1st ed.). New York, NY: Palgrave Macmillan. doi:10.1057/9780230601994

Shaffer, D. W., Squire, K. R., Halverson, R., & Gee, J. P. (2005). Video games and the future of learning. *Phi Delta Kappan*, *87*(2), 104.

Sheingold, K., & Hadley, M. (1990). *Accomplished teachers: Integrating computers into classroom practice*. New York, NY: Center for Technology in Education, Bank Street College of Education. Retrieved from http://www.eric.ed.gov/ERICWebPortal/contentdelivery/servlet/ERICServlet?accno=ED322900

Shulman, L. S. (1986). Paradigms and research programs in the study of teaching: A contemporary perspective. In Wittrock, M. C. (Ed.), *Handbook on Research on Teaching* (3rd ed., pp. 3–36). New York, NY: Macmillan.

Shulman, L. S. (1987). Knowledge and teaching: Foundations of the new reform. *Harvard Educational Review*, *57*(1), 1–23.

Simpson, E., & Stansberry, S. (2009). Video games and teacher development: Bridging the gap in the classroom. In Miller, C. T. (Ed.), *Games: Purpose and Potential in Education* (pp. 163–185). New York, NY: Springer. doi:10.1007/978-0-387-09775-6_7

Skoric, M. M., Teo, L. L. C., & Neo, R. L. (2009). Children and video games: Addiction, engagement, and scholastic achievement. *Cyberpsychology & Behavior*, *12*(5), 567–572. doi:10.1089/cpb.2009.0079

Snyder, T. D., & Dillow, S. A. (2011). *Digest of education statistics: 2010*. Washington, DC: U.S. Department of Education, Institute of Education Sciences, National Center for Education Statistics. Retrieved from http://nces.ed.gov/pubs2011/2011015.pdf

Sprague, D. (2004). Technology and teacher education: Are we talking to ourselves? *Contemporary Issues in Technology & Teacher Education*, *3*(4). Retrieved from http://www.citejournal.org/vol3/iss4/editorial/article1.cfm

Squire, K. (2005). Changing the game: What happens when video games enter the classroom. *Innovate: Journal of Online Education*, *1*(6), 25–49.

Steinkuehler, C., & Chmiel, M. (2006). Fostering scientific habits of mind in the context of online play. In *Proceedings of the 7th International Conference on Learning Sciences,* (pp. 723-729). New York, NY: International Society of the Learning Sciences.

U.S. Department of Education. (2000). *Teacher's tools for the 21st century: A report on teacher' use of technology*. Washington, DC: National Center for Education Statistics.

Wilson, L. (2009). *Best practices for using games & simulations in the classroom guidelines for k–12 educators*. New York, NY: SIIA Education Division.

Section 3
Curriculum Innovations

Chapter 9
Opening Both Eyes:
Gaining an Integrated Perspective of Geology and Biology

Renee M. Clary
Mississippi State University, USA

James H. Wandersee
Louisiana State University, USA

ABSTRACT

The focus of this chapter is an exploration of integrated geology and biology learning—from past to present. The chapter explains why active and integrated geological and biological learning became the lodestar of the authors' decade-long EarthScholars Research Group's research program. The authors argue that using an active and integrated geobiological pedagogical approach when teaching geology or biology provides natural opportunities for students to learn and do authentic scientific inquiry in a manner similar to how contemporary scientists conduct their work. The authors further review research that concerns the active, integrated geobiological science learning approach—in middle school, secondary, and college classrooms, laboratories, and field studies. The authors favor a gradual course transition to this pedagogy, while highlighting the advantages of adopting such an approach—both for teachers and students. Finally, the authors conclude the chapter with challenges and future directions in the design of active, integrated geobiological science learning environments.

INTRODUCTION

During the 2010 holiday season, a former student approached us in a crowded sporting goods store. It was practically a decade since she was registered in one of our large introductory geology courses, and yet she favorably recalled the experience, and sought us out to tell us so. When we asked what she remembered most about the course—and what she liked the most—she replied she enjoyed the petrified wood in the classroom (Clary & Wandersee, 2007). She also pleasingly volunteered the statement that our planet was 4.6 billion years old.

DOI: 10.4018/978-1-4666-2809-0.ch009

When we reflected back on that introductory geology course for non-science majors, we realized that this former student targeted not only what was successful with our class, but also the successful underpinnings of our entire research program. For ten years, our EarthScholars Research Group has advocated "opening both eyes" in every science class and informal learning experience, and immersing learners in the integrated and interdisciplinary nature of the sciences. Through state, national, and international field-based, image-based, and archival research, we seek to integrate geological and biological learning with the ultimate goal of improving public understanding of these sciences. EarthScholars' research has focused on traditional classrooms at the middle, high school, and college level, as well as non-traditional online and distance classroom settings, and informal educational environments.

In our research program, we stress that most biological and geological learning, and informal and formal science educational programs, are separated for artificial reasons. This separation is to the detriment of meaningful science learning. We seek to bridge informal and formal learning, and improve the articulation of the sciences, for a more scientifically literate population.

INTEGRATED GEOLOGICAL-BIOLOGICAL LEARNING: A HISTORICAL REFERENCE

Even within elementary grades, students often have definitive periods or days specifically dedicated to different subjects. By the time students reach middle school settings (US grades 6-8), their classes consist of separate courses within "life science" or "physical science" or "Earth science," with scientific content presented independently from mathematics instruction. However, the compartmentalization of scientific sub-disciplines in *educational settings* does not have a history that parallels that of *scientific investigations*.

The earliest analyses of our natural world were grouped under the general heading of "natural philosophy." Natural philosophy's encompassing nature resulted in an integrated approach to observed problems arising from nature, with mathematical expressions used as needed to explain and clarify the collected and recorded data. Chemistry, physics, and much of geosciences content can be subsumed under natural philosophy's earlier organization.

As scientific methodologies developed, the modern scientific approaches differed from natural philosophical ones in systematic observation and research. However, even with modern scientific approaches in the 1800s, scientific sub-disciplines were not separated in instruction, nor taught independently in schools. Before 1900, students in US classrooms generally studied an integrated science approach with chemistry, physics, and geosciences, with mathematics routinely included in the investigation and study of natural phenomena.

BENEFITS OF INTEGRATED SCIENTIFIC LEARNING: THE HUMAN CONSTRUCTIVIST VIEW

As firm supporters of the inclusion of the history and philosophy of science in science teaching (Matthews, 2003), we recognize that the most effective methods for science instruction do not necessarily involve new approaches or the latest technology. In our research and historical investigations, we concluded that the re-appropriation of some historical techniques can add value in modern science classrooms. For example, our research demonstrated that several historical visualization methods—including aquarium view graphics or scientific caricatures—can be used for effective teaching and assessment (Clary & Wandersee, 2005, 2010a). We further recognize that the older, more integrated approach in science classrooms more authentically represents the interdisciplinary nature in which contemporary scientists conduct research. We propose a return to the "natural philosophy" by which science instruction was historically conducted.

In 2002, we founded EarthScholars Research Group with the mission to promote interdisciplinary approaches in science classrooms. As a botanist and biology educator (Wandersee) and a geologist and geoscience educator (Clary), we actively research geological and biological integration in multiple educational settings, including traditional, informal, and online classrooms. Our research consistently demonstrates that an active, integrated geobiological approach provides natural opportunities for students (both formal and informal) to learn and conduct authentic scientific inquiry, similar to the approaches utilized by our contemporary scientists. Guiding our science educational research is the learning theory of human constructivism, which promotes inquiry based and active learning strategies that reflect the interdisciplinary approach of modern scientists in typical field and laboratory investigations.

Human Constructivism

Based upon the work of cognitive psychologist Ausubel (1963, 1968; Ausubel, et al., 1978) and the pioneering work of science education researcher Novak (1963), the learning theory of human constructivism postulates that humans construct meaning from their lived experiences when they integrate new concepts into their existing knowledge framework in new, non-verbatim ways. Therefore, in order for instruction to be effective and conceptual change to occur, learners must be able to anchor new knowledge and experiences into their existing cognitive structures. Ausubel (1968) stated that the most important factor was what the learner already knew when entering the science classroom; teachers should determine this existing knowledge and modify instruction accordingly. Because learners often possess scientifically inaccurate views of how certain natural phenomena operate, teachers have to identify these alternative conceptions (often termed misconceptions in the science education literature) and directly address them with instruction (Novak & Musonda, 1991; Wandersee, et al., 1994). Otherwise, instruction can be ineffective.

The science classrooms of teachers advocating the human constructivist approach can be noticeably different from other science classrooms. Instead of covering an entire textbook in the school year with a "mile-wide, inch-deep" approach, constructivists advocate a "less is more" curriculum in order to teach a selected number of important scientific constructs in a meaningful way. The constructivist approach advocates understanding over awareness, and meaning over memorization (Mintzes, et al., 1998, 2000). When students are aware of and monitor their own learning, meaningful learning can result (Novak, 1998; Novak & Gowin, 1984). This process, often termed "metacognition," involves both awareness and monitoring of the learning process *by the learner* (Gunstone & Mitchell, 1998). When the US National Research Council (2010) proposed the new US framework for K-12 science education, the selected core constructs, organizing themes, and vertical alignment across grade levels (with new concepts in each grade level building from previous years) echoed human constructivist guidelines.

Active Learning

Human constructivism supports hands-on, active learning strategies, in which students actively engage in scientific inquiry and build more powerful cognitive structures by connecting their experiences within their existing knowledge framework. In the 1990s, DeBoer (1991) recognized that one of the primary goals of science education since the mid-twentieth century was the inclusion of inquiry based activities in the classroom. Student-led inquiry investigations are still affirmed as important components of twenty-first century science classrooms (DeBoer, 2005), but not all inquiry classroom activities are student-driven (Asay & Orgill, 2010). However, researchers continue to affirm the value of active and student-centered learning (McConnell, et al., 2003; Lawrenz, et al., 2005; Michael & Modell, 2003). Hands-on, active learning more closely parallels scientific research processes, and can provide genuine investigative experiences for students (Felzien & Cooper, 2005; Hemler & Repine, 2006). The benefits of inquiry learning also include more positive *attitudes* toward the research process (Lord & Orkwiszewski, 2006). The value of active learning strategies is likewise confirmed in informal, field experiences (Elkins & Elkins, 2007), which can provide holistic experiences (Bernstein, 2004) and supply an environmental context for students (McLaughlin, 2005).

Interdisciplinary Science and the Nature of Science

In field investigations, learners are typically exposed to interdisciplinary science approaches during observation, sampling, and analysis. Because our natural world is not compartmentalized into separate chemical, physical, geological, and meteorological components, professional scientists, and researchers do not work in isolation within one scientific discipline. Therefore, in order to promote a more legitimate view of scientific endeavors, our constructivist guidelines promote the interdisciplinary approach to reveal the interconnectedness of the sciences, both within traditional classrooms and informal field excursions.

In addition to the interdisciplinary nature of science, instructors must counteract what Duschl (1990) termed "final form science." When scientific content is presented *only* within the current knowledge parameters, we reinforce the misconception that scientific understanding has always existed in this final form. When we incorporate the history of the science in the classroom, and reveal how scientific knowledge is restructured and modified as new knowledge is available, then we demonstrate the dynamic nature of the evolving body of scientific knowledge (Duschl, 1994). If we can successfully demonstrate to our students that science is not static, then our students can recognize that our scientific knowledge base evolves. With this acknowledgment and understanding, students can engage in what Langer (1997) described as "mindful learning."

The realization that science is not static has an additional benefit for our students: When we incorporate historical scientific controversies in the classroom, we also reveal to our students that misinformation reported in the scientific literature is ultimately corrected (Clary, et al., 2008). Therefore, our students are better able to understand the nature of science and recognize that it is through scientific replication that a scientific body of knowledge rectifies itself. With this more thorough understanding of how science operates, students are less likely to dismiss *all* science as inaccurate or ineffective when they encounter differing points of view, or accusations of scientific misconduct in modern controversial topics.

INTEGRATED GEOLOGICAL-BIOLOGICAL LEARNING IN K-12, COLLEGE, AND INFORMAL EDUCATIONAL SETTINGS

When we founded EarthScholars Research Group, we recognized the potential classroom benefits of an inquiry based, interdisciplinary approach, which incorporated both the life and Earth sciences. However, we also recognized that, with the exception of informal educational settings, K-12 science teachers must work within an existing educational system and cover specific content with mandated standards, competencies, or learning objectives. Therefore, we researched pedagogical integration in typical science classes (Earth science, biology, integrated) at the middle school and high school levels. We researched the incorporation of an interdisciplinary geobiological approach with the benefits of active learning and inquiry investigations *within* the existing educational system.

By carefully choosing topics that pique students' interest and serve as portals for several content strands of the US National Science Education Standards (National Committee on Science Education Standards and Assessment, 2006), we sought to document the learning benefits of an interdisciplinary, active learning approach, as well as facilitate its implementation within science classrooms. Since our research mission has been the integration of biological and geological content through active learning and inquiry approaches, we typically addressed Life Science and Earth/Space Science content strands, as well as the Science as Inquiry, History and Nature of Science, and Unifying Concepts and Processes in Sciences standards in our research approaches.

Our research investigated both middle school science classrooms (US grades 6-8) and high school science classrooms (US grades 9-12). We also extended our investigations into college geology and biology classrooms, and informal educational settings. These informal educational settings could be environments which extended the traditional or online classroom, and/or which served as educational sites for the general population.

Middle School Science Classrooms

Although some middle school science classrooms are divided into content according to grade level (e.g., Life Science in US grade 7, Earth Science in US grade 8), some of the US state curricular frameworks include an "integrated science" approach. However, the integration approach is not necessarily an interdisciplinary approach; instead, many middle school classrooms teach "integrated science" with separate units in biology, geology, chemistry, and physical science. Although students may receive instruction in several science content areas in an academic year, the content of each sub-discipline is still delivered independently from the other science content areas.

Interdisciplinary Science Portals

We began to research topics that could serve as portals through which to address both life and Earth science concepts *simultaneously*, in order to facilitate integrated science instruction at the middle school level. One of our earlier interdisciplinary portal investigations was amber, the mineraloid that both serves *as* a fossil and a repository *for* fossils. Through amber, several scientific concepts could be easily focused upon in middle school science classrooms, including a study of botany (properties of resin and tree physiology), entomology (identification of insects, and species percentages as represented in amber), organic evolution (extinct insect and tree species, and with more recent research, Cretaceous bird and dinosaur feathers (McKellar, et al., 2011)), ecology (environments represented by the trees producing resin and the included insects), chemistry (polymerization and desiccation reactions that produce amber from resin), and geology (geologic time, fossilization, paleoenvironment reconstruction) (Clary & Wandersee, 2009a). Amber served as a unique vehicle to introduce several scientific concepts; most middle school students were inherently interested in amber's potential for dinosaur cloning because of Crichton's (1990) novel, and its subsequent movie, *Jurassic Park*. When amber was used as a vehicle to extend learning beyond the science classroom, students were intrigued to learn that Czar Peter the Great built a complete room from amber, which was dismantled by the Nazis in World War II and never recovered.

Fortunately, amber offered a large number of topics through which science content could be delivered. A study of amber could be easily implemented in most middle school science classrooms, and then a selected number of topics in other science areas could be scaffolded onto the original investigation for an interdisciplinary approach. We concluded that amber could facilitate a more accurate representation of the interconnectedness of the scientific community (Clary & Wandersee, 2009).

Amber is not the only portal through which interdisciplinary science content can be delivered. With the current scientific and political interest in changing climate, we turned our investigation to the carbon cycle. While most students are well aware of the hydrologic cycle and its role in Earth systems, students appeared to encounter more difficulties with the carbon cycle, and demonstrated little understanding of how carbon moves through the various "spheres" of our planet (e.g., atmosphere, lithosphere, hydrosphere, biosphere). However, this important construct has several ramifications, and carbon literacy is needed so that our future citizens can make wise decisions about our planet.

Using the principles of human constructivism, we researched and investigated an interdisciplinary biological-geological approach that would build upon the incoming knowledge of middle school students. Although not readily familiar with a "carbon cycle," the students were typically familiar with one subset and variable within in: coal (Clary & Wandersee, 2010b). The formation of coal touches upon ancient environments, extinct plant species, sea level changes, and climates that were vastly different than the ones we experience today. Because coal is an inexpensive fossil fuel resource with relatively abundant reserves, we mine it for its energy value. However, our human extraction of coal and our consumption of it impact the Earth. From coal's formation, through its lithification, extraction, and burning as an energy resource, this subset of the carbon cycle efficiently addresses Earth System Science, and can be used effectively in the classroom in a variety of active learning strategies. Through classroom investigations of the different types (or ranks) of coal, their carbon content, and the implication of coal use in the human carbon footprint of the planet, teachers build upon students' pre-existing understanding of fossil fuels. Extending a coal cycle unit into biological and geological content standards can be accomplished in a variety of ways, with several free Web-based resources available to teachers. This integrated geobiological approach facilitates inquiry learning in the classroom, and better prepares students to evaluate the role of fossil fuels in the current discussions on climate change (Clary & Wandersee, 2010b).

Although the topics of coal and amber may appear to be obvious portals for integrated science learning, the majority of topics covered in middle school science classrooms can be extended beyond a single science content strand. For example, plants are typically incorporated into a classroom via the required unit on botany, but plants can be used to address content standards in other scientific arenas, including climate change, ecology, and geology (Clary & Wandersee, 2011a). Not only can plants be extended into other scientific disciplines, but most plant active learning and inquiry activities in middle school classrooms are relatively easy to incorporate (e.g., fruit investigations, floral scent experiments, pollinator identifications, exploration of natural dyes, plants' specific soil and pH requirements). In addition to providing opportunities for a pedagogically geobiological integrated science curriculum, there is another important reason that plants should be incorporated throughout the science curriculum: Although human existence depends heavily on plants, our earlier research revealed that humans typically do not notice the plants in their environments, but focus instead on the animals. We coined the term "plant blindness" to explain this phenomenon (Wandersee & Schlussler, 1999; Wandersee & Clary, 2006a). Because plants are important in our daily lives, students benefit from the realization of the plant interconnectedness of our world.

History and Philosophy of Science Inclusion

The historical development of scientific constructs provides opportunities for geobiological integration in middle school science classrooms, with an additional benefit of facilitating a more thorough understanding of the nature of science within our students. For example, the premier "soap opera" of paleontology reveals the ultimate competition in the discovery and naming of dinosaur species between Othneil Charles Marsh and Edward Drinker Cope (Clary, et al., 2008). While some teachers may be reluctant to incorporate the history of science when it illustrates that scientists blundered in double-naming species, erringly reconstructed extinct reptiles, and employed questionable tactics in their haste to outperform their rivals, our research revealed that these controversies *validate* the foundations of the scientific process. Ultimately, the scientific community identifies and corrects the mistakes of errant researchers.

Fortunately, active learning and inquiry-based activities need not be discarded when including the history and philosophy of science in the science classroom. Not only did practicing teachers acknowledge that the Cope-Marsh controversy would pique the interest of their students, but they further developed strategies for successful incorporation of the Great Dinosaur Feud in their classrooms (Clary, et al., 2008). The most popular strategy was a "trial by jury," or an analogous trial variation with a modern television show format (e.g. "Jerry Springer," "Judge Judy.") Teachers also developed a "rush to publication" activity, in which student groups had to describe and name new dinosaurs before their peer groups did so. When students recognized that the group that named the dinosaur group was the *only* group rewarded, the groups' descriptions and conclusions tended to become disorganized and sloppy in the rush to receive the reward. The teacher then introduced historical controversies, and addressed how Cope and Marsh conducted sloppy science in their attempts to out-perform their rivals.

Any historical investigation of dinosaurs can include multiple geobiological learning opportunities for students. Geologic history, extinction, evolution, endothermy, avian-dinosaur relationships, fossilization processes, and paleoclimatology are a few examples of the interdisciplinary opportunities afforded by dinosaur study. Because most students are inherently interested in dinosaurs, we researched variations on their potential inclusion in science classrooms.

Since children's dinosaur books are popular items, several of our middle school students have been exposed to multiple representations of dinosaurs. We opted to investigate whether student investigations of *historical* dinosaur representations might convey the changing nature of science (Clary & Wandersee, 2011b). In the 1800s, dinosaur remains were first recognized and described as extinct species. The early scientific representations that emerged from the fossils generally portrayed dinosaurs as slow, dim-witted lizards. Some fossils were inaccurately assembled, such as Mantell's *Iguanodon*: Mantell mistakenly placed the thumb spike of the dinosaur at the end of its nose! As more fossils emerged and more data were collected, the reconstructions of the dinosaurs were likewise modified to reflect the more current status of scientific knowledge. When students locate and analyze illustrations of the same dinosaur through various decades, the progression of dinosaur visualizations reveals that misconceptions in dinosaur science are corrected through additional research. The illustrations also facilitated an integrated scientific investigation of fossilization processes, extinction, populations, and ecosystems (Clary & Wandersee, 2011b).

In yet another historical investigation that potentially involves dinosaurs, we tapped into the enthusiasm—and at times disgust—that students bring into a classroom coprolite investigation

(Clary & Wandersee, 2011c, 2011d). William Buckland, eccentric minister and first reader in geology at Oxford University, began examining fossilized excrement in the 1820s. Although the topic was unusual, Buckland approached it in true scientific fashion. His research methods involved early attempts to recreate the anatomy of extinct organisms and the environmental conditions under which they lived; this illustrates sound, experimental methodology to our students (Duffin, 2006; Brook, 1993; Pemberton & Frey, 1991). Students are typically amazed to discover that Buckland had coprolites chemically analyzed for their phosphate content, and that he injected Roman cement into the intestines of local fish in order to recreate the spiral structures he observed in the coprolites that he collected in Lyme Regis, England.

Likewise, coprolites in the classroom can facilitate student research experiences. Similar to Buckland's investigations, student teams can create "coprolites" from a prehistoric animal by describing its diet, and choosing representative indigestible materials that would reflect the animal's eating habits. When these indigestible materials are baked into coprolite "cookies," they become the remains of a mystery organism for another team (Clary & Wandersee, 2011c). These inquiry and active strategies build upon students' original interest in scatological topics, and invite deeper study. We also have confirmed that scatological topics retain the interest of older, high school students, providing portals for interdisciplinary study and opportunities to address multiple US National Science Education Standards.

Secondary Science Classrooms

The integrated pedagogical approach we advocated and researched for the middle school classroom, including the implementation of interdisciplinary scientific portals, the use of active learning and inquiry based projects, and the incorporation of the history and philosophy of science, is also needed and beneficial in high school science classrooms. Many of the middle school activities that we developed can be adapted to a more rigorous high school setting and/or a more scientifically literate student population. In addition, many of the research findings are transferable into other educational settings.

Interdisciplinary Secondary Science

Many high school teachers lament the compartmentalization of knowledge by some of our students. It is not that unusual for students to ask, when faced with reading a logarithmic scale when analyzing earthquake events, or when discussing the physiology of extinct organisms, whether they need to "know" mathematics or biology for the Earth Science course! Although most high school courses are developed around specific scientific content (e.g., biology, chemistry, physical science, Earth science, physics), we maintain that the integration of material across scientific disciplines promotes scientific habits of mind within our student population.

In order to ascertain the incoming knowledge of our students, we designed and researched the Botanical Sense of Place and Geological Sense of Place writing templates (Wandersee, et al., 2006; Clary & Wandersee, 2006). Through the analysis of student responses, we learned that our students are most often influenced by the local landscape, although large magnitude events and unusual specimens may capture their attention and pique their interest. Therefore, when we researched sand as an interdisciplinary portal in the science classroom, we developed activities that built scientific content knowledge from *local* sand samples with which students were most familiar (Clary & Wandersee, 2011e). In our sand investigations, students face the task of procuring a naturally deposited sample of sand, and then reconstructing the sands' tectonic journey from its rock origins, through its weathering, and transportation, to its final deposition. When students first view their local sand samples under a microscope, many are amazed to find that there is an assortment of clasts, which detail a story that is unique to the sample (many students predicted a homogeneous collection of grains before microscopic examination). In addition to the obvious geological content that sand can address, biological concepts can be introduced, particularly when sand samples contain microfossils. Fortunately, there are several sand aficionados who collect sand samples from around the world; some collectors post assorted microphotographs of unique and diverse specimens. Teachers can extend sand comparisons from local areas to geographically wide spread, worldwide samples.

Similar to middle school classrooms, fossils can serve as a valuable interdisciplinary portal to address several scientific constructs in high school science classrooms (Clary & Wandersee, 2008a). We specifically developed the concept of marquee fossils as a portal through which high school teachers could construct meaningful learning, building upon the prior knowledge of their students. Outside of the science classroom, a marquee serves as a sign (often with chasing flasher lights) that draws attention to and identifies a facility. In the science classroom, the marquee fossil serves as a unique fossil with unusual characteristics that captures students' attention, and invites them in for further interdisciplinary study. Marquee fossils can be appropriated for geological, biological, and environmental concepts involved in the study of fossilized organisms, their paleoenvironments, and the subsequent changes in climate that the area has experienced since the organism lived (Clary & Wandersee, 2008a). Students can compare extinct marquee fossils to their extant relatives, create replicas of marquee fossils in the classroom, and recreate the environment of their local area at the time of the living marquee fossil. We concluded that marquee fossils inspire students to delve deeper into scientific content, and integrate it within other disciplines.

In addition to using sand and local fossils to facilitate interdisciplinary, inquiry based learning, we also researched the incorporation of another, locally available material. Most science classrooms have easy access to a supply of gravel that is used in construction or to line roadways. This inexpensive resource can also be used in investigations to foster critical thinking skills (Clary & Wandersee, 2012). Students can investigate weathering and erosion, transportation, composition, and physical properties of gravel (such as hardness and streak) to investigate the rock cycle, ecology, natural resources, and geologic time. Some gravels also contain fossils and can be used to illustrate changes in Earth history and evolution. Students can incorporate gravel into experiments and test which type of gravel works best in filtration, absorbs more solar energy, or is more durable in parking lots.

Similar to our middle school classrooms, our interdisciplinary investigations at the high school level included dinosaurs. High school students typically were still enamored with the topic, and relished the opportunity to selectively investigate a particular genus in more detail (Clary & Wandersee, 2011f). Although the Adopt-a-Dino project resulted in positive student responses and was interdisciplinary in nature, we continued to investigate ways in which to improve and maximize learning benefits. Because artists in the nineteenth century struggled with the depiction of aquatic organisms before the advent of parlor aquaria (Clary & Wandersee, 2005), we hypothesized that our students might encounter similar difficulties. However, when we required that students incorporate a paleoenvironmental reconstruction of their dinosaur, and include air, terrestrial, *and* water components, we found that student solutions exhibited much higher creativity than previous projects we assigned.

Inquiry and Active Learning in Secondary Classrooms

In addition to the incorporation of interdisciplinary topics to showcase the interconnectedness of scientific concepts, there remains value and need for active learning and inquiry based activities within high school science classrooms. The pedagogical approach we recommend and research, especially at the secondary level, involves *student-led* inquiry, in which students generate their own questions and design subsequent investigations (Clary & Wandersee, 2010c, 2011g). In order to effectively incorporate student-led inquiry, teachers must first tap into students' prior knowledge, and then link new content to their existing cognitive frameworks. In a "green root beer laboratory," the classroom engages in a fun activity of producing and bottling its own root beer. First, however, students review ecological principles and investigate where the root beer ingredients are produced, and how much transportation was involved in getting the products from their components' origins to the grocery shelves (Steffen, 2007; Clary & Wandersee, 2010c). Other active learning strategies in a green root beer laboratory include classroom calculations of the amount of plastic and/or energy saved when recycled plastic bottles are used. Finally, as the culminating activity, teachers allow students (or small student groups) to generate their own inquiry questions, design investigations to answer the questions, and conduct the necessary experimentation. Naturally, the teacher's role is to offer guidance, and to ensure that safety rules are followed and personal protective equipment is used.

Historical events and the history of science can also form the basis for eventual student-led inquiry activities in secondary science classrooms (Clary & Wandersee, 2011g). Contemporary volcanic eruptions fascinate students, and provide a natural opportunity to discuss the historical eruption of Krakatoa. In 1883, the news of this eruption traveled rapidly around the world by way of the first "Internet"—or wired telegraphy. In the Krakatoa investigation, students first participated in "curiosity starters" by reading authentic period news reports. Next, activities in the classroom engaged all five senses during the sensory priming stage of the investigation: Students touch and see pumice. They observe that, counterintuitive to predictions, this is one rock that floats! The sense of smell is accessed with safety matches in the classroom, since striking safety matches releases sulfur compounds that can be detected at very low concentrations. The sounds of actual eruptions are available on the Internet (Oregon State University, 2011); the *Krakatoa* video also recreated the sounds of the original eruption (Burke & Hall, 2005). Finally, commercial candies with trapped pockets of carbon dioxide (CO_2) gas engage students' sense of taste. Building from the curiosity starter readings and sensory priming activities, students form their own questions for subsequent investigations. We termed this student-led inquiry part of the project the "scientific wanderings" phase. In general, students have no difficulty generating questions for subsequent investigations, and the historic eruption of Krakatoa—one of the first well-documented natural disasters—offers multiple opportunities for interdisciplinary study. Possible investigations include animal behavior prior to eruptions (biology), the causes of volcanic eruptions (geology), plant and insect survival and recolonization after volcanic eruptions (ecology), and distances traveled by eruptions' sounds (physics).

College Science Classrooms

Research consistently demonstrated the rich benefits that can result from inquiry learning in K-12 science classrooms. Although it may appear intuitive that the university classroom should provide *more* inquiry opportunities for our future scientists, the reality is that the first university science courses that students encounter are often bereft of inquiry (Drew, 2011). Introductory science classes typically are delivered in large lecture settings with intense mathematical requirements; at least one professor has termed our science and engineering curriculum, the "math-science death march" (Drew, 2011).

The earliest research investigations of the EarthScholars Research Group attempted to counteract the introductory university, single science discipline, "sage on the stage" lecture classroom. In our early research of informal science settings, we encountered petrified wood. Fairly ubiquitous fossils, petrified wood samples offered an obvious portal through which multiple, interdisciplinary scientific constructs could be taught, with several of them identified by the Earth Science Literacy Initiative (2010) as "big ideas" in Earth Science. We researched and identified the science content areas that could be included with petrified wood study, and next hypothesized that an introductory Earth history classroom with integrated petrified wood study might result in positive student learning gains (Clary & Wandersee, 2007). Through our research, we developed and validated the Petrified Wood Survey (PWS) as a testing instrument that measured students' understanding of fossilization, geologic time, and evolution. Conducted over three semesters (N = 515), we confirmed that significant student gains were made over the semester in student understanding, regardless of whether petrified wood was directly incorporated in the classroom. However, when we exposed the treatment group to integrated petrified wood study (e.g., mini-laboratory comparisons of modern and fossilized wood, online discussions that compared and contrasted petrified wood/modern wood samples, independent research opportunities to investigate the origin and public display of petrified wood specimens), students performed significantly better on interdisciplinary science assessments ($\alpha = 0.05$). Gains were made in students' conceptual understanding of geologic time, evolution, and fossilization processes; the content area with the least gains was geochemistry. In particular, students still failed to recognize the role (or lack thereof) of oxygen in petrified wood's formation, and the origin of petrified woods' colors. However, when we analyzed the original scientific misconceptions that students exhibited at the beginning of the semester, it appeared that both weak and strong cognitive restructuring occurred within our students (Clary & Wandersee, 2007).

Informal Field Investigations in Online Environments

A rallying cry of many geologists is that "geology is best taught in the field." The human constructivist learning theory also promotes field investigations, especially those that facilitate interdisciplinary and active learning, inquiry approaches. It can be difficult to organize large introductory classes into field excursions, and even more problematic when students are enrolled in online courses, and separated by large geographic distances.

We designed an autonomous field investigation in an online paleontology class, and directed our students to locate fossil collecting sites, plan an investigation, procure fossils, and then identify the ones they collected (Clary & Wandersee, 2008b). Students downloaded a required logo and were directed to place it next to each fossil that they documented through photography. In general, students performed extremely well on this assignment, and there was a significant positive difference ($\alpha = 0.05$) between the scores of this assignment and previous laboratory projects. While instructor flexibility and adaptability were necessary, the students in online environments did not encounter more difficulties than students in traditional paleontology courses. By focusing upon fossils, the assignment also integrated the physiology of the organism, fossilization processes, and geological ranges for an authentic geobiological approach.

When some students encountered difficulties with field excursion assignments, we modified our expectations and researched whether informal learning environments—such as those offered by museums, university geology displays, and fossil parks—could also provide settings for meaningful learning experiences and integration of course material (Clary & Wandersee, 2009b). Since the local environment typically impacts students' sense of place the most (Clary & Wandersee, 2006; Wandersee, et al., 2006), we designed a fossil investigation that would include local informal educational sites. Our students, most of whom are in-service teachers, were required to investigate fossil specimens at a minimum of three informal sites. The interdisciplinary investigation was successful for our students. The three stable findings that emerged from our research were that the informal sites affirmed the local environment's importance and placed it in a larger context; the student-directed activities maximized learning; and the informal educational sites supplied an interdisciplinary "big picture" (Clary & Wandersee, 2009b).

Informal Learning Environments

Our research demonstrated that field investigations and informal educational sites support meaningful learning by providing students with interdisciplinary learning opportunities while affirming the importance of their local geographical area. However, informal education is important beyond the traditional classroom. The value of informal learning environments was well established before we began our interdisciplinary research investigations within them (McComas, 1996). Several researchers have examined the theoretical bases behind informal learning, the motivation of the informal learners, and assessment of learning at informal sites (Rennie & Johnson, 2004; Anderson, et al., 2003; Falk, 2001; Falk & Dierking, 2000; Meredith, et al., 1997; Orion & Hofstein, 1994). While it may seem obvious that informal education is the default learning style for an adult population, researchers have also demonstrated that even school-age students engage in informal learning more often than learning in traditional classroom environments (Falk & Dierking, 2002).

Our research program in informal learning environments promotes the bridging between informal sites and traditional, formal classrooms (Wandersee & Clary, 2012). Similar to the artificial compartmentalization between geology, biology, and other science disciplines, the separation of informal and formal learning has been detrimental to science learning. When active learning and inquiry-based investigations accompany informal education sites, learners are provided with authentic opportunities to engage in scientific activities. Our research program encourages geobiological investigations in informal sites that are integrated within traditional classrooms (Clary & Wandersee, 2008b, 2009b). It logically follows that authentic science course work calls for a field component (Wandersee & Clary, 2006b). By incorporating informal learning opportunities and field work in our science curriculum, we demonstrate to students that the instructors not only *teach* science, but can also *do* science. There exist numerous successes of fieldwork and informal learning environments that are coupled with traditional science classes (Drake, et al., 1997; Woltemade & Stanitski Martin, 2002; Haines & Blake, 2005; Clary, et al., 2005).

Beyond the traditional science classroom, informal sites can facilitate interdisciplinary science learning for those visitors who are unassociated with formal learning environments. Therefore, we also must consider the casual visitors to these informal science sites, and determine whether the opportunities for integrated geobiological instruction are optimized for meaningful learning. One of our informal research avenues has been that of fossil parks, or public sites that allow visitors, for a small fee, to collect and keep a limited number of fossils (Clary & Wandersee, 2011h). However, not all fossil parks—or informal educational sites—are equal. Following an in-depth investigation of the first three US fossil parks, we identified key variables and developed an optimized model for park development that would facilitate geobiological learning for visitors. Our research model concluded that the site needs to develop an informative pre-visit website; provide authentic *in situ* collection opportunities as opposed to spoil piles; supply—or at least allow—authentic collecting tools; be accessible for all visitors; install semi-permanent signage (as opposed to brochures) that visitors can access, hands-free, to aid in their fossil identification; and supply learning opportunities for an integrated geobiological field experience (Clary & Wandersee, 2011h). These informal sites should emphasize important constructs in science, including evolution, biodiversity, and geologic time. Safety considerations for visitors must also be incorporated within the design of the park. When correctly designed and implemented, informal sites can educate the general population for increased scientific literacy. However, we also concluded that the optimized fossil park could effectively serve as an outdoor laboratory for our secondary and college students (Singer, et al., 2005).

DISCUSSIONS AND RECOMMENDATIONS FOR IMPLEMENTING A GEOLOGICAL-BIOLOGICAL APPROACH

The EarthScholars Research Group program advocates integrated geological and biological content to facilitate a more holistic view of science within our students. Our guiding theory of human constructivism further advocates the incorporation of active learning and inquiry based projects in order to provide a more authentic scientific experience, and avoid what Schwab (1962) termed the "rhetoric of conclusions" in the science classroom.

Although our research has revealed and documented the benefits of the integrated geobiological approach, we propose that teachers facilitate a gradual transition to this pedagogical strategy. As with any approach, the human constructivist guidelines note that instructors must build upon the *pre-existing* knowledge of the learner. When students enter our classrooms with limited field experience, having been subjected to cookbook science investigations, the instructor must ascertain the prior knowledge and experiences of the student, and build instruction, as well as methods, accordingly. However, we definitely recommend a move from the cookbook strategy with pre-supplied laboratory equipment. In accordance with the "less is more" philosophy of human constructivism, students who were given *fewer* laboratory supplies developed and provided more creative solutions than those who were given typical laboratory materials (Jordan, et al., 2011).

Advantages for Teachers

When student-led inquiry investigations are incorporated within the science classroom, teachers may expect more motivated students. First, however, the background knowledge of students must be determined, and classroom investigations and instruction must be built accordingly. Students will then have the necessary knowledge and skills for further investigation. After classroom introduction, when students are allowed to choose the direction of their subsequent investigations, they have a vested interest in the process. Gowin (1981) noted that for successful science education to occur, it must scaffold upon an integration of thinking, *feeling*, and acting within a learner. When student interest can determine the direction of classroom inquiry, then the affective domain of the learner is addressed.

Advantages for Students

We propose that students also have much to gain with an active learning, interdisciplinary inquiry approach. When we bridge and integrate disciplines of science within the classroom, we expose students to a more holistic view of the way that science actually works, and in so doing, facilitate a more accurate understanding of the nature of science. The integrated, interdisciplinary approach can facilitate students' understanding of the Earth's complexity (Orr, 1994). In an integrated, inquiry based active learning classroom, students can participate in authentic research experiences that form the crux of scientific endeavors. They not only *learn* about science, but they actually *do* science. Although students will not leave the classroom having covered the entire textbook, teachers have facilitated an in-depth understanding of carefully selected, important scientific constructs. However, with sustained meaningful and mindful learning throughout their schooling, students integrate these constructs within their cognitive framework, and build upon them. Over time, students are able to continuously build more powerful knowledge structures.

CHALLENGES AND FUTURE DIRECTIONS OF INTEGRATED GEOLOGICAL-BIOLOGICAL LEARNING

We think that interdisciplinary geobiological instruction in middle, high school, and college science classrooms, coupled with an integrated, well-planned informal field experience, should be seen as a goal for science instructors. The active learning, inquiry based approach should be viewed as the core of geobiological understanding. Without including biological content in geological settings, or geological content in biological topics, we fail to explore a topic fully: we fail to "open both eyes" in a scientific investigation. Contemporary scientists research and attempt to gain knowledge by focusing upon all the variables within the system being studied. They cannot ignore the biological components of the Earth system when investigating weathering and erosion, nor can they ignore fossilization processes when reconstructing the physiology of extinct animals. Likewise, teachers must provide the necessary tools and experiences in order to reveal the interconnected nature of scientific inquiry, so that we encourage our students to open both eyes and experience a more authentic approach to science in the classroom. It is our hope that our research findings can provide guidance for educators seeking to incorporate an interdisciplinary science approach that capitalizes on informal science learning opportunities within traditional classrooms. We likewise hope that our research provides guidelines for effectively designing informal educational sites that maximize viewer understanding.

We caution, however, that each classroom is different, and educators must effectively ascertain the optimum teaching requirements within each new situation. Both underteaching and overteaching can have negative consequences: If instructors present too much content, students can become overwhelmed with information overload. On the other hand, if teachers fail to supply ample content in the classroom, or link it to students' existing knowledge structures, then students are unable to connect the new material in a meaningful way. Successful science instruction must locate an optimal teaching point in each, unique classroom.

Science educators must also be cautious in how formal and informal learning are supplied in a classroom, and how to balance classroom investigation and content delivery with individualized student-led inquiry investigations. Teachers must also work within pre-determined guidelines, and address state and national standards in their classrooms. Therefore, each class, each subject, and each topic will provide a unique challenge—as well as an opportunity—for the science educator. However, we encourage science educators to embrace the opportunities as well as the challenges. The public's scientific literacy is intricately linked with our efforts.

REFERENCES

Anderson, D., Lucas, K., & Ginns, I. (2003). Theoretical perspectives on learning in an informal setting. *Journal of Research in Science Teaching, 40,* 177–199. doi:10.1002/tea.10071

Asay, L., & Orgill, M. (2010). Analysis of essential features of inquiry found in articles published in *The Science Teacher*, 1998-2007. *Journal of Science Teacher Education, 21*(1), 57–79. doi:10.1007/s10972-009-9152-9

Ausubel, D. P. (1963). *The psychology of meaningful verbal learning*. New York, NY: Grune and Stratton.

Ausubel, D. P. (1968). *Educational psychology: A cognitive view*. New York, NY: Holt, Rinehart, and Winston.

Ausubel, D. P. Novak. J. D., & Hanesian, H. (1978). *Educational psychology: A cognitive view* (2nd ed.). New York, NY: Holt, Rinehart, and Winston.

Bernstein, S. N. (2004). A limestone way of learning. *The Chronicle Review, 50*(7).

Brook, A. J. (1993). The Reverend William Buckland, the first paleoecologist. *Biologist (Columbus, Ohio), 40*, 149–152.

Burke, R., & Hall, J. (2005). *Krakatoa*. [DVD]. PBS Home Video.

Clary, R. M., Gresham, D., Bases, F., Hamlin, E., Bergeron, N., & Petry, C. … Fischer, E. (2005). Sediment and water analysis adjacent to an active scrap yard and archived superfund site, Lafayette Parish, Louisiana. *Gulf Coast Association of Geological Societies Transactions, 55*, 89-100.

Clary, R. M., & Wandersee, J. H. (2005). Through the looking glass: The history of aquarium views and their potential to improve learning in science classrooms. *Science and Education, 14*, 579–596. doi:10.1007/s11191-004-7691-1

Clary, R. M., & Wandersee, J. H. (2006). A writing template for probing students' geological sense of place. *Science Education Review, 5*(2), 51–59.

Clary, R. M., & Wandersee, J. H. (2007). A mixed methods analysis of the effects of an integrative geobiological study of petrified wood in introductory college geology classrooms. *Journal of Research in Science Teaching, 44*(8), 1011–1035. doi:10.1002/tea.20178

Clary, R. M., & Wandersee, J. H. (2008a). Marquee fossils: Using local specimens to integrate geology, biology, and environmental science. *Science Teacher (Normal, Ill.), 75*(1), 44–50.

Clary, R. M., & Wandersee, J. H. (2008b). Earth science teachers' perceptions of an autonomous fieldwork assignment in a nationwide online paleontology course. *Journal of Geoscience Education, 56*, 149–155.

Clary, R. M., & Wandersee, J. H. (2009a). Amber: Use "tree tears turned to stone" to teach biology, ecology…and more! *Science Scope, 33*(3), 22–29.

Clary, R. M., & Wandersee, J. H. (2009b). Incorporating informal learning environments and local fossil specimens in earth science classrooms: A recipe for success. *Science Education Review, 8*. Retrieved from http://www.scienceeducationreview.com/open_access/index.html

Clary, R. M., & Wandersee, J. H. (2010a). Scientific caricatures in the earth science classroom: An alternative assessment for meaningful science learning. *Science and Education, 19*(1), 21–38. doi:10.1007/s11191-008-9178-y

Clary, R. M., & Wandersee, J. H. (2010b). Connect-the-spheres with the coal cycle. *Science Scope, 34*(2), 20–29.

Clary, R. M., & Wandersee, J. H. (2010c). The "green" root beer laboratory. *Science Teacher (Normal, Ill.), 77*(2), 25–28.

Clary, R. M., & Wandersee, J. H. (2011a). Our human-plant connection. *Science Scope, 34*(8), 32–37.

Clary, R. M., & Wandersee, J. H. (2011b). DinoViz: The history and nature of science through the progression of dinosaur visualization. *Science Scope, 34*(6), 14–21.

Clary, R. M., & Wandersee, J. H. (2011c). A coprolite mystery: Who DUNG it? *Science Scope, 34*(7), 32–42.

Clary, R. M., & Wandersee, J. H. (2011d). A "coprolitic vision" for earth science education. *School Science and Mathematics, 111*(6), 262–273. doi:10.1111/j.1949-8594.2011.00087.x

Clary, R. M., & Wandersee, J. H. (2011e). To see a scientific world in a grain of sand..... *Science Teacher (Normal, Ill.), 78*(5), 29–33.

Clary, R. M., & Wandersee, J. H. (2011f). Adopt-a-dino: Creative scientific visualization. *Science Teacher (Normal, Ill.), 78*(6), 36–41.

Clary, R. M., & Wandersee, J. H. (2011g). 1883 news report—Krakatoa erupts! A biology-geology integration inquiry. *Science Teacher (Normal, Ill.), 78*(9), 42–47.

Clary, R. M., & Wandersee, J. H. (2011h). Geobiological opportunities to learn at US fossil parks. In Feig, A., & Stokes, A. (Eds.), *Qualitative Inquiry in Geoscience Education Research (Vol. 474*, pp. 113–134). GSA Special Papers. doi:10.1130/2011.2474(09)

Clary, R. M., & Wandersee, J. H. (2012). Rock on! Using gravel to promote critical thinking in the classroom. *Science Teacher (Normal, Ill.), 79*(3), 42–46.

Clary, R. M., Wandersee, J. H., & Carpinelli, A. (2008). The great dinosaur feud: Science against all odds. *Science Scope, 32*(2), 34–40.

Crichton, M. (1990). *Jurassic park*. New York, NY: Knopf.

DeBoer, G. (1991). *A history of ideas in science education*. New York, NY: Teachers College Press.

DeBoer, G. (2005). Historical perspectives on inquiry teaching in schools. In Flick, L. B., & Lederman, N. G. (Eds.), *Scientific Inquiry and Nature of Science: Implications for Teaching, Learning, and Teacher Education* (pp. 17–36). Dordrecht, The Netherlands: Springer.

Drake, J., Worle, I., & Mehrtens, C. (1997). An introductory-level field-based course in geology and biology. *Journal of Geoscience Education, 45*, 234–237.

Drew, C. (2011, November 4). Why science majors change their minds (it's just so darn hard). *New York Times*. Retrieved from http://www.nytimes.com/2011/11/06/education/edlife/why-science-majors-change-their-mind-its-just-so-darnhard.html?pagewanted=1&_r=1&emc=eta1

Duffin, C. J. (2006). William Buckland (1786-1856). *Geology Today, 22*(3), 104–108. doi:10.1111/j.1365-2451.2006.00562.x

Duschl, R. A. (1990). *Restructuring science education: The importance of theories and their development*. New York, NY: Teachers College Press.

Duschl, R. A. (1994). Research on the history and philosophy of science. In Gabel, D. L. (Ed.), *Handbook of Research on Science Teaching and Learning* (pp. 443–465). New York, NY: Macmillian.

Earth Science Literacy Initiative. (2010). Earth science literacy principles: Big ideas and supporting concepts of earth science. *National Science Foundation*. Retrieved from http://www.earthscienceliteracy.org/es_literacy_6may10_.pdf

Elkins, J. T., & Elkins, N. M. L. (2007). Teaching geology in the field: Significant geoscience concept gains in entirely field-based introductory geology courses. *Journal of Geoscience Education, 55*(2), 126–132.

Falk, J. (2001). *Free choice science education: How we learn science outside of school.* New York, NY: Teachers College Press.

Falk, J., & Dierking, L. (2000). *Learning from museums: Visitor experiences and the making of meaning.* Walnut Creek, CA: Alta Mira Press.

Falk, J., & Dierking, L. (2002). *Lessons without limit: How free-choice learning is transforming education.* Walnut Creek, CA: Alta Mira Press.

Felzien, L., & Cooper, J. (2005). Modeling the research process: Alternative approaches to teaching undergraduates. *Journal of College Science Teaching, 34,* 42–46.

Gowin, D. B. (1981). *Educating.* Ithaca, NY: Cornell University Press.

Gunstone, R. F., & Mitchell, I. J. (1998). Metacognition and conceptual change. In Mintzes, J. J., Wandersee, J. H., & Novak, J. D. (Eds.), *Teaching Science for Understanding: A Human Constructivist View* (pp. 133–163). San Diego, CA: Academic Press.

Haines, S., & Blake, R. Jr. (2005). Field and natural science. *Journal of College Science Teaching, 34*(7), 28–31.

Hemler, D., & Repine, T. (2006). Teachers doing science: An authentic geology research experience for teachers. *Journal of Geoscience Education, 54,* 93–102.

Jordan, R. C., Ruibal-Villasenor, M., Hmelo-Siler, C. E., & Etkina, E. (2011). Laboratory materials: Affordances or constraints? *Journal of Research in Science Teaching, 48*(9), 1010–1025. doi:10.1002/tea.20418

Langer, E. J. (1997). *The power of mindful learning.* Reading, MA: Addison-Wesley.

Lawrenz, F., Huffman, D., & Appeldoorn, K. (2005). Enhancing the instructional environment. *Journal of College Science Teaching, 35*(7), 40–44.

Lord, T., & Orkwiszewski, T. (2006). Moving from didactic to inquiry-based instruction. *The American Biology Teacher, 68,* 342–345. doi:10.1662/0002-7685(2006)68[342:DTIIIA]2.0.CO;2

Matthews, M. R. (1994). *Science teaching: The role of history and philosophy in science.* New York, NY: Routledge.

McComas, W. (2006). Science teaching beyond the classroom: The role and nature of informal learning environments. *Science Teacher (Normal, Ill.), 73*(1), 26–30.

McConnell, D. A., Steer, D. A. N., & Owens, K. D. (2003). Assessment and active learning strategies for introductory geology courses. *Journal of Geoscience Education, 51*(2), 205–216.

McKellar, R., Chatterton, B., Wolfe, A., & Currie, P. (2011). A diverse assemblage of late cretaceous dinosaur and bird feathers from Canadian amber. *Science, 333*(6049), 1619–1622. doi:10.1126/science.1203344

McLaughlin, J. S. (2005). Classrooms without walls. *Journal of College Science Teaching, 35*(4), 5–6.

Meredith, J., Fortner, R., & Mullins, G. (1997). A model of affect in nonformal education. *Journal of Research in Science Teaching, 34,* 805–818. doi:10.1002/(SICI)1098-2736(199710)34:8<805::AID-TEA4>3.0.CO;2-Z

Michael, J., & Modell, H. I. (2003). *Active learning in secondary and college science classrooms: A working model for helping the learner to learn.* Mahwah, NJ: LEA Inc.

Mintzes, J. J., Wandersee, J. H., & Novak, J. D. (Eds.). (1998). *Teaching science for understanding: A human constructivist view.* San Diego, CA: Academic Press.

Mintzes, J. J., Wandersee, J. H., & Novak, J. D. (Eds.). (2000). *Assessing science for understanding: A human constructivist view*. San Diego, CA: Academic Press.

National Committee on Science Education Standards and Assessment. (1996). *National science education standards*. Retrieved from http://www.nap.edu/openbook.php?record_id=4962&page=R1

National Research Council. (2010). *A framework for science education. Preliminary Public Draft*. Washington, DC: Committee on Conceptual Framework for New Science Education Standards.

Novak, J. D. (1963). What should we teach in biology? *News & Views*, *7*(2).

Novak, J. D. (1998). The pursuit of a dream: Education can be improved. In Mintzes, J. J., Wandersee, J. H., & Novak, J. D. (Eds.), *Teaching Science for Understanding: A Human Constructivist View* (pp. 3–29). San Diego, CA: Academic Press.

Novak, J. D., & Gowin, D. (1984). *Learning how to learn*. Cambridge, UK: Cambridge University Press. doi:10.1017/CBO9781139173469

Novak, J. D., & Musonda, D. (1991). A twelve-year longitudinal study of science concept learning. *American Educational Research Journal*, *28*, 117–153.

Oregon State University. (2011). *Volcano sounds*. Retrieved from http://bit.ly/pWMB7m

Orion, N., & Hofstein, A. (1994). Factors that influence learning during a scientific field trip in a natural environment. *Journal of Research in Science Teaching*, *31*, 1097–1119. doi:10.1002/tea.3660311005

Orr, D. (1994). *Earth in mind: On education, environment, and the human prospect*. Washington, DC: National Academy Press.

Pemberton, S. G., & Frey, R. W. (1991). William Buckland and his coprolitic vision. *Ichnos*, *1*(4), 317–325. doi:10.1080/10420949109386367

Rennie, L., & Johnston, D. (2004). The nature of learning and its implications for research in learning from museums. *Science Education*, *88*, S4–S16. doi:10.1002/sce.20017

Schwab, J. (1962). The teaching of sicnece as inquiry. In Schwab, J., & Brandwein, P. (Eds.), *The Teaching of Science* (pp. 1–104). Cambridge, MA: Harvard University Press.

Singer, S. R., Hilton, M. L., & Schweingruber, H. A. (Eds.). (2005). *America's lab report: Investigations in high school science*. Washington, DC: National Academies Press.

Steffen, A. (Ed.). (2007). *Principle 1: The backstory*. Retrieved from http://www.worthchanging.com/archives/004534.html

Wandersee, J. H., & Clary, R. M. (2006). Fieldwork: New directions and examples in informal science education research. In Mintzes, J., & Leonard, W. (Eds.), *NSTA Handbook of College Science Teaching: Theory, Research, & Practice* (pp. 167–176). Arlington, VA: NSTA Press.

Wandersee, J. H., & Clary, R. M. (2006a). On seeing flowers: Are you missing anything? *The Human Flower Project*. Retrieved from http://www.humanflowerproject.com/index.php/weblog/on_seeing_flowers_are_you_missing_anything

Wandersee, J. H., & Clary, R. M. (2012). Envisioning a rainbow bridge: Eight research studies that reveal optimal opportunities to learn biology and geology at informal science education sites. In P. Kurtz & F. Ren (Eds.), *Proceedings of the 11th World Congress for Center for Inquiry-Transnational: Scientific Inquiry and Human Development*. Amherst, NY: Prometheus Books.

Wandersee, J. H., Clary, R. M., & Guzman, S. M. (2006). How-to-do-it: A writing template for probing students' botanical sense of place. *The American Biology Teacher*, *68*(7), 419–422. doi:10.1662/0002-7685(2006)68[419:AWTFPS]2.0.CO;2

Wandersee, J. H., Mintzes, J. J., & Novak, J. D. (1994). Research on alternative conceptions in science. In Gabel, D. (Ed.), *Handbook of Research on Science Teaching and Learning* (pp. 177–210). New York, NY: Macmillan.

Wandersee, J. H., & Schussler, E. (1999). Preventing plant blindess. *The American Biology Teacher*, *61*, 84–86. doi:10.2307/4450624

Woltemade, C., & Staniski-Martin, D. (2002). A student-centered field project comparing NEXRAD and rain gauge precipitation values in mountainous terrain. *Journal of Geocience Education*, *50*, 296–302.

Chapter 10
Promoting the Physical Sciences among Middle School Urban Youth through Informal Learning Experiences

Angela M. Kelly
Stony Brook University, USA

ABSTRACT

Numerous reform efforts in STEM education have been targeted towards increasing the number of qualified STEM professionals in the U.S., which necessitates promoting science participation among secondary and post-secondary students. Some novel designs have focused on the middle school years, when students tend to lose interest in science and formulate opinions on science self-identification. This chapter describes the effectiveness of developing informal physical science experiences for middle school students in underserved urban communities. Several cohorts of students have participated in inquiry-based physics and chemistry weekend classes that incorporated authentic applications from the urban setting, field visits to scientists' laboratories and museums, advanced educational technology tools, and learning complex scientific concepts. Participants reported significant improvements in their attitudes, knowledge, and appreciation of the physical sciences, suggesting that well designed constructivist physical science programs are potentially transformative in improving students' academic self-efficacy, confidence, and persistence in science, and positional advantage. The potential of early, rigorous experiences with the physical sciences is explored as a means for improving science participation and diversifying the ranks of future scientists.

DOI: 10.4018/978-1-4666-2809-0.ch010

INTRODUCTION

There has been much national discussion regarding the need to expand participation in STEM (Science, Technology, Engineering, and Mathematics) careers among underrepresented minorities. Recently, the National Academies reported the concern that "the scientific and technological building blocks critical to our global economic leadership are eroding at a time when many other nations are gathering strength" (National Academies of Sciences, 2007). This report and its follow-up have increased attention on improving K-12 mathematics and science education and addressing the relatively low number of underrepresented minorities in STEM fields (NAS, 2007, 2010a, 2010b). One variable that has affected accessibility and participation has been the relatively low enrollment in gateway elective sciences for high school students. Once students get to junior and senior year of high school, they often choose not to take chemistry and physics, two courses that are essential for their post-secondary success in college-level STEM coursework. Data have shown that African-American and Hispanic students are less like to take physics than their Caucasian and Asian counterparts (American Institute of Physics, 2011). The reasons for limited participation in pre-college physical sciences are complex. In some cases, students are steered away from taking courses that are intended for the academically elite. Others may not recognize the value of a strong foundation in the physical sciences for their career goals. Educators must seriously consider improving these enrollments to improve the science preparation of secondary students since they are the next generation in a modern technology-driven, global economy.

This essay will explore one promising solution for improving student enthusiasm for the physical sciences—making chemistry and physics concepts accessible earlier in a student's academic life. Specifically, the case will be made for middle school physical sciences in informal learning environments. Exploring the physical sciences is a critical experience for middle school students, though many middle school science curricula place higher value on life sciences. Meaningful physical science study at this level is a rarity, particularly in urban schools. Possible explanations include a scarcity of laboratory resources, inadequate teacher training, and a testing culture that focuses on literacy and mathematics. Consequently, this author initiated, developed, and taught a Saturday program for Bronx middle school students in physics, chemistry, and mathematics. The program was part of the Enlace Latino Collegiate Society at the Bronx Institute, which provided extracurricular academic opportunities for local students. The goal was to provide academic experiences that the students did not have in their neighborhood schools, with the hope that they would be better prepared for advanced science and mathematics in high school and, consequently, admission to selective colleges.

This chapter has three main components. First, the need for meaningful physical science instruction will be established by examining research in trends in physical sciences among American high school students, and looking at reasons why the physical sciences are important in STEM study. Secondly, the program structure of the physics and chemistry coursework at Bronx Institute will be described. Finally, empirical results from studies of the Bronx Institute participants will highlight the program's effectiveness and the potential for replication in larger contexts.

TRENDS IN PHYSICAL SCIENCES AMONG U.S. HIGH SCHOOL STUDENTS

There has been much recent concern about the quality and quantity of STEM education in the United States (NAS, 2007). The complex technological challenges facing our nation require diverse solutions that most likely will come from a diverse scientific workforce (American Chemical Society, 2008). Why is a diversified STEM workforce desirable and necessary? The scientific community will benefit from multiple voices and varied perspectives when deciding upon what questions to ask and what solutions are most appropriate for particular contexts. All scientists bring unique life experiences and cultural capital to their work, and equitable participation will ensure that all American interests are represented in the broader global scientific community. In a nation where 45% of all school-aged children are racial minorities (U.S. Census Bureau, 2012), this issue requires immediate and targeted solutions.

The preparation of American students in science and mathematics has not been at an optimal level to sustain and improve our scientific workforce. Echoing numerous reports in the media and the academic world, the recent issue of *Trends in International Mathematics and Science Study* suggested that American high school students were performing well below the levels of their global peers in science (National Commission on Mathematics and Science Teaching for the 21st Century, 2000). For example, of those American students who did enroll in the physical sciences, many performed poorly on tests when compared to their international peers (National Center for Education Statistics, 2004; NCMST, 2000). At the end of twelfth grade, American students scored nearly last among students in industrialized countries in both science and mathematics, outperforming only Cyprus and South Africa. Even when comparing Advanced Placement Physics students in the U.S. to advanced science students in other nations, the U.S. students scored second to last out of sixteen countries (NCES, 1998).

The reasons for this disappointing performance are somewhat ambiguous. U.S. twelfth graders were less likely to be enrolled in science and mathematics in their last year of high school, and a disproportionately small fraction of American high school students (37%) have taken at least one year of physics (AIP, 2011). Although 66% of graduating seniors in America were taking mathematics, the average in other participating countries was 79%. The same pattern was evident for science, with 53% of U.S. seniors enrolled compared to 67% for all TIMMS countries. International students spent more time on homework than American twelfth graders (NCES, 1998). In addition, many international students take multiple years of physics and chemistry, unlike the U.S. where one year is more typical for those who participate.

When looking at the U.S. as a whole, there has been some improvement in terms of increased enrollments in high school sciences, although serious inequities are still prevalent in urban schools. Transcript studies have indicated that between 1990 and 2000, the percentage of high school graduates completing a chemistry course rose from 45 to 63% and a physics course rose from 21.5% to 31% (National Science Board, 2006). Today, virtually all American students complete a biology course, while 54% complete chemistry and slightly more than one-third physics (Hehn & Neuschatz, 2006). However, these trends have not been replicated in urban areas. Many schools in low-income, under-resourced neighborhoods, both urban and rural, simply do not offer physics to students (Gollub & Spital, 2002). Research has shown that 55% of New York City's secondary schools did not offer any courses in physics; these schools enrolled approximately 23% of the secondary students in the city. Only 21% of NYC high school graduates have studied physics for at least one year (Kelly & Sheppard, 2008). Minority students have not made gains in pursuing engineering degrees at the undergraduate level in the past ten years and have remained underrepresented in physics, chemistry and computer science (National Science Foundation, 2007).

To illustrate ethnic disparities in preparation for participation, the NCES examined trends in a three-course sequence of biology, chemistry, and physics, and reported that 29% of White and 43% of Asian students completed the sequence, while 21% of African-American and 19% of Hispanic students did (NCES, 2009). Similarly, in 2008-2009, 52% of Asian students and 41% of White students enrolled in high school physics, compared to 25% of Black and Hispanic students (AIP, 2011). Research has suggested that variables related to disproportionate participation have included socioeconomic status (Carnevale & Strohl, 2010) and school racial/ethnic composition (Riegl-Crumb & Grodsky, 2010). In wealthier school districts, 47% of high school seniors enrolled in physics in 2008-2009, while in economically disadvantaged schools, 24% enrolled (AIP, 2011). Socioeconomic differences may have had a significant impact on advanced course taking, since educated parents have been more likely to intervene with school personnel to monitor academic progress and demand access to higher-level courses (Lareau & Horvat, 1999). More research is needed to explore additional explanations for low secondary physics and chemistry participation in predominantly minority urban schools.

Though the statistics revealing disparities are clear, students' reasons for choosing physical sciences, should they have the opportunity, were more ambiguous. When considering STEM fields in general, academic achievement has been shown to be a predictor of underrepresented students' interest (Barton, 1993), particularly their prior success in mathematics (Simpson, 2001). Students have also been influenced by their academic experiences in elementary and middle school science and mathematics (Eamon, 2004). Teacher quality at this level has been found to be a factor in promoting STEM interest, and for women and underrepresented minorities, a mentor, teacher, or counselor has frequently been cited as an important influence in their STEM persistence (U.S. Government Accountability Office, 2005; Whitten, et al., 2004). Other research has revealed that students may not have wanted to take physics and chemistry because they considered the subjects too abstract, intellectually rigorous, and lacking in exciting laboratory experiences and authenticity; a poorly regarded teacher was also cited as a negative factor (Woolnough, 1994).

Even if a student expressed interest, she may have had external barriers that prevented her from choosing the physical sciences. The existence of prerequisites has affected access; this can be in the form of formally required courses such as algebra II or trigonometry, which are "gateway" courses that may have prohibited certain tracks of students from having equal access to physical sciences (Lynch, 2001). As Arthur Eisenkraft pointed out (2010), Michael Faraday, one of the greatest experimentalists of the 19th century, might not have succeeded in today's physics classes because his mathematical preparation was so deficient (this weakness resulted in his creating the field concept). Today, a school administrator's judgment about the career path of a particular student has often determined what courses he/she could choose to take (DeLany, 1991; Vanfossen, Jones, & Spade, 1987). Physics and chemistry have typically been reserved for the college-bound, and even among those, for students who have planned to pursue science or engineering in college (Sadler & Tai, 2001). Therefore, a student's access can be severely constrained by past opportunities in math and science, their perceived career paths, and school-measured intellectual capabilities based on sometimes-flawed assessments.

Conversely, many stakeholders in physics education have rejected this elitist notion and supported the idea of "physics for all," suggesting that physical science should be open to broader participation among secondary students (American Association of Physics Teachers, 2002; Sheppard & Robbins, 2003). Teachers familiar with students' capabilities and cognitive development could initiate this option. By adjusting teaching strategies and adopting research-based curricula, all students can succeed and be challenged in physical science classrooms (Brewe, 2008; Eisenkraft, 2010; Kelly, 2011). The belief that every child can learn physical science is necessary for this to happen. Indeed, a strong foundation in middle school grades would facilitate learning at the high school level in the physical sciences.

In light of these enrollment and performance statistics, one can conclude that many American students will graduate from high school and college with a limited intellectual foundation in the physical sciences. Immediate innovations in secondary education, particularly in alternative settings, are necessary to promote physical science and engineering as attractive academic and career choices for American students. Such innovations should begin earlier in students' academic lives in order to nurture enthusiasm for science and inspire wonder of the natural world.

WHY ARE THE PHYSICAL SCIENCES IMPORTANT IN MIDDLE SCHOOL?

Research has suggested that physics is a necessary component of students' academic preparation for post-secondary STEM (Science, Technology, Engineering, and Mathematics) study (Tyson, Lee, Borman, & Hanson, 2007). If access to physics is limited and/or discouraged, it will negatively impact the future of the U.S. scientific and global enterprise (NAS, 2007, 2010a, 2010b), since fewer high school and college students will have the science background that is typically necessary to be successful in STEM majors (American College Testing, 2005). Among students with two or more advanced science courses in high school, 90% enrolled in a four-year college within two years of graduation, compared with 12% of students with no advanced science courses. The most commonly taken advanced science course was chemistry, though slightly more than one-half of high school students completed this course in 2009 (National Science Foundation, 2010).

Results from numerous studies have indicated that late elementary and early middle school is a crucial time in students' educational lives (Horizon Research, 2002). Unfortunately for many students, their first encounters with science during these years are not motivating and students easily become dissuaded from further study of science. Later in high school, many elect to complete only the minimum number of science courses required to graduate; this is particularly evident in high needs schools. Research has shown that in the Bronx, just 32% of high schools even offered physics. Fewer than one in twenty Bronx high school students took physics in 2011; this figure does not include the selective Bronx High School of Science, where most students commute from other boroughs (Kelly & Sheppard, 2009, 2010).

Consequently, when considering preparation for advanced science, it is helpful to examine conditions in earlier parts of the pipeline, specifically, to middle school preparation. For example, science in middle schools in New York City has often been neglected. Recent assessments indicated that 62% of NYC 8th graders scored below basic competency in science, compared to 38% of 8th graders nationally and 56% in other large cities; also, 44% of NYC 4th graders scored below basic (National Association of Educational Progress, 2009). The increase in 18% points between 4th and 8th grades suggests that science in city middle schools has been inadequate in preparing students for scientific achievement in an increasingly technological society. Inequitable access to high quality instruction has contributed to the achievement gap and disparate enrollment of underrepresented minorities in the city's science specialized schools (New York Times, 2006).

The opportunity to study advanced science in middle school may be severely limited by the availability of resources and qualified teachers. New York City's science education report *Lost in Space* found that more than one-third of the middle schools did not have science labs (Committee on Education, Council of the City of New York, 2004). Students in schools without labs scored 10 percentage points lower on the 8th grade standardized science exam, as seen in Figure 1 (New York City Coalition for Education Justice, 2007).

Without laboratories and appropriate resources for science instruction, it is difficult for students to learn science content and the habits of mind necessary for understanding their physical world.

Many middle school teachers lack the content background necessary to teach science effectively, resulting in a textbook-driven curriculum and rote individual work (Lee, 1995). Unprepared science teachers avoid complex scientific ideas, teach more biology than physical science, and avoid all but the simplest laboratory work (Appleton, 2007). In NYC middle schools, 20.6% of teachers are teaching out of their license area, compared to 14.3% if elementary school teachers and 18.9% of high school teachers (New York City Department of Education, 2006). Fewer than half of NYC middle school teachers have more than five years of teaching experience (NYCCEJ, 2007). Middle school teachers in the city will benefit from professional development in physical science content and teaching strategies to strengthen their students' science understandings (Committee on Education, Council of the City of New York, 2004). However, without immediate and sustained action to solve this problem, informal contexts can provide a mechanism for high needs to students to participate in high quality physical science instruction before they enter high school.

PROMOTING EARLY ACCESS TO THE PHYSICAL SCIENCES: A STANDARDS-BASED APPROACH

Earlier access to physics and chemistry is a means for greater participation and diversification in STEM post-secondary study and careers. This strategy is aligned with New York State Science Standards (New York State Education Department, 2009) as well as the Next Generations Science Standards (Achieve, Inc., 2012). The latter is a particularly important consideration since the first draft of the NGSS were just released and are expected to shape state science curricula guidelines and instruction over the next twenty years.

The New York State Standards provided the framework for the development of the Bronx Institute Enlace Program in Science and Mathematics. Although the program targeted middle school students, high school *Physical Setting* standards were used; this made the content more rigorous and explicitly physics- and chemistry-related. A second differentiating feature of this program was its emphasis on student-centered instruction and a focus on inquiry. Although the term *inquiry* can be ambiguous, this program was structured with the goal of giving students maximum autonomy and capital to explore scientific questions and generate their own means for finding solutions. The specific NYS standards that constituted the core of the program's scientific inquiry philosophy included the following:

1. Students will engage in scientific inquiry, where they will develop explanations of natural phenomena in a continuing, creative process.
2. Students will evaluate competing explanations and overcome misconceptions.
3. Students will be able to devise ways of making observations to test proposed explanations.
4. Students will be able to use various means of representing and organizing observations (e.g. diagrams, tables, charts, graphs, and equations) and insightfully interpreting the organized data.
5. Students will be able to apply statistical analysis techniques (percent error, standard deviation, line of best fit) when appropriate to test if chance alone explains the result.
6. Students will be able to assess correspondence between the predicted result contained in the hypothesis and the actual result, and reach a conclusion as to whether or not the explanation is supported.

Figure 1. Science performance among middle school students in New York City, based on science laboratory availability

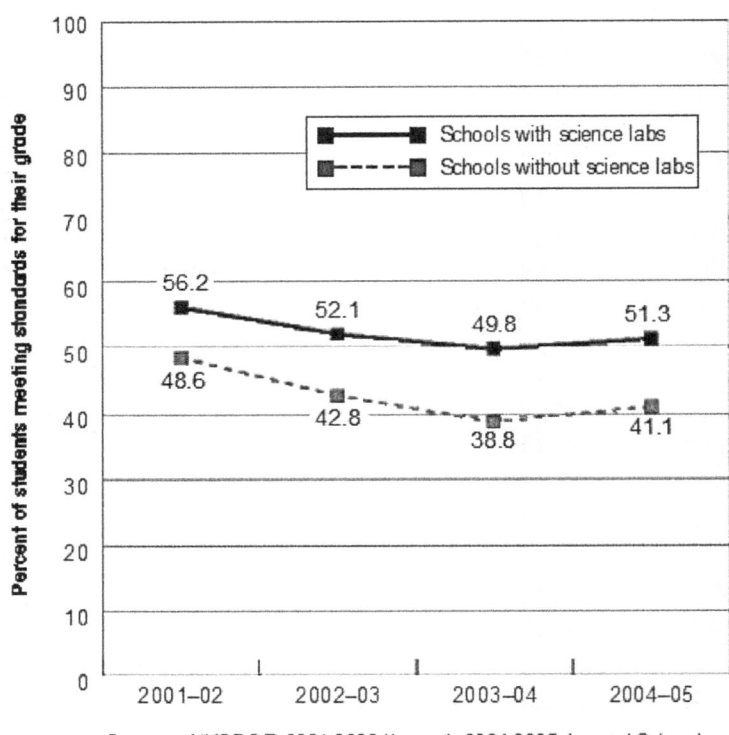

Sources: NYCDOE, 2001-2002 through 2004-2005 Annual School Reports; science lab information based on calls to schools and survey by NYCDOE

7. Students will be able to solve interdisciplinary problems using a variety of skills and strategies, including effective work habits; gathering and processing information; generating and analyzing ideas; realizing ideas; making connections among the common themes of mathematics, science, and technology; and presenting results.

Each week students performed hands-on, constructivist tasks to develop and strengthen their understanding of the physical sciences. For example, students engaged in inquiry-based learning by observing elevated subway trains and developing a method for estimating their speed, creating operational definitions for physical and chemical properties and changes based on prior knowledge (Kelly, 2012), and designing a classification scheme for pH using natural indicators and household substances.

When formulating content-based modules for the weekly lessons in physics and chemistry, it was necessary to incorporate mathematics so students might recognize its importance and usefulness in solving scientific problems. Again, the New York State Science Standards set the foundation for ways in which mathematics might be incorporated into the learning of science. By seamlessly combining science and mathematics instruction, students' skills in both areas were improved significantly. Select mathematics standards (NYSED, 2009) included the following:

1. Students will be able to graph independent and dependent variables, calculating slope and relating it to experimental data.
2. Students will be able to isolate unknown variables when presented with equations and given data.
3. Students will be able to use dimensional analysis to confirm algebraic solutions.
4. Students will use inductive and deductive reasoning to reach mathematical conclusions.
5. Students will be able to use scaled diagrams to represent and manipulate vector quantities.
6. Students will be able to extend their use of powers of ten notation to understand the exponential function and performing operations with exponential factors.
7. Students will be able to simplify calculations by using scientific notation.

Program participants were given graphing calculators to use in the program and keep for personal use in school and at home. They became proficient with this tool while completing tasks such as creating graphs of position-time for uniformly accelerated objects in Galileo's inclined plane experiment, using scientific notation in calculations of wavelength in Young's double slit experiment, and finding the average speed of subway cars with data collected in the neighborhood. They also learned the value of scale diagrams when adding vectors together—an alternative approach that does not require trigonometry, which was beyond the scope of the course. The concept of scale was further developed by the Scales of Space and Time exhibit at the American Museum of Natural History, which helped students visualize the incredible magnitudes of the size and history of the universe.

The Next Generation Science Standards (Achieve, Inc., 2012) were recently designed to improve science education by integrating science and engineering practices, disciplinary core concepts, and crosscutting concepts. The goal is to reduce the ambiguity of inquiry and prioritize the ability of students to design and implement scientific solutions to practical problems. Students are encouraged to engage in literacy practices as they formulate their understanding of science, which presents rich opportunities for students in high needs settings. The Bronx Institute program stressed these practices by encouraging discussion, reading and writing, and scientific discourse among students as they reached consensus on what data revealed.

STRUCTURE OF THE BRONX INSTITUTE PHYSICS AND CHEMISTRY COURSEWORK

Lehman College hosts the Bronx Institute, a non-profit organization that strives to promote equity and academic excellence in the education of Bronx students. The Institute develops educational programs that provide enrichment to increase college readiness, with the ultimate goal of giving urban youth the skills and experiences they need to have equitable access to competitive post-secondary institutions. The purpose of the enrichment program was to provide high quality introductory physics and chemistry education to low-income promising middle and high school students in the Bronx. Students experienced constructivist, inquiry-driven learning activities that helped them recognize the essential role of science in their everyday lives. Ultimately, they gained an appreciation for the physical sciences, with the promise of becoming more scientifically literate citizens, capable of making rationale decisions and having a positive impact in their own communities.

The program was an opportunity for informal physics and chemistry learning, and students did not receive school credit for attending. Seventh and eighth grade students were recruited from the community through secondary school-based guidance counselors. Students were invited to apply for the program if they had scored a 3 or 4 on the English Language Arts (ELA) assessment. Additionally, they had to complete an essay and participate in an interview to assess their motivation and sustained commitment. The majority of the students (98%) were underrepresented minorities (African American or Latino) who also qualified for free lunch at their schools.

The program included both physics and chemistry units over two separate semesters. Mathematics concepts were also integrated with the use of the latest technological tools. The program was structured to provide inquiry-based, challenging experiences for students during a 12-week cycle. Students attended two-hour classes, either in the morning or afternoon, had lunch with their peers and college counselors, and spent an additional two hours in a humanities or mathematics class. The communal cohort was instrumental in the success of the program; students moved together around the college campus, shared meals and subway rides, and regularly received college counseling from staff members.

Typical science classes started with short lectures, which described important physical science principles and how they related to other sciences (students were simultaneously enrolled in integrtaed science in their respective schools). The most important consideration during all lessons was to relate the topic to the everyday experiences of the students. Students' prior knowledge was elicited with questioning, and connections with new concepts were made explicit. Students worked in cooperative groups on hands-on activities that were designed for them to construct their own understandings through observations and experiences (Kelly, 2010a). The standards below outline sample physical science topics in which students were engaged.

Content Key Ideas for the Physical Setting

Key Idea 4.1: Energy exists in many forms, and when these forms change, energy is conserved. Students will observe and describe the transmission of various forms of energy.

1. Students will be able to describe and explain the interchange among potential energy, kinetic energy, and internal energy for mechanical systems.
2. Students will be able to predict velocities, heights, and spring compressions based on energy conservation.
3. Students will be able to determine the factors that affect the period of a pendulum.
4. Students will be able to compare the power developed when the same work is done at different rates.
5. Students will be able to construct simple and parallel circuits, and predict the behavior of light bulbs in these circuits.
6. Students will be able to map the magnetic field of a permanent magnet, indicating the direction of the field between the north-seeking and south-seeking poles.
7. Students will be able to use measurements to determine the resistance of a circuit element.
8. Students will be able to interpret graphs of voltage vs. current.

Key Idea 4.3: Students will explain variations in wavelength and frequency in terms of the source of the vibrations that produced them.

1. Students will be able to compare the characteristics of two transverse waves such as amplitude, frequency, wavelength, speed, period, and phase.
2. Students will be able to draw wave forms with various characteristics.
3. Students will be able to differentiate between transverse and longitudinal waves.
4. Students will be able to determine the speed of sound in air.
5. Students will be able to observe, sketch, and interpret the behavior of wave fronts as they reflect, refract, and diffract.

6. Students will be able to draw ray diagrams to represent the reflection and refraction of waves.
7. Students will be able to determine empirically the index of refraction of a transparent medium.

Key Idea 5.1: Energy and matter interact through forces that result in changes in motion. Students will explain and predict different patterns of motion of objects (e.g. linear and uniform circular motion, velocity and acceleration, momentum and inertia).

1. Students will be able to construct and interpret graphs of position, velocity, or acceleration vs. time.
2. Students will be able to determine and interpret slopes and areas of motion graphs.
3. Students will be able to determine the acceleration due to gravity near the surface of the Earth.
4. Students will be able to determine the resultant of two or more vectors graphically or algebraically.
5. Students will be able to draw scaled force diagrams using a ruler and a protractor.
6. SWBAT resolve a vector into perpendicular components both graphically and algebraically.
7. Students will be able to sketch the theoretical path of a projectile.
8. Students will be able to verify Newton's second law for linear motion.
9. Students will be able to verify conservation of momentum.

From these overarching content standards, topics were selected to best prepare students for standard secondary school course concepts in chemistry and physics. Although the standards provided the foundation, the program aimed for higher-level goals that established more rigorous learning and promoted students' sustained interest.

The program designed experiences that would help students recognize the interconnectedness of key ideas in chemistry, such as the classification of matter, the ways in which energy interacts with matter and is transformed, and how students might see the importance of chemistry in their everyday lives. The guiding questions in the introductory chemistry curriculum included the following:

1. How do matter and energy interact on a molecular level to produce observable physical and chemical changes?
2. How can classification schemes, such as the periodic table, allow us to predict the behavior of matter?
3. How does energy affect matter?
4. In what ways is an understanding of chemistry essential to become a scientifically literate citizen in a democratic society?
5. What are the characteristics of scientifically valid experimental design?
6. In what ways do I know and do science in my everyday life?
7. Why is an understanding of science essential for me to make personal decisions and actively contribute to our global society?
8. How is knowing chemistry important if I choose a career in science?

From these guiding questions, the chemistry course was developed by selecting major concepts, creating constructivist lab activities upon which students could further their prior knowledge, and integrating mathematics and technology in authentic ways. It was essential for students to spend the majority of class time engaged in collecting data and discussing results with the instructor/facilitator and their peers. Consequently, they could take their previous experiences and develop new ideas through experimentation and scientific reasoning. It should be noted that students used non-toxic chemicals in all experiments; when practical, household items were used so students could see the relevance of chemistry in things they use and see regularly. Table 1 summarizes the 12-week modules, which included field visits to the New York Botanical Garden and Brookhaven National Laboratory that complemented classroom learning and instruction. The rationale behind field visits to informal science sites is explained in further detail later in this chapter.

A similar process was utilized to produce the physics curriculum for the Bronx Institute program. Activities were selected that helped students envision themselves as future scientists. For example, they performed the same experiments that Galileo did when observing uniformly accelerated motion and the period of a pendulum. Students were required to keep laboratory notebooks of their experimental data, where they always began each lesson by writing a key question, and their primary task during the course of the class was to answer the question, using experimental data to justify their answers. As with chemistry, students

Table 1. Chemistry curriculum highlights

Session	Chemistry Component	Mathematics/Technology Component
1 Introduction to Matter	• Atomic structure (protons, neutrons, electrons) • Periodic table organization	• Positive and negative integers • Graphing calculator applications
2 Substance Identification	• Differentiating between physical and chemical properties • Relating properties of matter to periodicity	• iPod periodic table applications • Using Flip video cameras and downloading files
3 Phase Changes	• Thermodynamic equilibrium • Energy transfer during phase changes	• Making/analyzing graphs of phase changes • Using graphing calculators to collect and analyze temperature data
4 Acid/Base Chemistry	• Classifying substances according to pH • Making natural indicators	• Base 10 numbers • Logarithms
5 Field Visit NYBG	The purpose of this trip was to learn about how soil pH affects local plantlife. Students had a guided tour of the *New York Botanical Garden*, followed by structured activities and an inquiry-based experiment, where they collected soil and water samples were collected for data analysis (pH, nitrates, etc.) in the classroom. Students filmed parts of their trip with Flip video cameras, which were incorporated into their science documentaries as part of their culminating project.	
6 Water/Soil Analysis	• Water and soil sample analysis with Bronx River samples • Using remote sensing tools to determine dissolved oxygen in water samples	• Fractions and parts per million • Introduction to iMovie; creating projects using film from NYBG and Bronx River field visits
7 Chemical Reactions	• Replacement reactions • Combustion • Synthesis and decomposition	• Manipulating algebraic equations • Creating documentaries with iMovie
8 Energy Sources	• Nuclear chemistry • Environmental sustainability • LEED Green buildings – learning about the design of Lehman's new Science Center	• Balancing nuclear equations • iMovie documentaries
9 Field Visit Brookhaven	The purpose of this visit was to learn about research in new energy technologies at *Brookhaven National Laboratories*. Students visited with members of Hispanic Heritage Club to learn about their research and backgrounds, and tour the facilities.	
10 Molecular Structure	• Creating models of compounds • Using *Molecular Modeling* software	• iPod molecular modeling applications • iMovie documentaries
11 Things Are Made of…	• Composition of everyday materials (ceramics, construction materials, concrete, make-up, etc.) • Chemistry cooking experiments – how ingredients affect outcomes	• Using graphs and statistics to represent data • Filming experiments • iMovie documentaries
12	Sharing of Portfolios and iMovies (families invited)	

came to view science as means for revealing the intricacies of the physical world around them, a world in which are active participants and not passive observers. The guiding questions in the introductory physics curriculum included the following:

1. How can we describe moving objects utilizing principles of kinematics?
2. How do Newton's laws and the concept of force explain why interactions occur between objects?
3. How do electromagnetic forces explain interactions of matter and observable phenomena?
4. In what ways is an understanding of physics essential to become a scientifically literate citizen in a democratic society?
5. What are the characteristics of scientifically valid experimental design?
6. In what ways do I know and do science in my everyday life?
7. Why is an understanding of science essential for me to make personal decisions and actively contribute to our global society?
8. How is knowing physics important if I choose a career in science?

Table 2 summarizes the 12-week physics modules. The student-centered activities included a variety of cutting-edge technological tools, such as Vernier probeware, TI-84 graphing calculators, iPod Touches (see Kelly, 2011), and digital multimeters. The digital fluency that most children bring to school must be leveraged to facilitate data collection and the evaluation of evidence (Hsi, 2007). Kuech and Lunetta (2002) reported that real-time data sensors and graphs have engaged students in cognitive conflict and promoted conceptual growth; such tools allow students to rapidly evaluate hypotheses with multiple trials. The use of technology can also promote interest and help students see the relevance of physics concepts in their lives. By participating in the experiments and recording and discussing their observations, the students were able to make sense of important physics terminology and concepts. These exercises also gave them more confidence and background knowledge, so they could readily make connections to the new experiences they would learn about as the weeks progressed and scaffolded concepts became gradually more complex. The intent was to maximize learning and interest in science.

As part of the Bronx Institute Program, students engaged in field trips to regional science institutions – the American Museum of Natural History, the New York Botanical Garden, and Brookhaven National Laboratory. Experiences in informal environments for science learning have been typically characterized as learner-motivated, guided by learner interests, voluntary, personal, ongoing, contextually relevant, collaborative, nonlinear, and open-ended (Falk & Dierking, 2000; Griffin & Symington, 1998). Research evidence has shown that learning through field trips can be a valuable supplement and addition to classroom instruction (DeWitt & Storksdieck, 2008). Nevertheless, school constraints often have not been supportive of informal learning experiences, particularly when confronting the time constraints due to the pressure of standardized tests in middle and high school. Research has shown that, despite its potential, field trips have often been underused as learning experiences (DeWitt & Storksdieck, 2008).

Informal learning environments, such as museums, outdoor gardens, and science research centers, have demonstrated their success in making science accessible to a broad range of learners (Bell, Lewenstein, Shouse, & Feder, 2009). The designed spaces have allowed students to navigate freely through authentic scientific contexts, where they might choose experiences that fit their needs, interests, and educational agendas (Bell, et al., 2009; Cox-Peterson, Marsh, Kisiel, & Melber, 2003). Their learning has not only depended upon exhibit interactions within a safe, stable space, but

Table 2. Physics curriculum highlights

Session	Physics Component	Mathematics/Technology Component
1 Kinematics	• How can we describe moving objects utilizing principles of kinematics and deriving appropriate equations?	• Making/analyzing graphs of position vs. time • Calculating velocity from slope • Matching existing position-time and velocity-time graphs with motion
2 Uniform Acceleration	• How did Galileo observe and interpret the effect of mass on acceleration due to gravity?	• Interpreting graphs of position-time and position-time squared to make inferences about uniform acceleration
3 Projectiles and Vectors	• How is two-dimensional motion related to vector addition?	• Creating scale drawings of force vectors and using graphical methods to add and subtract vectors
4 Newton's Laws	• How do Newton's laws and the concept of force explain the interactions between objects?	• Using iPod Touch applications to predict the effects of force on motion (Kelly, 2011)
5 Momentum	• How is Newtons' third law evident in observations of collisions?	• Vernier photogates provide data on velocity before and after collisions
6 Mechanical Energy	• How is energy transformation evident with kinetic and potential energy.	• Calculating expended horsepower during climbing stairs
7 Amusement Park Physics	• How can we observe the conservation of mechanical energy in an everyday event?	• iPod Touch applications on roller coaster design
8 Waves and Sound	• How do sound waves propogate in different media?	• Base 10 numbers • Logarithms
9 Field Visit Vassar	The purpose of this visit was to learn about research in acoustics. Students visited with members of the physics faculty, participated in a college admissions session, experienced physics lab tours and acoustics demonstrations and experiments. The workshop was designed to solidify the students' understanding of acoustics fundamentals. After the acoustic demonstrations, the Bronx students worked in small groups as they rotated through three stations facilitated by physics undergraduates where they explored the transmission of sound (Bradley & Kelly, 2011).	
10 Light and Color	• How do we experience the different forms of light in our everyday lives?	• Measurement of visible wavelengths of light with a diffraction grating
11 Electricity	• How does are the electrical circuits constructed in my own home? What are the safety considerations?	• Using digital multi-meters to measure voltage, current, and resistance
12 Magnetism	• What do magnetic field lines tell us about the strength and direction of the magnetic field?	• Using magnetic field sensors to measure the fields around bar magnets • Using compasses to map a route through the neighborhood.

also upon the social and mediating factors that have allowed them to process their emotions and sensory responses (Rennie, Feher, Dierking, & Falk, 2003). Interactions with facilitators, scientists, teachers, and peers were essential when considering the value of the social setting in supporting science learning (Fadigan & Hammrich, 2004).

Table 3 summarizes one segment of a field visit to the American Museum of Natural History. As part of the chemistry module, students were directed to several permanent exhibits to answer questions about the physical and chemical characteristics of specific substances. The questions they explored at the Museum complemented their classroom learning and provided a unique context for them to see evidence of "micro," or atomic level, chemistry principles in a "macro" environment, with physical specimens and graphical representations.

Table 3. Exploration topics for American museum of natural history

Exhibit	Essential Questions
Planet Earth	1. What elements are necessary for life? 2. Give some examples of where carbon is located on Earth (look at carbon cycle). 3. Explain the water cycle. 4. What are the nine main elements that the Earth is made of? 5. Compare the densities of the Earth's core, crust, mantle, and oceans. 6. What are the heat-producing elements in the Earth's interior? 7. How can we tell the age of rocks? 8. Why can only Earth support life as we know it? 9. Explain how rocks form under different heat and pressures. 10. Why is the sea salty? 11. How is ozone produced? Why is it both good and bad?
Minerals & Gems	1. What are minerals? 2. How do minerals form? 3. Choose one or more of the displayed minerals and gems. Explain their composition in terms of their hardness and the atoms they contain. Where are these minerals/gems usually found?
Scales of Universe	1. Look at some examples of small things in the universe. Explain how to compare the size of the object in the photograph or model to the large sphere in the hall.
Journey to the Stars	1. What re the two main elements in stars? 2. How does the Sun affect the Earth? 3. How is the Sun similar to or different from other stars? 4. The Sun continuously blasts solar wind – what is this made of? 5. How does the Sun produce so much energy? 6. What are the stages of the life of a star?
Sea Rex iMax Show	1. Compare the size of these creatures to other living things. 2. What factors led to these creatures dying off? 3. What is the most fascinating thing you learned from this film?

RESEARCH AND RESULTS

Quantitative Data

The effectiveness of the informal physics and chemistry program was measured in several ways. Data were collected on student achievement, views, and attitudes towards science before and after their participation in the program. Quantitative and qualitative data suggested that students had significant increases in their knowledge of physical science, had improving attitudes towards science, and learned to appreciate the importance of science in their lives.

Preliminary data were collected on the physics portion of the course, using a modified *Force Concept Inventory* (FCI) with a pilot group of 20 students. The FCI is a set of validated conceptual physics questions that tests students' knowledge and application of basic mechanics (Savinainen & Scott, 2002). The performance among the students indicated significant gains. Table 3 and Figure 2 summarize the pre-post achievement gains for the overall group along with disaggregation by attendance, gender, and standardized math scores. As Table 4 shows, the largest gains observed for students with 100% attendance (18.8 percentage points), male students (15.9 points), and math level 3 students (16.1 points). The pre-post gains for the overall group tested using the t-test was significant ($t = 3.1$, $df = 19$, $p < .01$), with an effect size of 0.73 (medium). Figure 2 shows the pre-post scores for the overall group and subgroups (Kelly & Kennedy-Shaffer, 2011).

Data was collected to examine whether students' attitudes towards physics had changed as result of their participation in the program. Students were given four assessments to gauge their

Table 4. Physics content diagnostic scores, overall and within groups (Kelly & Kennedy-Shaffer, 2011)

Variable	N	Pre-Test Mean	Post-Test Mean	Pre-Post Gain
Students (all)	20	43.2	55.9	12.7
Students with 100% attendance	13	46.2	65.0	18.8
Female	12	47.7	58.3	10.6
Male	8	36.4	52.3	15.9
Math level 3 students	11	34.4	50.5	16.1
Math level 4 students	9	53.2	63.6	10.4

performance and their attitudes toward physics. During the first meeting of the Bronx Institute class, researchers administered a Likert-scale survey of students' attitudes toward physics and learning science in general (1=never; 2=rarely; 3=sometimes; 4=frequently; 5=always). Cronbach's alpha for a small group of survey respondents was 0.89. The survey statements included the following:

1. I like to solve problems.
2. I think about taking a lab science course in the future.
3. I see science as a collection of facts for me to learn.
4. I think about science outside of school.
5. I watch scientific programs on television.
6. I see myself in a career that requires me to use science.
7. I see science as a way to examine my surroundings.
8. Studying physics intimidates me.
9. I think physics is too difficult for me to learn.
10. I see physics as a powerful way to understand my world.
11. I think I need to study too much math for physics

Students rated statements using the Likert categorizations listed above. Because these responses were categorical, a Pearson chi-square test was used for analysis. The changes, though generally slightly increasing in the direction of more positive attitudes (notable statements are shaded), were not statistically significant. The exception for more positive attitudes towards science is question 3; more students considered science a collection of facts rather than a process for understanding the natural world. This was not intended but was something that could be addressed in future classes. Students' responses on the pre-survey and on the post-survey are displayed below (Table 5). The means were calculated by compiling all of the student responses, assigning 1 point for "never," 2 for "rarely," 3 for "sometimes," 4 for "frequently," and 5 for "always." The numerical values were reversed for negatively phrased questions (#8, 9,

Table 5. Students' attitudes toward science, pre- and post-survey

Statement	Pre-test mean	Post-test mean
1. I like to solve problems.	3.5	3.7
2. I think about taking a lab science course in the future.	3.6	3.7
3. I see science as a collection of facts for me to learn.	3.3	3.9
4. I think about science outside of school.	2.9	3.3
5. I watch scientific programs on television or the computer.	2.8	3.0
6. I see myself in a career that requires me to use science.	3.3	3.4
7. I see science as a way to understand my surroundings.	3.5	3.5
8. Studying physics intimidates me.	4.0	3.5
9. I think physics is too difficult for me to learn.	4.3	3.8
10. I see physics as a powerful way to understand my world.	3.3	3.6
11. I think I need to study too much math to learn physics.	3.7	3.6
COMPOSITE SCORE	37.6	38.4

Figure 2. Pre-test and post-test scores for students, overall and within groups (Kelly & Kennedy-Shaffer, 2011)

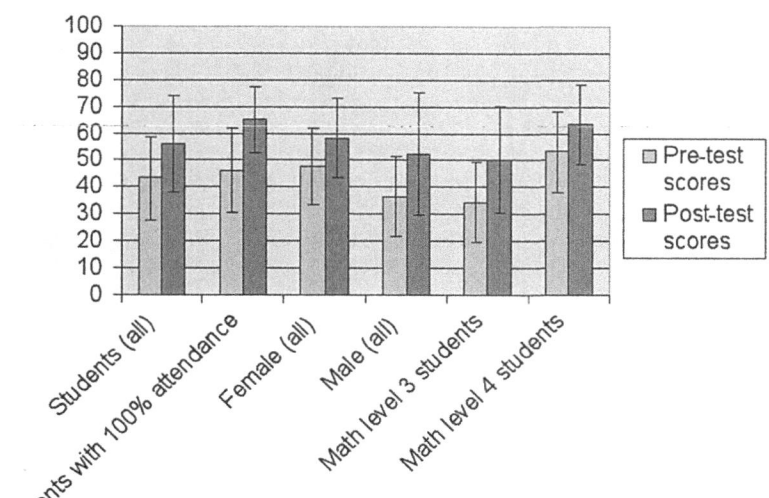

and 11). The composite score was computed by adding the means. The data suggest that the students' attitudes were moving towards an improved view of physics. Perhaps additional delayed post data would reveal statistically significant changes in students' attitudes towards the physical sciences; this is planned for future cohorts.

Qualitative Data

Qualitative data provided different insights into the program's effectiveness. While interviewing participants regarding their experiences with physical science, several themes emerged (see Figure 3). First, it was apparent that the authentic representation of scientific concepts in the physical science courses was an essential element for them to understand the relevance of physics and chemistry in their lives. Secondly, they viewed their success in physics and chemistry as an accomplishment, and this sense of satisfaction transformed their approach towards science from one of apprehension (passive) to one of confident participation (active). Third, their experiences in a rigorous student-centered, constructivist environment fostered collaboration with peers and a desire to give back to their communities. Finally, they recognized the positional advantage inherently resulting from physics and chemistry participation, and this educational capital influenced their academic aspirations (Kelly, 2010b).

Their engagement with physical science can be characterized as transformative, in that many approached the subject with trepidation, only to discover that they often found physics and chemistry to be their favorite sciences. Furthermore, their success in understanding key physical concepts gave them a sense of positional authority, where could access an elite intellectual domain through understandings they personally constructed in inquiry-based laboratory experiences. Unlike their school-based, typically passive science experiences, they were trusted and encouraged to learn through the active participation more closely aligned with authentic scientific practice. Future programs can look towards the positive experiences of these students in providing extracurricular opportunities for access to advanced science for urban students (Kelly, 2010b).

Figure 3. Thematic elements of physical science as a transformative experience (Kelly, 2010b)

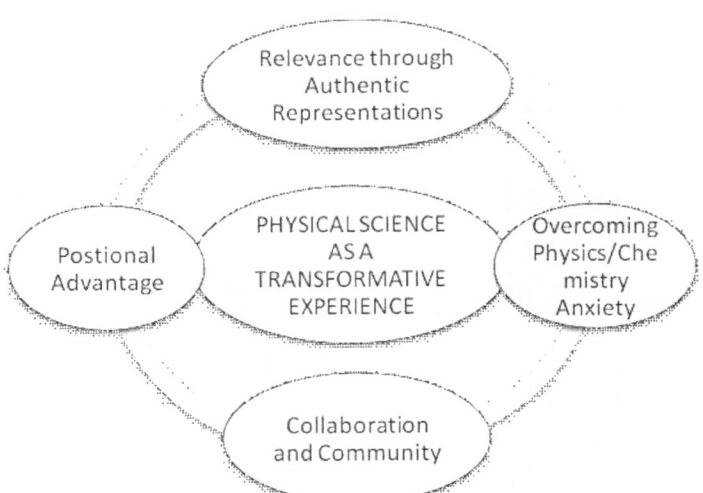

DISCUSSION AND IMPLICATIONS

Several lessons can be learned from creating informal physical science learning experiences for urban middle school students. First of all, teachers must have an unwavering belief that their students can be successful. By taking high school level standards and making them accessible to middle school students, one can communicate to the students that they are trusted and expected to reach higher levels of achievement. Students will rise to the occasion if given the opportunity. Secondly, making science come alive through student-centered instruction with exciting and practical laboratory activities is a clear path towards success. High-quality informal physical science teaching requires intensive planning, an awareness of what knowledge and misconceptions students bring to the classroom, and coordination of experiences whereby students construct understandings through observation. By creating an environment conducive to participation, collaboration, and risk-taking, students experienced active learning and could see themselves as potential future scientists (Kelly & Kennedy-Shaffer, 2011). Finally, this research suggests that informal settings have great potential for fostering interest in physical science among traditionally underrepresented students.

It takes years to transform schools, particularly in STEM education, and informal programs are ideally situated to address this immediate concern. Grant-funded initiatives like the Bronx Institute can develop innovative ways to approach access to physics and chemistry, and the empirical data from the program can inform broader reforms in similar contexts.

The urgent need for such early exposure to rigorous science study justifies novel approaches. The Bronx Institute created such a program to stimulate student interest in the physical sciences and cultivate valuable knowledge and skills for advancing their academic standing in high school. Students participating in an inquiry-based science program that seamlessly integrated mathematics and technology were taking advantage of an ideal opportunity to prepare themselves for active participation in STEM fields later in their lives. One student realized that colleges would look at her record favorably because she had taken a science course that most others had not, stating, "…it differentiates me from everybody else around me. It gives me a lot of choices for when I get older." Indeed, for many the experience was transformative and gave them greater confidence in their ability to succeed in what are typically considered elitist subjects. As one participant stated:

When I took the physics course, I felt that nobody's limited to anything in life. If you try hard enough to do anything you want, you could be able to do it. So it doesn't matter where you're from, what race you are, where you live, as long as you have the strength and the ability to do it, and you want to do it, then you're able to conquer anything that comes your way.

In the future, the Bronx Institute can serve as an example to other informal programs on how to reach more underserved students and better prepare them for success in secondary post-secondary STEM study. The program demonstrated that early exposure to rigorous and accessible physical science has great promise for improving student performance and interest in science. Their model can be scaled up for application in formal school settings with appropriate professional development, investment in classroom resources, and shifts in policy to promote better middle school science curricula. Institutionalizing early access to physical science in U.S. schools may transform the nation's pool of future scientists and engineers, and pave the way for global leadership with a diversified scientific community.

REFERENCES

Achieve, Inc. (2012). *The next generation science standards.* Retrieved May 18, 2012, from http://www.nextgenscience.org/next-generation-science-standards

American Association of Physics Teachers. (2002). *AAPT statement on physics first.* College Park, MD: AAPT. Retrieved October 11, 2011, from http://www.aapt.org/Resources/policy/physicsfirst.cfm

American Chemical Society. (2008). *Workshop on increasing participation of Hispanic undergraduate students in chemistry.* Washington, DC: ACS.

American College Testing. (2005). *Developing the STEM (science, technology, engineering, mathematics) education pipeline.* Retrieved August 31, 2011, from http://www.act.org/research/policymakers/pdf/ACT_STEM_PolicyRpt.pdf

American Institute of Physics. (2011). *Focus on underrepresented minorities in high school physics: Results from the 2008-09 nationwide survey of high school physics teachers.* College Park, MD: AIP Statistical Research Center. Retrieved October 11, 2011, from http://www.aip.org/statistics/trends/hstrends.html

Appleton, K. (2007). Elementary science teaching. In Abell, S. K., & Lederman, N. G. (Eds.), *Handbook of Research on Science Education* (pp. 493–535). Mahwah, NJ: Erlbaum.

Barton, P. E. (1993). *Hispanics in science and engineering: A matter of assistance and persistence.* Princeton, NJ: Educational Testing Service.

Bell, P., Lewenstein, B., Shouse, A. W., & Feder, M. A. (2009). *Learning science in informal environments: People, places, and pursuits.* Washington, DC: The National Academies Press.

Bradley, D. B., & Kelly, A. M. (2011). Promoting inclusiveness in acoustical physics. *Academic Exchange Quarterly, 15*(4), 88–93.

Brewe, E. (2008). Modeling theory applied: Modeling instruction in introductory physics. *American Journal of Physics, 76*(12), 1155–1160. doi:10.1119/1.2983148

Carnevale, A. P., & Strohl, J. (2010). How increasing college access in increasing inequality, and what to do about it. In Kahlenberg, R. D. (Ed.), *Rewarding Strivers* (pp. 71–190). New York, NY: The Century Foundation.

Committee on Education, Council of the City of New York. (2004). *Lost in space: Science education in New York City public schools.* New York, NY: Committee on Education, Council of the City of New York.

Cox-Petersen, A. M., Marsh, D. D., Kisiel, J., & Melber, L. M. (2003). Investigation of guided school tours, student learning, and science reform: Recommendations at a museum of natural history. *Journal of Research in Science Teaching, 40*(2), 200–218. doi:10.1002/tea.10072

DeLany, B. (1991). Allocation, choice, and stratification within high schools: How the sorting machine copes. *American Journal of Education, 99*(2), 181–207. doi:10.1086/443978

DeWitt, J., & Storksdieck, M. (2008). A short review of school field trips: Key findings from the past and implications for the future. *Visitor Studies, 11*(2), 181–197. doi:10.1080/10645570802355562

Eamon, M. K. (2004). Socio-demographic, school, neighborhood, parenting influences on the academic achievement of Latino youth adolescents. *Journal of Youth and Adolescence, 34*(2), 163–174. doi:10.1007/s10964-005-3214-x

Eisenkraft, A. (2010). Millikan lecture 2009: Physics for all: From special needs to olympiads. *American Journal of Physics, 78*(4), 328–337. doi:10.1119/1.3293130

Fadigan, K. A., & Hammrich, P. L. (2004). A longitudinal study of the educational and career trajectories of female participants of an urban informal science education program. *Journal of Research in Science Education, 41*(8), 835–860.

Falk, J. H., & Dierking, L. D. (2000). *Learning from museums*. Walnut Creek, CA: Altamira Press.

Gollub & Spital. (2002). Advanced physics in high schools. *Physics Today, 55*(5), 48–53. doi:10.1063/1.1485584

Griffin, J., & Symington, D. (1998). Moving from task-oriented to learning-oriented strategies on school excursions to museums. *Science Education, 81*(6), 763–779. doi:10.1002/(SICI)1098-237X(199711)81:6<763::AID-SCE11>3.0.CO;2-O

Hehn, J., & Neuschatz, M. (2006). Physics for all? A million and counting! *Physics Today, 59*, 37–43. doi:10.1063/1.2186280

Horizon Research. (2002). *2000 national survey of science and mathematics education: Status of elementary school science teaching*. Chapel Hill, NC: Horizon Research Inc.

Hsi, S. (2007). Conceptualizing learning from the everyday activities of digital kids. *International Journal of Science Education, 29*(12), 1509–1529. doi:10.1080/09500690701494076

Kelly, A. M. (2010a). Differentiating the underrepresented: Physics opportunities for Bronx high school students in a university setting. In H. Oluseyi (Ed.), *2009 American Institute of Physics Conference Proceedings Series: Vol. 1280: Joint Annual Conference of the National Society of Black Physicists and the National Society of Hispanic Physicists,* (pp. 176-181). Melville, NY: American Institute of Physics.

Kelly, A. M. (2010b). Transformative informal physics in the Bronx. *Academic Exchange Quarterly, 14*(1), 57–62.

Kelly, A. M. (2011). Teaching Newton's laws with the iPod Touch in conceptual physics. *The Physics Teacher, 49*(4), 202–205. doi:10.1119/1.3566026

Kelly, A. M. (2012). Introducing physical and chemical properties and changes. *Science Teacher (Normal, Ill.), 79*.

Kelly, A. M., & Kennedy-Shaffer, R. (2011). Teaching Newton's laws to urban middle school students: Strategies for conceptual understanding. *Journal of Curriculum and Instruction, 5*(1), 54–67.

Kelly, A. M., & Sheppard, K. (2008). Newton in the Big Apple: Access to physics in New York City. *The Physics Teacher, 46*(5), 280–283. doi:10.1119/1.2909745

Kelly, A. M., & Sheppard, K. (2009). Secondary physics availability in an urban setting: The relationship to academic achievement and course offerings. *American Journal of Physics, 77*(10), 902–906. doi:10.1119/1.3191690

Kelly, A. M., & Sheppard, K. (2010). The relationship between the urban small schools movement and access to physics education. *Science Educator, 19*(1), 14–25.

Kuech, R. K., & Lunetta, C. N. (2002). Using digital technologies in the science classroom to promote conceptual understanding. *Journal of Computers in Mathematics and Science Teaching, 21*(2), 103–126.

Lareau, A., & Horvat, E. M. (1999). Moments of social inclusion and exclusion: Race, class, and cultural capital in family-school relationships. *Sociology of Education, 72*(1), 37–53. doi:10.2307/2673185

Lee, O. (1995). Subject matter knowledge, classroom management, and instructional practices in middle school science classrooms. *Journal of Research in Science Teaching, 32*(4), 423–440. doi:10.1002/tea.3660320409

Lynch, S. (2001). Science for all is not equal to one size fits all: Linguistic and cultural diversity and science education reform. *Journal of Research in Science Teaching, 38*(5), 622–627. doi:10.1002/tea.1021

National Academies of Sciences. (2007). *Rising above the gathering storm: Energizing and employing America for a brighter economic future*. Washington, DC: The National Academies Press.

National Academies of Sciences. (2010a). *Expanding underrepresented minority participation: America's science and technology talent at the crossroads*. Washington, DC: National Academies Press.

National Academies of Sciences. (2010b). *Rising above the gathering storm, revisited: Rapidly approaching category 5*. Washington, DC: National Academies Press.

National Association of Educational Progress. (2009). *The nation's report card*. Retrieved January 30, 2012, from http://www.nationsreportcard.gov/science_2009/

National Center for Education Statistics. (1998). *Pursuing excellence: A study of U.S. twelfth grade mathematics and science achievement in international context*. Washington, DC: NCES.

National Center for Education Statistics. (2004). *Highlights from TIMMS*. Washington, DC: Office of Educational research and Improvement. Retrieved from http://nces.ed.gov/pubs99/1999081.pdf

National Center for Education Statistics. (2009). *Digest of education statistics: Percentage of public and private high school graduates taking selected mathematics and science courses in high school, by sex and race/ethnicity: Selected years, 1982 through 2005*. Washington, DC: U. S. Department of Education Institute of Education Sciences. Retrieved November 9, 2010, from http://nces.ed.gov/programs/digest/d09/tables/dt09_151.asp

National Commission on Mathematics and Science Teaching for the 21st Century (NCMST). (2000). *Before It's Too Late*. Washington, DC: Education Publications Center.

National Science Board. (2006). *Science and engineering indicators 2006*. Arlington, VA: National Science Foundation.

National Science Foundation, Division of Science Resources Statistics. (2007). *Women, minorities, and persons with disabilities in science and engineering: 2007*. Arlington, VA: National Science Foundation. Retrieved from http://www.nsf.gov/statistics/wmpd

New York City Coalition for Education Justice. (2007). *New York City's middle grade schools: Platforms for success or pathways to failure?* Retrieved January 12, 2012, from http://www.nyccej.org/117/new-york-citys-middle-grade-schools-platforms-for-success-or-pathways-to-failure

New York City Department of Education. (2006). *Annual school reports cards 2004-2005.* Retrieved January 12, 2012, from http://schools.nyc.gov/Accountability/data/AnnualSchoolReports/default.htm

New York State Education Department. (2009). *Physics/physical setting core standards.* Retrieved November 30, 2009, from http://www.emsc.nysed.gov/ciai/mst/pub/phycoresci.pdf

New York Times. (2006, August 18). *In elite new york high schools, a dip in blacks and hispanics.* Retrieved January 30, 2012, from http://www.nytimes.com/2006/08/18/education/18schools.html?pagewanted=all

Rennie, L. J., Feher, E., Dierking, L. D., & Falk, J. H. (2003). Toward an agenda for advancing research on science learning in out-of-school settings. *Journal of Research in Science Teaching, 40*(2), 112–120. doi:10.1002/tea.10067

Riegl-Crumb, C., & Grodsky, E. (2010). Racial-ethnic differences at the intersection of math course-taking and achievement. *Sociology of Education, 83*(3), 248–270. doi:10.1177/0038040710375689

Sadler, P. M., & Tai, R. H. (2001). Success in introductory college physics: The role of high school preparation. *Science Education, 85*(2), 111–136. doi:10.1002/1098-237X(200103)85:2<111::AID-SCE20>3.0.CO;2-O

Savinainen, A., & Scott, P. (2002). The force concept inventory: A tool for monitoring student learning. *Physics Education, 37*(1), 45–52. doi:10.1088/0031-9120/37/1/306

Sheppard, K., & Robbins, D. M. (2003). Physics was once first and was once for all. *The Physics Teacher, 41*, 420–424. doi:10.1119/1.1616483

Simpson, J. C. (2001). Segregated by subject – Racial differences in the factors influencing academic major between European Americans, Asian Americans, and African, Hispanic, and Native Americans. *The Journal of Higher Education, 72*(1), 63–100. doi:10.2307/2649134

Tyson, W., Lee, R., Borman, K. M., & Hanson, M. A. (2007). Science, technology, engineering, and mathematics (STEM) pathways: High school science and math coursework for postsecondary degree attainment. *Journal of Education for Students Placed at Risk, 12*, 243–270. doi:10.1080/10824660701601266

U.S. Census Bureau. (2012). *Statistical abstract of the United States, 2012.* Washington, DC: Government Printing Office. Retrieved May 18, 2012, from http://www.census.gov/compendia/statab/cats/education.html

U.S. Government Accountability Office. (2005). *Higher education: Federal science, technology, engineering, and mathematics programs and related trends.* Retrieved October 11, 2011, from http://www.gao.gov/new.items/d06114.pdf

Vanfossen, B. E., Jones, J. D., & Spade, J. Z. (1987). Curriculum tracking and status maintenance. *Sociology of Education, 60*(2), 104–122. doi:10.2307/2112586

Whitten, B. L., Foster, S. R., Duncombe, M. L., Allen, P. E., Heron, P., & McCullough, L. (2004). Like a family: What works to create friendly and respectful student-faculty interactions. *Journal of Women and Minorities in Science and Engineering, 10*(3), 229–242. doi:10.1615/JWomenMinorScienEng.v10.i3.30

Woolnough, B. (1994). Why students choose physics, or reject it. *Physics Education, 29*(6), 368–374. doi:10.1088/0031-9120/29/6/006

Chapter 11
Rooted in Teaching:
Does Environmental Socialization Impact Teachers' Interest in Science-Related Topics?

Lisa A. Gross
Appalachian State University, USA

Joy James
Appalachian State University, USA

Eric Frauman
Appalachian State University, USA

ABSTRACT

Research in Environmental Socialization (ES) and the impact of significant life experiences suggest that childhood play in outdoor environments shape later adult activities or career interests. Few studies have investigated how childhood experiences influence curricular interests of preservice and inservice teachers. This preliminary study examines what ES factors of teachers raised in rural and/or non-rural environments reveal about their interests in science topics and field-based learning opportunities. Results suggest that teachers growing up in rural areas were slightly less interested than non-rural teachers in field-based learning and expressed less experience with environmental education. Teachers with more ES experiences (e.g., played in the woods, built forts) expressed greater interest in science-related topics than those who had indicated fewer experiences. Rural teachers tended to have more ES experiences than non-rural teachers. The authors discuss how environmental socialization factors influence teacher preference for curricular programs specific to environmental and ecological topics and raise questions about the changing environmental socialization experiences of preservice and novice teachers.

DOI: 10.4018/978-1-4666-2809-0.ch011

ROOTED IN TEACHING: DOES ENVIRONMENTAL SOCIALIZATION IMPACT TEACHERS' INTEREST IN SCIENCE-RELATED TOPICS?

From the earliest years, children are actively engaged in their local surroundings. They explore their surroundings, discover new spaces, and construct personal places where their initial ideas and foundational knowledge are developed. This physical-world socialization begins in the home and over time, extends out into the local community. Each day offers new situations, events, and potential opportunities for furthering the child's understanding of the natural world. Over time, such experiences, combined with their social interactions with peers, family or local experts, contribute to what is known and how this knowledge is to be used (Piaget, 1954; Vygotsky, 1978; Wertsch, 1984).

Our students grow up in varying social communities, or lifeworlds (Lim & Calabrese Barton, 2006)—places where they seek out and interact with resources that enable cultural reproduction. Constructed around the physical, social, and emotional dimensions, an individual's lifeworld is the most basic knowledge-shaping mechanism. Too often, educational researchers take for granted how the context of one's upbringing has influenced her development into an adult social being.

Whether a student's lifeworld includes the sights and sounds of city life or country living, the extent in which knowledge has been learned and the ways in which it has been communicated to others has been shaped by formative interactions within her environment. The role of the teacher then, is to draw from this foundational knowledge, making connections to her prior experiences and interests as a means of developing more complex ideas and understanding. While this constructivist school of thought is prevalent in many teacher education programs, there seems to be an expanding crevasse between the experiences of today's youth and the appropriate pedagogical approaches used to bridge the gap toward understanding.

At present, we live in a day and time in which our nation's youth spend less time investigating and exploring their local surroundings. Richard Louv's (2005) book, *Last Child in the Woods*, describes how changes in our children's daily activities have influenced their connection with nature and the outdoors. Elementary teachers, as well as science teachers, must provide learners with experiences within and knowledge about their local surroundings. Lim and Barton (2006) propose the idea of community science, using lifeworld places as a context for connecting the science of the curriculum to the students' experiences occurring outside of the classroom and in the local community. To preserve the places that shape the knowledge of future generations requires a redirection and reconnection with our natural world.

Over the past two years, we have conducted surveys with pre-service and in-service educators within a six county regional area to assess interest in teaching ecological and environmental topics. Data collected were influential to the development of curricular programs and workshops offered through our university's outdoor education facility. While analyzing the responses by county, we detected patterns between the respondents' area of upbringing and their science topic interests. Upon further examination, we noted that the childhood activities indicated by the respondents seemingly correlated to their preferences in potential programs. In this chapter, we discuss our preliminary findings from the analysis of data and introduce how patterns in environmental socialization experiences may contribute to teachers' curricular choices and interests.

LITERATURE REVIEW

Sense of Place and Place-Identity

Shared among researchers of this study, is our interest in place and the development of a place-identity. Though our discipline areas and backgrounds differ (e.g. education and recreation management), we are aware of how our own identities and interests have been shaped by earlier interactions with our environment. Given this, we contend that the notion of *place* is significant in understanding self—that the meaning derived from past encounters have played into what we know, how we know it and how we communicate it to others. Relph (1976) describes how nearly every individual has "a deep association with and consciousness of the places where we were born and grew up, where we live now or where we have had particular moving experiences" (p. 43). From our own lived experiences, we are able to identify places where we feel welcome, comfortable and secure just as we are likely to avoid or have limited interactions within those places in which we are not.

The notion of place-identity includes the relationship between human action and physical space. For the social scientist, an individual's sense of place includes meanings and experiences of those who live in, visit, or imagine place as shaped by structural forces and/or past interactions (Massey & Jess, 1995). The formative experiences that occur in these spaces have a powerful impact on our understanding of self, our feelings of belonging, and our degree of attachment (Cresswell, 2004). This leads us to wonder how place plays into the statistical trends of the nation's professional teaching force. As members of the civilian workforce, elementary teachers are listed within the top fifteen occupational groups in the country (Bureau of Labor Statistics, 2010). Data reveal that most are employed in school systems similar to their upbringing—places that are familiar, comfortable and likely to promote the same cultural values and norms among group members.

Of the 3.4 million teachers, the demographic patterns have remained fairly consistent over the past four decades—the majority of the teaching force is white (71% city, 84% suburban, 90% rural) and female (76%). Eighty-four percent of this workforce is likely to remain at the same school in which they were originally employed (Chen, et al., 2011) with teacher turnover in rural schools much lower than the turnover rate found in urban school environments (Ingersoll, 2001). Given these stable demographic trends, we propose that the notion of place must be significant to this particular professional group and question if the foundational experiences of teachers have influenced their interests and attitudes toward curricular topics, specifically those associated with the natural sciences.

For the purposes of this study, we explore teacher interest in environmental and ecological concepts based on area of upbringing. In addition, we have employed environmental socialization measures to investigate if the earlier formative experiences of pre-service and in-service teachers' indicate curricular preferences.

Contexts of Environmental Socialization

As previously introduced, many formative learning experiences occur in informal settings; these include routine actions that occur in the home and extend to the boundaries of the local community. Much research has been conducted on the variation in learning opportunities, ranging from relatively formal and structured to entirely informal and ad hoc (Aubusson, Griffin, & Kearney, 2012, p. 1123). Because the formulation of an individual's ideas and thoughts are cumulative, personal, and influenced over time, the knowledge one possesses is quite difficult to trace. This poses numerous challenges to any educator, especially science teachers, as many misconceptions or alternative conceptions (Koch, 2009) are formed and reinforced early on. It is critical therefore, to consider the prior socialization experiences of our students and to develop further understanding in the context of their lifeworlds.

During childhood, one develops both positive and negative attitudes toward her local surroundings. "Childhood is also a period of socialization, of adult investment in the creation of socially relevant skills and beliefs, and motives" (Garbarino, 1989, p. 18). A child's appreciation of her natural environment develops through associations, with both peer and adults, occurring in particular places (Tanner, 1980; Peterson, 1982; Chawla, 1988; Chawla & Hart, 1988; Bixler, Carlisle, Hammitt, & Floyd, 1994). Bixler and Morris (2000) describe this as *environmental socialization*, a process that involves:

repeated experiences resulting in practical knowledge of the physical environment, conceptualization of self in terms of the environment in which rewarding actions take place, and the development of primary and ancillary skills and competencies that allow rewarding activities to be carried out efficiently (p. 67).

Environmental socialization contributes to the individual's understanding of her world and has the potential to impact future preferences, motivations, attitudes, and beliefs.

Research on environmental socialization has been conducted on individuals in less formal learning situations. Studies on socialization into conservation and outdoor recreation indicates that many outdoor activities and comfort in nature settings are learned during childhood (Sofranko & Nolan, 1972; Decker & Purdy, 1986; O'Leary, Behrens-Tepper, McGuire, & Dottavio, 1987; Scott & Willits, 1989). Participation in these outdoor recreational activities encompasses a basic knowledge of natural science and has the potential to cultivate interest in the natural world. This accumulation of childhood experiences in nature can lead to a body of knowledge and appreciation of the environment (Bixler, et al., 1994). Additionally, early childhood education research supports the idea that outdoor experiences are important to children's development and interest in outdoor activity (Cobb, 1993; Rivkin, 1995; Wilson, 1996; Bredekamp & Copple, 1997; Moore & Wong, 1997).

In past socialization studies for example, the roots of conservation behaviors were explored. Research on the significant life experiences of conservationists revealed how earlier formative experiences with one's environment led to interests and careers in professional conservation work. Dozens of studies in North America, Europe, and Africa point to the importance of childhood interactions with natural environments, the presence of environmentally concerned adults, and sometimes a negative experience involving environmental damage or loss (Tanner, 1980; Chawla, 1988; Chawla & Hart, 1988, Palmer, 1993). These findings mirror others reported in the outdoor recreation socialization literature.

The other striking parallel from the literature is the significance of an opportunity for repeated experiences over time. In a qualitative study of the socialization of paddle sport enthusiasts, the dominant theme emerging from the data was the reported frequency and variation in outdoor experiences noted by the informants. This accumulation of experiences not only cultivated the enthusiasts' competencies, interests and preferences, but also contributed to their comfort in the outdoors (Bixler & Morris, 1998; James, Bixler, & Vadala, 2010). Conservationists also reported an amassing of personally meaningful outdoor activities (Chawla, 1988), though the impact of these repeated experiences were less clear.

It seems plausible to argue that this ongoing environmental socialization process contributes to the development of an individual's place identity. As children explore their surroundings and engage in repeated experiences in various settings, they acquire skills, knowledge and interests that positively (or negatively) influence their comfort in these spaces. Readiness of the child and freedom from parental restrictions, as well as availability and access to natural environments also impact the extent and number of experiences (Vygotsky, 1978; Novak, 1998). Human development, social influences, and opportunities for multiple experiences have an effect on children's receptiveness to natural environments (Tanner, 1980; Peterson, 1982; Chawla, 1988; Bixler, et al., 1994; James, Bixler, & Vadala, 2010).

Human development influences an individual's readiness and ability to learn (Piaget, 1954; Arends, 2000). Three domains of human development affect an individual's readiness and learning: (1) physical the psychomotor abilities and physical growth, (2) cognition the mental processes of acquisition of knowledge and perception, and (3) psychosocial the personality, social and emotional development (Berger, 1988; Bee, 1997). Children's awareness of their environment and ability to interact socially is interdependent upon each domain and their particular stage of development. Thus, the developmental stage of children acts as a filter through the environmental socialization contexts of formal, non-formal, and informal learning.

While a positive attitude toward being in natural environments is important, being part of a supportive social group is also necessary (Cheek & Burch, 1976; Vygotsky, 1978). Many of the outdoor opportunities children have, such as playing outdoors, going camping, attending summer camp, or visiting parks, are primarily a function of being in a family that values these experiences (Bixler & Morris, 2000). Frequent engagements with supportive group members make repeated experiences within a physical environment possible; similarly, positive social interactions that take place before, during, and after such encounters shape the child's expanding understanding and respect for her surroundings (Vygotsky, 1978). This leads to the importance of the social and cultural influences.

Social influences that permeate a child's ability to become socialized into natural environment are parents, peers, and institutions. Each influence provides opportunities for introduction, extension, and enrichment of knowledge and awareness of natural environments as well as a socially supportive network. Parental values determine not only access to natural areas and opportunities for exploration or learning of natural areas, but also access to a socially supportive network (e.g., peers, mentors, and clubs). Interactions with peers in natural environments provide opportunities for learning of skills, support, and social norms. Institutions such as clubs or youth organizations (e.g., Girl Scouts, Boy Scouts, or Camp Fire Boys and Girls) offer educational opportunities to learn about the natural world as well as peers and mentors that have similar interests (Vygotsky, 1978; Hammerman, Hammerman, & Hammerman, 1994; Rivkin, 1997). Socialization into the outdoors occurs through parental values, peers, interest, social supportive network, and opportunity.

Finally, access to natural areas for people can be limited through location, cost, time, convenience, or familiarity. For some adults the priority of spending time in a natural setting is worth the effort, but for others the lack of accessibility, the lack of interest, or the lack of opportunity often prevents participation. Those who experience natural environments can develop a social network that provides an introduction, support, guidance, and instruction into outdoor leisure pursuits (Tanner, 1980; Chawla, 1988; Chawla & Hart, 1988; Palmer, 1993). This supportive social structure is integral to development of skills and competencies that enable enjoyment from natural environments (Bixler & Morris, 1998). In examining the environmental socialization experiences of potential and in-service teachers, it would seem likely that the locale of an individual's upbringing would influence access to natural areas and the networks available within particular settings.

In a recent study, James, Bixler and Vadala (2010) produced a developmental model from the data, describing natural history oriented environmental professionals. In each developmental stage, there are experiences contributing to the development of interest in natural history pursuits. Playing in nature, building forts, collecting critters or specimens (e.g., rocks, plants), pretense (e.g., playing Swiss Family Robinson) and getting dirty were found in early childhood experiences. Additional childhood experiences, of these nature-minded individuals, included caring for family pets, travel to different geographical areas (e.g., parks, farms), involvement in scouting, and going to residential camps. Sporting events were present but did not dominate the childhood experiences of the study participants. Each outdoor experience provided opportunities for the development of skills and competencies such as tolerance for weather extremes, wayfinding in unstructured environments, travel planning skills, and activity-related ancillary skills (Bixler & Morris, 2000). These competencies support nature-based interests that can lead to occupational choices, such as committed conservationists (Tanner, 1980; Chawla, 1988) and naturalists (James, Bixler, & Vadala, 2010) to personal interests, such as kayaking (Bixler & Morris, 2000) As evident thus far, the application of environmental socialization measures may offer additional insights for content areas preferences among teaching professionals as well as their beliefs and attitudes toward the teaching of content, specifically science-related concepts.

Shaping Teacher Beliefs and the Formal Schooling Process

Research in teacher education reveals how significant formal schooling experiences have been to those pursuing degrees in education (Brookhart & Freeman, 1992; Southerland & Gess-Newsome, 1999; Windschitl, 2003) as well as their beliefs and attitudes toward teaching (Freeman & Smith, 1997; Haim, 2003; Bleicher & Lindgren, 2005; Decker & Rim-Kaufman, 2008). According to Richardson (1996), three categories of experience influence the development of beliefs and knowledge about teaching and begin at different stages of one's career: personal schooling, experiences with schooling and instruction and formal knowledge. These experiences include earlier social and environmental interactions that contribute to one's foundational knowledge about teaching, shape beliefs about the profession, and influence interests in subject matter.

While much research has been conducted on the formative schooling experiences that contributed to teaching as a career choice, few studies have examined how one's upbringing has influenced curricular interests or confidence in teaching environmental or ecological concepts. Victor (1962) examined the reluctance of elementary school teachers to teach science as the result of inadequate science background. More recent studies have been conducted on preservice teachers efficacy in teaching science (Bleicher & Lindgren, 2005; Decker & Rim-Kaufman, 2008; Southerland & Gess-Newsome, 1999) or efficacy in teaching environmental concepts (Moseley, Reinke, & Bookout, 2002; Moseley & Utley, 2008; Powers, 2004.). Though such studies reveal similar findings to those conducted more than five decades ago, there has been no examination of teaching efficacy that has extended beyond formal education and into the informal experiences of one's upbringing. Only Soy's study of prospective teachers at the State College of Iowa (1967) touched on factors associated with choice of subject fields by geographic locale; she found prospective teachers who choose science, as an area of concentration were more likely to come from a farm background.

Making sense of science-related concepts in one's natural surroundings is influenced by motivation and choice. Learning processes are typically derived from exploration and inquiry into a problem, question, or a specific contextual phenomenon; such helps young people develop new ways of thinking, interpreting and engaging in inquiry. Tension between everyday learning and teachers' view of learning (His, 2007); need for school science to "acknowledge the potential of exploiting natural learning processes that operate when students occupy settings beyond the classroom. A reframing of what counts as legitimate learning in school science. Dewey (1938) asserted the need for teachers to incorporate the prior experiences and interests of students into meaningful, learning episodes. In fact, during the progressive education movement the idea of "learning by doing" included a more holistic, interdisciplinary approach that promoted learning *about* one's surroundings by exploring *within* those surroundings. Such a context was considered preparatory for future life experiences, shaping the individual's future actions within and toward it. Learning occurred in both formal and informal settings, extending beyond the four walls of the classroom. Teaching was a shared responsibility—between educators, parents, and members of the community.

As discussed by Louv (2005), there is a broadening disconnection between individuals and their local surroundings. Understanding the environment has become a sequence of mandated science courses regulated by a standardized curriculum. Teachers are expected to develop students' interests in curricular topics while simultaneously assessing the extent of knowledge learned along the way. For the teaching professional, personal experiences are inseparable from their own previous experiences in informal settings (as children) and their more formal experiences (as students). We must consider that the latter learning opportunities occur for seven hours a day and approximately 120 months of their most formative years. Through the socialization process, the school environment shapes and conditions their understanding of self, place identity and of course, their attitudes toward learning. It is during this time that our experiences with acquiring subject knowledge can be supported or resisted, emphasized, or ignored. Pedagogy plays heavily into how content knowledge is interpreted and acted upon. For the professional educator, all three categories shape their beliefs and attitudes toward teaching and their self-efficacy toward particular content.

Self-Efficacy and the Teaching of Science

Bandura's (1977) theory of social learning includes how the motivation to perform an action is more likely to result in a favorable outcome if the individual is confident that he can perform that action successfully. The construct of personal self-efficacy is situation-specific as well as subject-specific (Bandura, 1997). Beliefs, attitudes, and comfort play important roles in how individuals organize knowledge and information, process and respond. Studies have revealed that perceived academic efficacy is influential in career choice and development. It predicts academic grades, range of career options, and persistence and success in chosen fields (Bandura, 1997, p. 239). Perceived self-efficacy accounts for variations in these different intellectual aspects of occupational pursuits when past academic achievement, scholastic ability, and occupational interests are controlled (Lent, 1986, p. 87). Inseparable from an individual's socialization experiences is the efficacy that develops from these encounters occurring in specific places with particular group members.

Standardized curricula and accountability systems are realistic constraints of the classroom teacher, yet these same limiting structures are often a professional "scapegoat" for educators who do not feel comfortable or knowledgeable about a given topic or content area. Elementary teachers for example, make daily curricular choices based on district and state accountability systems. A majority of the school day time is devoted to content areas and skills that are assessed (reading, mathematics, writing) while other curricular areas (science, social studies) are skimmed over or avoided completely. While time may be a realistic constraint for some, others use these same accountability systems to avoid topics or curricular areas in which they lack interest or background experience. Similarly, we find secondary teachers in positions for which they are not qualified, prepared, or interested.

Fear of science, inadequate science background, lack of interest, difficulty of content and negative attitudes have been reasons cited by elementary teachers' in regard to their reluctance to teach science (Victor, 1962; Soy, 1967; Blosser & Howe, 1969; Westerback, 1982; Gabel, et al., 1987). Similar to Perkes notion of "sensed adequacy," self-efficacy has been used in research on science teaching and teacher education. A number of teacher-efficacy studies have focused on elementary teachers, and their ability to teach science (Freeman & Smith, 1997; Richardson & Liang, 2008). Such studies indicate a high correlation between sense of adequacy and one's preference to teach science.

PURPOSE OF THE STUDY

This collaborative study emerged from survey data collected for programming purposes for the university's environmental education facility. While categorizing teachers' interests in curricular programs and professional development, the researchers noted patterns from respondents given their earlier childhood experiences and the locale in which these interactions took place. This chapter examines the relationship between teachers' early environmental socialization and their indicated interests in curricular topics. More specifically, in that the survey data includes responses from teacher educators, we are interested in the relationships between the teacher's childhood locale (non-rural, rural), the activities she engaged in during these formative years, and her interests in science related programs as a teaching professional. The additive measures were taken from the James, Bixler and Vadala's (2010) Developmental Model for Natural History Oriented Environmental Professionals.

The primary research questions guiding the paper include:

1. Are their differences between teachers who grew up in a non-rural environment versus a rural environment and interest in field-based learning and environmental education?
2. Are their differences between teachers who grew up in a non-rural environment versus a rural environment and interest in teaching science related topics?
3. Are their differences between teachers who grew up in a non-rural environment versus a rural environment in environmental socialization?
4. How does environmental socialization based on growing up in a non-rural environment versus a rural environment relate to preference for teaching science related topics and interest in field based learning and environmental education?

Instrumentation

Content and construct validity of the instrument was assessed prior to distribution. Students enrolled in an undergraduate level recreation assessment course, two representatives from environmental education facility, and three faculty members representing two college departments (education and recreation) developed the initial survey instrument. In addition, this survey was analyzed by graduate level education students and then piloted for appropriateness of terminology, clarity of questions, and final edits. All individuals involved in the development of the instrument were asked to take the survey using an online survey instrument (SurveyMonkey.com) to determine the approximate average response time. The instrument was designed to take respondents less than 30 minutes. The survey instrument was comprised of 27 questions and included demographic information, environmental socialization measures, and field-based topic interest questions. Data were imported into SPSS, Version 14 for analysis.

Surveys were distributed through the university listserv to all undergraduate program student teachers in the teacher education program and their cooperating teachers during the spring 2008. The cover letter was developed into a webpage and embedded within the body of the e-mail message; the link to the survey was provided. Consent was evident if the participants chose to respond to the survey. An e-mail reminder that included the survey link was sent out again in late April. Around that same time, one representative of the research team attended a student teaching seminar and requested the completion of surveys if not done so at that time.

RESULTS

Demographic Profile

All respondents were local practitioners or student teachers in a six county region. Of the total sample, there were 88 usable surveys for this study. Forty-nine of the participants identified themselves as having grown up in a rural area (R); the remaining participants (n=39) indicated the area of their upbringing as non-rural (NR). One in seven (14.8%) of the sample population were male and were equally represented in each group (14.6% R, 15.4% NR). The average age of all respondents was 38.9, while the mean for rural teachers was 38.4 (standard deviation [s.d.] = 11.2) and 39.8 (s.d. = 10.2) for NR teachers. Only one participant was non-Caucasian. A crosstab analysis revealed that during childhood NR teachers moved more than expected while R teachers moved less than expected (chi-square = 17.9, p<.001, contingency coefficient = .413). The average number of years teaching for the full sample was 15.0, while rural teachers averaged 16.1 (s.d. = 8.9) and non-rural (NR) teachers 14.0 (s.d. = 8.3). There was not a statistically significant difference (p=.296) in reported years of teaching. A crosstab analysis did not reveal any unexpected associations (p=.889) when years of teaching were categorized based on five year intervals (i.e., 0-5, 6-10, 11-15, 16 or more). Similar percentages were found in each teaching category for both R and NR teachers.

Rural and Non-Rural Teachers: Interest in Field-Based Learning and Experience with Environmental Education

Rural and non-rural teachers were not statistically different (p=.485) regarding their "experience with environmental education." Both teacher groups indicated "none" to a "small amount" (R, mean=1.52; NR, mean=1.62) based on a 4-point Likert scale (1="none" to 4="I am very competent in EE"). Using a 4-point Likert scale (1="not at all interested" to 4="very interested"), an independent sample t-test revealed that rural teachers were slightly less interested (mean=2.67) but not statistically different (p<.05) than NR teachers (mean=3.03) in "providing students with field based experiences different from a typical classroom setting."

Rural and Non-Rural Teachers Interest in Field-Based Topics

Of the 10 science related topics suggested for curricular development or program implementation, crosstab analysis did not reveal any statistically significant associations (see Table 1). Few topics generated more than 40% interest with NR teachers more interested than rural teachers in 6 of 10 topics.

Rural and Non-Rural Teachers and Environmental Socialization

Teachers were asked to identify any of the twenty-five environmental socialization items they engaged in during their formative childhood years (e.g., "I played in the woods"). Six additive measures were created based on guidance from the environmental socialization literature (Bixler & James, 2005; James, Bixler, & Vadala, 2010). The six measures were labeled: Water-based (score range=0-3), Played in Nature (score range=0-7), Sports-based (score range=0-3), Camps/Groups/Scouts (score range=0-4), Parks/Travel (score range=0-5), and Critters (score range=0-3). Independent sample t-tests revealed only one statistically significant difference (p<.05); rural teachers had a higher group mean score than NR teachers on the "Critters" measure (see Table 2). In other words, rural teachers, on average, were more exposed to critters as a child than NR teachers. Rural teachers also had higher mean group scores for four of the remaining five measures. The "Camps/Groups/Scouts" measure was the exception.

Rural and Non-Rural Teachers: Interest in Science Based Topics Given Environmental Socialization Scores

Independent sample t-tests were run to analyze whether teachers who expressed interest in any of the 10 topics were more likely to have statistically significant higher mean scores from those with no interest. Tests were run across each of the six environmental socialization measures. For rural teachers, there were 23 group mean differences across topics for all six measures (see Table 3). Twenty of the 23 significant differences revealed that those with interest in a topic had a higher group mean score for the environmental socialization measure (see Table 3). Upon closer examination, six group mean differences for the "water-based" measure were revealed: forest ecology, mountain ecology, animal ecology, ponds and streams, night hikes, and town hall. Similarly, six differences were also found for "played in nature;" this included forest ecology, geology, mountain ecology,

Table 1. Crosstab analysis comparing rural (R) to non-rural (NR) teachers on interest in 10 science related topics

Topic	Rural Teacher Interest In	Non-rural Teacher Interest In	p-value
Forest Ecology	34.7%	35.9%	.907
Geology	28.6%	43.6%	.143
Mountain Ecology	36.7%	38.5%	.868
Animal Ecology	49.0%	53.8%	.650
Ponds and Streams	53.1%	46.2%	.520
Cherokee Culture	36.7%	35.9%	.935
Web of Life	41.0%	32.7%	.417
Mountain Bog	16.3%	20.5%	.613
Night Hike	22.4%	20.5%	.822
Town Hall	26.5%	38.5%	.233

Table 2. Independent sample t-test analysis comparing rural to non-rural teachers on environmental socialization measures

	Mean	Std. Deviation	p-value
Water-Based			.296
Non-rural	1.79	1.06	
Rural	2.04	1.12	
Played in Nature			.156
Non-rural	3.15	2.08	
Rural	3.79	2.09	
Sports-Based			.078
Non-rural	2.03	1.20	
Rural	2.43	.91	
Camps/Groups/Scouts			.434
Non-rural	1.05	1.02	
Rural	.88	1.03	
Parks/Travel			.239
Non-rural	2.53	1.68	
Rural	2.96	1.63	
Critters			.004*
Non-rural	1.18	.68	
Rural	1.69	.89	

Note: Water-based (score range=0-3), Played in Nature (score range=0-7), Sports-based (score range=0-3), Camps/Groups/Scouts (score range=0-4), Parks/Travel (score range=0-5), and Critters (score range=0-3). Asterisk (*) represents a statistically significant difference (p<.05) between groups.

Table 3. Non-rural (NR) and rural (R) teachers interest in topics given their environmental socialization mean scores

Measure Topic	Water-Based NR	Water-Based R	Played in Nature NR	Played in Nature R	Camps Groups Scouts NR	Camps Groups Scouts R	Parks Travel NR	Parks Travel R	Sports-Based NR	Sports-Based R	Critters NR	Critters R
Forest Ecology												
Interested	2.00	2.65**	4.21**	5.12**	1.35	1.35**	3.07	3.82**	2.07	2.41	1.36	1.94
Not interested	1.68	1.72	2.56	3.09	.88	.62	2.24	2.50	2.00	2.44	1.08	1.56
Geology												
Interested	2.06	2.43	3.82*	4.86**	1.00	1.36**	3.23**	3.71**	2.29	2.79*	1.29	2.00
Not interested	1.59	1.88	2.64	3.37	1.09	.69	2.00	2.66	1.82	2.29	1.09	1.57
Mtn. Ecology												
Interested	2.07	2.44**	3.87*	4.50*	1.13	1.11	3.20**	3.28	2.13	2.50	1.27	1.89
Not interested	1.62	1.81	2.71	3.39	1.00	.74	2.12	2.78	1.96	2.39	1.12	1.58
Animal Ecology												
Interested	1.90	2.37**	3.33	4.17	1.05	1.00	2.86	3.00	2.29	2.54	1.19	1.83
Not interested	1.67	1.72	2.94	3.44	1.06	.76	2.17	2.92	1.72	2.32	1.17	1.56
Ponds & Streams												
Interested	1.89	2.38**	3.50	4.35**	1.17	1.04	2.78	3.46**	2.06	2.61	.67	1.85
Not interested	1.71	1.65	2.86	3.17	.95	.69	2.33	2.39	2.00	2.22	.70	1.52
Web of Life												
Interested	1.94	2.00	3.81	4.00	1.12	1.37**	2.75*	3.63**	2.00	2.69	1.31	1.69
Not interested	1.69	2.06	2.69	3.97	1.00	.64	2.39	2.64	2.04	2.30	1.09	1.70
Mountain Bog												
Interested	2.25	2.37	4.75**	5.13**	1.25	1.37	3.50*	3.37	2.25	2.62	1.50	1.50
Not interested	1.68	1.97	2.74	3.54	1.00	.78	2.29	2.88	1.97	2.39	1.10	1.73
Night Hike												
Interested	1.75	2.73**	4.00	4.55	.75	1.37*	2.12	3.36	2.37	2.27	1.25	1.82
Not interested	1.81	1.84	2.94	3.58	1.13	.74	2.64	2.84	1.94	2.47	1.16	1.66
Cherokee Cult.												
Interested	1.57	2.28	3.43	4.44*	.93	1.00	2.79	3.39	1.93	2.39	1.36	1.78
Not interested	1.92	1.90	3.00	3.42	1.12	.81	2.40	2.71	2.08	2.45	1.08	1.64
Town Hall												
Interested	1.67	1.54	1.67	3.85	1.17	.46	2.60	2.92	2.00	2.46	1.13	1.31
Not interested	1.87	2.22*	1.87	3.78	.87	1.03*	2.50	2.97	2.04	2.42	1.21	1.83*

Note: Statistically significant mean differences between "Interested" and "Not interested" within each teaching group are represented in the Table with an asterisk (**p<.05, *p<.10). Water-based (score range=0-3), Played in Nature (score range=0-7), Camps/Groups/Scouts (score range=0-4), Parks/Travel (score range=0-5), Sports-Based (score range=0-3), and Critters (score range=0-3).

ponds and streams, mountain bog, and Cherokee culture. Five differences were found for "camps/groups/scouts": forest ecology, geology, web of life, night hike, and town hall. Four differences were revealed for the "parks/travel" measures: forest ecology, geology, ponds and streams, and web of life. One difference was found for both "sports-based" and "critters." The three statistically significant differences that revealed those with "no interest" having a greater mean score than those who expressed interest were linked to the "town hall" topic.

For non-rural teachers, there were eight statistically significant mean differences across topics for two of the measures ("played in nature" and "parks/travel"); there were no differences for the "water-based," "camps/groups/scouts," "sports-based," or "critters" measures. For the "played in nature" measure, four differences were found: forest ecology, geology, mountain ecology, and mountain bog. Similarly, four differences were revealed for "parks/travel." This included the topics of geology, mountain ecology, web of life, and mountain bog (see Table 3).

Beyond the findings of statistical significance, of the 60 comparisons performed for each teaching group, approximately eight of ten teachers who expressed an interest in a topic had higher mean scores than those who did not. Moreover, much was revealed from percentages indicated in Table 1; rural teachers who expressed interest in a topic tended to have greater mean scores than non-rural teachers who expressed an interest in a topic.

Rural Teachers and Non-Rural Teachers: Environmental Socialization and Correlation with Interest in Field-Based Learning and Experiences with Environmental Education

Two statistically significant and moderate positive correlations (p<.05) were found for rural teachers for the "sports-based" and "parks/travel" measures given their interest in field-based learning (see Table 4). In other words, as experience in sports based or parks/travel increased so did interest in field-based learning. Two statistically significant correlations were also found for non-rural teachers although they were for different measures: "water-based" and "played in nature." In this case, as experience in water-based or played in nature measures increased so did interest in field-based learning. Additionally, the strength of association for three of the environmental socialization measures was quite dissimilar when comparing teacher groups particularly for water-based, sports-based and critters.

Two statistically significant correlations were found for rural teachers given their experience with environmental education for "played in nature" and "parks/travel" while four statistically significant correlations were found for non-rural teachers: "water-based," "played in nature," "parks/travel," and "critters." Again, much as was found for interest in field based learning, the strength of association for water-based, sports-based and critters were quite dissimilar when comparing teacher groups.

DISCUSSION

We examined how teachers' environmental socialization experiences contributed to their expressed interest in particular science-related topics. More specifically, we looked at the relationships between the teachers' childhood locale (rural, non-rural), the activities they engaged in during these formative years and current interests in science-related programs as teaching professionals.

We found that teachers who grew up in rural areas were slightly less interested than non-rural

Table 4. Teacher interest in field-based learning (FBL) and experience with environmental education (EE) and correlation with environmental socialization

	Water-Based	Played in Nature	Sports-Based	Camps Groups Scouts	Parks Travel	Critters
Interest in FBL[1]						
Rural Teachers	.121	.269	.352*	.197	.302*	-.066
Non-rural	.350*	.385*	.172	.049	.208	.297
Experience with EE[2]						
Rural Teachers	.191	.375**	.067	.229	.345*	.230
Non-rural	.418**	.431**	.258	.280	.397*	.362*

Note: [1] Used a 4-point Likert scale where 1="not at all interested" and 4="very interested." [2] Used a 4-point Likert scale where 1="none" and 4="I am very competent in EE." *p<.05, **p<.01.

teachers in field-based learning and expressed less experience with environmental education. This did not fit with our initial expectations that rural teachers would be more likely to participate and express interest in field based opportunities. We would expect that rural teachers, situated in more open, less densely populated social environments, would be more likely to engage in outdoor activities. In this, we inferred that such environmental socialization experiences would result in higher levels of interest in field-based opportunities. Given this outcome, we questioned how the term "environmental education" had been interpreted and wondered if it may have conveyed a different meaning to teachers growing up in rural environments. We did find that regardless of teacher group, respondents had little to no experience with environmental education. This supports the findings of previous research studies conducted on teacher preparation and efficacy toward environmental education (Mastrilli, 2005; Mosely, et al., 2002; Mosely & Utley, 2008; Plevyak, Bendixen-Noe, Henderson, Roth, & Wilke, 2001, 2004; Powers, 2004).

Controlling for environmental socialization experiences we found the teacher's interest in science related program topics were not significantly different for rural and non-rural teachers. Our thought was that rural teachers would express greater interest than NR teachers on the majority of the topics. In fact the opposite occurred. Non-rural teachers expressed greater interest in 6 out of 10 the science related topics, although few topics generated more than 40% interest by either teaching group.

Soy (1967) investigated teachers' attitude toward science and found that negative attitudes had been developed throughout their earlier schooling experiences. In follow-up interviews, Soy found that the prospective teacher's content preference had been established prior to graduation from high school. Soy concluded that the prospective classroom teacher, who chose science as her content preference usually had more background in science and mathematics and in this, was more likely to come from a farm background than the prospective teacher who had chosen other subject areas. If this trend has been maintained over time, then it would seem reasonable to expect more interest in the science-related program topics from teachers who had indicated a rural upbringing. That was not the case given our findings.

Perkes (1975) probed relationships between teacher background and the individual's commitment and confidence to teach science. Results from her study indicated that previous frustration or dissatisfaction with science-related experiences significantly and negatively related to the individual's tendency to take science courses in college and influenced their sensed adequacy to teach science. Given that environmental socialization includes a greater accumulation of experiences that develops skills and competencies in the natural sciences, we believe these past experiences are influential in developing efficacy toward science content and may impact future teaching preferences for natural sciences. Findings in this preliminary study indicate rural teachers had more exposure to environmental socialization experiences for five of the six measures.

Upon examination of the environmental socialization experiences noted by rural teachers, the analysis indicated that there were many statistically significant differences between those who expressed interest in a topic versus those who did not. Each of the six environmental socialization measures revealed one or more mean differences. Rural teachers who expressed an interest in forest ecology, for example, had greater mean scores for four environmental socialization categorical measures (water-based, played in nature, camp/group scouts, and parks/travel) then those with a rural upbringing who expressed no interest in forest ecology. For the NR teachers on the other hand, only one significant difference was found (higher mean score in "played in nature") for those who expressed an interest in teaching forest ecology.

Across a number of topics, there were commonalities found among rural and NR teachers. As would be expected, those teachers with more environmental socialization experiences tended to be more interested in science-related topics. Teachers who expressed interest in topics tended to have greater environmental socialization experiences. When examining teachers' environmental socialization and interest in field-based learning, there were significant relationships found for rural teachers concerning "sports-based" and "parks/travel" measures. The more interested the teacher was in field-based learning, the more likely she was to have had experiences in sports and with travel. In contrast, NR teachers interested in field-based learning were more likely to have indicated a greater number of experiences with "water-based" activities and "playing in nature." Additionally, while teachers in this study had little experience with environmental education, there were a couple significant positive correlations found for rural teachers' environmental socialization as linked to "playing in nature" and "travel." As experiences with environmental education increased so did the likelihood that the individual played in nature and traveled as a child. For non-rural teachers, significant positive correlations were found for playing in nature, travel, water-based activities and playing with critters. Intuitively each of these findings makes sense as it would be expected teachers with greater ES experiences (particularly those associated with experiencing the natural world) would seek opportunities to learn more about their immediate surroundings, whether field-based or in the context of a formal learning environment.

IMPLICATIONS

Results from this study indicate the impact of childhood experiences and the significance of the learning environment. It is important that teachers include, regardless of their efficacy toward particular topics, opportunities for students to explore and interact with their natural world. As described in previous studies (Tanner, 1980; Peterson, 1982; Chawla, 1988; Chawla & Hart, 1988; Bixler, Carlisle, Hammitt, & Floyd, 1994), early environmental experiences that support nature-based interests can lead to career/vocation choices. Investigating preservice teachers' past Environmental Socialization (ES) experiences could offer insight into their sensed adequacy toward content or their preferences in instructional planning and delivery, specifically incorporating ES experiences into undergraduate experiences to foster preservice comfort in teaching outdoors as well as science.

We found that teachers from a non-rural environment tended to express more interest in topics that were typical of those brought up in a rural environment. Whereas the teacher with a rural childhood was less interested in field-based learning, we wonder if this lack of interest is due to their familiarity with the topics from earlier ES experiences within their rural environment. Thus implying teacher preparation programs should not only consider the preservice teacher's childhood ES experiences and surrounding environment but determine ways to incorporate ES competencies into curriculum. Examples of ES competencies that could be incorporated into curriculum include dressing for weather, outdoor classroom management, multiple outdoor teaching practice sessions in different natural locals as well as getting hands dirty handling bugs/worms or dirt. Fostering these ES competencies might encourage comfort in teaching science content both in and out of the classroom as well as field-base learning or environmental education.

Not only is building ES competencies important in undergraduate preservice curriculum but research in ES indicates a social supportive network, knowledgeable experts, multiple experiences in natural settings and access to natural environments not only is important to children but young adults. Through ES competency building in curriculum, preservice teachers are exposed to "experts" in the field and have opportunity to access natural areas multiple times. Thus creating comfort with the natural environment, which may foster additional comfort in teaching outdoors.

One other implication of environmental socialization is to be a self-reflective teacher—understanding how past experiences influence choices in content, how one teaches and one's proclivity for field-based learning. By acknowledging one's ES influences, upbringing, and future, self-reflection offers the pre-service teacher insight to overcome their own discomfort in science content. Why does this matter? While student learning is of utmost importance, ultimately our nation can be impacted environmentally as well as economically through a lack of available competent science teachers.

In a study conducted by Phenice and Griffore (2003), the researchers explored how environmental education played into young children's perceptions of their place in relation to their natural world. Described as the ecopsychological self, this natural sense of self-identity connects well to environmental socialization and the notion of place. Our findings support the need for further investigations of early schooling experiences and how one's interactions within a given environment shape individuals understanding of self. We must not lose sight of the significance of such experiences, as present day instructional decisions will surely impact future attitudes, interests and actions.

Finally, we assert the need for community science (Adams, 2012), a pedagogical structure that allows for localized ways of enacting science. "Community science is contextualized science—science shaped by people's sense of place" (p. 1167). As a potential area of science education, an examination of students' place attachment, identity development, and academic efficacy toward science is essential.

LIMITATIONS AND FUTURE RESEARCH

In future studies, we intend to identify and define rural, suburban, and urban categories to fit the state census description of such locales. We believe this will help us identify the childhood activities within a specific type of environment. In addition, we intend to include attitudinal measures specific to the teaching of science on the survey instrument. Since the population sample was limited to a group of student teachers and their cooperating teachers during an academic semester, the findings of this study will provide baseline data for comparative purposes in the future.

Next, there is a need for further study of the teaching professional's beliefs and identity as influenced by environmental socialization. Teacher preparation programs generally include a greater number of methods courses for the teaching of language arts and mathematics than in science, thus influencing how preservice teachers perceive the importance of science in the elementary curricula. We intend to explore further how environmental socialization play into preservice teachers' course of study choices, especially those in elementary programs who often must select an area of concentration within the general curricula.

Finally, we conclude with questions for future studies. How does environmental socialization play into an individual's content interest during the formal schooling process? Is there a relationship between environmental socialization and self-efficacy toward the teaching of particular content knowledge? How does environmental socialization play into the elementary teacher's self-efficacy to science-related content? Several studies (Mosely & Utley, 2008; Mosely, Reinke, & Bookout, 2002; Zint, 2002) have been conducted on this topic, but without regard to environmental socialization. Perhaps through the addition of this lens, science educators will have a better understanding of factors influencing self-efficacy in the teaching of science content, especially among elementary teachers who are responsible for addressing all curricular areas.

REFERENCES

Adams, J. (2012). Community science: Capitalizing on local ways of enacting science in science education. In Fraser, B., Tobin, K., & McRobbie, C. (Eds.), *The Second International Handbook of Science Education*. Dordrecht, The Netherlands: Springer. doi:10.1007/978-1-4020-9041-7_77

Arends, R. I. (2000). *Learning to teach*. Boston, MA: McGraw Hill.

Aubusson, P., Griffin, J., & Kearney, M. (2012). Learning beyond the classroom: Implications for school science. In Fraser, B., Tobin, K., & McRobbie, C. (Eds.), *The Second International Handbook of Science Education* (pp. 1123–1134). Dordrecht, The Netherlands: Springer. doi:10.1007/978-1-4020-9041-7_74

Bandura, A. (1997). *Self-efficacy: The exercise of control*. New York, NY: Freeman Publications.

Bee, H. (1997). *The developing child* (8th ed.). New York, NY: Longman.

Berger, K. S. (1988). *The developing person through the life span* (2nd ed.). New York, NY: Worth Publishers.

Bixler, R. D., Carlisle, C. L., Hammitt, W. E., & Floyd, M. F. (1994). Observed fears and discomforts among urban students on school field trips to wildland areas. *The Journal of Environmental Education*, *26*, 24–33. doi:10.1080/00958964.1994.9941430

Bixler, R. D., & Morris, B. (2000). Factors differentiating water-based wildland recreationists from nonparticipants: Implications for recreation activity instruction. *Journal of Park and Recreation Administration*, *18*(2), 54–72.

Bleicher, R. E., & Lindgren, J. (2005). Success in learning science and preservice science teaching self-efficacy. *Journal of Science Teacher Education*, *16*, 205–225. doi:10.1007/s10972-005-4861-1

Bredekamp, S., & Copple, C. (Eds.). (1997). *Developmentally appropriate practice in early childhood programs*. Washington, DC: National Association for the Education of Young Children.

Brookhart, S., & Freeman, D. (1992). Characteristics of entering teaching candidates. *Review of Educational Research*, *62*(1), 37–60.

Chawla, L. (1988). Children's concern for the natural environment. *Children's Environments Quarterly*, *5*, 13–20.

Chawla, L., & Hart, R. (1988). Roots of environmental concern. In Lawrence, D., Habe, R., Hacker, A., & Sherrod, D. (Eds.), *People's Needs/Planet Management: Paths to Coexistence* (pp. 15–18). Pomona, CA: Environmental Design Research Association.

Cheek, N. Jr, & Burch, W. Jr. (1976). *The social organization of leisure in human society*. New York, NY: Harper and Row.

Chen, C., Sable, J., Mitchell, L., & Liu, F. (2011). *Documentation to the NCES common core of data public elementary/secondary school universe survey: School year 2009–10 (NCES 2011-348)*. Washington, DC: National Center for Education Statistics. Retrieved May 16, 2011 from http://nces.ed.gov/pubsearch/pubs.info.asp?pubid=2011348

Cobb, E. (1993). *The ecology of imagination in childhood* (2nd ed.). Dallas, TX: Spring Publications.

Cresswell, T. (2004). *Place: A short introduction*. Malden, MA: Blackwell Publications.

Decker, D. J., & Purdy, K. G. (1986). Becoming a hunter: Identifying stages of hunting involvement for improving hunter education programs. *Wildlife Society Bulletin, 14*, 474–479.

Decker, L., & Rim-Kaufman, S. (2008). Personality characteristics and teacher beliefs among preservice teachers. *Teacher Education Quarterly, 35*(2), 45–62.

Dewey, J. (1938). *Experience and education.* London, UK: Collier-Macmillian.

Freeman, C., & Smith, D. (1997). *Active and engaged? Lessons from an interdisciplinary and collaborative college mathematics and science course for preservice teachers.* Paper presented at the Annual Meeting of the American Educational Research Association. Chicago, IL.

Garbarino, J. (1989). An ecological perspective on the role of play in child development. In Bloch, M. N., & Pellegrini, A. D. (Eds.), *The Ecological Context of Children's Play* (pp. 16–34). Norwood, NJ: Ablex Publishing Corporation.

Haim, E. (2003). Inquiry-events as a tool for changing science teaching efficacy beliefs of kindergarten and elementary school teachers. *Journal of Science Education and Technology, 12*(4), 495–501. doi:10.1023/B:JOST.0000006309.16842.c8

Hammerman, D. R., Hammerman, W. M., & Hammerman, E. L. (1994). *Teaching in the outdoors.* Danville, IL: Interstate Publishers, Inc.

James, J. J., Bixler, R. D., & Vadala, C. (2010). From play in nature, to recreation then vocation: A developmental model for natural history-oriented environmental professionals. *Children, Youth and Environments, 20*(1), 231–256.

Koch, J. (2010). *Science stories: Science methods for elementary and middle school teachers* (4th ed.). Belmont, CA: Wadsworth Cengage Learning.

Lent, R. W., Brown, S., & Larkin, K. (2012). Comparison of three theoretically derived variables in predicting career and academic behavior: Self-efficacy, interest congruence, and consequence thinking. *Journal of Counseling Psychology, 34*(3), 293–298. doi:10.1037/0022-0167.34.3.293

Lim, M., & Calabrese Barton, A. (2006). Science learning and a sense of place in an urban middle school. *Cultural Studies of Science Education, 1*(1), 107–142. doi:10.1007/s11422-005-9002-9

Louv, R. (2005). *Last child in the woods: Saving our children from nature-deficit disorder.* Chapel Hill, NC: Algonquin Books.

Massey, D., & Jess, P. (Eds.). (1995). *A place in the world? Places, culture and globalization.* Oxford, UK: Oxford University Press.

Mastrilli, T. (2005). Environmental education in Pennsylvania's elementary teacher education programs: A statewide report. *The Journal of Environmental Education, 36*(3), 22–30. doi:10.3200/JOEE.36.3.22-30

Moore, R., & Wong, H. (1997). *Natural learning: Rediscovering nature's way of teaching.* Berkeley, CA: MIG Communications.

Moseley, C., Reinke, K., & Bookout, V. (2002). The effect of teaching outdoor environmental education on pre-service teachers' attitudes toward self-efficacy and outcome expectancy. *The Journal of Environmental Education, 34*(1), 9–15. doi:10.1080/00958960209603476

Moseley, C., & Utley, J. (2008). An exploratory study of preservice teachers' beliefs about the environment. *The Journal of Environmental Education, 39*(4), 15–30. doi:10.3200/JOEE.39.4.15-30

Novak, J. D. (1998). *Learning, creating, and using knowledge: Concept maps as facilitative tools in schools and corporations.* Mahwah, NJ: Lawrence Erlbaum Associates Publishers.

O'Leary, J. T., Behrens-Tepper, J., McGuire, F. A., & Dottavio, F. D. (1987). Age of first hunting experience: Results from a nationwide recreation survey. *Leisure Sciences*, *9*, 225–233. doi:10.1080/01490408709512164

Palmer, J. (1993). Development of concern for the environment and formative experiences of educators. *The Journal of Environmental Education*, *24*, 26–30. doi:10.1080/00958964.1993.9943500

Perkes, V. (1975). Relationships between a teacher's background and sensed adequacy to teach elementary science. *Journal of Research in Science Teaching*, *12*(1), 85–88. doi:10.1002/tea.3660120112

Peterson, N. (1982). *Developmental variables affecting environmental sensitivity in professional environmental educators*. (Unpublished Master's Thesis). Southern Illinois University. Carbondale, IL.

Phenice, L., & Griffore, R. (2003). Young children and the natural world. *Contemporary Issues in Early Childhood*, *4*(2), 167–171. doi:10.2304/ciec.2003.4.2.6

Piaget, J. (1954). *The construction of reality in the child*. New York, NY: Basic Books. doi:10.1037/11168-000

Plevyak, L., Bendixen-Noe, M., Henderson, J., Roth, R., & Wilke, R. (2001). Level of teacher preparation and implementation of EE: Mandated and non-mandated EE teacher preparation states. *The Journal of Environmental Education*, *32*(2), 28–37. doi:10.1080/00958960109599135

Powers, A. (2004). Teacher preparation for environmental education: Faculty perspectives on the infusion of environmental education into pre-service methods courses. *The Journal of Environmental Education*, *35*(3), 3–11.

Relph, E. (1976). *Place and placelessness*. London, UK: Pion.

Richardson, V. (1996). The role of attitudes and beliefs in learning to teach. In Sikula, J. (Ed.), *Handbook of Research on Teacher Education* (2nd ed., pp. 102–119). New York, NY: Macmillan.

Rivkin, M. (1995). *The great outdoors: Restoring children's right to play outside*. Washington, DC: National Association for the Education of Young Children.

Scott, D., & Willits, F. K. (1989). Adolescent and adult leisure patterns: A 37-year follow up study. *Leisure Sciences*, *11*, 323–335. doi:10.1080/01490408909512230

Sofranko, A. J., & Nolan, M. F. (1972). Early life experiences and adult sports participation. *Journal of Leisure Research*, *4*, 6–18.

Southerland, S., & Gess-Newsome, J. (1999). Preservice teachers' views of science teaching as shaped by images of teaching. *Science Education*, *83*(2), 131–151. doi:10.1002/(SICI)1098-237X(199903)83:2<131::AID-SCE3>3.0.CO;2-X

Soy, E. (1967). Attitudes of prospective elementary teachers toward science as a field of specialty. *School Science and Mathematics*, *67*, 507–520. doi:10.1111/j.1949-8594.1967.tb15233.x

Tanner, T. (1980). Significant life experiences: A new research area in environmental education. *The Journal of Environmental Education*, *11*(4), 20–24. doi:10.1080/00958964.1980.9941386

United States Department of Labor. (2010). *Occupational employment statistics*. Retrieved from http://www.bls.gov/oes/current/largest_occs.htm

Victor, E. (1962). Why are our elementary school teachers reluctant to teach science? *Science Education*, *46*(2), 185–192. doi:10.1002/sce.3730460231

Vygotsky, L. S. (1978). *Mind in society*. Cambridge, MA: Harvard University Press.

Wertsch, J. V. (1984). The zone of proximal development from a comparative and organismic point of view. In Rogoff, B., & Wertsch, J. V. (Eds.), *New Directions for Child Development*. San Francisco, CA: Jossey-Bass.

Wilson, R. (1996). *Starting early: Environmental education during the early childhood years*. Columbus, OH: ERIC Clearinghouse for Science, Mathematics, and Environmental Education.

Windschitl, M. (2003). Inquiry projects in science teaching teacher education, what can investigative experiences reveal about teacher thinking and eventual classroom practice? *Science Education, 87*(1), 112–144. doi:10.1002/sce.10044

Zint, M. (2002). Comparing three attitude-behavior theories for predicting science teachers' intentions. *Journal of Research in Science Teaching, 39*(9), 819–844. doi:10.1002/tea.10047

Chapter 12
Analysis of Discourse Practices in Elementary Science Classrooms using Argument-Based Inquiry during Whole-Class Dialogue

Matthew J. Benus
Indiana University Northwest, USA

Morgan B. Yarker
University of Iowa, USA

Brian M. Hand
University of Iowa, USA

Lori A. Norton-Meier
University of Louisville, USA

ABSTRACT

This chapter discusses an analysis of discourse practices found in eight different elementary science classrooms that have implemented the Science Writing Heuristic (SWH) approach to argument-based inquiry. The analysis for this study involved examining a segment of whole-class talk that began after a small group presented its claim and evidence and ended when the discussion moved on to a new topic, or when a different group presented. The framework for the analysis of this whole-class dialogue developed through an iterative process that was first informed by previous analysis, review and modification of other instruments, and notable anomalies of difference from this data set. Each classroom was then rated using the Reform Teaching Observation Protocol (RTOP), which provided a score for the extent to which the teacher was engaged with reform-based science teaching practices. Our analysis shows that elements of whole-class dialogue in argument-based inquiry classrooms were different across varying levels of RTOP implementation. Overall, low level RTOP implementation (little evidence of reformed-

DOI: 10.4018/978-1-4666-2809-0.ch012

based practice) had a question and answer format during whole class talk that rarely included discourse around scientific reasoning and justification. Higher levels of RTOP implementation were more likely to be focused on student use of scientific evidence to anchor and develop a scientific understanding of "big ideas" in science. These findings are discussed in relation to teacher professional development in argument-based inquiry, science literacy, and the teacher's and students' grasp of science practice.

INTRODUCTION

For quite some time now, reform documents have set the benchmarks for science literacy by stressing the importance of using inquiry-based approaches in the science classroom (AAAS, 1993; NRC, 1996, 2012). Yet, what drives and supports an inquiry-based approach in the classroom is still up for discussion and currently being thoroughly examined through research. Kuhn (1991) gives one good idea of what it means to actually be engaged in scientific inquiry, stating "Scientific inquiry is fundamentally a knowledge building process in which explanations are presented to the community so they can be critiqued, debated, and revised" (1991, p. 4). Deriving Kuhn's definition of scientific inquiry leads us to understand that to engage in argument-based inquiry means to use evidence in support of one's claim, which can help to focus and drive dialogues of critique and consensus (Duschl & Osborne, 2002; Osborne, Erduran, & Simon, 2004). Dialogue "is about a shared inquiry, a way of thinking and reflecting together. It is not something you do to another person. It is something you do with people" (Isaacs, 1999, p. 9) and dialogic conversation can be a way to generate knowledge (Alexopoulou & Driver, 1997; Ford, 2008; Kelly & Green, 1998; Schein, 1993). While the research community recognizes the value of dialogic communication in inquiry-based learning approaches, research that examines communication patterns found in classrooms using argument-based inquiry is still needed. This chapter will explore the talk patterns within whole-class dialogue in fifteen elementary classrooms that utilize argument-based inquiry. The two research questions guiding this study are (1) what are the key factors that contribute to whole-class discussion found in classrooms using argument-based inquiry and (2) how do these key factors develop across levels of implementation of argument-based inquiry?

ARGUMENTATION IN PRACTICE

Argumentation is conversational dialogue. This can be reasoned through by exploring the meanings of the words dialogue and argumentation. The word dialogue is from two Greek roots: dia and logos. This roughly translates to be "meaning flowing through" (Isaacs, 1993, p. 25). Conversation, which begins all dialogues, means "to turn together" (Isaacs, 1993, p. 35). Argumentation then is a conversational dialogue where meanings can flow through and turn together within those engaged. Argumentation "initiates change, it transforms the significance of material, it enables reflection and action, it brings divergent voices together in interaction..." (Mork, 2005, p. 18; referencing Costello & Mitchell, 1995). It is worth noting the difference in argumentation versus argument. Argumentation by many researchers is considered a discourse process (e.g. Jimenez-Aleixandre & Erduran, 2008; Osborne, et al., 2004) while argument is centering around producing or influencing a particular outcome (Cavagnetto, 2010; Toulmin, 1958). In this study, we focus on argumentation.

Scientific argumentation is about evaluating and critiquing the construction of scientific claims, evidence, and explanation (Duschl, Schweingruber, & Shouse, 2007). Once reasoned through by being constructed and deconstructed, knowledge is claimed because it carries with it tentative, but stable, evidence that is deemed trustworthy by the community of science (Gross, 1990). In actual practice, scientists engage in the process of argumentation to understand "why" (Duschl, 1990) their ideas matter. As these scientists engage in argumentation, they modify their ideas by making or critiquing claims supported with evidence (NRC, 2012). The net effect of their engagement further strengthens their understanding within their domain of study.

SCIENTIFIC ARGUMENTATION IN CLASSROOMS

This process of argumentation can and should also be a critical component of instruction in our science classrooms (Driver, Asoko, Leach, Mortimer, & Scott, 1994; Driver, Newton, & Osborne, 2000; Duschl & Osborne, 2002; Kuhn, 1993). In particular, "why" questions can help to engage students into the core practices found in communities of scientists. The reason for this is that "why" questions set up conversations around claims, data, and evidence, all of which are core essential components of scientific argumentation (Brinker & Bell, 2008). The process of engaging in scientific argumentation in the science classroom is centered around what Lemke (1990) calls "learning to talk science." This does not mean a science classroom is dominated by the teacher talking science (unfortunately this is still widely being practiced today (see Crawford, 2005)) but rather the classroom community with the teacher engaged in an ongoing conversational dialogue of scientific argumentation.

Over the last decade there have been several studies focusing on dialogic elements of scientific argumentation in science classrooms (e.g., Cavagnetto, Hand, & Norton-Meier, 2010; Driver, Newton, & Osborne, 2000; Duschl & Osborne, 2002; Hogan, Nastasi, & Pressley, 1999; Kelly & Duschl, 2002; Martin & Hand, 2009; Martins, Mortimer, Osborne, Tsatsarelis, & Jiménez-Aleixandre, 2001; Naylor, Keogh, & Downing, 2007; Simon, Erduran, & Osborne, 2006; von Aufschnaiter, Erduran, Osborne, & Simon, 2008; Zohar & Nemet, 2002). Without question, these studies imply, suggest, and/or stress that it takes time and skill to engage students in some element of conversational dialogue in order to aid them in their construction of knowledge.

Classrooms that engage in scientific argumentation also engage in "activity, reflection, and conversation" (Fosnot, 1989, p. 29). In later writings Fosnot states, "The learners [rather than the teacher] are responsible for defending, providing, justifying, and communicating their ideas to the classroom community. Ideas are accepted as trustworthy only insofar as they make sense to the community and thus rise to the level of 'taken-as-shared,'" (Fosnot, 1996, p. 29-30). Being responsible for nurturing this "taken-as-shared" experience where meanings can flow through and turn together (Isaacs, 1993) is a significant responsibility of the teacher. McNeill (2009) states, "The teacher can be essential for helping students explain phenomena, justify the claims they are making, and debate the strength of alternative explanation" (p. 259). Scientific argumentation in full classroom practice is thought to "transforms the common monologic discourse of the school science classroom" (Osborne, 2007, p. 12) that is often wrought with "monolithic paths of logic or pre-ordained discovery" (Yerrick, 2000, p. 814), which is lead or voiced by the teacher.

SUSTAINED UNDERSTANDING AND PEDAGOGICAL PRACTICE OF SCIENTIFIC ARGUMENTATION

Without doubt, transforming schools and classrooms toward the practice of scientific argumentation is slow to happen because the science curriculum and instruction is often orchestrated around predetermined investigations that serve to verify what is already known (Duschl, 1990; Lemke, 1990). Scientific argumentation in classrooms certainly will draw on expert understandings, but at its core is the ability to engage in the meta-thinking skills and practices around the evaluation, critique, and construction of evidence as discussed in documents such as the NSES (NRC, 1996), A Framework for K-12 Science Education: Practices, Crosscutting Concepts, and Core Ideas (NRC, 2012), Taking Science to School (Duschl, Schweingruber, & Shouse, 2007), and Benchmarks for Science literacy (AAAS, 1993). Certainly, classrooms that engage in high levels of scientific argumentation would entail individuals coherently constructing ideas for others so meaning can flow in and through others as understandings turn and evolve within the community of learners. To do this, teachers need sustained professional development in the ways to approach and practice scientific argumentation. They also need ongoing support that reassures them that the strength of what one knows comes from one's ability to coherently construct and critique one's understanding with self and others.

Unfortunately, sustained professional development around what is scientific argumentation and how one learns is scarcely noted in the literature. When Newton et al. (1999) surveyed 14 experienced science teachers, the teachers reported that they needed more professional development time to confidently manage and facilitate elements of argumentation.

There are two examples of well-researched professional development programs on argumentation; one is called Evidence and Argument in Science (IDEAS) (Osborne, Erduran, & Simon, 2004) from King's College in London, England, and the second is called the Science Writing Heuristic (SWH) (Hand, Wallace, & Yang, 2004; Keys, Hand, Prain, & Collins, 1999) from the University of Iowa, USA. Both research programs have produced numerous reports and guides on implementing argumentation in science classrooms (Norton-Meier, Hand, Hockenberry, & Wise, 2008; Osborne, Erduran, & Simon, 2004b) as well as having produced peer-reviewed papers (Cavagnetto, Hand, & Norton-Meier, 2010; Hand, Wallace, & Yang, 2004; Keys, Hand, Prain, & Collins, 1999; Martin & Hand, 2009; Simon, Erduran, & Osborne, 2006; von Aufschnaiter, Erduran, Osborne, & Simon, 2008) on its effects towards enhancing student learning through engagement in elements of scientific argumentation.

Professional development work in Iowa (e.g. Martin & Hand, 2009; others in press) has shown that shifting classroom activities to include elements of science argumentation not only takes time and practice (at least 18 months) but also that the teacher needs to hold a sustainable understanding that student learning happens through ongoing engagement in scientific argumentation. The professional development work in London (Osborne, Erduran, & Simon, 2004) acknowledged the ways in which the strength of a teacher's initial understanding of argumentation "determines their development, particularly in the short term" (p. 31). In a two-year study of 8th grade science teachers, related to the initial study, Simon, Erduran, and Osborne (2006) found that their extended period of professional development helped to adapt classroom practice toward the use of argumentation.

In classroom practice, talking and reading are more likely to occur than writing. This is not to suggest that writing is not important, but rather to point out that writing and especially revising one's writing is fostered by additional conversations and readings. The literature in science argumentation regarding student learning has mostly been focused on the evaluation of the written processes/products of arguments (e.g. Bell & Linn, 2000; Kelly, et al., 2000; McNeill, 2009; McNeill, Lizotte, Krajcik, & Marx, 2006; Sampson, Grooms, & Walker, 2011; Schweizer & Kelly, 2005; von Aufschnaiter, Erduran, Osborne, & Simon, 2008: Sandoval & Reiser, 2004; Zohar & Nemet, 2002) rather than analysis of the depth and nature of conversations leading up to and following student writing.

Yore, Bisanz, and Hand (2003) point out that elements of Toulmin's Argument Pattern (TAP) (often used to study writing samples) are nearly impossible to pattern and sequence in detail because of the speed in which conversation happens. Understanding whole classroom dialogue goes well beyond the notions of two or more people talking. Approaches to dialogue are many and Burbules and Bruce (2001) reminds us that "no single approach holds the patent on dialogue" (p. 1112). However, there are certainly factors that help to characterize dialogic approaches in scientific argumentation. Effort is needed to provide insight into the extent to which students understand the depth of the question(s), evidence, ideas, and extent of interactions consistent with practicing science argumentation.

ELEMENTS OF CONVERSATIONAL DIALOGUE

The role of questioning, how the class reasons and uses evidence, and how classrooms conversationally "turn together" and integrate ideas seem to be elements strongly supported in the literature and most recently in the framework for K-12 science education (NRC, 2012). Below are five key factors that help to guide whole-class dialogue in classrooms practicing scientific argumentation.

Issues of Questioning

Questioning is essential to scientific argumentation. Boyd and Rubin (2006) talks about contingent questioning and how it takes students from where they are and launches them further. These questions are generally authentic (Nystrand, Gamoran, Kachur, & Prendergast, 1997) and stay well within reach of the student's zone of proximal development (Vygotsky, 1986). These questions are also more likely to be voiced by students to further clarify, elaborate, and/or extend ideas because the teacher in dialogically organized instruction is thought to ask fewer questions and encourage more conversational talk among the class (Boyd & Rubin, 2006, p. 144).

Depth of Exchange

Exchanging ideas takes time. As classroom talk turns together, Boyd and Rubin (2006) also suggests that students must have critical turns of talk that are linguistically extended for 10 seconds or more. These extended turns help in formulating and connecting our ideas for others as well as one's self (Scott & Ametller, 2007; Wells & Mejía Arauz, 2006). Extended exchanges discourage a back-and-forth conversation volley and tend to give more ideas and insights that sustain thinking together (Issacs, 1999). In these sorts of exchanges, ideas from others are expanded, modified, and/ or reframed.

Classroom Interactions

Classrooms that turn together interact in ways that are dynamic and open in the classroom community (Burbules & Bruce, 2001). There is an ongoing openness to different points of view. The community spends time "probing and revoicing each other's thinking" (Varelas & Pineda, 1999, p. 26) as a common understanding is constructed. Viewpoints of others are not only included but also acknowledged during dialogue (Scott, Mortimer, & Aguiar, 2006).

Evidence-Based Ideas

Justification of a claim requires evidence if we expect the classroom community to be aligned to the social practices of scientists (Kuhn & Reiser, 2006). In classroom practice, Kuhn (1991) suggests that some do not feel that claims need to be justified with evidence. Some do not understand what counts as evidence (Sadler, 2004). However, the social aspect of scientific argumentation requires that claims and evidence not only be presented, but subjected to ongoing discussion and critique (Kuhn & Reiser, 2006, p. 5).

Conversational Pattern

A dialogically organized conversation around one idea is "relatively unpredictable" (Boyd & Rubin, 2006, p. 146) and negotiated by the classroom on the fly (Boyd & Rubin, 2006; Nystrand, et al., 1997). Classroom roles are "neither distinct nor stable" (Burbules & Bruce, 2001, p. 19) and all can have a voice in the conversation. Turns of talks are structurally coherent and socially engaging (Boyd & Rubin, 2006). The conversation varies in density and quality, and has an ebb and flow as ideas build (Varelas & Pineda, 1999).

METHODOLOGY AND PARTICIPANTS

Data used for this study were chosen from an existing collection of K-5 classroom videos. This collection of videos comes from classrooms whose teachers have undergone professional development in argument-based inquiry from one to six consecutive years. The overarching focus of the professional development is to help these in-service teachers understand that learning is a process of negotiation, and that by using argument-based inquiry in their science classrooms, "big ideas" in science may build and develop through on-going classroom negotiation. Their annual Professional Development (PD) consisted of five days of summer workshops and at least five visits from PD staff in their classroom during the school year. The teachers were not given any curriculum to teach. However, they were supported through PD during the academic year to embed argument-based inquiry into their existing classroom curriculum. While engaged in the PD process the teachers used the Science Writing Heuristic (SWH) (Keys, et al., 1999) in their classrooms, which we view as one approach to establishing a classroom community that practices argument-based inquiry.

As part of the research component of the PD, teachers were asked to record a video of their classrooms. This was a lesson of their choosing. For this particular study, we choose only video segments of whole-class conversation after a small group of students presented their written claim and evidence about a concept in science. We identified fifteen classroom videos that met this criterion, and a total of fifty-six segments from these classrooms. We used no more than five segments from any one classroom video. A segment began after small group presentation of claim and evidence ended and whole-class conversation about the presentation began. The segment ended when the discussion moved on to a new topic or a new presenting group.

Initially we did a comparative analysis (Charmaz, 2002; Dey, 1999; Glaser & Strauss, 1967), within and between transcript segments in order to establish codes and patterns. During this process, we explored several published instruments and research products that could further inform our coding and analysis. In particular, we found that Section V "Discourse Factors" of Marshall et al. (2009) instrument helped to guide and shape some of our early thinking as we grounded our descriptions in our K-5 classroom.

After many rounds of revision, informed by our data set as well as the literature, we settled on three levels of criteria and five whole-class dialogue factors. The criteria ranges from level 1 (little to no dialogic interaction) to level 3 (high dialogic interaction) we settled on the five key factors to be: conversational patterns, complexity of questions, evidence-based ideas, depth of idea exchange, and classroom interaction. Table 1 represents the whole class dialogue factors used in this study.

Once the framework was established, each segment was coded with a dialogue level for each of the five key dialogue factors. The first two authors of this study coded the segments individually and then discussed the differences in ratings until a consensus was reached for each of the segments in this study.

Table 1. Whole class dialogue factors

Factor	Level One	Level Two	Level Three
Complexity of Question	In most cases, questions were asked to explain explicit knowledge.	In most cases, questions ask students to explain their comprehension of ideas.	In many cases, questions challenged students to explain, reason through, and/or justify.
Depth of Idea Exchange	Even if opportunities existed, ideas were rarely discussed beyond initial response.	Ideas were discussed for several turns of talk but were usually limited to comparing/checking/understanding some smaller element of the "big idea."	Ideas were discussed over many turns of talk to help understand many elements/viewpoints of the "big idea."
Classroom Interactions	Students did not ask a student to justify reasoning or evidence. Teacher may occasionally ask for justification and/or reasoning.	Occasionally student(s), and often the teacher, asked follow-up responses that required student(s) to justify reasoning or evidence.	Students often, and teacher may or may not as often, ask follow-up responses that required students to justify reasoning or evidence.
Evidence-based Ideas	There is little discussion of the claim/evidence presented.	There is some discussion of the claim/evidence presented.	There is extensive discussion of the claim/evidence presented.
Conversational Pattern	Student conversation not well connected to previous turns of talk, with very short conversations about student ideas. Generally Q&A format.	Student conversation at least occasionally is connected to previous turns of talk. Some medium-length conversation occurs about a student idea.	Student conversation was consistently integrated with previous turns of talk. Lengthy discussions occur about a student idea.

Additionally, every video used in this study was scored by experienced scorers using a modified version (Martin & Hand, 2009) of the Reform Teaching Observation Protocol (RTOP) (Sawada & Piburn, 2000). Using this protocol gave us a sense of how well each classroom engaged in "reformed" (Sawada, et al., 2000) instruction. The quantitative RTOP score was then categorized into qualitative descriptions; i.e., high, medium, and low RTOP scores to indicate high, medium, and low implementation of reformed instruction.

RESULTS

Analyses of the data lead us to compare the criteria-based level of dialogue (i.e., level 1, level 2, and level 3) with the classroom RTOP ranking (i.e., low, medium, and high). The findings of this analysis are reported by RTOP score. For each level, we provide an overall summary of implementation and then provide more detailed descriptions of key factors noting what makes a level different than another. We also provide an actual segment of a classroom transcript for each level to further illustrate our findings.

SUMMARY OF LOW IMPLEMENTATION

In low implementation classrooms, we did not see any instances of what we would call level 3 talk. *Conversational patterns* were mostly very structured question/answer situations and generally led by the teacher with fractured connections to previous turns of talk. The teacher's goals seem to be to check student responses to *low-complexity questions* and compare them with his or her own ideas about the concept; therefore there was little talk about *evidence-based ideas*. The teacher and students rarely asked follow-up questions, which hindered the *depth of idea exchange* and tempered whole *classroom interaction* (see Figure 1).

Figure 1. Frequency of whole-class talk categories for low SWH implementation classroom segments

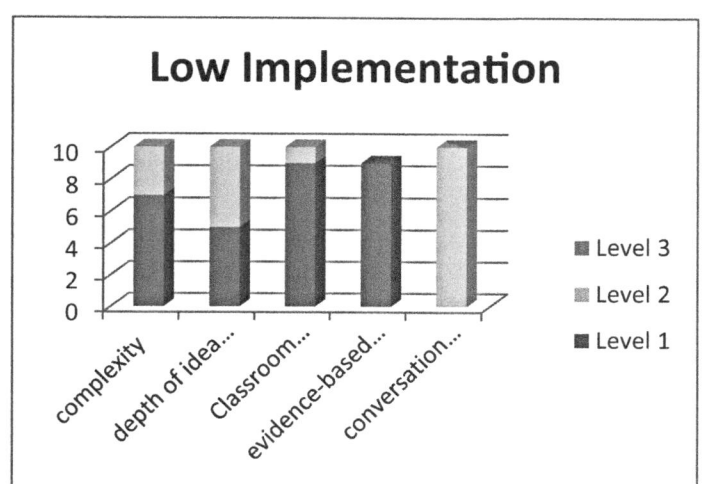

Questioning and Use of Evidence

Quality of the question was generally quite low with seven segments containing mostly level 1, three segments containing mostly level 2, and no segments containing mostly level 3; however, there were a few instances of higher level questions being asked that required students to explain their conception. However, regardless of the question type being asked, student responses tended to foster a recalling of some piece of knowledge that the questioner accepted without any follow-up or elaboration. This lack of critique of student response suggests that *evidence-based ideas* were not well developed, because nine segments were level 1, and there were no level 2 or level 3. Questions that only required a response for a particular piece of content with no follow-up (see Tables 2 and 3) did not encourage ideas to be reasoned through by the classroom community, hindering *classroom interactions*. Additionally ideas being discussed were not grounded in the relationship between claim and evidence. This limited the amount of reasoning during classroom interaction.

Depth of Idea Exchange and Conversational Pattern

Depth of idea exchange was generally very low, with five segments at level 1, five segments at level 2, and no segments at level 3. Even when multiple turns of student talk were present, opportunities for students to reason through their ideas seemed limited. In each turn of student-student talk there were too few words and a lack of sufficient follow-up for a developed idea exchange. Overall, *conversational patterns* were generally very structured in low implementation classrooms, with all ten segments at level 2, and no segments at level 1 and level 3.

Example of Low Implementation

In Table 2, a group is presenting on the United States' third manned mission to the moon (Apollo 13) whose mission was aborted due to a malfunction. The teacher is asking questions about the spaceship malfunction and receiving responses from the students. When Kenny elaborated on an idea the teacher did not take the opportunity to discuss Kenny's idea beyond his initial response. This seems to indicate that the teacher was more

Table 2. Example of level one whole class dialogue

Line	Person	Words Spoken
119	Teacher	What do they need oxygen for?
120	Joyce	To breathe.
121	Teacher	So if the oxygen was leaking, is that a problem?
122	Students	Yep.
123	Teacher	Ok
124	Kenny	Cuz it was going into space.
125	Teacher	Ok.
126	Joyce	They shut down the command module… down to save battery power so they could get back to Earth. They lived in the lunar module.
127	Teacher	What's the lunar module?
Grade 4, 025-04-01-03-2009		

interested in checking to see if the class agreed with her own idea rather than understanding the students' ideas.

In Table 3, a fourth-grade class is discussing animal adaptations. Sarah asks a question that seemed to request further explanation of the groups claim. Instead, the students responded by repeating their claim with no explanation. Neither the teacher nor students challenged the group's response; rather, the teacher allowed the conversation to move on by asking for other questions. We consider these two examples to be coded as a *missed opportunity* that seems to establish an environment with limited opportunity for students to discuss their ideas beyond the initial response.

SUMMARY OF MEDIUM LEVEL IMPLEMENTATION

In medium ranked classrooms, teachers and students both are asking questions with some level of complexity. When a level 2 or level 3 *complex question* is asked, there is generally some level of *in-depth idea exchange* that stems directly from *evidence-based ideas*. There tends to be less idea exchange when the questions are less complex, because these questions are rarely focused on the relationship of evidence to the claim. Conversation patterns are generally structured by the teacher, or not as neatly structured if the students are leading the discussion; however, in either case, there are some instances of *classroom interaction* that involve many students sharing their ideas about the claim and evidence.

Questioning, Evidence, and Idea Exchange

In medium level classrooms, whole class dialogue that supported *deep idea exchange* was developing. In these classrooms, the student turns of talk tended to show *evidence of ideas* at least occasionally being responded to by other students and consistently being listened to and responded to by the teacher. At least for the teacher, their turns of talk usually reflected a response to one or more turns of student talk. *Question complexity* by both students and teacher tended to vary. Follow-up to question responses varied from no elaboration to extended dialogue. When questions asked were about understanding what something meant, there was often extensive classroom interaction among

Table 3. Example of level one whole class dialogue

Line	Person	Words Spoken
	Jack	What was the best animal adaptation? Claim: the best animal adaptation is the tail for a possum. For a alligator, fish and a possum…
209	Teacher	Sara?
210	Sara	I didn't get your claim.
211	Jack	The best animal adaptation is the possum's tail.
212	Teacher	Any other comments? Sheri?
213	Sheri	I think they did pretty good as far as their claim.
Grade 4, 027-03-01-05-2008		

students over many turns of talk (see Tables 4 and 5). In the questions and ideas found in these turns of talk, it seems as though students at this level were starting to work towards being able to argue for their ideas. We noticed that when questions were more focused on a call or need to have evidence, talk tended to stop after a response was given (see Table 4 and Figure 2).

Conversational Interactions and Evidence

In each of the medium implementation classrooms, *conversation patterns* tended to be conversational and most often monitored and/or mediated by the teacher. There were a few cases where the teacher was not actively involved in the discussion, which sometimes led to conversations where students were not practicing "listen first" skills. Student conversations in these situations would have flair-ups of multiple simultaneous conversations that mainly focused on a group's claim and/or evidence. In other cases, the teacher was not actively involved in the conversation because the students were able to maintain the discussion on their own, focusing clearly on exchanging ideas (see Table 4). Overall, what we begin to see emerging is the fine balance of how to conversationally express, listen, respond, and negotiate through discussion of *evidence-based ideas*. At this level, it seems as though these elements sometimes get out of balance, thus making an expression of an idea more exuberant, listening first more challenging, and/or reacting rather than responding more likely. In this way, conversation at this level is sometimes bumpy, because of the varied use and quality of evidence and critique.

Example One of Medium Implementation

In Table 4, the presenting group is discussing the deciduous forest ecosystem. The rest of the class groups have been exploring other ecosystems.

Table 4. Example of level two whole class dialogue

Line	Person	Words Spoken
333	Maria	Um do you know uh, if like, Iowa is in the deciduous forest?
334	Group Member	No. It has a map in there but it didn't have the United States. Lance?
Grade 6, 010-01-01-01-2007		

Maria asks if Iowa was in the deciduous forest. This question first suggests that Maria was listening and thinking about what was presented. Second, she asked a question that either called for a factual response or a speculation of what might be. In this case, the presenting group said no and said that their source did not have the United States in it. In terms of listener understanding, it is murky if no meant, "I don't know" or "No, it does not." The response was not grounded in evidence but lack of information. More importantly, the lack of follow-up by Maria, other students, or the teacher we see as a missed opportunity to have an evidence-based conversation, which would have explored and reasoned through the possibility of a deciduous forest in Iowa.

Example Two of Medium Implementation

In Table 5, the class was discussing evidence about the characteristics of objects that have a high or low auditory pitch. In this segment, one group claimed that a large jar generated a low pitch and a small jar generated a high pitch. Danny asked if they thought it was possible for a large jar also to generate a high pitch, because they had experienced that during their experimentation. The students discuss for several turns of talk the reasons that the two groups heard different things, which eventually lead to discussion about water levels in the jars. In this example and characteristic to this level, follow-up type questions tended to either seek clarification (see Table 5 lines 33-35) or to further challenge an idea (see Table 5

Table 5. Example of level two whole class dialogue

Line	Person	Words Spoken
31	Danny	Uh do you also think that larger objects can make high pitch because the jar the jar the big jar, that one also made a high pitch?
32	Alyssa	In the jar, it was like the one with no water in it made a high sound and the one almost full made a low sound. And that's what you meant with the jars.
33	Danny	How full was the jar?
34	Alyssa	It's the one over there.
35	Luke	So I think it differs with how much water you put in each jar. Cause one group, cause one group (mumble) cause one jar was overflowing.
36	Hunter	Yeah
37	Peyton	But the one with the strings had a shorter string, the higher the pitch was and the longer the string the lower it was.
38	Luke	But we are talking about the jars.
39	Peyton	I know but you said it might affect them not a lot and I'm saying cause that the jars have a little bit of water and a lot and with the strings that has a little shorter string makes a higher pitch and the longer one has a lower pitch.
40	Luke	But what our group did is that the shorter string had a really high sound and the longer ones had the same. All was the same it had a high sound.
41	Peyton	That's not what we got.
42	Luke	I'll try it.

Grade 5, 012-10-08-04-2009

line 39). When an idea was challenged, it often brought in other students to engage in the developing negotiation. In this example, because the students spent several turns of talk sharing their ideas, they were able to develop an explanation as to the reasons that two groups came up with two different results; this in turn helped them to negotiate the concept of pitch.

In Table 5, line 37 when Peyton was making an attempt to establish a relationship of length of string to amount of water, Luke refocused the conversation back to the jars. While we see this refocus as acceptable, the detail that Peyton's outside evidence was brought into the conversation and not used then or later in the class session points out the ways in which ideas when exchanged at this level are more likely to be situated in only the group's evidence and do not take in a broader array of related evidence. Although this example clearly has a particular group arguing for their own ideas over extended and developed turns of talk, we consider this, too, a missed opportunity of reasoning through commonalities in and between groups as the class works toward understanding the properties of sound.

SUMMARY OF HIGH LEVEL IMPLEMENTATION

In high level classrooms whole class dialogue that supported student exchange of ideas was consistently evident. Questioning consistently supported a well-developed conversation that involved students reasoning through and justifying their claims and evidence (see Figure 3).

Complexity of Questions and Interactions

For classrooms at high implementation, the complexity of the question seems to be less important

Figure 2. Frequency of whole-class talk categories for medium SWH implementation classroom segments

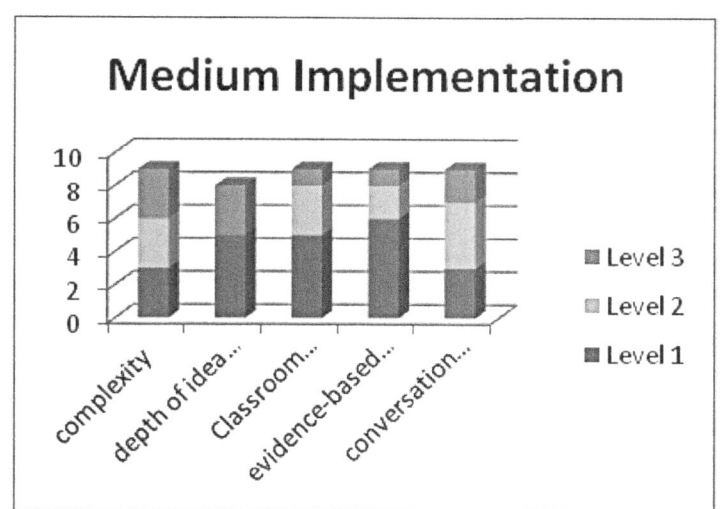

Figure 3. Frequency of whole-class talk categories for high SWH implementation classroom segments

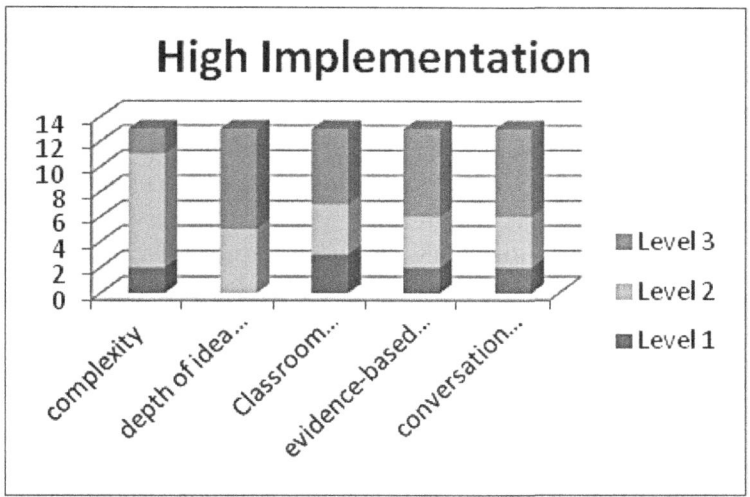

than the ways in which the response of the question is carried out. Questions that begin these conversations are often focused on the presenting groups' claim and evidence. The teacher and the students in these classrooms usually ask follow-up questions or make follow-up statements to any type of question that is asked. This questioning practice encourages student ideas to be reasoned through over many turns of talk.

Deep Exchanges of Ideas and Evidence

Follow-ups to questions in these classrooms were important to foster the depth of ideas and evidence. Follow-up questions by students generally clarified or challenged an ongoing idea. If the teacher stepped into these exchanges, the teacher often asked a student how he/she felt about the presenting group's ideas. The explanation by the classmate often included their own personal ideas or the ideas of their group that often stimulated more questions and/or lengthier negotiations. During these deep idea exchanges, "big ideas" were driven by more than one group's claim and evidence. Not only were the presenting groups ideas circulated but so were the ideas held by other prepared groups and individuals.

Conversational Pattern

Communication in high implementation classrooms was very conversational. Conversations were about building upon student's own evidence-based ideas. Most of the time, students were able to maintain the discussion on their own, focusing the "big idea" with minimal guidance from the teacher (see Table 6). When the teacher was involved with the conversation, often it was to further challenge and/or scaffold student-generated ideas. Sometimes when the teacher was not involved in conversation over many turns of student talk, the conversation became intense because student's ideas were being challenged. On these occasions, the teacher was able to remind students to exhibit "listen first" skills, be respectful and/or thoughtful in responding. In this way, the teacher did not need to manage or control the conversation, but rather helped in extending, patching, and/or consolidating ideas in process.

Example of High Implementation

Table 6 is a second grade classroom that is discussing what is inside a seed. Several members of the class are challenging the presenting group's claim that there is a seed inside a seed. In line 49 JoAnn asks "How you can see a seed in a seed?" JoAnn remains persistent in denying support (line 49, 53, 57, 61, 64, and 67) that a seed is inside a seed. Over seventeen turns of talk happen because the "how" evidence appears to be lacking for JoeAnn. Then in line 68, Erin provides further support for JoeAnn's point and further explains with a related line of support.

Eventually after thirty more turns of talk (not included here), their discussion of observational data is left unresolved. The last presenting group of this session thinks that there are lungs inside the seed. A similar discussion ensues like the first. In the end, the teacher asks the students to recap their unresolved questions, reminds them that the posters they presented will stay in front of the room (this is to insure the ideas remain public), and they will be exploring, in the next class session, nonfiction books to see what they "say about whether seeds are alive or not and if there's something inside the seeds" (line 205).

In this sample excerpt we provide from Table 6 only a few questions were asked. However these questions and the ensuing discussion were about students reasoning through and justifying one group's evidence. The teacher encouraged students to listen to each other and to draw other students into the conversation. The effect of whole class conversation helped to build ideas and viewpoints over many turns of talk. In the end of this session none of the ideas were resolved, but many ideas were expressed and challenged to guide their future look into seeds and plants.

DISCUSSION

This study provides us with a snapshot of one moment in time for each of these fifteen classrooms. All of these classrooms were engaged in an authentic attempt to incorporate argument-based inquiry (Forman, Barnhart, Deafenbaugh, & Ewing, 2010). In each case, students presented their claim and evidence and some level of classroom conversation ensued. The conversations that happened within each classroom varied. However, the overall evaluation for any one classroom fell into one of the three levels fairly consistently. Dialogical patterns tend to be localized (Pratt, 1987) and this was generally the case for this data

Analysis of Discourse Practices in Elementary Science Classrooms

Table 6. Example of level three whole class dialogue

Line	Person	Words Spoken
45	JoAnn	I disagree that you could see a seed in a seed.
46	Nellie	Why?
47	JoAnn	It just seems weird that-
48	Nellie	I saw a seed inside my seed.
49	JoAnn	It's different how you can see a seed in a seed, and I think that's make believe or something.
50	Cindy	It's not make believe, JoAnna.
51	JoAnn	I'm still talking, you know.
52	Teacher	In other words, what she's really saying to you is that what? You need to let her finish, right? Remember what we talked about in discussion, Nellie. You've got to listen first, and then she needs to have her chance, and then you can have hers.
53	JoAnn	I think it's make believe because I have never seen a seed inside a seed before, and that's kind of weird how a seed can be in a seed, and I just don't really think it's true.
54	Nellie	It's true. If you got the seed that I got, then you would be able to see a seed.
55	JoAnn	Well, our group saw a stem in it, and….
56	Nellie	And we saw a seed.
57	JoAnn	A seed in a seed. I just don't think that's true.
58	Cindy	JoAnn, the lima bean is sort of a seed on the outside, and then, JoAnn, on the inside and hanging out partly was a stem.
59	JoAnn	Well you guys didn't say anything about the stem in there.
60	Nellie	Well, we didn't want to. And there's a seed in a seed.
61	JoAnn	I just don't think that can happen.
62	Cindy	Sometimes it can.
63	Nellie	Yep. Sometimes it can't, sometimes it can.
64	JoAnn	I just don't get it.
65	Nellie	Sometimes it can. Erin.
66	Teacher	Let's go to someone else, and we can come back to you, JoAnn.
67	JoAnn	Are you sure that the little seed you saw wasn't a plant?
68	Erin	I kind of disagree with you saw a seed inside of a seed, because the plant has to be inside of the seed, and the seed is supposed to be part of the plant, and the plant is supposed to be growing from the seed, and that little thing…
69	Nellie	[interrupting] I think that worms grow out of that little seed.
70	Erin	That's what I'm trying to say. That little ball probably had the plant inside of it, so it's probably not the seed, really, it's probably just the shell to keep the seed from cracking and breaking all the way through it so what I think she's trying to say- it's a coat that's really hard that looks like a seed, but it's not really a seed, it's actually a coat. The stem from...
71	Teacher	What do you think about that? She's saying there's a coat. Nellie, listen to what she's saying. What do you think-
72	Cindy	I kind of agree with that.
73	Teacher	You need to tell her.
74	Cindy	I kind of agree with that because that outer skin of it is kind of a shield to keep it from getting…the dirt from…
75	Erin	I'm not talking about that. I'm talking about…
76	Cindy	I know. But Erin, the skin on the outside probably protects the seed from inside the lima bean from getting dirty and cracked. I think it's probably trying to keep it together so then the seed won't go out.
Grade 2, 015-08-01-01-2008		

set. For example, no segment rated at Low-Level Implementation (LLI) had evidence of level 3 talk, and rarely did a High-Level Implementation (HLI) classroom have elements of level 1 talk. Medium-Level Implementation (MLI) saw no single level of talk dominate. In all cases, the teacher, by virtue of using this approach, which supports student talk, provided their own version of support toward classroom conversation (Eichinger, et al., 1991; Osborne, 2001).

From our perspective, the teacher was both the navigator and lynchpin in HLI that held together overall classroom discourse. As navigator, the teacher insured that the axis of conversation was maintained and developed around the "big idea" in science. The teacher acted as a lynchpin by insuring the five factors of whole-class dialogue did not fly off the axle and ground the conversation.

In high-level classrooms the sophistication of idea exchange, classroom interaction, and the sustained use of evidence to guide conversation was more monitored than sustained by the teacher. In these high level classrooms there was what Burbules and Bruce (2001) calls a "reciprocal relation" in which " prerogatives of questioning, answering, commenting, or offering reflective observations on the dynamic are open to all participants" (Burbules & Bruce, 2001, p. 19). The classroom community often took questions as starting points for conversation. These questions were more likely to come from students, but a teacher's question was just as likely to be acted upon in a similar fashion. The quality of questions in HLI was not critical in fostering sustained turns of student talk. In HLI the students and sometimes the teacher helped to integrate previous turns of talk. The teacher was most likely to encourage uncontested underrepresented evidence to be brought in to conversation.

In this sense, conversation in HLI classrooms had a "natural flow of conversation" (Schein, 1993, p. 30) and students were afforded opportunities for "sustained expression" (Dillon, 1981, p. 16) of their thoughts. Throughout the HLI classrooms, we saw in varying degrees flare-ups, pauses, wandering, focus and refocus, unidirectional flow, multi-voices, ignoring, and pressing (Scardamalia & Bereiter, 1991; Schein, 1993; Scott & Ametller, 2007; vanZee, Iwasyk, Kurose, Simpson, & Wild, 2001; Windschitl, 2004). These were not examples of behavioral classroom management in crisis but rather examples of what the National Science Education Standards call "reasoned arguments" (NRC, 1996, p. 50). For these reasoned arguments teachers set up expectations which required students "to explain and justify their understanding, argue from data and defend their conclusions, and critically assess and challenge the scientific explanations of one another" (NRC, 1996, p. 50). The argumentation found in these classrooms was certainly with its ebbs and flows—as are most all engaged conversations. However, this was consistent with McNeill's et al. (2006) findings in that we also saw, even in HLI classrooms students struggling to reason through how to use or what counts as evidence during argument-based inquiry.

In low-level classrooms we saw many textbook descriptions of IRE or IRF cycles (Mehan, 1979) or what Isaacs (1999) calls a "back-and-forth volley" (p. 365). Isaacs sees this as a good example of "people thinking alone, the ping-pong nature of conversation dominating - not dialogue" (Isaacs, 1999, p. 365). These LLI classrooms also had a dominance of knowledge level questions, asked by both teacher and students, which were adequately addressed but unlike HLI did not sustain further turns of talk beyond the IRF/IRE cycle.

In HLI we saw "genuine questions" (Scott, 1998, p. 66) in which the response could not follow a typical IRE or IRF pattern because it served as a "thinking device" (p. 66) for the community of learners. In these cases, genuine questions led to extensive turns of talk supported by evidence. Regardless of questioning, in LLI we saw very few sustained student-student turns of talk that included evidence. In this way, Kuhn (1991) is correct. LLI Classrooms we analyzed were poor at bringing together their claims and evidence and conversation did not naturally flow. In HLI we consistently saw elements of discourse around the claim and evidence that seemed "natural," "coordinated," "constructed," and "reasoned" (Kuhn, 1991; McNeill, et al., 2006; Osborne, Erduran, & Simon, 2004). Although LLI is not the ideal, we did see elements of Driver et al.'s (2000) notion of dialogic argumentation beginning to emerge, but differences in ideas were not given sufficient space for critique, consensus, and agreement because the conversation was truncated. In HLI, claims were supported by evidence and questions were asked that challenged ideas that were being expressed. There were attempts to persuade or convince classmates though expressions of doubt, alternate viewpoints, and the recognition that some concepts were still not well understood by the classroom community (Driver, et al., 2000; Newton, Driver, & Osborne, 1999).

The mere existence of MLI classrooms provides recognition that inquiry-based approaches are part of a dynamic process for the teacher and student, PD team, and whole learning community. MLI classrooms can be seen to be either classrooms in transition or classrooms that have settled into a form of practice that was not stable in low level or high-level practices. Duschl and Osborne (2002) points out that "normal classroom discourse is predominantly monologic" (p. 55). We are comfortable in suggesting that the MLI classroom is not usually monologic and just beginning to disengage from monologic activities. In MLI classrooms we see "change both in the structure of classroom activities and the aims that underlie them" (Duschl & Osborne, 2002, p. 55). In MLI those changes are well in place as evidenced by the increasing likelihood of student-student dialogue over more complex questioning and a greater withholding of teacher feedback, evaluation, and forced directions. The most obvious change from LLI to MLI that we see is in the form of questions. Knowledge recall questions are replaced with more complex questioning that cannot be responded to over one turn of talk. Additionally, the teacher is supportive of letting the idea develop before moving the conversation along. During MLI it seems as though the students are beginning to construct arguments for each other rather than just for the teacher or not at all (Berland & Reiser, 2009). However, as Wertsch (1979) suggests, students cannot internalize argumentative discourse until they can make counter arguments for themselves. The major difference between MLI and HLI is that in HLI we usually hear persistent evidence-based whole-class dialogue and in MLI the dialogue generally emerges and fades. While much more research is needed to properly understand the role and predominance of MLI classrooms in our study, the very existence of an MLI classroom indicates that PD can help teachers move beyond a low level implementation of argument-based inquiry; however it is unclear if all classrooms will reach a sustained high level of whole-class dialogue while engaging in argument-based inquiry. In summary, one can most easily differentiate LLI from MLI by the emergence of more "genuine questions" (Scott, 1998, p. 66) that are given sustained support through the "teacher's uptake of student ideas" (Ametller & Scott, 2009, p. 14; also see Collins, 1982). More importantly, student uptake of these questions over several turns of conversational talk begins to emerge. In HLI classrooms, what is consistently noticed is that students always seemed to have something more to say to each other (Eichinger, et al., 1991) and that classroom dialogue collaboratively built understanding as the dialogue progressed (Wells & Mejía Arauz, 2006). The whole-class dialogue in these HLI classrooms was left unsettled, yet it seemed satisfying to the classroom participants. Students seemed intrigued by others reasoned claims, evidence, and ideas. In LLI classrooms when the conversation ended or moved on, it seemed like it was "won" or "settled" and the discussion of one's idea was a moderately interactive spectator event rather than an opportunity to build knowledge with each other.

CONCLUSION

The implications of this research indicate that there are specific dialogical characteristics that occur in the science classroom that can contribute to highly effective implementation of argument-based inquiry and reformed teaching practices. In our discussion, we attempted to create a picture of what is happening across each level of implementation and how existing research also supports our evidence. The focus in the discussion was on the classroom's practice. However, we fully understand that it is ultimately the teacher that gives the authority for more than monologic instruction in any given classroom. "In learning to talk science we must buy into, and learn to work with, the conceptual tools, epistemological framing, ontological perspectives, and forms of reasoning of the scientific community" (Scott, 1998, p. 74). Without question, the first person to begin the process of engaging the classroom in the practices of an authentic scientific community must be the teacher. Maintaining a steadfast approach to helping students of elementary age to develop reasoned scientific arguments takes time, courage, and ongoing PD to fully understand and implement. This research is only scratching the surface of the ways in which classrooms engage in argument-based inquiry. Many research questions surface from this study. For example, how are students' writing samples influenced through engagement in varying levels of whole class dialogue in argument-based inquiry? What influences why some classrooms develop into HLI and others do not? How successful are students in classrooms that practice the highest level of whole class dialogue? These questions and many others are needed to understand more fully the overall impact on student learning from classrooms engaged in argument-based inquiry.

REFERENCES

Alexopoulou, E., & Driver, R. (1997). Small group discussions in physics: Peer interaction modes in pairs and fours. *Journal of Research in Science Teaching*, *33*(10), 1099–1114. doi:10.1002/(SICI)1098-2736(199612)33:10<1099::AID-TEA4>3.0.CO;2-N

American Association for the Advancement of Science. (1993). *Benchmarks for scientific literacy*. Washington, DC: AAAS.

Ametller, J., & Scott, P. (2009). *Classroom discourse, dialogic teaching and teacher professional development in science education: A selective review of the literature*. Retrieved April 16, 2010, from https://www.ntnu.no/wiki/download/attachments/8324755/Final+draft_lit_review_Leeds.doc

Bell, P., & Linn, M. C. (2000). Scientific arguments as learning artifacts: Designing for learning from the web with KIE. *International Journal of Science Education*, *22*, 797–817. doi:10.1080/095006900412284

Berland, L. K., & Reiser, B. J. (2009). Making sense of argumentation and explanation. *Science Education*, *93*(1), 26–55. doi:10.1002/sce.20286

Boyd, M., & Rubin, D. (2006). How contingent questioning promotes extended student talk: A function of display questions. *Journal of Literacy Research*, *38*(2), 141–159. doi:10.1207/s15548430jlr3802_2

Bricker, L. A., & Bell, P. (2008). Conceptualizations of argumentation from science studies and the learning sciences and their implications for the practices of science education. *Science Education*, *92*(3), 473–498. doi:10.1002/sce.20278

Burbules, N. C., & Bruce, B. C. (2001). Theory and research on teaching as dialogue. In Richardson, V. (Ed.), *Handbook of Research on Teaching* (4th ed., pp. 1102–1121). Washington, DC: American Educational Research Association.

Cavagnetto, A., Hand, B. M., & Norton-Meier, L. (2010). The nature of elementary student science discourse in the context of the science writing heuristic approach. *International Journal of Science Education, 32*(4), 427–449. doi:10.1080/09500690802627277

Cavagnetto, A. R. (2010). Argument to foster scientific literacy: A review of argument interventions in K-12 science contexts. *Review of Educational Research, 80*(3), 336–371. doi:10.3102/0034654310376953

Charmaz, K. (2002). Qualitative interviewing and grounded theory analysis. In Gubrium, J., & Holstein, J. A. (Eds.), *Handbook of Interview Research: Context and Method* (pp. 675–694). Thousand Oaks, CA: Sage Publications.

Costello, P. J. M., & Mitchell, S. (1995). Introduction: Argument: Voices, text and contexts. In Costello, P. J. M., & Mitchell, S. (Eds.), *Competing and Consensual Voices: The Theory and Practice of Argument*. Clevedon, UK: Multilingual Matters Ltd.

Crawford, T. (2005). What counts as knowing: Constructing a communicative repertoire for student demonstration of knowledge in science. *Journal of Research in Science Teaching, 42*(2), 139–165. doi:10.1002/tea.20047

Dey, I. (1999). *Grounding grounded theory: Guidelines for qualitative inquiry*. San Diego, CA: Academic Press.

Dillon, J. T. (1981). To question and not to question during discussion: Non-questioning techniques. *Journal of Teacher Education, 32*(15), 15–20. doi:10.1177/002248718103200605

Driver, R., Asoko, H., Leach, J., Mortimer, E., & Scott, P. (1994). Constructing scientific knowledge in the classroom. *Educational Researcher, 23*(7), 5–12.

Driver, R., Newton, P., & Osborne, J. (2000). Establishing the norms of scientific argumentation in classrooms. *Science Education, 84*(3), 287–312. doi:10.1002/(SICI)1098-237X(200005)84:3<287::AID-SCE1>3.0.CO;2-A

Duschl, R. (1990). *Restructuring science education: The importance of theories and their development*. New York, NY: Teachers College Press.

Duschl, R., & Osborne, J. (2002). Supporting and promoting argumentation discourse. *Studies in Science Education, 38*, 39–72. doi:10.1080/03057260208560187

Duschl, R. A., Schweingruber, H. A., & Shouse, A. W. (Eds.). (2007). *Taking science to school: Learning and teaching science in grades K-8*. Washington, DC: National Academies Press.

Eichinger, D. C., Anderson, C. W., Palinscar, A., & David, Y. M. (1991). *An illustration of the roles of content knowledge, scientific argument, and social norms in collaborative problem solving*. Paper presented at the American Educational Research Association. Chicago, IL.

Ford, M. (2008). Disciplinary authority and accountability in scientific practice and learning. *Science Education, 92*(3), 404–423. doi:10.1002/sce.20263

Forman, E., Barnhart, B., Deafenbaugh, L., & Ewing, M. (2010). *Fostering constructive criticism in a high school biology classroom: Understanding the social dynamics of argumentation*. Paper presented at the Annual Meeting of the National Association for Research in Science Teaching. Philadelphia, PA.

Fosnot, C. T. (1989). *Enquiring teachers, enquiring learners: A constructivist approach to teaching*. New York, NY: Teachers College Press.

Fosnot, C. T. (1996). Constructivism: A psychological theory of learning. In Fosnot, C. T. (Ed.), *Constructivism: Theory, Perspectives, and Practice* (pp. 8–33). New York, NY: Teachers College Press.

Glaser, B. G., & Strauss, A. L. (1967). *The discovery of grounded theory: Strategies for qualitative research.* New York, NY: Aldine. doi:10.1097/00006199-196807000-00014

Gross, A. G. (1990). *The rhetoric of science.* Cambridge, MA: Harvard University Press.

Hand, B., Wallace, C. S., & Yang, E. M. (2004). Using the science writing heuristic to enhance learning outcomes from laboratory activities in seventh grade science: Quantitative and qualitative aspects. *International Journal of Science Education, 26,* 131–149. doi:10.1080/0950069032000070252

Hogan, K., Nastasi, B. K., & Pressley, M. (2000). Discourse patterns and collaborative scientific reasoning in peer and teacher-guided discussions. *Cognition and Instruction, 17,* 379–432. doi:10.1207/S1532690XCI1704_2

Isaacs, W. (1999). *Dialogue and the art of thinking together.* New York, NY: Currency/Doubleday.

Isaacs, W. H. (1993). Taking flight: Dialogue, collective thinking, and organizational learning. *Organizational Dynamics, 22*(2), 24–39. doi:10.1016/0090-2616(93)90051-2

Jiménez-Aleixandre, M., & Erduran, S. (2008). Argumentation in science education: An overview. In Erduran, S., & Jiménez-Aleixandre, M. (Eds.), *Argumentation in Science Education: Perspectives from Classroom-Based Research* (pp. 3–27). Dordrecht, The Netherlands: Springer.

Kelly, G., Chen, C., & Prothero, W. (2000). The epistemological framing of a discipline: Writing science in university oceanography. *Journal of Research in Science Teaching, 37*(7), 691–718. doi:10.1002/1098-2736(200009)37:7<691::AID-TEA5>3.0.CO;2-G

Kelly, G., & Duschl, R. (2002). *Toward a research agenda for epistemological studies in science education.* Paper presented at the Annual Meeting of the National Association for Research in Science Teaching. New Orleans, LA.

Kelly, G., & Green, J. (1998). The social nature of knowing: Toward a sociocultural perspective on conceptual change and knowledge construction. In Guzzetti, B., & Hynd, C. (Eds.), *Perspectives on Conceptual Change* (pp. 145–182). Mahwah, NJ: Erlbaum.

Keys, C., Hand, B., Prain, V., & Collins, S. (1999). Using the science writing heuristic as a tool for learning from laboratory investigations in secondary school. *Journal of Research in Science Teaching, 36*(10), 1065–1084. doi:10.1002/(SICI)1098-2736(199912)36:10<1065::AID-TEA2>3.0.CO;2-I

Kuhn, D. (1991). *The skills of argument.* Cambridge, UK: Cambridge University Press. doi:10.1017/CBO9780511571350

Kuhn, D. (1993). Science as argument: Implications for teaching and learning scientific thinking. *Science Education, 77*(3), 319–337. doi:10.1002/sce.3730770306

Kuhn, L., & Reiser, B. J. (2006). S*tructuring activities to foster argumentative discourse.* Paper presented at the American Educational Research Association. San Francisco, CA.

Lemke, J. (1990). *Talking science: Language, learning and values.* Norwood, NJ: Ablex.

Marshall, J. C., Horton, B., Smart, J., & Llewellyn, D. (2009). *EQUIP: Electronic quality of inquiry protocol*. Retrieved from http://iim-web.clemson.edu/wp-content/uploads/2009/02/equip-2009.pdf

Martin, A. M., & Hand, B. (2009). Factors affecting the implementation of argument in the elementary science classroom: A longitudinal case study. *Research in Science Education, 39*(1), 17–38. doi:10.1007/s11165-007-9072-7

Martins, I., Mortimer, E., Osborne, J., Tsatsarelis, C., & Jiménez-Aleixandre, M. P. (2001). Rhetoric and science education. In Behrendt, H., Dahncke, H., Duit, R., Gräber, W., Komorek, M., Kross, A., & Reiska, P. (Eds.), *Research in Science Education - Past, Present, and Future* (pp. 189–198). Dordrecht, The Netherlands: Kluwer Academic Publishers.

McNeill, K. L. (2009). Teachers' use of curriculum to support students in writing scientific arguments to explain phenomena. *Science Education, 93*(2), 233–268. doi:10.1002/sce.20294

McNeill, K. L., Lizotte, D. J., Krajcik, J., & Marx, R. W. (2006). Supporting students' construction of scientific explanations by fading scaffolds in instructional materials. *Journal of the Learning Sciences, 15*(2), 153–191. doi:10.1207/s15327809jls1502_1

Mehan, H. (1979). *Learning lessons: Social organization in the classroom*. Cambridge, MA: Harvard University Press.

Mork, S. M. (2005). Argumentation in science lessons: Focusing on the teacher's role. *NorDiNa, 1*(1), 17–30.

National Research Council. (1996). *National science education standards*. Washington, DC: National Academy Press.

National Research Council. (2012). *A framework for K-12 science education: Practices, crosscutting concepts, and core ideas*. Washington, DC: The National Academies Press.

Naylor, S., Keogh, B., & Downing, B. (2007). Argumentation and primary science. *Research in Science Education, 37*, 17–39. doi:10.1007/s11165-005-9002-5

Newton, P., Driver, R., & Osborne, J. (1999). The place of argumentation in the pedagogy of school science. *International Journal of Science Education, 21*(5), 553–576. doi:10.1080/095006999290570

Norton-Meier, L., Hand, B., Hockenberry, L., & Wise, K. (2008). *Questions, claims and evidence: The important place of argument in children's science writing*. Portsmouth, NH: Heinemann.

Nystrand, M., Gamoran, A., Kachur, R., & Prendergast, C. (1997). *Opening dialogue: Understanding the dynamics of language and learning in the English classroom*. New York, NY: Teachers College Press.

Osborne, J. (2007). Towards a more social pedagogy in science education: the role of argumentation. *Revista Brasileira de Pesquisa em Educação em Ciências, 7*(1). Retrieved July 12, 2010, from http://www.fae.ufmg.br/abrapec/revista/index.html

Osborne, J., Erduran, S., & Simon, S. (2004). Enhancing the quality of argumentation in school science. *Journal of Research in Science Teaching, 41*(10), 994–1020. doi:10.1002/tea.20035

Osborne, J. F. (2001). Promoting argument in the science classroom: A rhetorical perspective. *Canadian Journal of Science. Mathematics and Technology Education, 1*, 271–290. doi:10.1080/14926150109556470

Osborne, J. F., Erduran, S., & Simon, S. (2004b). *Ideas, evidence and argument in science: In-service training pack, resource pack and video*. London, UK: Nuffield Foundation.

Pratt, M. L. (1987). Linguistic utopias. In Fabb, N., Attridge, D., Durant, A., & MacCabe, C. (Eds.), *The Linguistics of Writing: Arguments between Language and Literature* (pp. 48–66). New York, NY: Methuen.

Sampson, V., Grooms, J., & Walker, J. (2011). Argument-driven inquiry as a way to help students learn how to participate in scientific argumentation and craft written arguments: An exploratory study. *Science Education*, 95(2), 217–257. doi:10.1002/sce.20421

Sandoval, W. A., & Reiser, B. J. (2004). Explanation-driven inquiry: Integrating conceptual and epistemic scaffolds for scientific inquiry. *Science Education*, 88(3), 345–372. doi:10.1002/sce.10130

Sawada, D., & Piburn, M. (2000). *Reformed teaching observation protocol (RTOP)*. Technical Report No. IN00-1. Tempe, AZ: Arizona Collaborative for Excellence in the Preparation of Teachers.

Sawada, D., Piburn, M., Turley, J., Falconer, K., Benford, R., Bloom, I., & Judson, E. (2000). *Reformed teaching observation protocol (RTOP) training guide*. ACEPT Technical Report No. IN00-2. Tempe, AZ: Arizona Collaborative for Excellence in the Preparation of Teachers.

Scardamalia, M., & Bereiter, C. (1991). Higher levels of agency for children in knowledge-building: A challenge for the design of new knowledge media. *Journal of the Learning Sciences*, 1(1), 37–68. doi:10.1207/s15327809jls0101_3

Schein, E. H. (1993). On dialogue, culture, and organizational learning. *Reflections: The SoL Journal*, 4(4), 27–38. doi:10.1162/152417303322004184

Schweizer, D. M., & Kelly, G. J. (2005). An investigation of student engagement in a global warming debate. *Journal of Geoscience Education*, 53(1), 75–84.

Scott, P. (1998). Teacher talk and meaning making in science classrooms: A Vygotskian analysis and review. *Studies in Science Education*, 32, 45–80. doi:10.1080/03057269808560127

Scott, P., & Ametller, J. (2007). Teaching science in a meaningful way: Striking a balance between 'opening up' and 'closing down' classroom talk. *The School Science Review*, 88(324), 77–84.

Scott, P. H., Mortimer, E. F., & Aguiar, O. G. (2006). The tension between authoritative and dialogic discourse: A fundamental characteristic of meaning making interactions in high school science lessons. *Science Education*, 90, 605–631. doi:10.1002/sce.20131

Simon, S., Erduran, S., & Osborne, J. (2006). Learning to teach argumentation: Research and development in the science classroom. *International Journal of Science Education*, 28(2-3), 235–260. doi:10.1080/09500690500336957

Toulmin, S. (1958). *The uses of argument*. Cambridge, UK: Cambridge University Press.

vanZee, E. H., Iwasyk, M., Kurose, A., Simpson, D., & Wild, J. (2001). Student and teacher questioning during conversations about science. *Journal of Research in Science Teaching*, 38, 159–190. doi:10.1002/1098-2736(200102)38:2<159::AID-TEA1002>3.0.CO;2-J

Varelas, M., & Pineda, E. (1999). Intermingling and bumpiness: Exploring meaning making in the discourse of a science classroom. *Research in Science Education*, 29(1), 25–49. doi:10.1007/BF02461179

von Aufschnaiter, C., Erduran, S., Osborne, J., & Simon, S. (2008). Arguing to learn and learning to argue: Case studies of how students' argumentation related to their scientific knowledge. *International Journal of Science Education*, 45(1), 101–131.

Vygotsky, L. S. (1986). *Thought and language*. Cambridge, MA: MIT Press.

Wells, G., & Mejía Arauz, R. (2006). Dialogue in the classroom. *Journal of the Learning Sciences*, *15*, 379–428. doi:10.1207/s15327809jls1503_3

Wertsch, J. V. (1979). From social interaction to higher psychological processes: A clarification and application of Vygotsky's theory. *Human Development*, *22*(1), 1–22. doi:10.1159/000272425

Windschilt, M. (2004). *What types of knowledge do teachers use to engage learners in "doing science"? Rethinking the continuum of preparation and professional development for secondary science educators*. Paper presented at the Meeting High School Science Laboratories: Role and Vision. Washington, DC: National Academy of Sciences.

Yerrick, R. (2000). Lower track science students' argumentation and open inquiry instruction. *Journal of Research in Science Teaching*, *37*(8), 807–838. doi:10.1002/1098-2736(200010)37:8<807::AID-TEA4>3.0.CO;2-7

Yore, L. D., Bisanz, G. L., & Hand, B. M. (2003). Examining the literacy component of science literacy: 25 years of language arts and science research. *International Journal of Science Education*, *25*(6), 689–725. doi:10.1080/09500690305018

Zohar, A., & Nemet, F. (2002). Fostering students' knowledge and argumentation skills through dilemmas in human genetics. *Journal of Research in Science Teaching*, *39*(1), 35–62. doi:10.1002/tea.10008

Section 4
Evaluation and Assessment Issues

Chapter 13
Next Generation Science Assessment:
Putting Research into Classroom Practice

Edward G. Lyon
Arizona State University, USA

ABSTRACT

The recent release of science education documents such as A Framework for K-12 Science Education: Practices, Crosscutting Concepts, and Core Ideas (National Research Council, 2012) marks the transition into a new generation of science education. This transition necessitates a close look at how pre-college science teachers will assess a diverse group of students in ways that are consistent with science education reform. In this chapter, the authors identify current research in science assessment and employ assessment coherence, assessment use, and assessment equity as guiding principles to address the challenges of putting science assessment research into classroom practice. To exemplify these challenges, they describe a study where a research instrument designed to measure scientific reasoning skills was translated into a high school science classroom assessment. The goal of this chapter is to stimulate conversation in the science education community (researchers, assessment developers, teacher educators, administrators, and classroom teachers) about how to put science assessment research successfully into practice and to describe what next steps need to be taken, particularly around assessing diverse student populations.

DOI: 10.4018/978-1-4666-2809-0.ch013

NEXT GENERATION SCIENCE ASSESSMENT: PUTTING RESEARCH INTO CLASSROOM PRACTICE

Assessment has always been a topic of interest in science education, but its importance is perhaps even more critical now, as pre-college science teachers prepare the next wave of students for economies of the 21st century. The importance of assessment rests on the premise that what *is assessed* often drives what *is taught*—thus, to equip students with the scientific knowledge, skills, and habits of mind called for in science education means that science teachers should also *assess* desirable knowledge, skills, and habits of mind. Furthermore, the general role of educational assessment in society has changed considerably over the last few decades due to shifts in how educational researchers view measurement, learning, and the role of education (Shepard, 2000). One outcome of this shift is an emphasis on using assessment for formative purposes, meaning that "activities undertaken by teachers—and by their students in assessing themselves—[provide] information to be used as feedback to modify teaching and learning activities" (Black & Wiliam, 1998, p. 140). This description emphasizes the function and not the form of the assessment; to be formative, "feedback needs to contain an implicit or explicit recipe for future action" (Wiliam & Leahy, 2007, p. 31). Therefore, classroom assessment can serve an instructional role, making it even more important to understand in light of classroom realities.

By science *classroom assessment*, I am referring to the process of finding out what students know and can do in the context of what is being taught in the science classroom. Ideally, assessment informs the teacher and the students themselves, so that decisions could be made—giving a grade or other evaluative mark, diagnosing the students' prior knowledge, or deciding what next steps to take in instruction. In this chapter, rather than answer an empirical research question or review a body of literature, I aim to stimulate a conversation by addressing the challenges of putting science assessment research into classroom practice. I begin by highlighting some of the major reforms and innovations in science assessment over the last twenty years to demonstrate where we are in terms of research. I then describe three assessment principles that can guide how researchers examine the practice of assessing in science classrooms as well as guide science teachers themselves in terms of how they assess. Finally, I will describe my participation in a classroom-based research study that exemplifies the challenges faced when putting science assessment research, centered on assessing scientific reasoning, into classroom practice. The intended audience is primarily science teachers who have the daunting task of translating research into practice and the science education researchers who seek to both study and support science teachers in this endeavor. However, anyone invested in science education—curriculum/assessment developers, administrators, and policy makers—can benefit from the discussion that follows.

WHAT TO ASSESS IN SCIENCE CLASSROOMS? A STARTING POINT

The essence of any science classroom assessment is to elicit and be able to interpret some desirable information about what students are being, or will be, taught. If the desirable information is whether students know the definition of the word "adaptation," then the assessment could ask students to write out the definition of adaptation, or have them select from a list of definitions. However, the next generation of science education is not about preparing students to recite definitions, but rather preparing scientifically literate students. Scientific literacy has been conceptualized in a variety of ways, primarily referring to being "knowledgeable, learned, and educated in science" (Phillips & Norris, p. 224), a conceptualization that, unfortunately, often overlooks reading and writing in science—a fundamental sense of literacy. I appropriate the term scientific literacy broadly to encompass how various science education documents and frameworks describe what is important to learn in science education.

A premise for this chapter is that what is being assessed should resemble what is deemed important to learn in science, thus assessments should inform the teacher about student progress toward becoming scientifically literate. Organizations and committees have published seminal documents, such as *Benchmarks for Science Literacy* (American Association for Advancement in Science, 1993), *National Science Education Standards* (National Research Council, 1996), and *Science Education in Europe: Critical Reflections* (Osborne & Dillon, 2008), which represent the aspirations of scientists and science educators. For instance, the United States' *National Science Education Standards,* or NSES, (NCR, 1996) defines scientific literacy as "the knowledge and understanding of scientific concepts and processes required for personal decision making, participation in civic and cultural affairs, and economic productivity" (p. 22). The NSES consists of eight broad strands—related to scientific content, processes, technology, personal/social perspectives, and unifying themes. More recently, the National Research Council (2012) released *A Framework for K-12 Science Education: Practices, Crosscutting Concepts, and Core Ideas* that drew on new research in the learning sciences and science education to "[articulate] a broad set of expectations for students in science" (p. 1). According to the report, "By the end of the 12th grade, students should have gained sufficient knowledge of the practices, crosscutting concepts, and core ideas of science and engineering to engage in public discussions on science-related issues, to be critical consumers of scientific information related to their everyday lives, and to continue to learn about science throughout their lives" (p. 9). A noticeable difference in this framework, which will inform the next generation of science standards in the United States, is a shift from scientifically inquiry to scientific practices, such as engaging in argument from evidence. Moreover, the new framework focuses on *core* ideas, in both science and engineering, such as "from molecules to organisms: Structures and processes," rather than a laundry list of scientific concepts to learn. Finally, beyond the next generation of science standards (informed by the new framework and currently being developed), common core language arts and literacy standards are already written and include literacy across the subject areas. Thus, *literacy in science* is being instantiated as an important part of scientific literacy.

The emphasis on core ideas and scientific practices is present in several large-scale assessments, both in the United States and internationally. For example, The College Board® recently revised the Advanced Placement (AP) Biology curriculum—a high school course in the United States, usually taken as a second biology course, which prepares students to take an examination and receive college credit. The current AP Biology curriculum calls for students to master a list of core biological ideas (e.g., the process of evolution drives the diversity and unity of life) and scientific practices (the student can work with scientific explanations and theories). The curriculum goes so far as to point out details that are *beyond* the scope of the course and the AP examination, such as "specific steps, names of enzymes and intermediates of the pathways for capturing and storing free energy" (The College Board, 2011).

Two large-scale assessments intend to measure science literacy (or achievement more generally) employ a matrix, in which scientific practices and content are integrated, or crossed. The National Assessment of Educational Progress (NAEP) developed a science framework that consists of *content areas* (life science, physical sciences, and earth and space sciences) and *scientific practices* (identifying science principles, using science principles, using scientific inquiry, using technological design). By integrating both domains into a matrix, each assessment item addresses a content area and a scientific practice (see NAEP, 2009). In a similar vein, the Programme for International Student Assessment, or PISA (OECD, 2009) characterizes four interrelated aspects—context, knowledge, competencies, and attitudes—once again addressing how aspects of scientific literacy intersect each other. PISA is intended to assess specific competencies (identifying scientific issues, explaining phenomena scientifically, and using scientific evidence); however, the framework developers recognized that each assessment item would occur within a particular context (either person, social, or global) and incorporate scientific knowledge (physical systems, living systems, earth and space systems, or technology systems). The PISA framework explicitly mentions that they are assessing the competencies (application of knowledge) and not the knowledge itself, given the variety of curricula students from across the world would bring while completing the assessment.

In summary, an examination of frameworks and large-scale assessments (assuming they should guide what is taught in science classrooms) demonstrates that the next generation of science education learning should focus on scientific practices and core ideas as opposed to a list of discrete concepts. Furthermore, attempts should be made to integrate practices with core ideas and cross cutting concepts. What kinds of tasks can meet the challenge of assessing scientific practices, in the context of core ideas, in science classrooms? What about culturally and linguistically diverse science classrooms?

REFORM AND INNOVATIONS IN SCIENCE ASSESSMENT TASKS

Selected-response (e.g., multiple-choice, true/false) have been a staple of science assessments both at the classroom level and on larger scale assessments for some time and primarily, though not always, draw on a behaviorist view of learning. New research about how students learn, such as the view that students learn by actively constructing knowledge, set the stage for concept mapping, which resembles how knowledge might be organized in a student's mind since the task is analogous to the schematic representations that individuals mentally construct (Ausubel, Novak, & Hanesian, 1978). Put simply, concept mapping involves student modeling of the interrelationship among a series of concepts by organizing the written concepts, often referred to as "nodes," schematically, and connecting the nodes with a linking phrase (Novak, 1990). Because concept maps can reveal a students' conceptual understanding schematically, they are useful as assessment tools to promote metacognitive strategies and provide feedback to address alternative conceptions. However, even concept maps, like selected-response items, are limited in that they do not resemble, and thus cannot solicit competencies associated with, scientific practices. On the other hand, a different type of assessment—science performance assessment—may be better suited for this task.

Science performance assessment, which gained momentum in the 1990s, is a subset of broader categories of assessment, commonly referred to as "alternative assessment" (Gipps, 2008) "authentic assessment" (Wiggins, 1989), or "contextual assessment" (Klassen, 2006). According to Johnson, Penny, and Gordon (2009), performance assessment involves "examinees demonstra[ting] their knowledge and skills by engaging in a process or constructing a product" (p. 2). A subtle nuance to performance assessment that distinguishes them from other alternative assessments is their authenticity, meaning that they use real life or simulated scenarios so that the *desirable* performance or product is more directly observed (Stiggins & Bridgeford, 1982). For example, since scientists engage in scientific arguments, then the assessment should call for students to do the same—instead of a selected-response item that tries to represent the practice. Science performance assessment also has the potential to be integrated seamlessly with instruction, since they often resemble instruction—particularly in an inquiry-based instruction. For example, the Science Teacher Enhancement Project-unifying the Pikes Peak region (STEP-uP) focused on the improvement of elementary students' science learning by aligning curriculum, instruction and assessment coupled with a comprehensive program of teacher professional development. Since inquiry was a focus of the already developed curriculum, project members designed science performance assessments that resembled the actually activities taught during the unit (Kuerbis & Mooney, 2008). In one such unit, *Food Chemistry* (National Science Resource Center, 1994), students engage in activities in which they test different types of nutrients—sugar, starch, fat, and protein. Thus, the science performance assessment calls for students to apply their knowledge of how to conduct food tests to a novel situation—recommending a "space snack" for astronauts. Thus, in the context of physical science, students engage in several different scientific practices.

While science performance assessment has been touted as better eliciting what students can do, as opposed to what they can recall, researchers have questioned their technical quality, especially on a large scale. Shavelson, Baxter, and Pine (1991) found that the variation of science performance assessment tasks and methods limits generalizability. For instance, teachers may ask students to follow a list of instructions that results in performing an exothermic reaction. Alternatively, a teacher may ask them to design a set of procedures and carry out an exothermic reaction. The variety of tasks begs the question, what exactly is being assessed? Moreover, the scoring method influences the inferences made about what students know and can do. Will teachers observe students as they carry out the procedures? Will teachers only look at the final product? Shavelson et al. suggest that science performance assessments need to (a) go beyond factual recall, (b) actively involve students (e.g., hands on), (c) utilize computer technology, (d) reflect development in cognitive research, and (e) align with curricular reform. These recommendations reflect why science performance assessments were initially developed and researched, but science education researchers are only now beginning to incorporate some of these recommendations and are in the process of learning even more about the use of science performance assessments with diverse students.

Recently, I have been part of a research team that examined science performance assessment through the lens of what students have to do with language, which we refer to as the assessment's language demands. We have documented the range of ways students participate and communicate in science performance assessment as well as the plethora of written texts they encounter. We argue that while challenging to second language learners (referred to as *English Learners* in this chapter); with proper support, the language demands could actually be opportunities for ELs, by providing them with multiple ways to access science content, engage in scientific inquiry and discourse, and develop English proficiency (Shaw, Bunch, & Geaney, 2010; Lyon, Bunch, & Shaw, 2012). I bring up this research to point out that researchers can move beyond examining the technical quality of assessments and look for other ways in which to analyze them that may have implications for classroom practice—particularly with diverse learners.

Finally, assessment developers have taken assessment, including performance assessment, into the 21st century by integrating technology as a way to elicit complex skills emphasized in science education, efficiently evaluate student responses, and even provide automated feedback to students, thus serving a formative role in the classroom. For example, WestEd, an educational laboratory in the Western United Sates, has been developing and researching computer-based science assessments in projects such as SimScientists (Quellmalz & Pellegrino, 2009). There are several advantages of computer-based assessments. For one, they can use multiple modalities and representations, which can help students demonstrate complex practices like manipulating given data and testing predictions based on data manipulation. Perhaps even more valuable, computers can respond to what students input and automatically give them new tasks, helpful hints, or feedback. Computers can also keep written records of student work, which teachers can access. SimScientists science assessments demonstrate high psychometric quality and hold promise for becoming part of balanced state science assessment systems (Quellmalz, Timms, Silberglitt, & Buckley, 2012). Although computers have been used extensively in science instruction, we are just scratching the surface of determining how to take advantage of computes while assessing. This is certainly a line of intriguing and valuable research.

Research is moving in a direction that can help deliver assessment tasks that answer the call of next generation science education. However, that does not mean that research can be put directly into practice without further considerations. As we enhance next generation science assessments, we also need to consider a host of challenges science teachers will face while employing assessment in the realities of a dynamic and diverse science classroom. The next section outlines three guiding principles—*assessment coherence, assessment use, and assessment equity*—informed by research on educational assessment, science teaching, and science learning that can inform the translation of science assessment research into practice.

GUIDING CLASSROOM ASSESSMENT PRINCIPLES

Assessment Coherence

The first guiding principle, assessment coherence, can help us examine whether assessments at the classroom level are in fact aligned with what and how science is being taught. According to *Knowing What Students Know: The Science and Design of Educational Assessment* (Pellegrino, Chudowsky, & Glaser, 2001), while the purpose of assessment can vary from measuring achievement to evaluating programs to assisting learning, any assessment involves three core features, cognition, observation, and interpretation, which should be designed as a coordinated whole. The coherence among these core features is represented as the assessment triangle model. In the model, the cognition vertex refers to theories of how students learn and represent knowledge. Ultimately, the assessor makes observations of what students know via some assessment task. Thus, the task should not only align with the topic of interest (and in classrooms the learning objectives), but should also draw on theoretically underpinnings of learning—including what *is important* to learn. I have thoroughly described what science education documents and frameworks have deemed important to learn, so the question is whether or not the teacher's learning objectives align with science education tenets and to what extent the assessment tasks allow teachers to observe competencies, such as argument from evidence, that are aligned with their learning objectives. Furthermore, given the complexities of interpreting scientific practices and core ideas as opposed to discrete factual recall, criteria should be clearly laid out that they also align with the task and the theoretical underpinnings.

Assessment Use

The second guiding principle, assessment use, can help us examine the function of assessment in classroom instruction. As already described, science education emphasizes that assessment be used for formative, and not just evaluative purposes. When using assessment formatively, teachers can engage in informal or formal assessment practices. Informal formative assessment involves eliciting, recognizing, and using information and relates to reasoning "on the fly" as opposed to planned reasoning. For instance, while teaching about energy conservation, the teacher may "elicit" student understanding by having them show a thumbs up if they feel they understand the concept and a thumbs down if they do not. The teacher "recognizes" the thumbs-up, thumbs-down signal, and uses the information to decide how to proceed in the lesson. Formal formative assessment involves gathering information, interpreting information, and acting upon information. For instance, a teacher may construct a physics problem to gather information about students' progress in solving energy conservation problems, use set criteria to interpret student responses, and use the information to provide students with additional instructional support, if needed (Ruiz-Primo & Furtak, 2007). Although both informal and formal practices are valuable, I focus on the latter when discussing classroom challenges given the nature of the assessment I will be discussing in a later section.

Another important aspect of formative assessment is to embed formal assessments into instruction. The model of a cycle is helpful to represent the relationship of assessment to instruction (Bell & Cowie, 2001; McMillan, 2007). Conceptualizing assessment as a cycle aligns with the view that its role is to support learning, since learning does not culminate with the assessment, but rather happens continuously. Assessment used in this way serves as a road map to guide both teachers and students where to go, thus formative assessment informs both teaching and learning (Gipps, 1994). A formative assessment cycle consists of the following steps: (a) deciding what students should know, (b) finding out what they do know, and (c) using assessment information to help student make progress toward the learning goal.

Assessment Equity

The final guiding principles, assessment equity, can help us look at issues of fairness and access to science learning. Given that much of the literature has focuses on these issues through the lens of language and assessment, it is helpful to focus on equity for second language learners, referred to as English Learners (ELs) in the United States. Many factors, including students' language proficiency and culture influence student performance on assessment (Abedi & Lord, 2001; Solano-Flores & Nelson-Barber, 2001). Even attitudes during the assessment and related tasks influence the student, affecting future learning (Bandura, 1997). Teachers can take steps to modify the language of assessment, ideally without compromising the rigor of the content (Siegel, 2007). The latter point about rigor is particularly important in terms of equity so that ELs' access to the next generation of science education is not hindered by a perception that they will be unable to engage in complex practices or literacy tasks, such as argument from evidence. Furthermore, scholars argue that engaging ELs in scientific discourse and inquiry can in fact promote science learning and help them develop English proficiency (Stoddart, Pinal, Latzke, & Canaday, 2002).

CHALLENGES TO PUTTING SCIENCE ASSESSMENT RESEARCH INTO CLASSROOM PRACTICE

Needless to say, it takes considerable effort to design assessments that align with and provide meaningful information about students' scientific literacy, let alone use such inform to further enhance students' scientific literacy and ensure that the assessment addressed equity for diverse learners. Yet, this is just what researchers, teacher educators, explicitly or implicitly, are calling upon science teachers to do. The remainder of this chapter focuses on a particular research study that illuminates the challenges of putting assessment into practice. Thus, it is the experience, and not the outcomes, that I hope can stimulate conversation around science classroom assessment.

Research Context

The context is a pilot study I participated in as part of the Assessing Scientific Inquiry and Leadership Skills (AScILS) Project—a project developed to study high school and undergraduate programs that promote entry into and success in biomedical research careers, particularly by underrepresented minority students (Chemers, 2006). While testing a conceptual model of science support program effects, the team had developed a set of paper and pencil simulations to measure scientific reasoning skills, broadly defined, with a future goal to "evaluate education needs and to develop skills" in science classrooms (Chemers, 2003). As an initial step toward this goal, AScILS project members invited me to pilot the simulations in my high school Advanced Placement (AP) Biology classroom, which I was teaching while beginning a doctorate in science education. The students in the United States typically take AP Biology after a first year of high school Biology. Students have the option at the end of the school year to take the AP Biology examination and earn college credit if they pass the examination.

Although several simulations were developed by the AScILS team, it was decided that I would implement the "Malaria Epidemic Simulation," or MES. The MES is centered on the following problem: identify which of four geographic zones is most prone to an outbreak of malaria twenty years into the future. To solve this problem, individuals read background information about malaria, the importance of its containment, and how malaria spreads, including four factors that increase the risk of malaria epidemics—climate, urbanization, agricultural development, and population sensitivity. For the first three risk factors, numerical data are provided in tabular form for each geographical zone from 1950 to 2000. Empirical data are purposely lacking for population sensitivity to provide students the opportunity to note such an omission as a limitation as well as make reasonable inferences about its potential impact.

For each risk factor, individuals write (a) what information they used to arrive at the conclusion, (b) their explanation (how the information led to their conclusion), (c) assumptions (what they are taking for granted), and (d) limitations (additional information they would like to have). Students then choose which zone is most at risk of a malaria epidemic in the period from 2000-2020 and write a final conclusion.

The main purpose of piloting the MES was to document the extent to which the assessment could be productively used in the classroom—meaning that it elicited desirable aspects of students' scientific reasoning, could be scored by a teacher, and could be used to inform the teacher about how to support students' scientific reasoning. By exploring how the assessment worked in a real classroom, the team hoped to improve upon the simulation design and use in classroom contexts, prior to implementing larger scale pilot and efficacy studies. For the purposes of this chapter, the situation provides a unique look into the translation process from research to practice, especially since I was the one primarily responsible for making the translation and documenting the challenges associated with it. My goal is to highlight the challenges encountered during this translation process, which can stimulate conversation around how to overcome challenges, and research that can provide empirical support for practices to overcome the challenges.

Research Method

Sources from multiple perspectives were used to document the challenges of implementing the MES. In addition to using student performance on the MES, the teacher and university researchers collaborated to design questionnaires—with both Likert-scale items and open-ended items—that elicited the students' attitudes toward the MES and related tasks. Questionnaires were not designed to measure particular constructs; therefore, we performed no reliability analysis. Instead, they served to provide broad feedback to the teacher and university researchers that could be used to shed light on experience from the student perspective. Throughout the implementation of the MES and related tasks, I kept written field notes to document (a) what happened, (b) what worked well, and (c) any challenges faced. Conversations between me and other project team members were recorded and became data itself as I reflected on my experience and the university researchers learned about the experience. This dialogue yielded a more organic and dynamic reflection than would be accomplished through a semi-structured interview.

Analysis occurred through several stages. First, the project team analyzed student information (MES scores and questionnaire responses). I originally scored student responses, during the instructional unit; then another project team member scored each response at a time after the instructional unit. The team then analyzed student responses and Likert-scale questionnaire items descriptively and identified qualitative patterns to three open-response items: what students learned, enjoyed, and disliked about the unit as a whole. Next, the team reviewed and discussed field notes, identified teaching decisions and justifications in the field notes, and categorized decisions/justifications based upon Shaw's (2005) 3-D framework, which relates to the process of classroom assessment. In the 3-D framework, *Design* refers to justifications for the design of the individual simulation and subsequent tasks. *Delivery* refers to justifications related to the delivery of the individual simulation and subsequent tasks. Finally, *Decisions* refers to how the evidence from the individual simulation and subsequent tasks was interpreted (scored), reported, and otherwise used to guide or revise instruction. This framework served as an organizational tool for the teacher and university researchers to interpret the challenges faced throughout each stage of using the MES. Finally, drawing on my field notes, student data, and conservations with the university researchers, I wrote several iterations of research memos (Corbin & Strauss, 1990) that were reviewed with university researchers, who wrote responses to the research memos. The dialogue, both oral and written, between parties became essential for the study given the covert nature of the assessment process (Mavrommatis, 1997), which the university researchers may not have even observed if they were present in the classroom. Finally, research memos were translated into written narratives that captured the challenges faces, reported next according to the three assessment dimensions: coherence, use, and equity.

Assessment Coherence Challenges

The project team first went through extensive measures to design a task that matched the construct of interest and could interpret the construct. Consistent with the guiding principle of coherence, an important first step was deciding *what* should be assessed. The AScILS research team chose scientific reasoning as a focal construct, because it (a) was identified as a skill addressed in all the science support programs studied by the project and (b) is an important aspect of scientific literacy, according to science education documents. Scientific reasoning in the simulation related mostly to the practice of engaging in argument from evidence. Along with the simulations, the team developed rubrics based on three dimensions of scientific reasoning—use of data, clarity of the argument, and assumptions/limitations (see Table 1). Research team members scored responses in prior studies through several iterations, refining the rubrics as needed.

Although the simulations drew on literature and went through an extensive design process, I had to decide whether or not one of the designed simulations was aligned with the curricular context of my class. AP Biology is intended to prepare students for the AP examination by following a curriculum established by The College Board®. If the simulation—both in terms of scientific practices and content—bore no relevance to the

Table 1. Criteria for assessed dimensions of scientific reasoning

	1 Novice	2 Apprentice	3 Proficient	4 Advanced
Use of data	Makes little use of the available data; lacks detail and/or accuracy	Makes use of some data	Makes use of key data available in an accurate manner	Makes use of nearly all data available in a detailed and accurate way
Clarity	Linkage of evidence to claim is largely absent and/or confusing	Response shows some linkage of evidence to claim	Shows clear and coherent linkage of evidence to claim	Shows detailed and comprehensive linkage of evidence to claim
	1 Novice	2 Proficient	3 Advanced	
Assumptions/ Limitations	Irrelevant, inappropriate, inaccurate, ambiguous, circular and/or re-statements/affirmations of data provided in the text	Relevant, appropriate, and refer to actual omissions/ inconsistencies in the text	Relevant, appropriate, and refer to actual omissions/ inconsistencies in the text in a sophisticated manner and/or consider alternative explanations	

Note. A zero was given for blank or unintelligible responses.

curriculum or learning objectives of my class, the simulation would not be appropriate. Not too surprisingly, the context of the MES (epidemic spread of malaria) did relate to such AP Biology topics such as evolutionary biology, ecology, and diversity of organisms.

Science teachers putting externally developed assessment tasks into practice should consider the alignment of the task with what they wanted students to learn—i.e., the learning objectives. The learning objectives may be teacher developed, but most likely determined through established standards or frameworks, as previously described. I had taught my students about aspects of scientific reasoning and assessed them primarily through written laboratory reports and essay exams. Yet, I did not use the specific competencies—use of data, clarity, and assumptions/limitations—that guided the MES. Thus, students were not given the opportunity to learn how they would assessed. Although this would be a major obstacle for using the MES to evaluate what students had learned in my courses, it would not be an obstacle if I used the MES to guide, instead of merely evaluative, learning. I will discuss how I *used* the MES in the next sub-section.

Another challenge was alignment with instruction. The project team had designed the MES drawing on scientific reasoning literature, particular around argumentation and explanation. Literature demonstrates how scientific reasoning is intimately connected with subject matter when learning science (Sandoval & Millwood, 2005; Zohar & Nemet, 2002), once again consistent with frameworks that integrate content and practices. My students had been learning AP Biology through a routine set of classroom procedures and tasks. According to a survey given to students after the simulation, the students, on average, reported that the MES only was between a little and somewhat similar to other tasks done in class. This suggests students would have to navigate the cognitive and language demands of the task (e.g., its structure and expectations) in addition to demonstrating what they know—information that could confirm that the MES alone was not aligned with the instructional context. As will be later described, I supplemented the MES with additional activities better aligned with how students learned science throughout the year.

Once I was confident the task did in fact match the curricular context and could be supplemented to match the instructional context, I had to consider the interpretation of the task. I relied heavily on the project team's assessment design and prior studies to decide whether the tasks did in fact match the construct of interest—scientific reasoning. Moreover, since I had never explicitly assessed my students' scientific reasoning before, I decided not to make any changes to the dimensions being scored (use of data, clarity, assumptions/ limitations), nor the criteria on the rubric itself. Of importance here is that the MES was not meant to assess conceptual understanding, nor was that how I wanted to use the MES. In fact, the MES was designed so that students were provided with all of the background information that they needed in order to complete the task, consistent with the notion that scientific practices occur within the *context* of science content.

In order to make sense of the scoring rubric, I received an abbreviated training (around two hours) from the second author, who led the development of the rubric used to score the simulations. As is usually the case in the classroom, I was the only person who initially scored the assessments. Although this might decrease the reliability of the scores, it takes a concerted effort to collaborate with other teachers or university researchers to score responses in a timely fashion. As it turned out, with very little training on how to use the rubrics, I was able to score students' responses quickly (within a week) and consistently with a more experienced scorer from the project team ($r = .80$, using pearson's correlation).

Assessment Use Challenges

Coherence, Use, and even Equity principles coalesced when deciding exactly *how* I would implement the MES. For one, because students had not been exposed directly to the content being assessed nor the format of simulation, I decided to use the MES diagnostically at the beginning of a unit rather than at the end. This way, the MES could serve as an instructional tool to help students refine their scientific reasoning skills. Using it for- matively generated new challenges—namely how to incorporate the MES into an instructional unit.

I collaborated with the AScILS team to plan an instructional unit divided into 4 instructional periods of various lengths (due to the school's bell schedule), occurring over 3 different instructional days. During the first instructional period (90 minutes), students would individually complete the MES to diagnose students' initial scientific reasoning, thus providing a baseline from which to refine the skills. During the second instructional period (90 minutes) students collaborated in previously established groups of four to (a) discuss each risk factor and understand the reasoning behind each student's written response, and (b) prepare a 7-10 minute presentation on an assigned risk factor. Students had engaged in these participant structures throughout the year—once again addressing how students learn throughout the year. The third instructional day was split into two separate periods. During this day, I led a lecture on the elements of scientific reasoning being assessed on the MES, drawing students' attention to the rubric criteria. Thus, a *coherence* aspect (establishing evaluative criteria) allowed me to *use* the assessment for instructional support. Throughout the lecture, I displayed actual student responses and the class discussed the quality of the responses as operationally defined by the rubrics. Following the lecture, students engaged in a routine assignment called "test corrections" in which students made revisions to their original MES responses based on the score from the researcher teacher written on a score sheet with the evaluative criteria. I encouraged students to work collaboratively on this assignment, as done throughout the year. Through test corrections, I aligned assessment elements with the instructional context and was able to use the assessment for instructional support—specifically a form of self-assessment. Even though I was able to provide feedback in the form of scores, I was not able to give more detailed feedback to students about their scientific reasoning. This was only one class and the reality of scoring and providing feedback to four or five classes seems unimaginable. However, I utilized several self-assessment opportunities that curtailed this challenge. By putting groups of students together to discuss the simulation, they heard peer responses that could be compared to his or her own response. Moreover, the test corrections assignment gave students an additional means to self-assess their original response, use the scoring sheet, and revise their response. Despite these successes, students would benefit from detailed feedback from someone with more expertise in the domain than their peers. Perhaps this is where university researchers could assist, in scoring and providing feedback that takes the burden off of the teacher who could focus on how to use the information to inform teaching and learning.

I was presented with challenges associated with assessment use even while interpreting student responses. As mentioned earlier, I was able to provide reliable scores to my students, but as written in a research memo "[t]he most critical aspect of scoring the assessments...involved trustworthiness. As a teacher, assigning a '3' must not be arbitrary, but mean something" (research memo, dated 8/19/2008). By understanding what it means to be proficient at "using data" or having "clear conclusions" the students could better understand not just how to go from a "3" to a "4," but what it takes to refine their scientific reasoning. During the AScILS project, the university researchers had experimented with both a holistic (one score) and a dimensional (multiple sub-scores) scoring system. With this pilot, we all saw the instructional value of dimensional approach, since it provided more nuanced feedback to myself and the students via the scored rubric.

Although I implemented tasks, anchored around the MES, to support student refinement of scientific reasoning, I did not have opportunities to use the assessment information to interpret patterns in students' reasoning—at least not until the summer—and modify instruction to provide individualized support. The challenge of providing opportunities to learn was stifled by time constraints, but also how I could modify teaching to support students since I had little experience in the kind of supports students might need if they do not demonstrate proficient reasoning. This was perhaps the most difficult challenge and a place where technology could provide tremendous help. If the simulation was completed through a computer program, then tasks could be structured in a way where students received instant feedback and provide me with instant information, which I can then use in a more efficient and timely manner.

Assessment Equity Challenges

Not only was it challenging to adapt assessment to the curricular and instructional context, it was challenging to adapt assessment to *who* is in the class. I needed to carefully consider both the MES and related tasks to (a) ensure the students had opportunities to demonstrate what they know and (b) benefited from the assessment. After reviewing the structure of the simulation, I decided not to make any modifications because I believed the language and vocabulary used was appropriate for his class, and felt that the tasks could be accomplished reasonably well by my students. These decisions were based on my experience working with this individual class as well as previous AP Biology classes and supported ad-hoc by student responses to the questionnaires. 17 out of 23 students thought that the instructions for the MES were very clear and 22 out of 23 thought the language was very understandable. However, every class is different and language, from the instructions to the rubric, might need to be modified depending on students' language proficiency and reading level. Indeed one drawback to the study was that it did not occur within a linguistically diverse classroom, an important next step.

Issues of language are not restricted to vocabulary and grammar. As mentioned in relation to performance assessment, language also relates to how students participate and communicate—in a small group versus individually? Presenting or interpreting information? My students enjoyed collaborating with others during the group discussion, a finding supported by my students' perception that collaboration in science is important and that they enjoyed working on the MES as a group. As a teacher, I could attend to this feedback by expanding collaborative opportunities, such as letting groups of students discuss a common risk factor (e.g. urbanization) and then instructing groups to present their results to each other. Finally, as a class, we could then discuss the cumulative evidence. One potential benefit of including additional activities like group discussions and presentation is to provide another opportunity for students to demonstrate what they know and build on what they know while learning science. For diverse students, the additional interactions may also open doors for them, but only if appropriate for the particular student context and if properly supported so that they can be successful.

Did the assessment benefit students? The most common items students reported that they learned from the unit related to the actual learning goals of the unit—the process and importance of scientific reasoning, specifically using data, identifying assumptions/limitations, and drawing clear and logical conclusions. In addition, some students indicated that they did not just learn strategies (predicting data, linking evidence to claims, etc.), but also the metacognitive aspects: "I do not know how to use scientific reasoning any better…but now I …have more of a grasp on what I am doing" (Emma, italics original). Some students better understood why you must articulate your reasoning clearly and the value of assumptions/limitations. Moreover, some students learned and appropriated new terms for the skills they already knew: "I learned some new terms like 'interpolation and extrapolation'" (Elise). These findings suggest the value of the MES as an instructional tool to refine scientific reasoning skills and understand the importance of scientific reasoning. Furthermore, the MES and corresponding activities provided students with opportunities to engage in scientific discourse—both in talk and in writing. For linguistically diverse classrooms, these opportunities coupled with attention to language development can be particularly powerful.

Summary of Major Findings

By piloting the Malaria Epidemic Simulation in a high school classroom and collecting student and teacher perspectives about the role of the assessment in the classroom, I along with other members of the study team, identified many of challenges faced while translating a research instrument into a productive science classroom assessment. In depth analysis of technical quality (i.e., validity, reliability, and bias) of science assessments is an important, and arguably necessary, step while designing quality science assessments. However, our analysis was guided theoretically by other considerations that more aptly appeal to the realities of classroom practice—most notably alignment to standards and curricular frameworks, the role of the assessment in support student learning, and equity considerations. The Malaria Epidemic Simulation was embedded within a curricular unit on scientific reasoning and used to diagnose student conceptions around as well as anchor instructions about scientific reasoning. The students reported that they understood the directions of the assessment, learned from it, and even enjoyed it. Yet, the study's relevance to this chapter lies in the description of challenges while translating the MES into a productive assessment, which can inform how research, teachers, and other put research of the next generation of science assessments into classroom practice.

CONCLUDING REMARKS

The specific purpose of this chapter was to stimulate a conversation by addressing the challenges of putting next generation science assessment research into practice. I began by highlighting some of the major reforms and innovations in science assessment over the last twenty years to demonstrate where we are in terms of research. I described three assessment principles that can guide how researchers examine the practice of assessing in science classrooms as well as guide science teachers themselves in terms of how they assess. I used the guiding principles to focus on a research study that exemplifies the challenges faced when putting science assessment research, centered on scientific reasoning, into classroom practice.

In science education, we are in position to advance our understanding about what happens when the next generation of science assessments are put into practice. We have developed a range of tasks that can elicit students' scientific literacy, consistent with seminal science education documents. We have the potential to integrate technology and draw on emerging theories of how students learn science. Science education researchers has also advanced how assessment, even complex ones, can provide feedback to students and address the role of language in assessment. However, the description of the AScILS pilot study exemplifies the many challenges that still need to be overcome when putting research in the realities of science classrooms. This is important given that most advances in science assessment will occur outside of the classroom; however, the classroom is where many of the assessments constructed would ideally like to be enacted.

The study described in this chapter took an instrument originally designed for research purposes and translated into a science classroom assessment. The experience illuminated many of the differences between a classroom context and an out of classroom context, summarized in Table 2. For one, teachers and science education researchers need to consider the assessment's purpose, which alters the role of the teacher and students as well as what is done with the assessment information. Since the university researchers were only using the MES for evaluative purposes, they did not have to consider feedback or how the assessment fit into the instructional context. Decisions to use the particular assessment overlapped, but also differed. Instead of considering what would be learned in support programs, I had to consider what was important for my class. Researchers and assessment developers may struggle if what they

Table 2. Comparison of issues between AScILS project and my (the teacher's) use

Issue	AScILS Project's Use	Teacher's Use
Primary purpose	Assess scientific reasoning skills in order to test a conceptual model related to the effects of science support program activities.	Assess scientific reasoning skills in order to ground follow-up learning about scientific reasoning.
Decision to use based on...	Analysis that scientific reasoning was addressed by the science support programs being studied.	Support for scientific reasoning in *National Science Education Standards* (NCR, 1996, 2000) and science education literature (e.g., Jiménez-Aleixandre & Erduran, 2007).
	Content thought to be engaging to students across a broad range of disciplines.	Content fit with Collegeboard® AP Biology curriculum outline.
Procedure for implementation	Standardized across programs; at the beginning and end of programs, at a time convenient to the programs.	Reflected curricular needs, teacher's routine pedagogical practice, and school schedule.
Time constraints	Determined by programs.	Determined by school schedule and curricular demands.
Scoring purpose	Efficient, reliable, and valid across hundred of responses for test of the hypotheses.	Efficient and valid for individual student feedback.
Scoring process	Multiple scorers, blind to pre/post condition, scoring taking place months to years after response collected.	Single scorer, with quick turnaround demand for use in instruction.

feel are important differs from what the teacher feels is important. Another important distinction, corresponding with the different assessment functions, was the process of implementation. The university researchers were more interested in standardized, uniform procedures whereas I was more concerned about how the assessment could be used formatively—what was the instructional value. Both university researchers and myself encountered time constraints when implementing the assessment, albeit different kinds of constraints. Finally, the process of scoring differed. I had to score alone, within a week, and my goal was to focus on individual feedback. The university researchers employed multiple scorers, which took place months after the assessment and were more interested in evidence to support their research hypotheses.

I conclude this chapter by displaying a set of pertinent questions when attempting to implement science classrooms assessments. These questions come directly from the challenges identified in the pilot study. Although not an exhaustive list, they can hopefully lead to fruitful conversations as the next generation of science assessments are put into classroom practice in facilitate scientific literacy for all students.

ACKNOWLEDGMENT

This research was supported by Grant Number R01GM071935 from the National Institute of General Medical Sciences. The content is solely the responsibility of the author and does not necessarily represent the official views of the National Institute of General Medical Sciences or the National Institutes of Health. I wish to acknowledge the other AScILS Classroom Pilot project team member, Dr. Jerome Shaw, Dr. Barbara Goza, and Dr. Beth Jaworski for their contribution, as well as my 2007-2008 AP Biology class for their participation in the study.

REFERENCES

American Association for the Advancement of Science. (1993). *Benchmarks for science literacy*. Washington, DC: American Association for the Advancement of Science.

Ausubel, D. P., Novak, J. D., & Hanesian, H. (1978). *Educational psychology: A cognitive view*. New York, NY: Holt Rinehart and Winston.

Bandura, A. (1997). *Self-efficacy: The exercise of control*. New York, NY: Freeman.

Bell, B., & Cowie, B. (2001). *Formative assessment and science education*. Dordrecht, The Netherlands: Kluwer Academic Publishers.

Black, P., & Wiliam, D. (1998). Inside the black box: Raising standards through classroom assessment. *Phi Delta Kappan*, *80*(2).

Bunch, G. C., Shaw, J. M., & Geaney, E. R. (2010). Documenting the language demands of mainstream content-area assessment for English learners: participant structures, communicative modes, and genre in science performance assessments. *Language and Education*, *24*(3), 185–214. doi:10.1080/09500780903518986

Chemers, M. (2006). *Understanding undergraduate and high school science support programs: A theory-driven, interdisciplinary, multi-method approach*. Paper presented at the Annual Meeting of the American Evaluation Association. Portland, OR.

Chemers, M. M. (2003). *Assessing scientific inquiry and leadership skills: Research plan*. Santa Cruz, CA: University of California Santa Cruz.

Chinn, C. A., & Malhotra, B. A. (2002). Epistemologically authentic inquiry in schools: A theoretical framework for evaluating inquiry tasks. *Science Education*, *86*(2), 175–218. doi:10.1002/sce.10001

College Board. (2011). *AP biology curriculum framework*. New York, NY: The College Board.

Gipps, C. V. (1994). *Beyond testing: Towards a theory of educational assessment*. London, UK: Routledge.

Johnson, R. L., Penny, J. A., & Gordon, B. (2009). *Assessing performance: Designing, scoring, and validating performance tasks*. New York, NY: The Guilford Press.

Klassen, S. (2006). Contextual assessment in science education: Background, issues, and policy. *Science Education*, *90*(5), 820. doi:10.1002/sce.20150

Kuerbis, P. J., & Mooney, L. B. (2008). Using assessment design as a model of professional development. In Coffey, J., Douglas, R., & Stearns, C. (Eds.), *Assessing Science Learning: Perspectives from Research and Practice* (pp. 409–426). Arlington, VA: National Science Teachers Association Press.

Lyon, E. G., Bunch, G. C., & Shaw, J. M. (2012). Navigating the language demands of an inquiry-based science performance assessment: Classroom challenges and opportunities for English learners. *Science Education*, *96*(4). doi:10.1002/sce.21008

Martiniello, M. (2008). Language and the performance of English-language learners in math word problems. *Harvard Educational Review*, *78*(2), 333–368.

Mavrommatis, Y. (1997). Understanding assessment in the classroom: Phases of the assessment process—The assessment episode. *Assessment in Education: Principles. Policy & Practice*, *4*(3), 381–400.

McMillan, J. H. (2007). *Formative classroom assessment: Theory into practice*. New York, NY: Columbia University.

McNeill, K. L., Lizotte, D. J., Krajcik, J., & Marx, R. W. (2006). Supporting students' construction of scientific explanations by fading scaffolds in instructional materials. *Journal of the Learning Sciences*, *15*(2), 153–191. doi:10.1207/s15327809jls1502_1

Millar, R., & Osborne, J. (1998). *Beyond 2000: Science education for the future*. London, UK: King's College.

National Center for Education Statistics. (2009). *Science framework for the 2009 national assessments of educational progress*. Washington, DC: National Center for Education Statistics.

National Research Council. (1996). *National science education standards*. Washington, DC: National Academy Press.

National Research Council. (2012). *A framework for K-12 science education: Practices, crosscutting concepts, and core ideas*. Washington, DC: National Academy Press.

Norris, S. P., & Phillips, L. M. (2003). How literacy in its fundamental sense is central to scientific literacy. *Science Education*, *87*(2), 224–240. doi:10.1002/sce.10066

Novak, J. D. (1990). Concept mapping: A useful tool for science education. *Journal of Research in Science Teaching*, *27*(10), 937–949. doi:10.1002/tea.3660271003

OECD. (2009). *Pisa 2009 assessment framework – Key competencies in reading, mathematics, and science*. Washington, DC: OECD.

Osborne, J., & Dillon, J. (2008). *Science education in Europe: Critical reflections*. London, UK: Nuffield Foundation.

Pellegrino, J. W., Chudowsky, N., & Glaser, R. (2001). *Knowing what students know: The science and design of educational assessment*. Washington, DC: National Academy Press.

Quellmalz, E. S., & Pellegrino, J. W. (2009). Technology and testing. *Science, 323*(5910), 75. doi:10.1126/science.1168046

Ruiz-Primo, M. A., & Furtak, E. M. (2007). Exploring teachers' informal formative assessment practices and students' understanding in the context of scientific inquiry. *Journal of Research in Science Teaching, 44*(1). doi:10.1002/tea.20163

Sandoval, W. A., & Millwood, K. A. (2005). The quality of students' use of evidence in written scientific explanations. *Cognition and Instruction, 23*(1), 23–55. doi:10.1207/s1532690xci2301_2

Shaftel, J., Belton-Kocher, E., Glasnapp, D., & Poggio, J. (2006). The impact of language characteristics in mathematics test items on the performance of English language learners and students with disabilities. *Educational Assessment, 11*(2), 105. doi:10.1207/s15326977ea1102_2

Shavelson, R. J., Baxter, G. P., & Pine, J. (1991). Performance assessment in science. *Applied Measurement in Education, 4*(4), 347–362. doi:10.1207/s15324818ame0404_7

Shavelson, R. J., Young, D. B., Ayala, C. C., Brandon, P. R., Furtak, E. M., & Ruiz-Primo, M. A. (2008). On the impact of curriculum-embedded formative assessment on learning: A collaboration between curriculum and assessment developers. *Applied Measurement in Education, 21*(4), 295–314. doi:10.1080/08957340802347647

Shaw, J. (2005). Getting things right at the classroom level. In Herman, J. L. H. (Ed.), *Uses and Misuses of Data for Educational Accountability and Improvement: The One Hundred Fourth Yearbook of the National Academy Society for the Study of Education, Part 2* (pp. 340–357). Chicago, IL: National Society for the Study of Education. doi:10.1111/j.1744-7984.2005.00036.x

Shaw, J. M., Bunch, G. C., & Geaney, E. R. (2010). Analyzing language demands facing English learners on science performance assessments: The Sald framework. *Journal of Research in Science Teaching, 47*(8), 909–928.

Shepard, L. A. (2000). The role of assessment in a learning culture. *Educational Researcher, 29*(7), 4–14.

Siegel, M. A. (2007). Striving for equitable classroom assessments for linguistic minorities: Strategies for and effects of revising life science items. *Journal of Research in Science Teaching, 44*(6). doi:10.1002/tea.20176

Solano-Flores, G., & Nelson-Barber, S. (2001). On the cultural validity of science assessments. *Journal of Research in Science Teaching, 38*(5), 553–573. doi:10.1002/tea.1018

Stoddart, T., Pinal, A., Latzke, M., & Canaday, D. (2002). Integrating inquiry science and language development for English language learners. *Journal of Research in Science Teaching, 39*(8), 664–687. doi:10.1002/tea.10040

White, B. Y., & Frederiksen, J. R. (1998). Inquiry, modeling, and metacognition: Making science accessible to all students. *Cognition and Instruction, 16*(1), 3–118. doi:10.1207/s1532690xci1601_2

Wiggins, G. (1989). A true test: Toward more authentic and equitable assessment. *Phi Delta Kappan, 70*(9), 703–713.

Wiliam, D., & Leahy, S. (2007). A theoretical foundation for formative assessment. In McMillan, J. H. (Ed.), *Formative Classroom Assessment* (pp. 29–42). New York, NY: Teachers College Press.

Zohar, A., & Nemet, F. (2002). Fostering students' knowledge and argumentation skills through dilemmas in human genetics. *Journal of Research in Science Teaching, 39*(1), 35–62. doi:10.1002/tea.10008

APPENDIX

Questions for Translating Science Assessment Research into Classroom Practice

Principle	Question for Classroom Practice
Assessment coherence	• Is the assessment task aligned with instructional goals? (including standards or other school/national guidelines) • Is the assessment task aligned to the ways in which students are taught or will be taught science? • How can student work be efficiently (and reliably) interpreted? Are the criteria used to interpret aligned to the instructional goals?
Assessment Use	• How will the assessment be embedded into instruction? • How will feedback be provided in a timely and effective manner? • How will decision be made about next steps based upon assessment information?
Assessment Equity	• What is the student context? How can the assessment be responsive to the diverse needs of students? • What do students have to do with language in the assessment? How can students be supported so they successfully navigate the language demands of assessment and still be exposed to challenging literacy tasks in science?

Chapter 14
A Tool for Analyzing Science Standards and Curricula for 21st Century Science Education

Danielle E. Dani
Ohio University, USA

Sara Salloum
Long Island University, USA

Rola Khishfe
American University of Beirut, Lebanon

Saouma BouJaoude
American University of Beirut, Lebanon

ABSTRACT

Twentieth century curricula are no longer sufficient to prepare students for life and work in today's diverse, fast-paced, technologically driven, and media saturated world of the 21st century. This chapter presents a new framework for analyzing science standards and curricula to determine the extent of alignment with 21st Century essential understandings and skills. The Tool for Analyzing Science Standards and Curricula (TASSC) was developed using the conceptual frameworks proposed by the Partnership for 21st Century Skills, the Organization for Economic Co-Operation and Development, and the typology of knowledge proposed by Jurgen Habermas. Development of TASSC relied on an iterative process of refinement, testing, and discussions resulting in an instrument with three sections and related rating scales: content, skills, and additional curricular components. TASSC was piloted using middle school science standards and curricula in the context of two US states (Ohio and New York) and two Arab countries (Lebanon and Qatar). The analysis procedure and individual case study results are presented and discussed in the chapter.

DOI: 10.4018/978-1-4666-2809-0.ch014

INTRODUCTION

For more than twenty years, reforms in science education have been calling for the preparation of scientifically literate students. However, standards and curricula promoting 20th Century conceptualizations of scientific literacy (see American Association for the Advancement of Science, 1993; Bybee, 1997; National Research Council, 1996) may not be sufficient to prepare students for the global economy of the 21st century. Today's students live in a world that is extremely fast-paced, constantly changing, increasingly culturally diverse, technologically driven, and media saturated (Wan & Gut, 2011). The kinds of skills students need to be prepared for of the 21st century are different from those needed 20 years ago (P21 Skills, 2009). To prepare for this *"second Renaissance period"* (Treadwell, 2011), we need to bring what we teach and how we teach into the 21st Century. Essential understandings and skills that are necessary for success in today's world must be included.

As Nations and States around the world revise existing curricula or develop new standards, a framework and instrument for analyzing science standards and curricula and determining the extent of alignment with 21st Century essential understandings and skills is necessary. BouJaoude (2010) created such a framework for the analysis of education programs (FAEP, see Appendix 1). In preparation for a National Association for the Research in Science Teaching symposium in 2011, the authors of this chapter decided to use FAEP to analyze and evaluate the standards and curricula of US states and several countries from around the world. After determining that FAEP was not adequate for analyzing science standards and curricula, the authors decided to develop a new framework. This chapter describes the development of this new framework entitled Tool for Analyzing Science Standards and Curricula (TASSC). The chapter begins with a discussion of the conceptual basis that supports the call for 21st Century essential understandings and skills. The chapter ends with a presentation of case studies using TASSC in multiple contexts.

CONCEPTUAL BASIS

The essential understandings and skills necessary for a college and career ready populace have been posited by several organizations. For example, the Partnership for 21st Century Skills (P21) proposed the Framework for 21st Century Learning. Similarly, the Organization for Economic Cooperation and Development (OECD[1]) proposed a set of competencies needed for a successful life and well-functioning 21st Century society. BouJaoude (2010) additionally used Habermas' types of knowledge (1971; as cited by Cranton, 2002) as he developed FAEP. These conceptual frameworks are described in this section.

Framework for 21st Century Learning

The partnership for 21st century skills (2009) developed a framework for 21st century learning delineating a set of learning outcomes that specify the knowledge, skills, expertise, and literacies needed for success in work and life. In addition to the core subjects of science, the framework identifies a set of 21st century interdisciplinary themes that, when integrated into core subjects, will result in much higher levels of learning. The themes consist of global awareness, financial, economic, business, and entrepreneurial literacy, civic literacy, health literacy, and environmental literacy.

The framework divides the necessary skills into three types, (a) learning and innovation, (b) information, media, and technology literacies, and (c) life and career skills. Learning and Innovation skills include the four C's: creativity and innovation, critical thinking and problem solving, communication, and collaboration. Life and career skills include flexibility and adaptability, initiative and self-direction, social and cross-cultural skills, productivity and accountability, and leadership and responsibility. The Partnership for 21st Century Skills further defines these outcomes in their P21 Framework Definitions (2009).

OECD's Key Competencies for the 21st Century

OECD (2005) proposed three broad categories of key competencies: Using tools interactively, interacting in heterogeneous groups, and acting autonomously. Today's world is reliant on a variety of tools ranging from physical ones, such as information technology, to socio-cultural ones such as the use of language. For this reason, individuals need to be able to use this wide variety of tools for interacting effectively with the environment and they need to understand them well enough to adapt them for their own purposes. Because today's world is also increasingly interdependent, individuals need to be autonomously able to engage with others of different backgrounds. Finally, individuals in today's world must be able to take responsibility for managing their own lives and situating their lives in the broader social context and act autonomously.

Reflective skills are central to OECD's competency framework. They involve the ability to deal with change effectively, learn from experience, and take a critical stance toward issues. Such a critical stance will only occur in an appropriate learning environment where individuals are not satisfied with applying what they learn blindly, but rather learn to be dynamic and responsive actors in society.

Habermas' Types of Knowledge

Scholars such as Merizow (1997) and Cranton (1997, 2001, 2002) have utilized Habermas' (1971) conceptualizations of human interests and ensuing knowledge to outline knowledge types within the context of adult learning. Habermas (1971; as cited by Cranton, 2001, 2002) proposes three types of knowledge: Instrumental, communicative, and emancipatory. According to Habermas, knowledge types are acquired in response to three basic human interests. A technical interest to explain and manipulate our environment leads to *instrumental knowledge*. Instrumental knowledge is objective and empirically derived allowing for prediction and control of the environment and to an understanding of how the world works (Cranton, 1997). Our practical interest to live within social settings leads us to acquire *communicative knowledge*[2]. Such knowledge is acquired through language and involves knowledge of societal and cultural norms (at the micro, macro, and global levels). Validity of communicative knowledge is determined by "consensus within a group and a sense of rightness or morality" (Cranton, 2001, p. 12); for example, agreed-upon norms and knowledge in one culture may not hold in another culture. Finally, our emancipatory interests lead to *emancipatory knowledge*. Emancipatory knowledge is motivated by an interest in growth, self-awareness, and autonomy. In addition, emancipatory knowledge springs from an interest to free ourselves from the oppression of unquestioned social norms, values, and beliefs (Cranton, 2001). It is developed through critical reflection be it on self or the society.

Habermas' three types of knowledge are deemed important in our global age since:

The essential learning required to prepare a productive and responsible worker for the twenty-first century must empower the individual to think as an autonomous agent in a collaborative context rather than to uncritically act on the received ideas and judgments of others. Workers will have to become autonomous, socially responsible thinkers (Merizow, 1997, p. 8).

A similar argument can be made for teaching and learning at the secondary school level. In order to be compatible with a highly technological and global age, students need to develop instrumental and communicative knowledge. Concurrently, they need to develop emancipatory forms of knowledge that would allow them to become autonomous independent thinkers who critically reflect on their own thinking, values and beliefs and those of the surrounding culture.

TOOL FOR ANALYZING SCIENCE STANDARDS AND CURRICULA (TASSC)

The Tool for Analyzing Science Standards and Curricula (TASSC, see Appendix 2) was developed using an iterative process of refinement, testing, and discussions over a period of two years. TASSC consists of 13 items distributed over the three sections of content, skills, and additional curricular components. Each section and its individual rating scale will be described below. The process began with the development of a framework to analyze programs in terms of 21st century skills (BouJaoude, 2010). The various steps in the development of TASSC are described in the following sections.

A Framework for Analyzing Education Programs

Saouma BouJaoude (2010) developed and used the Framework for Analyzing Education Programs (FAEP, see Appendix 1) to analyze the secondary educational systems in England, Germany, Finland, Japan, and the International Baccalaureate in order to derive lessons that might be helpful in improving secondary education in the Arab States. BouJaoude's (2010) framework is based on an alignment and critical examination of the proposals of the Partnership for 21st Century Skills and the OECD using Habermas' types of knowledge (described above) and Prensky's (2001) discussion of student thinking and ways of processing knowledge. According to Prensky, digital "Natives" process knowledge differently from "Immigrants." "Natives" are used to receiving information fast, like to parallel process and multi-task, prefer their graphics before their text rather than the opposite, and function best when networked. Conversely, "Immigrants" have learned—and so choose to teach—slowly, step-by-step, one thing at a time, and individually.

BouJaoude's (2010) analysis suggests that advocates of the recommendations of P21 emphasize "Instrumental Knowledge" with some emphasis on "Communicative Knowledge." Prensky's recommendations, on the other hand, seem to focus more on student characteristics and methods of acquiring knowledge and less on the types of knowledge needed by all learners. The competencies advocated by OECD, however, move more closely to what Habermas advocates but with a preference for the communicative and the emancipatory knowledge. BouJaoude concludes that it might be productive to adopt an integrative approach to determine the knowledge, skills, and dispositions needed for students to live and work in the twenty-first century. This integrative approach would not neglect any of the three types of knowledge but rather integrates the competencies advocated by different groups. Moreover, such an approach focuses on preparing knowledgeable, autonomous, and reflective critical thinkers who (a) are capable of using information to solve academic and everyday life problems, (b) use technology meaningfully, (c) interact well with others in work and other social settings, (d) reflect on their actions and knowledge to improve the quality of their lives, and (e) construct knowledge through research. This integrative approach rooted the development of FAEP and was subsequently adopted in the making of TASSC.

Development of the TASSC

As mentioned previously, TASSC was developed after it was determined that the FAEP was not adequate for analyzing science standards and curricula. The development of TASSC began with revisiting the conceptual base used to develop FAEP. In a first step, the framework for 21st Century Skills and the OECD key competencies were respectively aligned with Habermas' types of knowledge, and compared to determine areas of similarities and difference. Only areas of similarities that aligned with Habermas' instrumental, communicative, and emancipatory types of knowledge were used in TASSC. For example, a change that was made to reflect Habermas' work was separating the FAEP item "developing information and communication skills" into two distinct items.

Next, FAEP was revisited in light of the results of the analysis performed in step 1. Four main changes were made:

- Distinguishing between knowledge and skills, which are subsumed under Core Academic Content by creating separate sections to address each in the instrument,
- Creating a third section of the instrument to include additional curricular components such as teaching and learning activities, assessment, and overall educational systems,
- Further defining of the Core Academic Content as content knowledge and 21st century themes, and
- Including creativity and innovation skills in the skills section of the instrument.

In the Core Academic Content section, content knowledge consists of the unifying concepts and processes of science, physical science, life science, earth and space science, understandings about scientific inquiry, understandings about science and technology, science in the personal and social perspectives, and the history and nature of science. Twenty-first Century Themes include environmental literacy, global awareness, economic literacy, civic literacy, and health literacy. Conversely, Skills consist of information and communication, creativity and innovation, thinking and problem solving, reflective skills, autonomy, and self-direction.

Then, the rating scales were revisited and further defined. The skills rating scale used in the NRC (2010) report about the intersection of science education and 21st century skills was adopted because it assessed the degree to which a skill is present, complete, explicit, and/or optional. The Skills Rating Scale consists of the following levels:

- **4: Strong whole skill:** The skill is found almost in its entirety in the curriculum in a strong form likely to produce high levels of performance if the curricular goals are met.
- **3: Weak whole skill:** The skill is found almost in its entirety in the curriculum in a weak form, either because it is made optional or described vaguely.
- **2: Strong component skill:** Only one or two components of the larger skill are found in the curriculum, but those elements are met to a high degree.
- **1: Weak component skill:** Only one or two components of the larger skill are found in the curriculum, and even then only a weak form, either because they are made optional or described vaguely or are implicit in the activities of a listed curriculum.
- **0: None:** The skill is completely absent.

The skills rating scale was used as a model for the development of a new content rating scale. The new rating scale was designed to assess the degree to which core academic content and 21st century themes are present, complete, explicit, integrated, and/or optional. The Content Rating Scale consists of the following levels:

- **4: Comprehensive High Level:** The content is found almost in its entirety and explicitly described in curriculum activities in a manner likely to produce high levels of learning if the curricular goals are met.
- **3: Comprehensive Low Level:** The content is found almost in its entirety and is

vaguely described or made optional in curriculum activities.
- **2: Partial High Level:** Only one or two content themes are explicitly described and integrated in curriculum activities.
- **1: Partial Low level:** Only one or two content themes are vaguely described or implicit in the activities of the curriculum.
- **0: None:** The content themes are completely absent.

Finally, the rating for the additional curricular components section consisted of "1" to represent "mostly present" or "0" to represent "mostly absent."

In short, the development of TASSC was cooperatively accomplished by the authors through an iterative process involving the coalescing and separation of categories based on conceptual and empirical considerations. The 13 items of the instrument are partly based on how the group conceptualized the categories and the distinct nature of each. For example, to account for Prensky's (2001) notions of digital natives and immigrants, the item targeting overall educational systems was further defined using the following prompts:

- Students are perceived as active learners.
- Curriculum supports a rich technological environment (availability of rich graphics and interaction with technological tools and media), and
- Curriculum affords utilization of technological tools allowing for parallel processing.

The process also involved using versions of the instrument to analyze segments of curricula and noting which categories yielded useful distinctions and which were redundant. For example, an earlier version of the TASSC combined interpersonal and self-directional skills in one item. Conceptually, the authors considered these skills to be distinct from each other, a notion that was supported by preliminary analyses. As a result, these two skills were separated. Furthermore, self-directional skills were found to overlap with autonomy and were therefore compiled together as one item in the final version to avoid redundancy. The result was an instrument that was detailed enough to give an adequate and clear picture of the strengths and limitations of the standards and core curricula, while at the same time succinct so as to be practical to use.

CASE STUDIES USING TASSC

TASSC was used to analyze the standards and curricula of two US states (Ohio and New York) and two Arab countries (Lebanon and Qatar). Middle school rather than high school was chosen as the focus of the analysis for two reasons. First, middle school science is more unified and it is also required for all students. At the high school level, students have a choice of several science electives; for example, fewer students in urban settings take physical science electing to satisfy science graduation requirements with life and earth science courses. Second, middle school years are instrumental in developing student attitudes and dispositions towards science. This section of the chapter describes how the analysis process was accomplished and then presents the individual case studies.

Analysis Procedure

The standards and curricular analysis process used in the individual case studies was conducted in multiple steps and at different levels. First, standard and curricular documents and websites were first read in their entirety, including introductions, prefaces, and appendices, to discern *content*, *skills*, and *additional curricular component features* as outlined in the framework. Then, the first level of analysis involved color-coding key statements according to the broad categories of Content, Skills and Other. In some instances, cross coding was allowed but it was indicated which category was more prominent. An example of a cross coding instance from the New York City standards is:

- **M1.1a:** Identify independent and dependent variables.
- **M1.1b:** Identify relationships among variables including: direct, indirect, cyclic, constant; identify non-related material.

The above performance indicators require knowledge and understanding of the concepts of dependent and independent variables and therefore were designated as both process skills and content. In this case, Skills was more prominent.

For the second level of analysis, each item in the framework was assigned a code (see Appendix 3). Next, *Content, Skills,* and *Other* statements identified in the level one analysis were coded and separated into categories based on the items of the framework. Frequencies were calculated for each code. No rating was given initially for the first two levels of analyses. Table 1 presents an example of level 2 coding for the following State of Ohio grade 6 Model Curriculum statement:

Make a geological map of the local community. Use existing geologic data, historic geologic data, and field explorations to analyze types of formations that are present. Use the finished map to evaluate possible land and resource uses. Present map and recommendations to an authentic audience.

For the third level of analysis (see Table 1), statements within each level two *Content* and *Skill* coding category were re-examined and separated into subcategories based on the degree to which each statement presented (a) explicit, implicit, integrated, or optional content, or (b) explicit, implicit, optional, and/or vague skill. The subcategories were derived from the TASSC rating scales described previously. Frequencies were calculated for each subcategory. Finally, a rating for each of the TASSC items was holistically determined using the frequencies calculated in the level 2 and level 3 analyses.

The Case of the Ohio Standards and Model Curriculum

The Ohio Department of Education (ODE) began a standard revision process in 2009 in response to then Governor Strickland's (2009) proclamation that preparing students to be informed decision makers in the global economy of the 21st century is an Ohio state imperative. In addition to revising its standards, ODE has developed a Web-based Model Curriculum, of which 21st century skills are an essential component:

Table 1. Analysis levels of curriculum statements

Level 1	Level 2		Level 3
Content	SC		
	CL	Use the finished map to evaluate possible land and resource uses	Implicit, integrated
Skills	INF	Use existing geologic data, historic geologic data and field explorations	Explicit, required
	THP	to analyze types of formations that are present	Explicit, required
	COM	Present map and recommendations to an authentic audience	Explicit, required
Other	21CON	Make a geological map of the local community	
	TR	Use existing geologic data, historic geologic data and field explorations AND Make a geological map of the local community	

Throughout all grades, PreK-12, [21st century] skills are explicitly addressed through scientific inquiry and applications. The integration of skills and content aligns with the Framework for 21st Century Learning. (ODE, 2010)

Structure of the Ohio Middle Level Standards and Model Curriculum

The revised standards for grades 6-8 are organized around Strands and Topics for each grade level. Strands represent the disciplines of life science, physical science, and earth and space science. Topics consist of the main focus for each strand and provide the basis for further defined Content Statements, which delineate the science content to be learned. Strands are connected by Themes that progressively increase in complexity. The integration of strands and topics within each grade are illustrated by Grade Level Connections. Finally, the standards describe the inquiry skills and applications that must be developed by all students during grades 6-8.

The Model Curriculum is a Web-based resource that illustrates how the material in the Content Statements may be taught by providing content elaborations, expectations for learning, classroom examples, and instructional strategies and resources. A noteworthy characteristic of the Model Curriculum consists of the Cognitive Demands that are used to identify expectations for learning. At the most basic level, students are expected to recall accurate science. At the next level, students are expected to interpret and communicate science concepts. At the third level, students are expected to demonstrate science knowledge. At the highest level of cognitive demand, students are expected to design technological and engineering solutions using science concepts.

Results

Results of the TASSC analysis of the Ohio grades 6-8 revised standards and model curriculum are presented in this section for core academic content, skills, and additional curricular components.

Core Academic Content

Core academic content consists of science content and 21st century themes. Results of the analysis process indicate that the science content component of the core academic content is mostly addressed in the description of the content strands, topics, content statements, themes, grade band connections, and the inquiry skills and applications. Strands, topics, and content statements explicitly delineate earth and space science, life science, and physical science essential understandings for each grade level. Furthermore, unifying concepts and processes are given prominent positions in grade band themes and strand connections. For example, the 7th grade band theme is "order and organization," and the strand connections relate to "systems." Similarly, understandings about and abilities to do scientific inquiry are explicitly included and integrated throughout the middle grade standards as evidenced by, among others, the following statements:

- Identify questions that can be answered through scientific investigations; and
- Recognize and analyze alternative explanations and predications.

Other content components, including science in the Personal and Social Perspectives (PSP) and the History and Nature of Science (HNOS), are addressed in the model curriculum rather than the standards. Statements aligned with these components were found in the sections of content elaborations, instructional strategies, and expectations for learning as illustrated by the following examples:

It is important to provide the background knowledge regarding how scientists know about the structure and composition of the interior of Earth (without being able to "see" it). [HNOS implicit, required – 8th grade content elaboration]

In order to fully grasp plate movement, students must investigate Earth's history and the evidence used in determining plate movement...Interpreting paleomagnetic data for the geologic periods demonstrates how scientists have determined where the plates were over time. [HNOS vague, optional – 8th grade instructional resources]

Analyze and evaluate scientific tradeoffs (environmental, projected research required to move from current knowledge to application) for use of microbes to produce alternative energy or clean-up environmental spills. [PSP implicit, required – 6th grade expectations for learning]

Even though the content statements identified in the content elaboration and expectations for learning sections are required, references to the HNOS and PSP content are vague or implicitly made. Moreover, content statements identified in the instructional resources section of the model curriculum seem optional because the section in question is intended as "additional support and information for educators...not intended to be a prescriptive list of lessons."

The 21st century themes of the core academic content are also addressed in the model curriculum. Results of the analysis process indicate that environmental and civic literacy are the most frequently and explicitly integrated themes in the Ohio model curriculum. Some examples of statements targeting these two themes include:

Research cloning in the food industry. Select one practice and determine whether or not it is an environmentally healthy practice. Justify your position with scientific evidence. [EnvL explicit, integrated – 8th grade expectations for learning]

Develop, test, and evaluate plans outlining a specific method to reduce storm water flow at a specific site in the local community (for example a housing construction project or the school parking lot). Present findings/plan to school administrators or local government. [CL, explicit, integrated – 7th grade expectations for learning]

The themes of economic literacy and global awareness are addressed less often in the Ohio model curriculum. In most instances, references to these themes are vague or made optional. For example:

Ethanol, a plant product, is used in place of fossil fuels. Discuss the pros and cons of using biomass products such as ethanol versus traditional fossil fuels. [EcL vague – 7th grade expectations for learning]

Examining student-based (classroom data) soil sample results can be a good way to compare soil types by regions. The GLOBE program can allow a connection to other classrooms, but can also be used when to analyze data beyond the local area to draw conclusions about specific criteria for soil formation. [GA optional – 6th grade instructional resources]

The theme of health literacy was not addressed in the Ohio standards and model curriculum. In conclusion, the Ohio standards and model curriculum core academic content was present at a partially high level because only two of the 21st century themes were explicitly described and integrated into curriculum activities.

Skills

Results of the analysis process indicate that several 21st century skills are emphasized in the Ohio science standards. Namely, inquiry and its application incorporates the skills of thinking, problem solving, communication, creativity, and innovation. Of these skills, thinking, problem solving, and communication are explicitly included and integrated throughout the middle grade standards and model curriculum as evidenced in Table 1 and, among others, the following statements:

- Design and conduct a scientific investigation;
- Analyze and interpret data;
- Develop descriptions, models, explanations and predictions;
- Think critically and logically to connect evidence and explanations;
- Communicate scientific procedures and explanations.

Creativity and innovation skills are also present in the Ohio standards and model curriculum both explicitly and implicitly in the practice of inquiry. An example of a statement that explicitly supports the development of creativity and innovation skills in the 7th grade Earth and Space Science expectations for learning is:

Design, build, and test a buoy that can sample water temperatures (or other water quality test, such as pH or turbidity levels) of a local lake, pond, pool, or stream. Deploy the buoy and collect/analyze data. Compare and discuss results with the class.

Finally, no statements explicitly or implicitly describing student engagement with reflection, autonomy, and self-direction skills were identified.

Additional Curricular Components

Results of the analysis of the additional curricular components described in TASSC indicate that the Ohio revised standards and model curriculum for grades 6-8 support rich technological environments (see Table 1) and include teaching and learning activities set in real-world examples, applications, and experiences. For example, the following recommended instructional strategy and resource suggest an activity set in a local context that invites investigation of a real-world issue:

Choosing local issues that involve water and conducting field studies and research about the movement of water and/or contamination can lead to deeper understanding of how the cycles work. An example could be researching acid mine drainage problems in southeastern Ohio. The Monday Creek Organization provides research and data for southeastern Ohio and acid mine drainage cleanup efforts.

Furthermore, the standards and model curriculum account for the different ways of student knowing through an attention to learning progressions described in content elaborations. For example, the following excerpt from the 6th grade model curriculum highlights the past and future learning of students with respect to rocks:

Prior concepts related to rocks: PreK-2: Objects have physical properties, properties of objects can change, and Earth's nonliving resources have specific properties. Grades 3-5: rocks and soil have characteristics, soil contains pieces of rocks, rocks form in different ways, and objects are composed of matter and may exhibit electrical conductivity and magnetism.

Future Application of Concepts: Grades 7 and 8: Sedimentary, metamorphic, and igneous environments, and the history of Earth (including the changing environments) from the interpretation of the rock record are studied. High school: The formation of elements, chemical bonding, and crystal structure are found in the Physical sciences. In 11/12th grade Physical Geology, depositional environments, volcanics, characteristics of rocks, and mineralogy are explored at depth.

The model curriculum also accounts for active student learning through references to misconceptions. Resources that describe common misconceptions in core content are provided in the instructional strategies and resources section. For example, one such reference for the 6th grade life science model curriculum states, "The Annenberg Media series Essential Science for Teachers: Life Science Session 1: Children's Ideas provides greater insight to misconceptions children hold about the origin of living things."

Finally, the Ohio model curriculum proposes the use of appropriate tools and approaches to measure students' performance on twenty-first century content and skills. This focus on assessment is evident in the description of the Cognitive Demand expectations for learning described in the section addressing structure of the curriculum. Examples of these expectations for learning are provided in the previous two sections.

The Case of the New York State Standards

New York City (NYC) schools are among the nation's most diverse schools with large percentages of minorities, recent immigrants, and English Language learners (NYC Department of Education-NYC DOE[3]). NYC Public schools follow the New York State (NYS) Core Curriculum[4]. The NYC DOE website gives the following general statement on the goal of science instruction:

[Providing students] with opportunities to interact directly with the natural world and gives them the skills they need to explain the world that surrounds them. The City's approach allows students to practice problem-solving skills, develop positive science attitudes, learn new science content, and increase their scientific literacy (NYC DOE).

Structure of the NYS Science Core Curriculum

The NYS core curriculum outlines seven Standards for Science, Mathematics, and Technology, of which five are designated as applying specifically to science. Table 2 lists the standards and explanatory statements that outline learning expectations for each.

The science core designates Standards 1, 2, 6, and 7 as the process skills standards and Standard 4 as the science content standard with two main strands: The Living Environment and The Physical Setting. For reasons not explicit in the document, Strand 5, Technology (Students will apply technological knowledge and skills to design, construct, use, and evaluate products and systems to satisfy human and environmental needs) is not designated for science and consequently is not included in the science core curriculum document. Each standard is further expanded by a number of key ideas. Then each key idea is expanded with

Table 2. NYC science specific standards

Standard	Expectation
Standard 1: Analysis, Inquiry, and Design	Students will use mathematical analysis, scientific inquiry, and engineering design, as appropriate, to pose questions, seek answers, and develop solutions.
Standard 2: Information Systems	Students will access, generate, process, and transfer information using appropriate technologies.
Standard 4: Science	Students will understand and apply scientific concepts, principles, and theories pertaining to the physical setting and living environment and recognize the historical development of ideas in science.
Standard 6: Interconnectedness: Common Themes	Students will understand the relationships and common themes that connect mathematics, science, and technology and apply the themes to these and other areas of learning.
Standard 7: Interdisciplinary Problem Solving	Students will apply the knowledge and thinking skills of mathematics, science, and technology to address real-life problems and make informed decisions.

one or two levels of associated performance indicators. For example Standard 1(Analysis, Inquiry, and Design) subsumes the following key idea and subsequent performance indicators:

Key Idea 2: Beyond the use of reasoning and consensus, scientific inquiry involves the testing of proposed explanations involving the use of conventional techniques and procedures and usually requiring considerable ingenuity.

- **S2.1:** Use conventional techniques and those of their own design to make further observations and refine their explanations, guided by a need for more information.
    - **S2.1a:** demonstrate appropriate safety techniques
    - **S2.1b:** conduct an experiment designed by others
    - **S2.1c:** design and conduct an experiment to test a hypothesis
    - **S2.1d:** use appropriate tools and conventional techniques to solve problems about the natural world, including:
        - Measuring
        - Observing
        - Describing
        - Classifying
        - sequencing

Depending on its scale and expansiveness, a standard may be divided into broad themes that include several Key Ideas. For example standard 1(Analysis, Inquiry, and Design) incorporates the themes of Mathematical Analysis, Scientific Inquiry, and Engineering Design.

Results

The analysis results for Standards 1, 2, 4, and 7 will be presented below to demonstrate the use of TASSC on the curriculum document.

Core Academic Content

Standard 4 is the standard that outlines the science concepts and principles to be covered at the middle school level for New York State students. The major unifying concepts and processes of science are covered at comprehensive high level in standard 4, especially as they pertain to Living Environment and Physical setting.

Standard 1 also addresses science content and was assigned a rating of 'partial high level' coverage for Science Content for understandings about scientific Inquiry and its features. This is somewhat expected, as the standard itself explicitly cites 'inquiry' as a focus in its statement and includes 'scientific inquiry' as one of its themes. Therefore there are several key ideas and performance indicators that incorporate understandings about inquiry, for example, "The central purpose of scientific inquiry is to develop explanations of natural phenomena in a continuing, creative process" (Key Idea). Other performance indicator examples include:

- **S1.2c:** Differentiate among observations, inferences, predictions, and explanations
- **S3.1:** Design charts, tables, graphs, and other representations of observations in conventional and creative ways to help them address their research question or hypothesis.

Of the 21st Century Themes, content within Environmental literacy is covered at a more comprehensive level than Health literacy. For example, two of the seven key ideas in the Living Environment explicitly include aspects of Environmental Literacy, which is covered at comprehensive high level:

- **KI 6:** Plants and animals depend on each other and their physical environment.
- **KI 7:** Human decisions and activities have had a profound impact on the physical and living environment.

Health Literacy is covered at a partial-high level in Key Idea 5 in Living Environment. The component with the strong presence involves the role of nutrients and diet as is apparent in the following performance indicator:

- **5.2:** Describe the importance of major nutrients, vitamins, and minerals in maintaining health and promoting growth, and explain the need for a constant input of energy for living organisms.

No reference is made to global awareness and only minimal references are made to Economic and Civic literacy. For example, the following key idea presents a vague reference to civic literacy issues:

- **3.2:** Describe applications of information technology in mathematics, science, and other technologies that address needs and solve problems in the community.

Skills

Communication, thinking, and problem solving skills are explicitly integrated in the NYS science curriculum, specifically for scientific inquiry. For example, standard 1 presents a strong skill for the items 'thinking and problem solving skills' and 'communication skills' as the latter relates to presenting work associated with scientific inquiry or sharing understandings. Similarly, developing information skills are at a strong skill level in standard 2, clearly requiring that students manage and evaluate information from different sources. The following performance indicators illustrate this finding:

- **1.4c:** Use the collected data to communicate a scientific concept
- **3.1:** Use graphical, statistical, and presentation software to present projects to fellow classmates.

Autonomy, self-direction, creativity, and innovation skills are included but in weak form and mostly as they pertain to carrying out scientific inquiry. Autonomy skills are covered by the expectations that students carry out aspects of scientific inquiries 'independently.' Creativity and innovation skills are implicit to inquiry processes such as formulating hypotheses and designing experiments.

Additional Curricular Components

The strong focus on inquiry in Standard 1 supports students as active learners. Perceptions of students as active learners was evident by the strong emphasis in standard 1 on themes and key ideas under analysis, inquiry, and design, including "students are expected to use inquiry process skills to learn about the world." Parallel processing, on the other hand is not explicitly addressed in any place in the curriculum document.

The NYS core curriculum emphasizes teaching and learning in 21st Century teaching contexts. For example in standard 7, the notion of using content and skill to address everyday problems is explicitly stated in all Key Ideas. Furthermore, two performance indicators of standard 1 mention explicitly solving "real world problems" (M3.1) and the "investigation of situations of general social interest" (T1.1), therefore this item was designated as mostly absent.

Technologically rich environments are also supported in the NYS core curriculum, specifically in standard 2. Standard 2 is concerned with Information Systems and states that "Students will access, generate, process, and transfer information using appropriate technologies." Upon examining all the key ideas within the standard, technologically rich context was strongly emphasized and present.

Finally, assessment appropriate to a 21st Century context is not explicitly addressed in the standards. However, in the appendix of the NYS

Core Curriculum it is explicitly mentioned that 20-25% of the exam grade is on "real world problems," with this being the only reference made to assessment in middle school science. Vagueness about assessment is somewhat disconcerting considering the way by which the 'tested curriculum' (Cuban, 1995) percolates to classroom practices of science teachers.

The Case of Lebanon

Lebanon underwent educational reform in 1998 in order to meet the demands and requirements of the present society, and adequately prepare students for the future. The preparation of the resultant Lebanese curriculum[5] was inspired by the rapid expansion of science and technology in the present century, the new international and global tendencies towards science teaching, and the important role in all aspects of our everyday life. The intermediate (middle level) stage of the Lebanese curriculum consists of grades 7-9. Grade 6 is presently compulsory for all Lebanese students, and all students are provided with a common content until grade 10.

Structure of the Lebanese Science Curriculum

The focus of science teaching in the Lebanese science curriculum is defined by the overall general objectives. Some examples of the overall general objectives include:

- Develop learners' intellectual and practical scientific skills.
- Deepen learners' awareness in the ability of humans to understand, invent, and create.
- Understand the nature of science and technology, their historical development, and their impact on human thought.
- Insure that learners have acquired the facts, concepts, and principles that are essential to understand natural phenomena.
- Acquire knowledge and accordingly behave about issues of health, environment, and safety practices.
- Realize that some natural resources can be depleted and advocate for the role of science in sustaining these resources.
- Encourage learners to use scientific knowledge and skills in novel everyday life situations.
- Emphasize the role of scientists in the advancement of human kind.
- Encourage learners to be open to the ideas of scientists from different cultures and to their contributions to the advancement of science.
- Encourage learners to work independently and cooperatively to solve scientific problems.
- Make learners aware of career possibilities in different science related areas.

The subjects that are addressed in the Lebanese intermediate science curriculum are physics, chemistry, life, and earth sciences. The analysis was performed on all components of the intermediate science curriculum including the introduction for each of the three subjects, objectives of each subject, instructional objectives, and activities of each subject. The objectives of each subject are conceptual (technical and methodological) and advance a relationship between teaching and evaluation. The analysis was also performed on the overall general objectives of science education.

Results

The results of the TASSC analysis of the Lebanese Intermediate science curriculum are presented below as core academic content, skills, and additional curricular components.

Core Academic Content

The core academic content is composed of the science content and the 21st century themes. The TASSC analysis showed that the standards are highly comprehensive and strong for developing the science content for the earth and life sciences, chemistry, and physics.

As for the 21st century themes, the themes of the environmental and health literacy are emphasized in the introduction of the earth and life sciences curriculum with statements such as:

The program deals with a global perspective about health and the environment. This helps the student to maintain his health and the health of others, to develop a responsible behavior towards the environment, and to comprehend scientific messages transmitted by media.

The theme about the environment literacy is further carried along in several objectives of the earth and life sciences, including:

Allow the student to identify the essential biological, physical, and geological components of the environment and to understand their relations that initiate the development of an open and a responsible behavior toward environmental problems: management of media and resources, preservation of species, and risk prevention.

Particularly for Grade 8, there is a chapter in the content that targets the environment, "earth and the environment." That chapter deals with the concrete geodynamic manifestations of the earth activity, leading to the study of changes starting from rocks to landscape that constitutes the life framework for humans. The chapter further addresses the study of geology with a focus on the relationship between man and earth in order to develop among the students a sense of a long-time responsibility towards the environment.

Similarly within the chemistry curriculum, there is a separate chapter "chemistry and the environment" at the grade 9 level. It aims at providing students with a better understanding of the role of chemistry in causing and solving environmental problems. Environmental application and problems are presented as (a) dilution of toxic wastes and pollution addressed at grade 7, (b) acid rain, antacids, fertilizers addressed at grade 8, (c) pollution problems of air, water, and soil addressed at grade 9.

Likewise, developing a responsible attitude towards the environment is manifested as a theme in the instructional objectives within the physics curriculum. However, it is not clear how that is addressed in the instructional activities and guidelines.

Skills

The TASSC analysis showed that the skills for developing creativity and innovation, as well as the reflective skills are completely absent in the Lebanese curriculum. Furthermore, the curriculum is not very explicit with respect to developing information, communication, and autonomy skills. For example, the introduction of the life and earth sciences curriculum emphasizes the importance of communication skills by indicating that the program needs to focus on teaching scientific processes in a problem situation to allow the learner to acquire communication skills. However, the presence of communication skills is not reflected in the instructional objectives and activities of the curriculum.

The skill of developing autonomy is apparent in the introduction of the life and earth sciences curriculum as noted in the following statement, "The curriculum fosters the development of autonomy and responsibility of personal behavior in learners that are characteristics of a free citizen." Autonomy is further addressed in the instructional objectives of life and earth sciences curriculum as follows:

Develop in students a scientific attitude and motivate them to acquire a better autonomy.

This statement affirms that autonomy is strongly adopted in the curriculum. Yet and again, this is only addressed in the introduction and instructional objectives in only the life and earth sciences curriculum.

Promoting thinking and problem-solving skills is considered a critical component in the Lebanese Intermediate curriculum. It is explicitly present in the general science objectives and also emphasized as one of the instructional objectives in the life and earth sciences curriculum for the intermediate grades:

Permit students to acquire scientific processes, specifically by developing an experimental approach and problem solving activities

Additional Curricular Components

Results of the analysis of the additional curricular components described in TASSC indicate that the Lebanese intermediate curriculum vaguely describes the criteria of assessment and learning the academic content through real-world examples, applications, and experiences. For example, assessment is emphasized in the introduction of the life and earth sciences curriculum as "evaluation must not be limited to recalled information but should also assess the skills required to be achieved." However, it is not clear how that is manifested in terms of the assessment at the level of instructional activities or guidelines.

Furthermore, the Lebanese curriculum was found to account for the different ways of student knowing to various degrees. First, the curriculum makes no mention of experiences that allow for parallel processing. Second, it only references students as active learners. For example, the description in the introduction section of the life and earth sciences curriculum presents the learner as an "active participant" and it is also indicated that the use of different approaches encourages students to "construct" their knowledge themselves.

Third, the curriculum explicitly supports a technologically rich environment. For example, the introduction of the life and earth sciences curriculum states, "The program tends to establish a relationship between technology and society with which the student has strong ties." Technology is also referred to in the instructional objectives of the life and earth sciences curriculum, where it is indicated that there is a need to progressively and continually initiate in students "a scientific knowledge, technical performances and actual research in science and technology." Moreover, it is explicitly stated in the instructional objectives of the chemistry curriculum that students should "perceive the close relationship among chemistry, technology, and industry." This is also referred to in the instructional objectives of the physics curriculum about having students understand the relation between physics, technology, and society.

The Case of Qatar

A new reform initiative, *Education for a New Era*, was launched in Qatar in 2005. This initiative is built on four principles:

- **Autonomy:** Allowing an innovative approach for schools and teachers to meet the needs of students and parents within the curriculum standards framework.
- **Accountability:** Implementing an objective and transparent assessment system to hold school leaders, teachers, and parents responsible for the students' success.
- **Variety:** Encouraging different kinds of schools and instructional programs.
- **Choice:** Allowing parents to select the school that best fits their child's needs and seeking their input on important schools decisions.

Structure of the Educational System

The curriculum of the Qatari preparatory (middle level) cycle emphasizes basic literacy and numeracy skills and is follows the *Early Years Education Foundation Curriculum*. The *Foundation Curriculum* is believed to be future-oriented in attending to the qualities and characteristics needed in young citizens in 2020 in the State of Qatar with the goal of preparing the Qatari youth to compete in the 21st century global economy. The curriculum is also aligned with the vision of *Education for a New Era,* which aims for schools to educate students who will (a) make a positive contribution to society by being literate and numerate, (b) be effective communicators in Arabic and English, (c) be creative and critical thinkers, (d) have the dispositions and strategies to explore and inquire, and (e) work well in teams and demonstrate pride in being Qatari.

The Qatari curriculum for grades 7-9 includes the subjects of inquiry, life science, materials, earth and space, and physical processes. TASSC analysis was performed on the following components of the Qatari curriculum: overall general goals of science, summary of students' performance for each of the five subjects, and the instructional objectives and activities of each subject at grades 7, 8, and 9. It is important to note that the Qatari standards are explicit in outlining the different instructional activities, questions, and examples for each of the subjects.

Results

The results of the TASSC analysis of the Qatari preparatory curriculum are presented below as core academic content, skills, and additional curricular components.

Core Academic Content

The TASSC analysis showed that the standards are highly comprehensive and strong for developing the science content. As for the 21st century themes, attention to the environment is noted in the overall aims of the Qatari science standards (Supreme Education Council, 2002) as follows:

Students should (a) have a sound and systematic knowledge of important scientific facts, concepts and principles, and possess the skills needed to apply these in new and changing situations in a range of personal, domestic, industrial and environmental contexts, (b) recognize the importance of the application of scientific knowledge in the modern world and be aware of the moral, ethical, social and environmental implications

Further, one of the topics of the life science subject addresses living things and their environment. This topic is addressed in grade 7 and 8 but applications and examples related to the environment seem to be scarce. That is not the case for grade 9, where environmental literacy is a focus along with a relation to the economy of Qatar. For example, the life science curriculum states the expectation that students discuss the economic dimension of genetic engineering and its implications. According to the materials subject, grade 9 students need to "explain the importance of fossil fuels to the economy of Qatar."

Global awareness is manifested in the Qatari standards in grade 9 in the context of the materials subject. For example, students need to explain the causes of global warming and discuss various proposals about the removal of carbon dioxide from the atmosphere as proposed under the Kyoto protocol. Another instance is represented by having students "make a study of the consequences of acid rainfall in some other parts of the world (e.g., northern Europe)."

Health literacy is represented at various instances in the life science curriculum for grades 7 and 8 only. For example, students in grade 7 need to know the importance of good nutrition and hygiene to the health of babies. Students in grade 8 study the effects of smoking on health and the problems of diabetes and obesity, as follows:

Know that smoking damages the lungs and reduces the efficiency of gas exchange.

Know the symptoms, causes, and problems of diabetes and obesity. Chart the statistics on the frequency of diabetes and obesity in Qatar and compare with other countries.

Skills

The TASSC analysis showed that developing information and communication skills is explicitly addressed and described in the Qatari standards for grades 7-9. Actually, one of the topics in the subject of scientific inquiry is processing and communicating information, which is addressed in grades 7 to 9.

The development of creativity and innovation skills is also attended to in the standards. Having students develop the ability to work creatively is noted to be critical in the overall aims of the Qatari standards, and is explicitly reflected in the instructional activities described for grades 8 and 9, as shown in the following examples:

Construct a card game to illustrate what eats what in a range of environments (life science, grade 8)

Design a thermal solar panel and investigate its effectiveness (materials, grade 9)

Build model bridges from simple materials such as mud bricks (high compressive strength but low tensile strength), spaghetti (good tensile strength but easily broken), and cotton (flexible and good tensile strength). Use different kinds of bridge construction (e.g. arch, cantilever, girder, suspension) that make the best use of the properties of the materials. Test the bridges to destruction (physical processes, grade 9)

Developing problem-solving skills among students is considered important as indicated in the overall aims of the Qatari standards. And that is clearly reflected in the instructional activities of the curriculum of grades 7, 8, and 9.

Developing the interpersonal and self-directional skills, the reflective skills, and autonomy do not appear to be manifested in the Qatari standards.

Additional Curricular Components

Based on the TASSC analysis, results show that the instructional activities in the Qatari curriculum harness a 21st century context in the use of many real-world examples, applications and experiences at grades 7, 8, and 9. Following are some examples of the instructional activities at grade 7 targeting the subjects about materials and earth and space:

Collect information on the distillation of seawater to provide drinking water for Qatar (materials)

Show, using a flow-chart, how Qatar gas is used (earth and space).

More examples, which target real-world applications and experiences, are evident from the instructional activities at grade 8 across the subjects of materials and physical processes:

Discuss the occurrence of uncombined metals at the bottom of the reactivity series and methods used to recover alluvial gold. Link this natural low reactivity with the use of these metals in jewelry (materials)

Study the formation of mirages on a hot day, noting how they can be explained by refraction (physical processes)

Still other examples are found from the instructional activities for grade 9 in the subjects of life science, materials, and physical processes:

Collect information on the frequency of common inherited disorders in Qatar (life science).

Make a study of the cooling of industrial processes in Qatar, including a visit to a cooling plant such as the one at the Ras Laffan (materials).

Furthermore, Qatari science assessment seems to use appropriate tools and approaches to measure students' performance on twenty-first century content and skills. Science assessments have three objectives (Supreme Education Council, 2002):

- Knowledge and understanding; application of knowledge and understanding, analysis and evaluation of information.
- Scientific enquiry skills and procedures.
- The weightings of the assessment objectives change at various grade levels.

As students' scientific proficiency and experience develop, greater emphasis is placed on assessing the application of knowledge to solve problems in new situations.

Finally, the Qatari standards were found to vaguely account for students' ways of knowing to various degrees. First, there seems to be no indication that the curriculum affords the utilization of technological tools allowing for parallel processing. Second, the perception that students are active learners is implicit in the strong emphasis on the doing and understanding about scientific inquiry. Third, the standards somewhat support a technologically rich environment. The effective use of information and communications technology in the pursuit and communication of science, as well as the relationship between technology and society is emphasized in the Qatari science standards. However, that is not clearly addressed in the instructional activities of the curriculum.

DISCUSSION

In the previous section, individual case studies present the results of using TASSC to analyze the standards and curricula of Ohio, NYS, Lebanon, and Qatar. In this section, we summarize the findings from the four case studies with respect to core academic content, skills, and additional curricular components. We end this section with a discussion of next steps necessary in the development and use of TASSC.

Core Academic Content

Science content is found almost in its entirety in Ohio and NYS standards and curricula. It is explicitly described in curriculum activities in a manner likely to produce high levels of learning if the curricular goals are met. Specifically, unifying concepts and processes of science, inquiry, earth and space, life, and physical science content were integrated in the targeted US standards.

Twenty-first century themes are present in Ohio and NYS standards and curricula at partially high levels. Environmental Literacy is integrated in the activities of both US standards and curricula. Additionally, civic and health literacy are explicitly integrated in the Ohio and NYS curricula respectively. Other 21st century themes are vaguely referenced (economic literacy) or absent (e.g. global literacy in NYS curriculum and health literacy in Ohio).

Similarly, the science content is found in its entirety in the Lebanese and Qatari curricula. The central theme of the environment permeates all the topics covered in middle level curricula. However, the manifestation of that theme varies within the different subjects. Of the remaining 21st century themes, global and health literacy are explicitly integrated in the Qatari standards while only health literacy is alluded to vaguely in the Lebanese curriculum.

Skills

Twenty-first century skills are present in the Ohio and NYS standards and curricula to varying degrees. Communication, thinking, and problem solving skills are present as strong skills in both states. Creativity and innovation are present in both state documents as weak skills in the NYS standards and strong skills in the Ohio ones. Reflective skills, autonomy, and self-direction are present in the NYS curriculum as weak skills. Reflective skills, autonomy and self-direction seem to be absent from the Ohio standards.

Interestingly, all 21st century skills, with the exception of reflection and autonomy, are explicitly integrated in the Qatari curriculum. Reflection and autonomy are not addressed. Conversely, only thinking and problem solving skills are explicitly integrated in the Lebanese curriculum. Information, Communication, and autonomy skills are vaguely alluded to in the introductory comments only. Creativity, innovation, and reflective skills seem to be absent from the Lebanese curriculum.

Additional Curricular Components

Teaching and learning activities in the Ohio and NYS standards and curricula involve a twenty-first century context and promote the learning of academic content through real-world examples, applications and experiences. These Curricula also support a rich technological environment through the availability of rich graphics and interaction with technological tools and media. Activities within the Ohio and NYS standards and curricula account for students as active learners, but do not explicitly attend to students' ability to parallel process. However, while the assessments advocated in the Ohio standards and model curriculum seem to promote appropriate tools and approaches to measure students' performance on twenty-first century content and skills, the NYS standards only vaguely address assessment.

The majority of additional curricular components of interest in this study are vague or absent from the Lebanese and Qatari curricula. Only vague references in the general statements of both curricula address students as active learners or appropriate tools and approaches for the assessment of 21st century content and skills. Neither attends to students' ability to parallel process. While the Qatari curriculum consistently promotes the learning of academic content through real-world examples, applications and experiences, such recommendations are absent from the Lebanese curriculum. Conversely, the Lebanese curriculum explicitly supports technologically rich environments. References to such environments are vague in the Qatari curriculum.

Next Steps for TASSC Development Process

Even though TASSC was successfully used to analyze the standards and curricula described in the four case studies presented in this report, further refining of the instrument is warranted. First, TASSC needs to be revisited to remove additional areas of overlap that were identified in the development of the case studies. For example, statements depicting thinking and problem solving skills in the standards and curricula were also coded as instances of active learning in context.

Next, attention to the rating scale is needed. For example, additional instructions are necessary to make determinations about the extent themes and skills are explicit. Furthermore, considerations about how many times an instance of a particular theme or skill needs to be identified in the text to warrant a particular rating.

Finally, the content section of the instrument can benefit from the further definition of the science content. Such a definition will allow the instrument to provide more discriminating data by providing a rating for each content category. Similarly, skills can be further subdivided or operationalized for greater accuracy. For example, communication skills could be divided into five component skills as described by Schunn in the 2010 National Research Council report (NRC, 2010):

- Select key pieces of a complex idea to express in words, sounds, and images, in order to build shared understanding,
- Social perceptiveness,
- Persuasion and negotiation,
- Instructing, and
- Service orientation.

Such operationalization allows us to determine if a skill is present in its entirety warranting a rating of "whole," or if parts of the skill are present warranting a rating of "component." This operationalization may also allow us to distinguish between the general education components of each skill and more subject-specific ones. For example, persuasion and negotiation are science specific components of the communication skills that are represented in the science education community as argumentation skills.

CONCLUSION AND IMPLICATIONS

The use of TASSC to analyze the standards and curricula of Ohio, NYC, Lebanon, and Qatar provided useful information for the evaluation of these curricula with respect to their incorporation of 21st century essential understandings and skills. It is evident from the case studies that these states are at various points in their journey to meet the imperative of preparing students for the 21st century. The use of TASSC allows reformers and curriculum designers to identify areas for further development and continuous improvement.

The case studies also illustrate that TASSC can be used in multiple contexts. The intent of TASSC development is not the creation of an instrument for rank-ordering states with respect to the degree to which they incorporate 21st century essential understandings and skills. The use of TASSC is intended to be formative; results can enhance and inform the standard and curriculum development process. Finally, TASSC is most useful for the evaluation of standards, curricula and long-term planning efforts at the local levels.

REFERENCES

American Association for the Advancement of Science. (1993). *Benchmarks for science literacy: Project 2061*. Oxford, UK: Oxford University Press.

BouJaoude. S. (2010). *Competencies and educational structures needed to prepare secondary students for the 21st century*. Paper presented at a Symposium on Secondary Education Organized by the Ministry of Education in the United Arab Emirates in Cooperation with the Arab Bureau of Education for the Gulf States. Dubai, UAE.

Bybee, R. (1997). *Achieving scientific literacy: From purposes to practices*. Portsmouth, NH: Heinemann Educational Books.

Cranton, P. (1997). Higher education: A global community. *New Directions for Teaching and Learning, Winter, 72*, 5-9.

Cranton, P. (2001). Interpretive and critical evaluation. *New Directions for Teaching and Learning, 88*, 11–18. doi:10.1002/tl.33

Cranton, P. (2002). Teaching for transformation. In Ross-Gordon, O. M. (Ed.), *Contemporary Viewpoints on Teaching Adults Effectively: New Directions for Adult and Continuing Education*. San Francisco, CA: Jossey-Bass.

Habermas, J. (1971). *Knowledge and human interests*. Boston, MA: Beacon.

Mezirow, J. (1997). Transformative learning: Theory to practice. *New Directions for Adult and Continuing Learning, 74*, 5–12. doi:10.1002/ace.7401

National Research Council. (1996). *The national science education standards*. Washington, DC: National Academy Press.

National Research Council. (2010). *Exploring the intersection of science education and 21st century skills: A workshop summary*. Washington, DC: National Academies Press Ohio Department of Education. (2010). *Science education standards revision and model curriculum development*. Retrieved from http://www.ode.state.oh.us/GD/Templates/Pages/ODE/ODEDetail.aspx?page=3&TopicRelationID=1705&ContentID=76585

OECD. (2005). *Definition and selection of key competencies*. Retrieved from www.oecd.org/dataoecd/47/61/35070367.pdf

Partnership for 21st Century Skills. (2009). *Framework for 21st century learning*. Retrieved from http://www.p21.org/index.php?option=com_content&task=view&id=254&Itemid=119

Prensky, M. (2001). *Digital natives, digital immigrants*. Retrieved from http://www.marcprensky.com/writing/Prensky%20-%20Digital%20Natives,%20Digital%20Immigrants%20-%20Part1.pdf

Strickland, T. (2009). *Reforming Ohio's education system for the 21st century: Governor Ted Strickland's education reform and funding plan*. Retrieved from http://www.conversationoneducation.org/information/

Supreme Education Council. (2002). *Science standards*. Retrieved February 16, 2012, from http://www.english.education.gov.qa/section/sec/education_institute/cso/science_standards

Treadwell, M. (2011). Whatever happened? In Wan, G., & Gut, D. (Eds.), *Bringing Schools into the 21st Century* (pp. 7–40). New York, NY: Springer. doi:10.1007/978-94-007-0268-4_2

Wan, G., & Gut, D. (Eds.). (2011). *Bringing schools into the 21st century*. New York, NY: Springer. doi:10.1007/978-94-007-0268-4

ENDNOTES

[1] http://www.oecd.org
[2] Habermas coined this knowledge as practical knowledge and Merizow (1997) called it communicative knowledge.
[3] http://schools.nyc.gov/Academics/Science/default.htm
[4] http://www.p12.nysed.gov/ciai/cores.html#MST
[5] http://www.crdp.org/crdp/all%20curriculum/Sciences/Sciences_english/science_english.pdf

A Tool for Analyzing Science Standards and Curricula

APPENDIX 1

Framework for Analyzing Education Programs

Item	Evident			Not evident
1. The program focuses on:				
• helping students understand core academic content at high levels				
• developing information and communication skills				
• developing thinking and problem-solving skills				
• developing interpersonal and self-directional skills				
• developing reflective skills				
• developing autonomy				
2. The program encourages the development of social skills (living peacefully with others, cooperation, empathy...)				
3. Teaching and learning seem to be implemented in a twenty first century context (learning academic content through real-world examples, applications and experiences)				
4. Assessment seems to use appropriate tools and approaches to measure students' performance on twenty-first century content and skills.				
5. Systems seem to consider the fact that students think and process information differently from their predecessors (accounting for students' ways of knowing)				

APPENDIX 2

Tool for Analyzing Science Standards and Curricula (TASSC)

	Rating
The Standards/Curriculum focuses on:	
Content	
1. Helping students understand core academic content at high levels	
a. Science Content	
b. 21st Century Themes (Environmental literacy, global awareness, economic literacy, civic literacy, and health literacy	
Skills	
2. Developing information skills	
3. Developing communication skills	
4. Developing creativity and innovation skills	
5. Developing thinking and problem-solving skills	
6. Developing reflective skills	
7. Developing autonomy and self-directional skills	
Other	
8. Teaching and learning activities involve a twenty first century context (learning academic content through real-world examples, applications and experiences)	
9. Assessment seems to use appropriate tools and approaches to measure students' performance on twenty-first century content and skills.	
10. Systems consider the fact that students think and process information differently from their predecessors (accounting for students' ways of knowing)	
a. Students are perceived as **active learners** (learning theories)	
b. Curriculum supports a rich technological environment (availability of rich graphics and interaction with technological tools and media)	
c. Curriculum affords utilization of technological tools allowing for parallel processing	

A Tool for Analyzing Science Standards and Curricula

APPENDIX 3

Level 2 Analysis Codes

Content	Code
1. Helping students understand core academic content at high levels	
a. Science Content	SC
b. **21ˢᵗ Century Themes:** Environmental literacy Global awareness Economic literacy Civic literacy Health literacy	EnvL GA EcL CL HL
2. Teaching and learning activities involve a twenty first century context (learning academic content through real-world examples, applications and experiences)	**21Con**
3. Assessment seems to use appropriate tools and approaches to measure students' performance on twenty-first century content and skills.	**AsP**
4. Systems consider the fact that students think and process information differently from their predecessors (accounting for students' ways of knowing)	**IP:--**
a. Students are perceived as **active learners** (learning theories)	**AL**
b. Curriculum supports a rich technological environment (availability of rich graphics and interaction with technological tools and media)	**TR**
c. Curriculum affords utilization of technological tools allowing for **parallel processing**	**PP**
Skills	
1. Developing **information skills**	Inf
2. Developing **communication skills**	Com
3. Developing **creativity and innovation skills**	CInn
4. Developing **thinking and problem-solving skills**	ThP
5. Developing **autonomy** and **self-directional skills**	ASD

Chapter 15
Measuring and Facilitating Highly Effective Inquiry-Based Teaching and Learning in Science Classrooms

Jeff C. Marshall
Clemson University, USA

ABSTRACT

For the last decade or so there has been a huge push to incorporate best practice into the classroom. For science, this includes bringing effective inquiry-based instruction into all classrooms as a means to engage the learner. However, all inquiry instruction is not equal in terms of improving student achievement and conceptual development. This chapter explores how four critical constructs to learning (curriculum, instruction, discourse, and assessment) can be effectively measured and then used to guide more effective instructional practice. The Electronic Quality of Inquiry Protocol (EQUIP) is an instrument that can be used to measure and then to frame the discussion regarding the quality of inquiry-based instructional practice. Specifically, this chapter provides an overview of EQUIP, details the reliability and validity of EQUIP, shares a sample lesson that is analyzed using EQUIP, explores ways that EQUIP can help with teacher transformation relative to inquiry instruction, and addresses the relationship of EQUIP scores and student achievement data. There is a very high correlation between teacher performance on EQUIP and the ensuing student growth noted during an academic year.

DOI: 10.4018/978-1-4666-2809-0.ch015

MEASURING AND FACILITATING HIGHLY EFFECTIVE INQUIRY-BASED TEACHING AND LEARNING IN SCIENCE CLASSROOMS

Bringing high-quality inquiry-based instructional practice into science classrooms has continued to be central to reform efforts for the last several decades (Bransford, Brown, & Cocking, 2000; National Academy of Sciences, 2007; National Research Council, 1996, 2000, 2012; National Science Teachers Association, 1998). However, merely increasing the quantity of self-reported inquiry instruction is insufficient (Marshall, Horton, Igo, & Switzer, 2009); the quality of inquiry instructional practice must be at such a level that teachers are effective in facilitating rigorous, standards-based, inquiry-based learning. Success in achieving this goal has been largely inconsistent at best in programs across the country. Definitions of inquiry-based instruction may vary somewhat, but clear direction has been given to defining and exemplifying inquiry-based instruction (NRC, 1996, 2000). Even though consistency can be found among many of the definitions and agreement is found in the desire for reform that includes inquiry-based instruction, the implementation remains inconsistent.

As science education looks ahead to the next decade or so of science instruction, *A Framework for K-12 Science Education: Practices, Crosscutting Concepts, and Core Ideas* (NRC, 2012) has begun to pave the way for a new vision of teaching and learning that is more intentional and more integrated. This framework which serves as the predecessor for the *Next Generation Science Standards* (due out soon) makes clear that inquiry forms of instruction need to integrate cross disciplinary concepts and core ideas into the learning. No longer is it sufficient to teach inquiry as a stand-alone unit and then proceed to learning "the content." Inquiry helps provide the context for learning major concepts and ideas.

We know that successful inquiry-based instruction is often the result of numerous professional development experiences. For more experienced teachers, these experiences are necessary because transformation of practice is needed to move from prior more teacher-centered paradigms to a more student-centered, constructivist approach where students build on prior knowledge through a series of science learning experiences. For neophyte teachers, inquiry instruction should not be a new concept, but support is often needed to help achieve a significantly inquiry-centered environment. For these beginning teachers, they often have to overcome many of their college experiences, which were typically solely confirmatory experiences as well as overcoming peer, departmental, and/or school structures that may model learning that is counter to inquiry.

Capps, Crawford, and Constas (2012) conducted an analysis of professional development programs in an effort to see how well aligned to best practices many of the programs currently are. Their findings suggest that most of the analyzed programs generally align with the recommended features of effective professional development (Darling-Hammond & McLaughlin, 1995; Garet, Porter, Desimone, Birman, & Yoon, 2001; Loucks-Horsley, Stiles, Mundry, Love, & Hewson, 2010). The structural features of effective professional development experiences include significant interactions that include extended support through the academic year while providing authentic experiences for teachers. Further, core features of such professional development include many or all of the following: coherence, lesson development, modeled inquiry experiences, reflection, transference of new skills, and content knowledge development. Our work has shown during the past 5 years, that teachers who are engaged in our professional development model associated with Inquiry in Motion are able to raise student achievement higher than the virtual comparison group of similar students (Marshall, Horton, & Edmondson, 2007; Marshall & Horton, 2009).

These changes in teacher practice bring about the most significant change in student achievement after two or more years. During the first year, teaching practice is often inconsistent and unstable as new instructional approaches are used, and the data suggest that during year two that as teachers become more consistent in implementing guided forms of inquiry-based instruction that learning goes up significantly.

In helping to facilitate highly successful inquiry-based teaching and learning, two components are essential: 1) a measure is needed that provides an understanding of the degree of success experienced and tracks any progress made, and 2) a plan is needed for how to successfully implement inquiry-based instruction. Regarding the metrics, numerous instruments exist that monitor and measure teacher effectiveness and include the following:

- *Inside the Classroom Observational Protocol* (Horizon Research, 2002)
  - **Target:** Provides a global view of effective classroom practice;
  - **Possible limitation:** Very long and complex measure—lacks targeting specific aspects of instruction.
- *Reformed Teaching Observation Protocol (RTOP)* (Sawada, et al., 2000)
  - **Target:** Provides measure of constructivist practice in math and science classrooms;
  - **Possible limitations:** The non-descript Likert scale makes interpreting the value of individual and holistic scores difficult, indicators within constructs do not hold together for factor analysis, and entire lesson sequence needs to be observed for accurate score to be determined.
- *Electronic Quality of Inquiry Protocol (EQUIP)* (Marshall, Horton, & White, 2009; Marshall & Horton, 2009; Marshall, Smart, & Horton, 2010)
  - **Target:** Provides a measure of inquiry-based instruction in math and science through the use of a descriptive rubric;
  - **Possible limitation:** Targets specific type of instructional practice.
- *Science Teacher Inquiry Rubric (STIR)* (Beerer & Bodzin, 2003)
  - **Target:** Inquiry instruction aligned to the *National Science Education Standards (NSES)*;
  - **Possible limitations:** All components of inquiry defined by NSES are not present in every lesson, and instrument will be quickly antiquated because it is targets a narrow (NSES) view of inquiry.
- *Science Management Observation Protocol (SMOP)* (Sampson, 2004)
  - **Target:** Measures classroom management issues that support effective science instruction;
  - **Possible limitations:** Classroom management is necessary but not solely sufficient to predict successful instructional practice; so, while informative, this instrument seems to need to be coupled with another instrument for solid understanding of successful instruction to be understood.
- *Secondary Science Teaching Analysis Matrix (STAM)* (Adams & Krockover, 1999)
  - **Target:** Move novice teachers toward more student-centered instructional strategies;
- *Expert Science Teaching Educational Evaluation Model (ESTEEM)* (Burry-Stock & Oxford, 1994)
  - **Target:** Excellence in science teaching from a constructivist framework; and
- **Teacher efficacy scales (Riggs & Enochs, 1990)—target:** Predict likelihood of reform.

Despite all the available metrics, initial research showed that two instruments, the RTOP and the EQUIP, provided solid measures for assessing the degree to which inquiry-based instruction is being facilitated in K-12 science classrooms. These two instruments stand out for several reasons: (1) both of these instruments provide measures that target inquiry-based instruction in math and science classrooms, (2) both have validity studies to support their work, and (3) the other instruments seem too general (e.g., consider all elements of effective practice), too granular (e.g., consider one aspect of instruction such as classroom management), or too complex (e.g., necessary to use multiple rubrics over multiple days). Further, by attending any science education research conference, it seems clear that RTOP is one of the most often used instruments when studying inquiry instruction and has been seen as the leader in this area over the past several years. A comparative study between RTOP and EQUIP identified many similarities and differences, but the conclusions report that EQUIP has higher inter-rater reliability, provides broader utility (effective and meaningful for both researchers and practitioners), is more targeted to inquiry-based instruction (not the more general constructivist approaches), provides a more comprehensive understanding of teaching performance (macro and micro look at specific teaching aspects), and includes a descriptive rubric to allow teachers and leaders to target tangible goals (Marshall, Smart, Lotter, & Sirbu, 2011).

EQUIP is a valid instrument that was developed, tested, and refined over the past four years and is increasingly being seen as a solid and comprehensive tool for tracking inquiry-based instruction (Marshall, et al., 2010). The framework for EQUIP originated from joining components of many existing frameworks in an attempt to provide a means for studying teachers' transformation toward greater quantity and quality of inquiry-based instruction (Horizon Research, 2002; Llewellyn, 2007; Sampson, 2004; Sawada, et al., 2000).

Although inquiry is not the only student-centered instructional strategy, it is a critical strategy that should be part of every science classroom (Bransford, et al., 2000; NRC, 1996, 2000). For clarity, I use an established definition of inquiry, set forth by *NSES*.

Inquiry is a multifaceted activity that involves making observations; posing questions; examining books and other sources of information to see what is already known; planning investigations; reviewing what is already known in light of experimental evidence; using tools to gather, analyze, and interpret data; proposing answers, explanations and predictions; and communicating the results. Inquiry requires identification of assumptions, use of critical and logical thinking, and consideration of alternative explanations (NRC, 1996, p. 23).

Various nuances of inquiry are further detailed in the *NSES* and in other research documents and publications (Karplus, 1977; Llewellyn, 2002, 2007; NRC, 2000; NRC, 2012), but the essence of scientific inquiry is clear—students critically and systematically engage in examining, interpreting, and analyzing questions regarding the world around them, and then communicate their findings, providing convincing arguments for their conclusions.

EQUIP OVERVIEW

So how do we assess the quality of the inquiry that teachers lead in their classrooms? How can this assessment be used to improve teacher performance and ultimately student achievement? This chapter provides an overview of the EQUIP instrument, shares the reliability and validity information that EQUIP is based on, offers an example of instruction matched to the scoring, and then shares how

EQUIP provides a predictive indicator of student performance (both content and process). This protocol, which we have named the Electronic Quality of Inquiry Protocol (EQUIP), can be used as a snapshot to measure the quality of inquiry on several indicators for a given class or as a guide that outlines specific areas for teachers to target for growth. EQUIP provides a reliable and valid resource to measure the quality of inquiry that is being facilitated within classrooms. We researched, developed, and refined EQUIP over a period of three years.

Good teachers use many different instructional methods throughout a day, a week, and a year. EQUIP is not designed for all situations; it specifically focuses on the factors associated with the quality of inquiry-based instruction being facilitated by teachers, not with other methods that may be used in the classroom. The complete EQUIP instrument can be accessed from www.clemson.edu/iim. The first three sections of EQUIP are intended mainly for researchers; whereas the remaining sections of EQUIP are useful for both researchers and practitioners alike.

Section 1 contains basic demographic information and descriptive lesson information such as objectives and standards. Section 2 is used to rate five-minute snapshots on several issues such as student attention and cognitive level. Section 3 is used for field notes that teachers may or may not want to use, depending on their goals for individual growth. The column concerning "Classroom Notes of Observation" is for the evaluator to indicate as objectively as possible what has transpired in the class. The "Comments" column is for the evaluator to express her/his interpretation of what is going on; consequently it is more subjective.

Sections 4-8 of EQUIP are critical for researchers and practitioners and should be completed at the culmination of the observed lesson. The quantitative data from these sections of EQUIP include 19 indicators and five different composites. The 19 indicators are divided into the following four constructs:

- **Instruction (5 items):** Section 4.
- **Discourse (5 items):** Section 5.
- **Assessment (5 items):** Section 6.
- **Curriculum (4 items):** Section 7.

After scoring each indicator, five composite scores are generated—Section 8 (one for each construct plus an overall score for the lesson). Each indicator and composite score can range from 1-4 (Level 1 = Pre-inquiry, Level 2 = Developing Inquiry, Level 3 = Proficient Inquiry, and Level 4 = Exemplary Inquiry). The composite scores are based on the essence for that composite rather than an average of all indicators in the composite. I encourage teachers to avoid becoming defensive about the ratings; it is more important to understand why a score falls into a specific level and what can be done to advance to a higher level in the future then to argue over a particular score.

Once a benchmark measurement is determined, teachers can begin, individually or in teams, to chart the growth and target areas where improvement is desired. The descriptive nature of EQUIP helps teachers move from the "I know it when I see it" to an understanding of the specific aspects that were or were not effective. The insights gained when using EQUIP can provide a foundation for developing a plan that will ultimately improve instruction and student learning. See Table 1 for uses of EQUIP.

EQUIP RELIABILITY AND VALIDITY INFORMATION

EQUIP was created in response to a need for a reliable and valid instrument to assess the quantity and quality of inquiry in K-12 math and science classrooms (Marshall, et al., 2010). None of the other protocols met our specific needs for guiding teachers as they plan and implement inquiry-based instruction and for assessing the quantity and quality of inquiry instruction. EQUIP's structure provides both a formative and summative means

Table 1. Various uses of EQUIP paired with intended audience

Use of EQUIP	Teacher	Instructional Leader	Researchers
Establish benchmarks and then chart growth over time	X	X	X
Work with teachers to target growth in performance		X	X
Reflect back upon a lesson	X		
Videotape lesson and then complete protocol either alone or with peers during a replay of the lesson	X	X	X
Complete the instrument while observing another teacher's class	X	X	
Guide conversations with a teacher or team of teachers		X	X

to study inquiry-based instruction in K-12 science and math classrooms. EQUIP was specifically designed to (1) evaluate teachers' classroom practice, (2) evaluate PD program effectiveness, and (3) guide reflective practitioners as they try to increase the quantity and quality of inquiry. Though EQUIP is designed to measure both quantity and quality of inquiry instruction, the reliability and validity issues associated with only the quality of inquiry are addressed below.

Face validity: Five science education researchers, four math education researchers, and two doctoral students in Curriculum and Instruction from four universities helped assess the face validity of EQUIP. Further, two measurement experts with knowledge of instrument development assessed the instrument structure. To guide face validity conversations, we posed the following questions. Does EQUIP seem well-designed way to assess the quality of inquiry? Does it seem as though it will provide reliable measures? For the content specialists, does it maintain fidelity to the discipline (math/science)? Does each indicator, along with descriptor, provide a critical measure that seamlessly progresses from non-inquiry to exemplary inquiry? Finally, does a Level 3 descriptor provide an accurate benchmark representation of proficiency for a given indicator? Through a series of face-to-face meetings, email communication, and phone conversations, each indicator with the accompanying descriptor was scrutinized until both educational researchers and individuals conducting measurements in the field achieved consensus. Negotiation and refinements centered on balancing what theory suggested with what was consistently measurable.

Internal consistency: EQUIP indicators were examined for internal consistency using Cronbach's Alpha (α) for 102 class observations used for field-testing. The α-value ranged from .880-.889, demonstrating strong internal consistency. Further, the indicators that comprise the instrument hold together well in both science and mathematics settings.

Inter-rater reliability: Cohen's Kappa (κ) was used to determine inter-rater reliability from 16 paired observations. Using the Landis and Koch (1977) interpretative scale, κ scores averaged .6 and thus fall between moderate and substantial agreement. Further, since EQUIP items fall along a continuum and not an absolute discrete scale, it is not surprising that Kappa scores were not close to 1.0, which would show total agreement for all items. For these 16 paired observations, the coefficient of determination, r^2, was .856, indicating a strong collective agreement between raters. Specifically, 85.6% of Observer B's assessment is explained by Observer A's assessment and vice versa.

Content and construct validity: Once face validity and high reliability had been established, content validity was examined to provide a deeper analysis of the validity surrounding the instrument.

In assessing content validity, we are essentially asking: How well does EQUIP represent the domain it is designed to represent? In this instance, EQUIP was designed to represent components associated with the quality of inquiry, as defined by the research literature. In order to establish content validity, the primary constructs measures in EQUIP were aligned with key literature associated with inquiry-based instruction. Since only the indicators that remain in the model will be justified with research literature, content validity and construct validity are addressed together.

In evaluating construct validity, a Confirmatory Factor Analysis (CFA) using Structural Equation Modeling (SEM) was ran with model trimming used to eliminate any indicators that did not contribute significantly to each construct. The resulting more parsimonious model, trimmed the 26 total indicators to 19 (five for Instruction, four for Curriculum, five for Discourse, and five for Assessment).

Final EQUIP model: The five indicators (with the theory and research to justify) that comprise the Instructional Factors include: (1) *Instructional Strategies* (Abell & Lederman, 2007; Bransford, et al., 2000; Chiappetta & Koballa, 2006; National Research Council, 2000), (2) *Order of Instruction* (Abell & Lederman, 2007; Biggs, 1996; Bybee, et al., 2006), (3) *Teacher Role* (Lampert, 1990; Mortimer & Scott, 2003; National Research Council, 1996; van Zee, Iwasyk, Kurose, Simpson, & Wild, 2001), (4) *Student Role* (Cobb, Wood, & Yackel, 1990), and (5) *Knowledge Acquisition* (Chinn & Brewer, 1998; Mortimer & Scott, 2003). Note that all four constructs that frame the EQUIP has been thoroughly discussed and validated in prior work (Marshall, 2009). The descriptive rubric used to measure all five Instructional Factor indicators along with the 14 other indicators is found at www.clemson.edu/iim.

The four indicators that were identified in the CFA that comprised the Curriculum Factor construct, along with literature to support its inclusion, include: (1) *Content Depth* (Schmidt, McNight, & Raizen, 2002; Wiggins & McTighe, 1998), (2) *Learner Centrality* (Donovan & Bransford, 2005; Knowles & Brown, 2000; NBPTS, 2000; NRC, 1996), (3) *Integration of Content and Investigation* (Llewellyn, 2002, 2007; Luft, Bell, & Gess-Newsome, 2008; NRC, 2000), and (4) *Organizing and Recording Information* (Marzano, Pickering, & Pollock, 2001).

The five tightly aligned indicators identified as part of the Discourse construct include: (1) *Questioning Level* (Krathwohl, 2002; Vygotsky, 1978),(2) *Complexity of Questions* (Chin, 2007), (3) *Questioning Ecology* (Morge, 2005; Mortimer & Scott, 2003), (4) *Communication Pattern* (Kelly, 2007; Lemke, 1990; Moje, 1995), and (5) *Classroom Interaction* (Lampert, 1990; van Zee, et al., 2001).

The final construct, Assessment, contains the following five indicators: (1) *Prior Knowledge* (Bransford, et al., 2000; Chambers & Andre, 1997), (2) *Conceptual Development* (Driver, Squires, Rushworth, & Wood-Robinson, 1994), (3) *Student Reflection* (Mezirow, 1990; White & Frederiksen, 1998, 2005; Wiggins & McTighe, 1998), (4) *Assessment Type(s)* (Black, Harrison, Lee, Marshall, & Wiliam, 2004; Black & Wiliam, 1998), and (5) *Role of Assessing* (Bell & Cowie, 2001; Stiggins, 2005; Stigler & Hiebert, 1999).

Three indicators were trimmed from the original model even though their importance in quality instruction can be easily justified. It was determined that these three items are too difficult to measure accurately by a single item during a single class period observation and include the following items: (1) *Teacher Content Knowledge*, (2) *Meaningful Context*, and (3) *Fundamental Ideas*.

Although absolute parameters for SEM do not readily exist, the values and justifications for the model include the following: χ^2 is significant p < .001, $\chi^2/df \leq 2$ indicates reasonable fit (Kline, 2005), RMSEA of .1 is on the threshold of reasonable fit (Browne & Cudeck, 1993), SRMR < .1 is considered favorable (Kline, 2005), and the computerized fit index, CFI, of > .90 is considered a good fit (Hu & Bentler, 1999). The four-construct model, 19-indicator model, provides a good-fitting model that also is solidly supported by the literature base regarding effective inquiry instruction.

SAMPLE LESSON ANALYZED USING EQUIP

In the following pages, the factors of Instruction, Discourse, Assessment, and Curriculum are discussed in more depth as I illustrate the application of EQUIP. Again, the indicators for these factors are all assessed at the end of the observational period. The example comes from a physical science lesson framed by the essential question *"What factors affect the motion of an object?"* (Portions of the example and discussion that follows were published in *The Science Teacher* (Marshall, Horton, & White, 2009)). In the observation from which the example is drawn, the teacher provided teams of 3 or 4 students with mousetrap racer kits and challenged them to create a mousetrap racer that would go 5 meters the fastest, but would stop before it had traveled 6 meters. This competition incorporated process skills (e.g., asking good scientific questions, collecting meaningful data, analyzing results), and conceptual ideas (e.g., speed, motion, force, conservation of energy) from science, math, and engineering.

Instruction: Table 2 shows two of five indicators that comprise the factor of Instruction. I discuss the ratings for the science observation for the example described above. Because the indicators are associated with the same factor, there are connections among them. However, these connections are not absolute; there are sufficient distinctions among the indicators so that the levels often vary considerably even within the same factor.

Because the teacher provided the vehicle assembly instructions before students had sufficient time to think through their own creation and because she stopped and lectured about the terminology associated with motion, the *Instructional Strategies* earned a Level 2 inquiry rating. Had the teacher provided more opportunities for input of student ideas throughout the investigation, then the quality of the inquiry would have been at least Level 3.

The teacher did, however, achieve a Level 3 inquiry rating for *Order of Instruction* because the lesson engaged the students in exploring concepts before the teacher explained them, and students were involved in explaining their conceptual ideas to the teacher and their peers.

Discourse: Discourse measures the classroom climate and interactions relating to inquiry instruc-

Table 2. Sample of EQUIP instructional indicators associated with inquiry-based instruction

Indicator Measured	Pre-Inquiry (Level 1)	Developing Inquiry (Level 2)	Proficient Inquiry (Level 3)	Exemplary Inquiry (Level 4)
Instructional Strategies	Teacher predominantly lectured to cover content.	Teacher frequently lectured and/or used demonstrations to explain content. Activities were verification only.	Teacher occasionally lectured, but students were engaged in activities that helped develop conceptual understanding.	Teacher occasionally lectured, but students were engaged in investigations that promoted strong conceptual understanding.
Order of Instruction	Teacher explained concepts. Students either did not explore concepts or did so only after explanation.	Teacher asked students to explore concept before receiving explanation. Teacher explained.	Teacher asked students to explore before explanation. Teacher and students explained.	Teacher asked students to explore concept before explanation occurred. Though perhaps prompted by the teacher, students provided the explanation.

tion and learning. Two of five indicators associated with this factor are shown in Table 3.

As the lesson progressed, the teacher provided challenging, higher-level questions (e.g., How did your results compare with those from other groups?) as students presented their findings, which resulted in a Level 3 inquiry rating for *Questioning Level*. However, once students responded to the higher-level questions, the quality of the interactions dropped as the teacher followed-up responses with only low-level probes (e.g., How did you find the second point on the graph?). This resulted in a rating of Level 2 for *Classroom Interaction*. The teacher could raise this score by following-up student responses with more thought-provoking questions such as, "Why was the slope calculated by group 2 larger than the slope calculated by group 1? What does that slope tell us?"

Assessment: Five indicators are used to measure the Assessment factor relating to instructional practice. Two of the indicators are shown in Table 4.

Because the teacher did not attempt to assess or take into consideration the prior knowledge students possessed, the lesson earned a Level 1 inquiry rating for *Prior Knowledge*. A short pretest, a KWL chart, or even a discussion concerning what students already knew may have revealed strengths or, on the other hand, some misconceptions regarding motion that should be addressed. The teacher also fell short on *Conceptual Development*. When formative assessments are implemented throughout the lesson, student-learning increases. By making the lesson more prescribed than necessary, critical thinking was minimized. This resulted in a Level 2 rating for this indicator.

Table 3. Sample of EQUIP discourse indicators associated with inquiry-based instruction

Indicator Measured	Pre-Inquiry (Level 1)	Developing Inquiry (Level 2)	Proficient Inquiry (Level 3)	Exemplary Inquiry (Level 4)
Questioning Level	Questioning rarely challenged students above the remembering level.	Questioning rarely challenged students above the understanding level.	Questioning challenged students up to application or analysis levels.	Questioning challenged students at various levels, including at the analysis level or higher; level was varied to scaffold learning.
Classroom Interaction	Teacher accepted answers, correcting when necessary, but rarely followed-up with further probing.	Teacher or another student occasionally followed-up student response with further low-level probe.	Teacher or another student often followed-up response with engaging probe that required student to justify reasoning or evidence.	Teacher consistently and effectively facilitated rich classroom dialogue where evidence, assumptions, and reasoning were challenged by teacher or other students.

Table 4. Sample of EQUIP assessment indicators associated with inquiry-based instruction

Indicator Measured	Pre-Inquiry (Level 1)	Developing Inquiry (Level 2)	Proficient Inquiry (Level 3)	Exemplary Inquiry (Level 4)
Prior Knowledge	Teacher did not assess student prior knowledge.	Teacher assessed student prior knowledge but did not modify instruction based on this knowledge.	Teacher assessed student prior knowledge and then partially modified instruction based on this knowledge.	Teacher assessed student prior knowledge and then modified instruction based on this knowledge.
Conceptual Development	Teacher encouraged learning by memorization and repetition.	Teacher encouraged product- or answer-focused learning activities that lacked critical thinking.	Teacher encouraged process-focused learning activities that required critical thinking.	Teacher encouraged process-focused learning activities that involved critical thinking that connected learning with other concepts.

When students are challenged to defend their solutions to scientific questions, a Level 3 or 4 rating is appropriate.

Curriculum: The EQUIP includes four indicators associated with various Curriculum issues related to inquiry instruction. These indicators are tied directly to what is experienced by students, not what appears in a text or notes. *Organizing and Recording Information* is one of several areas in which teachers can provide students with different levels of scaffolding—thus differentiating instruction. The goal is to challenge all students to their highest level while not overly frustrating anyone. For instance, one student with a learning disability may need the structure that a graphic organizer provides, whereas an ESL student may need more visuals to help decode the language barriers. We should always strive to help students progress to a level where less direct assistance is needed. By doing so, we will have encouraged and helped to develop habits of lifelong learning. To earn Level 4 on this and other indicators, teachers should consider the various needs of *all* students in their class. Two of the Curriculum indicators, *Integration of Content and Investigation* and *Organizing and Recording Information*, are displayed in Table 5.

The *Integration of Content and Investigation* earned a Level 3 inquiry rating because the investigation almost continually integrated concepts such as speed vs. time graphs and conservation of energy into the student investigations. *Organizing and Recording Information* was scored at Level 2 because the teacher provided little opportunity for the students to determine how the data should be collected and organized. When data sheets are provided with the headings and axes already labeled, which is what happened during this observation, students are deprived of a rich opportunity to think about how to collect, organize, and convey meaning from the data. By having the opportunity to organize and record information as they see fit, students think more deeply and more critically about the concepts being investigated (e.g., how many trials are needed? Is speed the independent or dependent variable, and why?). Had the teacher provided this opportunity, the rating for this indicator would have risen to a Level 3 or 4.

IMPROVING QUALITY OF INQUIRY TEACHING

After each of the indicators associated with the four factors has been assessed, in Section 8 of EQUIP an overall, or holistic, rating is determined for each factor. Again, this holistic rating is not necessarily the mean of the indicators, but is the Level that best captures the essence of the lesson. Though it may seem that approaching the rating this way would make this section overly subjective, we have found that our inter-rater reliability, or consistency between different raters, is quite high.

Table 5. Sample of EQUIP curriculum indicators associated with inquiry-based instruction

Indicator Measured	Pre-Inquiry (Level 1)	Developing Inquiry (Level 2)	Proficient Inquiry (Level 3)	Exemplary Inquiry (Level 4)
Integration of Content & Investigation	Lesson either content-focused or activity-focused but not both.	Lesson provided poor integration of content with activity or investigation.	Lesson incorporated student investigation that linked well with content.	Lesson seamlessly integrated the content and the student investigation.
Organizing & Recording Information	Students organized and recorded information in prescriptive ways.	Students had only minor input as to how to organize and record information.	Students regularly organized and recorded information in non-prescriptive ways.	Students organized and recorded information in non-prescriptive ways that allowed them to effectively communicate their learning.

Once the instrument has been completed and the current state of inquiry instruction is established, the next step is to improve the quality of inquiry. Though establishing the benchmark may bring about some change just by having specific aspects of instructional practice brought to the teacher's attention, the goal is to become more intentional and explicit by developing an action plan of next steps. It is normal to desire to improve everything that ails our instruction all at once. However, such a course of action often leads to frustration and undue anxiety; effective change is usually incremental.

I recommend that teachers focus on one specific indicator that they wish to improve upon during the next lesson or unit of study. Once the desired growth has been achieved, then it is time to tackle another indicator. After four indicators relating to inquiry instruction have been improved, perhaps one from each of the factors Instruction, Discourse, Assessment, and Curriculum, the teacher should strive to maintain that level of performance before undertaking more improvements. If teachers work together and note common areas for growth, it may make sense to work on certain indicators together. This shared approach provides a support structure to exchange thoughts and ideas.

If current practice falls largely in Level 1, then it makes sense to begin reading about constructivist approaches to learning and inquiry-based methods of teaching, looking for examples of lessons and instruction. Many articles from NCTM journals (e.g. *The Science Teacher*, *Science Scope*), along with journals from other professional organizations, provide these, along with many innovative ideas that can be of immense value to teachers who wish to modify their practice. In addition, teachers may seek out any one of many professional development institutes that provide opportunities to experience inquiry learning firsthand. Generally, a Level 2 performance suggests that a teacher is familiar with getting students engaged and active, but the lessons tend to be more prescriptive, with students having only limited opportunities to develop the ideas for themselves. Additionally, instruction is still heavily teacher-focused. At Level 3, the teacher has demonstrated a student-centered inquiry-learning environment that actively engages students in investigations, questioning, and explanations. The role of the teacher remains vital (as it does at all levels), but she now functions more as a facilitator who scaffolds learning experiences than as a giver of facts and knowledge.

I do not expect that any one lesson would merit a Level 4 for all indicators or even for all factors. In fact, I have yet to see such a lesson, and I have seen some amazing lessons. Further, the point is not to make every instructional moment a Level 3 or higher; rather, the goal is to help teachers become more intentional about their practice. By making teachers aware of what high quality inquiry practice entails, I believe they will be more likely to implement it successfully when it is their desired instructional approach.

EQUIP provides teachers with a concrete way to reflect on their own teaching practice as they strive to lead inquiry-based learning experiences in their classroom. Inquiry instruction is challenging to implement well, but, when done effectively, learning is clearly evident with *all* students and at *all* ability levels.

EQUIP AND STUDENT ACHIEVEMENT

The value of EQUIP is partially predicated on the fact that it provides a reliable, valid metric to measure of inquiry-based instruction. In and of itself, this is valuable, but the ultimate goal is to find an instrument that helps predict student achievement based on teacher performance. Thus, the challenge becomes finding the most appropriate dependent measure for student achievement. Specifically, what metric would provide valuable student achievement growth data with minimal testing bias?

The Measures of Academic Progress (MAP) from Northwest Evaluation Association (NWEA) was selected as the metric to study student growth in science classrooms. MAP, a reliable and valid assessment (NWEA, 2004) used by schools in 48 states, is an adaptive test that provides either more or less challenging items, depending on students' success or failure on previous questions. Further, because it is aligned with state and national science standards (NWEA, 2005), MAP can pinpoint students' current level of achievement. Students are assessed both in the fall and in the spring; hence, growth during the majority of the academic year can be readily determined. MAP also allows success to be studied with various ability levels, thus providing a means to research possible effects on the achievement gap.

The MAP test provides a mechanism to measure the growth of students in both science content and science process during the course of a given academic year for elementary through early high school grades. Further, in a collaborative effort with NWEA (funding by an NSF grant), I was able to compare the growth of the students in my study group to similar students from around the state who also took the test, the Virtual Comparison Group (VCG). The VCG is comprised of students who: 1) are the same race and gender, 2) took the test during the same test window (both pre- and post-), and 3) attend a school with a similar free and reduced lunch status (a measure of poverty in the school). Each of the students in the intervention is compared to 21-51 like students from the VCG.

MAP inherent strengths are several. First, because test items are aligned to state science standards, it has high predictive validity when compared to other state assessments (Cronin, Kingsbury, Dahlin, Adkins, & Bowe, 2007; NWEA, 2005). Second, performance of a given teacher's students can be studied at both in terms of science content and process knowledge scores without requiring an additional test. Third, because it is adaptive, MAP provides a broader, more robust sample of the entire domain than a fixed-form test does (NWEA, 2003). Finally, since the districts that I work with already use MAP, no additional testing is necessary to obtain a reliable and valid measure of student performance.

With EQUIP as the predictor measure (teacher performance) and MAP scores as the dependent measure (student achievement), a correlation was analyzed to see if teacher performance relative to inquiry-based instruction could partially predict student achievement growth during the academic year. The results seemed mixed at first, but then a clear pattern emerged. Specifically, it was found that EQUIP could not predict for student growth in content knowledge or process skills during the first year of involvement in transformation of practice—no significance found. This lack of significance was attributed to several reasons: 1) teacher performance is unstable during times of transformation (larger variation of practice), 2) teachers tend to "perform" for observers (evaluators) during the first year often providing what they think is expected but which is often atypical of most of their instruction, and 3) inquiry teaching is an inconsistent part of everyday practice.

During the second year of involvement with teachers through a long-term professional development training program, it was found that teachers' changes in practice became more solidified and consistent. In addition to instruction becoming more consistent relative to inquiry, EQUIP became a powerful predictor of the growth seen in students during the academic year. Specifically, the average overall lesson score earned on the EQUIP for a teacher over the course of four observations, was able to explain approximately 36% of variance in the student growth. These data will continue to be analyzed as more teachers become involved in the second year portion of the program. However, the initial two years of data indicate a very power metric that could be used to improve student achievement in both science content and process. It would be expected that inquiry-based instruction increases students' performance on science process knowledge, but these initial results suggest that the effect is equally high for science content knowledge as well.

IMPLICATIONS

In an effort to help teachers move beyond an "I know it when I see it" mentality regarding inquiry, I propose the use of EQUIP as an instrument to help guide teacher practice to greater quantity and quality of inquiry-based instruction. The evidence suggests that when high-quality, proficient inquiry (Level 3 on EQUIP) becomes a consistent portion of instruction then the growth in student achievement exceeds a Virtual Comparison Group (VCG) of typical instructional practice. It is infrequent in educational research to find strong predictive indicators of things that when done by the teacher in the classroom result in improved student achievement.

Further, it is exciting to confirm that when inquiry is implemented well in the classroom that student achievement exceeds the comparison group on process skill knowledge (e.g., interpreting a graph or designing a study). However, these findings also show that we do not need to just "teach to the test" (e.g., tell, rehearse, memorize), because as students improve process knowledge, it is also possible for them to improve their content knowledge at a rate that also significantly exceeds the VCG.

Obviously, EQUIP is not a single panacea for all that ails our education system, but it does provide a clear, descriptive means to guide the transformation of instructional practice so that student learning is greatly increased via inquiry-based forms of instruction. Because of the complex, multifaceted nature of inquiry instruction, it has been very challenging to develop a protocol that assesses the quality of inquiry instruction in a valid and reliable manner. EQUIP seems to have met that challenge and was designed to (1) evaluate teachers' classroom practice, (2) evaluate PD program effectiveness, and (3) provide a tool to guide reflective practitioners as they strive to increase the quantity and quality of inquiry that they lead in their classrooms (Marshall, Horton, & White, 2009). The culminating four-construct (Instruction, Curriculum, Discourse, and Assessment) EQUIP is a reliable and valid instrument that meets these goals.

ACKNOWLEDGMENT

This material is based upon the work supported by the National Science Foundation under Grant #DRL-0952160 and from a grant from the South Carolina Commission on Higher Education under the auspices of the EIA Teacher Education Centers of Excellence Grant Program. Any opinions, findings, and conclusions or recommendations expressed in this material are those of the author(s) and do not necessarily reflect the views of the National Science Foundation.

REFERENCES

Abell, S. K., & Lederman, N. G. (2007). *Handbook of research on science education*. Mahwah, NJ: Lawrence Erlbaum Associates.

Adams, P. E., & Krockover, G. H. (1999). Stimulating constructivist teaching styles through use of an observation rubric. *Journal of Research in Science Teaching*, 36(8), 955–971. doi:10.1002/(SICI)1098-2736(199910)36:8<955::AID-TEA4>3.0.CO;2-3

Beerer, K., & Bodzin, A. (2003). *Science teacher inquiry rubric (STIR)*. Retrieved April 25, 2007, from http://www.lehigh.edu/~amb4/stir/stir.pdf

Bell, B., & Cowie, B. (2001). The characteristics of formative assessment in science education. *Science Education*, 85, 536–553. doi:10.1002/sce.1022

Biggs, J. (1996). Enhancing teaching through constructive alignment. *Higher Education*, *32*(3), 347–364. doi:10.1007/BF00138871

Black, P., Harrison, C., Lee, C., Marshall, B., & Wiliam, D. (2004). Working inside the black box: Assessment for learning in the classroom. *Phi Delta Kappan*, *86*(1), 9–21.

Black, P., & Wiliam, D. (1998). Assessment and classroom learning. *Assessment in Education*, *5*(1), 7–74. doi:10.1080/0969595980050102

Bransford, J. D., Brown, A. L., & Cocking, R. R. (2000). *How people learn: Brain, mind, experience, and school*. Washington, DC: National Academies Press.

Browne, M. W., & Cudeck, R. (1993). Alternative ways of assessing model fit. In Bollen, K. A., & Long, J. S. (Eds.), *Testing Structural Equation Models* (pp. 136–162). Beverly Hills, CA: Sage.

Burry-Stock, J., & Oxford, R. (1994). Expert science teaching edcuational evaluation model (ESTEEM): Measuring excellence in science teaching for professional development. *Journal of Personnel Evaluation in Education*, *8*(3), 267–297. doi:10.1007/BF00973725

Bybee, R. W., Taylor, J. A., Gardner, A., Scotter, P. V., Powell, J. C., & Westbrook, A. (2006). *The BSCS 5E instructional model: Origins, effectiveness, and applications*. Colorado Springs, CO: BSCS.

Capps, D. K., Crawford, B. A., & Constas, M. A. (2012). A review of empirical literature on inquiry professional development: Alignment with best practices and a critique of the findings. *Journal of Science Teacher Education*, *23*(3), 291–318. doi:10.1007/s10972-012-9275-2

Chambers, S. K., & Andre, T. (1997). Gender, prior knowledge, interest and experience in electricity and conceptual change text manipulations in learning about direct current. *Journal of Research in Science Teaching*, *34*(2), 107–123. doi:10.1002/(SICI)1098-2736(199702)34:2<107::AID-TEA2>3.0.CO;2-X

Chiappetta, E. L., & Koballa, T. R. J. (2006). *Science instruction in the middle and secondary schools: Developing fundamental knowledge and skills for teaching* (6th ed.). Upper Saddle River, NJ: Pearson Perrill Prentice Hall.

Chin, C. (2007). Teacher questioning in science classrooms: Approaches that stimulate productive thinking. *Journal of Research in Science Teaching*, *44*(6), 815–843. doi:10.1002/tea.20171

Chinn, C. A., & Brewer, W. F. (1998). Theories of knowledge acquisition. In Fraser, B. J., & Tobin, K. (Eds.), *International Handbook of Science Education* (pp. 97–113). London, UK: Kluwer Academic Publishers.

Cobb, P., Wood, T., & Yackel, E. (1990). Classrooms as learning environments for teachers and researchers. In Davis, R. B., Maher, C. A., & Noddings, N. (Eds.), *Constructivist Views of the Teaching and Learning of Mathematics* (pp. 125–146). Reston, VA: NCTM. doi:10.2307/749917

Cronin, J., Kingsbury, G. G., Dahlin, M., Adkins, D., & Bowe, B. (2007). *Alternate methodologies for estimating state standards on a widely-used computerized adaptive test*. Paper presented at the National Council on Measurement in Education. New York, NY.

Darling-Hammond, L., & McLaughlin, M. W. (1995). Policies that support professional development in an era of reform. *Phi Delta Kappan*, *92*(6), 81–92.

Donovan, M. S., & Bransford, J. D. (2005). *How students learn: History, mathematics, and science in the classroom*. Washington, DC: National Academies Press.

Driver, R., Squires, A., Rushworth, P., & Wood-Robinson, V. (1994). *Making sense of secondary science: Research into children's ideas*. London, UK: Taylor & Francis Ltd.

Garet, M. S., Porter, A. C., Desimone, L., Birman, B. F., & Yoon, K. S. (2001). What makes professional development effective? Results from a national sample of teachers. *American Educational Research Journal, 38*(4), 915–945. doi:10.3102/00028312038004915

Horizon Research. (2002). *Inside the classroom interview protocol*. Retrieved from http://www.horizon-research.com/instruments/clas/cop.php

Hu, L., & Bentler, P. M. (1999). Cutoff criteria in fix indexes in covariance structure analysis: Conventional criteria versus new alternatives. *Structural Equation Modeling, 6*(1), 1–55. doi:10.1080/10705519909540118

Karplus, R. (1977). Science teaching and the development of reasoning. *Journal of Research in Science Teaching, 14*, 169–175. doi:10.1002/tea.3660140212

Kelly, G. J. (2007). Discourse in science classrooms. In Abell, S. K., & Lederman, N. G. (Eds.), *Handbook of Research on Science Education*. Mahwah, NJ: Lawrence Erlbaum Associates.

Kline, R. B. (2005). *Principles and practice of structural equation modeling* (2nd ed.). New York, NY: Guilford Press.

Knowles, T., & Brown, D. F. (2000). *What every middle school teacher should know*. Portsmouth, NH: Heinemann.

Krathwohl, D. R. (2002). A revision of Bloom's taxonomy: An overview. *Theory into Practice, 41*(4), 212–218. doi:10.1207/s15430421tip4104_2

Lampert, M. (1990). When the problem is not the question and the solution is not the answer: Mathematical knowing and teaching. *American Educational Research Journal, 27*(1), 29–63.

Landis, J. R., & Koch, G. G. (1977). The measurement of observer agreement for categorical data. *Biometrics, 33*, 159–174. doi:10.2307/2529310

Lemke, J. L. (1990). *Talking science: Language, learning, and values*. Norwood, NJ: Ablex.

Llewellyn, D. (2002). *Inquiry within: Implementing inquiry-based science standards*. Thousand Oaks, CA: Corwin Press.

Llewellyn, D. (2007). *Inquiry within: Implementing inquiry-based science standards in grades 3-8* (2nd ed.). Thousand Oaks, CA: Corwin Press.

Loucks-Horsley, S., Stiles, K. E., Mundry, S., Love, N., & Hewson, P. W. (2010). *Designing professional development for teachers of science and mathematics* (3rd ed.). Thousand Oaks, CA: Corwin Press.

Luft, J., Bell, R. L., & Gess-Newsome, J. (2008). *Science as inquiry in the secondary setting*. Arlington, VA: National Science Teachers Association.

Marshall, J. C. (2009). *The creation, validation, and reliability associated with the EQUIP (electronic quality of inquiry protocol): A measure of inquiry-based instruction*. Paper presented at the National Association of Researchers of Science Teaching Conference. Washington, DC.

Marshall, J. C., Horton, B., & Edmondson, E. (2007). *4E x 2 instructional model*. Retrieved from http://www.clemson.edu/iim

Marshall, J. C., Horton, B., Igo, B. L., & Switzer, D. M. (2009). K-12 science and mathematics teachers' beliefs about and use of inquiry in the classroom. *International Journal of Science and Mathematics Education, 7*(3), 575–596. doi:10.1007/s10763-007-9122-7

Marshall, J. C., Horton, B., & White, C. (2009). EQUIPping teachers: A protocol to guide and improve inquiry-based instruction. *Science Teacher (Normal, Ill.), 76*(4), 46–53.

Marshall, J. C., & Horton, R. M. (2009). *Developing, assessing, and sustaining inquiry-based instruction: A guide for math and science teachers and leaders*. Saarbruecken, Germany: VDM Publishing House Ltd.

Marshall, J. C., Smart, J., & Horton, R. M. (2010). The design and validation of EQUIP: An instrument to assess inquiry-based instruction. *International Journal of Science and Mathematics Education, 8*(2), 299–321. doi:10.1007/s10763-009-9174-y

Marshall, J. C., Smart, J., Lotter, C., & Sirbu, C. (2011). Comparative analysis of two inquiry observational protocols: Striving to better understand the quality of teacher facilitated inquiry-based instruction. *School Science and Mathematics, 111*(6), 306–315. doi:10.1111/j.1949-8594.2011.00091.x

Marzano, R. J., Pickering, D. J., & Pollock, J. E. (2001). *Classroom instruction that works: Research-based strategies for increasing student achievement*. Alexandria, VA: ASCD.

Mezirow, J. (1990). *Fostering critical reflection in adulthood: A guide to transformative and emancipatory learning*. San Francisco, CA: Jossey-Bass.

Moje, E. B. (1995). Talking about science: An interpretation of the effects of teacher talk in a high school classroom. *Journal of Research in Science Teaching, 32*(4), 349–371. doi:10.1002/tea.3660320405

Morge, L. (2005). Teacher-pupil interaction: A study of hidden beliefs in conclusion phases. *International Journal of Science Education, 27*(8), 935–956. doi:10.1080/09500690500068600

Mortimer, E. F., & Scott, P. H. (2003). *Meaning making in secondary science classrooms*. Maidenhead, UK: Open University Press.

National Academy of Sciences. (2007). *Rising above the gathering storm: Energizing and employing America for a brighter economic future*. Washington, DC: National Academies Press.

National Board for Professional Teaching Standards. (2000). *A distinction that matters: Why national teacher certification makes a difference*. Greensboro, NC: Center for Educational Research and Evaluation.

National Research Council. (1996). *National science education standards*. Washington, DC: National Academies Press.

National Research Council. (2000). *Inquiry and the national science education standards: A guide for teaching and learning*. Washington, DC: National Academies Press.

National Research Council. (2012). *A framework for K-12 science education: Practices, crosscutting concepts, and core ideas*. Washington, DC: The National Academies Press.

National Science Teachers Association. (1998). *Standards for science teacher preparation*. Retrieved from http://www.nsta.org/main/pdfs/nsta98standards.pdf

Northwest Evaluation Association. (2003). *Technical manual*. Lake Oswego, OR: Northwest Evaluation Association.

Northwest Evaluation Association. (2004). *Reliability and validity estimates: NWEA achievement level tests and measure of academic progress*. Retrieved from http://www.nwea.org

Northwest Evaluation Association. (2005). *NWEA reliability and validity estimates: Achievement level tests and measures of academic progress.* Lake Oswego, OR: Northwest Evaluation Association.

Riggs, I. M., & Enochs, L. G. (1990). Toward the development of an elementary teacher's science teaching efficacy belief instrument. *Science Education, 74*(6), 625–637. doi:10.1002/sce.3730740605

Sampson, V. (2004). The science management observation protocol. *Science Teacher (Normal, Ill.), 71*(10), 30–33.

Sawada, D., Piburn, M., Falconer, K., Turley, J., Benford, R., & Bloom, I. (2000). *Reformed teaching observation protocol (RTOP).* Technical Report No. IN00-01. Phoenix, AZ: Arizona State University

Schmidt, W. H., McNight, C. C., & Raizen, S. A. (2002). *A splintered vision: An investigation of U.S. science and mathematics education.* Retrieved from http://imc.lisd.k12.mi.us/MSC1/Timms.html

Stiggins, R. (2005). From formative assessment to assessment FOR learning: A path to success in standards-based schools. *Phi Delta Kappan, 87*(4), 324–328.

Stigler, J. W., & Hiebert, J. (1999). *The teaching gap: Best ideas from the world's teachers for improving education in the classroom.* New York, NY: The Free Press.

van Zee, E. H., Iwasyk, M., Kurose, A., Simpson, D., & Wild, J. (2001). Student and teacher questioning during conversations about science. *Journal of Research in Science Teaching, 38*(2), 159–190. doi:10.1002/1098-2736(200102)38:2<159::AID-TEA1002>3.0.CO;2-J

Vygotsky, L. (1978). *Mind in society: The development of higher psychological processes.* Cambridge, MA: Harvard University Press.

White, B. Y., & Frederiksen, J. R. (1998). Inquiry, modeling, and metacognition: Making science accessible to all students. *Cognition and Instruction, 16*(1), 3–118. doi:10.1207/s1532690xci1601_2

White, B. Y., & Frederiksen, J. R. (2005). A theoretical framework and approach for fostering metacognitive development. *Educational Psychologist, 40*(4), 211–223. doi:10.1207/s15326985ep4004_3

Wiggins, G., & McTighe, J. (1998). *Understanding by design.* Alexandria, VA: ASCD.

Compilation of References

AAAS. (1990). *Science for all Americans*. Oxford, UK: Oxford University Press.

Abd-El-Khalick, F. (2001). Embedding nature of science instruction in preservice elementary science courses: Abandoning scientism, but..... *Journal of Science Teacher Education, 12*, 215–233. doi:10.1023/A:1016720417219

Abd-El-Khalick, F. (2005). Developing deeper understandings of nature of science: The impact of a philosophy of science course on preservice science teachers' views and instructional planning. *International Journal of Science Education, 27*, 15–42. doi:10.1080/0950069041 0001673810

Abd-El-Khalick, F. S., & Akerson, V. L. (2004). Learning about nature of science as conceptual change: Factors that mediate the development of preservice elementary teachers' views of nature of science. *Science Education, 88*, 785–810. doi:10.1002/sce.10143

Abd-El-Khalick, F., & Akerson, V. L. (2004). Learning as conceptual change: Factors that mediate the development of preservice elementary teachers' views of nature of science. *Science Education, 88*(5), 785–810. doi:10.1002/sce.10143

Abd-El-Khalick, F., Bell, R. L., & Lederman, N. G. (1998). The nature of science and instructional practice: Making the unnatural natural. *Science Education, 82*, 417–436. doi:10.1002/(SICI)1098-237X(199807)82:4<417::AID-SCE1>3.0.CO;2-E

Abd-El-Khalick, F., & Lederman, N. G. (2000). Improving science teachers' conceptions of the nature of science: A critical review of the literature. *International Journal of Science Education, 22*, 665–701. doi:10.1080/09500690050044044

Abell, S. K., & Lederman, N. G. (2007). *Handbook of research on science education*. Mahwah, NJ: Lawrence Erlbaum Associates.

Achieve, Inc. (2012). *The next generation science standards*. Retrieved May 18, 2012, from http://www.nextgenscience.org/next-generation-science-standards

Adams, J. (2012). Community science: Capitalizing on local ways of enacting science in science education. In Fraser, B., Tobin, K., & McRobbie, C. (Eds.), *The Second International Handbook of Science Education*. Dordrecht, The Netherlands: Springer. doi:10.1007/978-1-4020-9041-7_77

Adams, P. E., & Krockover, G. H. (1999). Stimulating constructivist teaching styles through use of an observation rubric. *Journal of Research in Science Teaching, 36*(8), 955–971. doi:10.1002/(SICI)1098-2736(199910)36:8<955::AID-TEA4>3.0.CO;2-3

Adams, W. K., Paulson, A., & Wieman, C. E. (2009). What levels of guidance promote engaged exploration with interactive simulations? In *Proceedings of PERC*. PERC.

Aguillard, D. (1999). Evolution education in Louisiana public schools: A decade following Edwards v. Aguillard. *The American Biology Teacher, 61*(3), 182–188. doi:10.2307/4450650

Aikenhead, G. (2004). Science-based occupations and the science curriculum: Concepts of evidence. *Science Education, 89*(2), 242–275. doi:10.1002/sce.20046

Aikenhead, G. J., & Jegede, O. J. (1999). Cross-cultural science education: A cognitive explanation of a cultural phenomenon. *Journal of Research in Science Teaching, 36*(3), 269–287. doi:10.1002/(SICI)1098-2736(199903)36:3<269::AID-TEA3>3.0.CO;2-T

Ainley, M. D. (1993). Styles of engagement with learning: Multidimensional assessment of their relationship with strategy use and school achievement. *Journal of Educational Psychology, 85*, 395–405. doi:10.1037/0022-0663.85.3.395

Akerson, V. L., & Abd-El-Khalick, F. S. (2003). Teaching elements of nature of science: A year long case study of a fourth grade teacher. *Journal of Research in Science Teaching, 40*, 1025–1049. doi:10.1002/tea.10119

Akerson, V. L., Abd-El-Khalick, F., & Lederman, N. G. (2000). Influence of a reflective explicit activity-based approach on elementary teachers' conceptions of nature of science. *Journal of Research in Science Teaching, 37*, 295–317. doi:10.1002/(SICI)1098-2736(200004)37:4<295::AID-TEA2>3.0.CO;2-2

Akerson, V. L., & Hanuscin, D. L. (2007). Teaching nature of science through inquiry: Results of a 3-year professional development program. *Journal of Research in Science Teaching, 44*, 653–680. doi:10.1002/tea.20159

Akerson, V. L., Morrison, J. A., & McDuffie, A. R. (2006). One course is not enough: Preservice elementary teachers' retention of improved views of nature of science. *Journal of Research in Science Teaching, 43*, 194–213. doi:10.1002/tea.20099

Akerson, V. L., Townsend, J. S., Donnelly, L. A., Hanson, D. L., Tira, P., & White, O. (2009). Scientific modeling for inquiring teachers network (SMIT'N): The influence on elementary teachers' views of nature of science, inquiry, and modeling. *Journal of Science Teacher Education, 20*, 21–40. doi:10.1007/s10972-008-9116-5

Akerson, V. L., & Volrich, M. L. (2006). Teaching nature of science explicitly in a first grade internship setting. *Journal of Research in Science Teaching, 43*, 377–394. doi:10.1002/tea.20132

Akindehin, F. (1988). Effect of an instructional package on preservice science teachers' understanding of the nature of science and acquisition of science-related attitudes. *Science Education, 72*, 73–82. doi:10.1002/sce.3730720107

Alexopoulou, E., & Driver, R. (1997). Small group discussions in physics: Peer interaction modes in pairs and fours. *Journal of Research in Science Teaching, 33*(10), 1099–1114. doi:10.1002/(SICI)1098-2736(199612)33:10<1099::AID-TEA4>3.0.CO;2-N

American Association for the Advancement of Science. (1993). *Benchmarks for science literacy.* Washington, DC: American Association for the Advancement of Science.

American Association of Physics Teachers. (2002). *AAPT statement on physics first.* College Park, MD: AAPT. Retrieved October 11, 2011, from http://www.aapt.org/Resources/policy/physicsfirst.cfm

American Chemical Society. (2008). *Workshop on increasing participation of Hispanic undergraduate students in chemistry.* Washington, DC: ACS.

American College Testing. (2005). *Developing the STEM (science, technology, engineering, mathematics) education pipeline.* Retrieved August 31, 2011, from http://www.act.org/research/policymakers/pdf/ACT_STEM_PolicyRpt.pdf

American Institute of Physics. (2011). *Focus on underrepresented minorities in high school physics: Results from the 2008-09 nationwide survey of high school physics teachers.* College Park, MD: AIP Statistical Research Center. Retrieved October 11, 2011, from http://www.aip.org/statistics/trends/hstrends.html

Ames, C. (1992). Achievement goals and the classroom motivational climate. In Schunk, D. H., & Meece, J. L. (Eds.), *Student Perceptions in the Classroom* (pp. 327–348). Hillsdale, NJ: Lawrence Erlbaum.

Ames, C., & Archer, J. (1988). Achievement goals in the classroom: Students' learning strategies and motivational processes. *Journal of Educational Psychology, 80*, 260–267. doi:10.1037/0022-0663.80.3.260

Ametller, J., & Scott, P. (2009). *Classroom discourse, dialogic teaching and teacher professional development in science education: A selective review of the literature.* Retrieved April 16, 2010, from https://www.ntnu.no/wiki/download/attachments/8324755/Final+draft_lit_review_Leeds.doc

Anderson, C. (2007). Perspectives on science learning. In Abell, S., & Lederman, N. (Eds.), *Handbook of Research on Science Education* (pp. 3–30). Mahwah, NJ: Lawrence Erlbaum.

Anderson, C. A. (2004). An update on the effects of playing violent video games. *Journal of Adolescence, 27*(1), 113–122. doi:10.1016/j.adolescence.2003.10.009

Anderson, C. A., & Bushman, B. J. (2001). Effects of violent video games on aggressive behavior, aggressive cognition, aggressive affect, physiological arousal, and prosocial behavior: A meta-analytic review of the scientific literature. *Psychological Science*, *12*(5), 353–359. doi:10.1111/1467-9280.00366

Anderson, D., Lucas, K., & Ginns, I. (2003). Theoretical perspectives on learning in an informal setting. *Journal of Research in Science Teaching*, *40*, 177–199. doi:10.1002/tea.10071

Anderson, R. D., & Mitchener, C. P. (1994). Research on science teacher education. In Gabel, D. L. (Ed.), *Handbook of Research on Science Teaching and Learning* (2nd ed., pp. 3–44). New York, NY: McMillan.

Andriessen, J., Baker, M., & Suthers, D. (2003). *Arguing to learn: Confronting cognitions in computer-supported collaborative learning environments*. Berlin, Germany: Springer.

Annetta, L. A. (2008). Video games in education: Why they should be used and how they are being used. *Theory into Practice*, *47*(3), 229–239. doi:10.1080/00405840802153940

Annetta, L. A., Cheng, M., & Holmes, S. (2010). Assessing twenty-first century skills through a teacher created video game for high school biology students. *Research in Science & Technological Education*, *28*(2), 101–114. doi:10.1080/02635141003748358

Annetta, L. A., Mangrum, J., Holmes, S., Collazo, K., & Cheng, M.-T. (2009). Bridging realty to virtual reality: Investigating gender effect and student engagement on learning through video game play in an elementary school classroom. *International Journal of Science Education*, 31.

Annetta, L. A., Minogue, J., Holmes, S. Y., & Cheng, M.-T. (2009). Investigating the impact of video games on high school students' engagement and learning about genetics. *Computers & Education*, *53*(1), 74–85. doi:10.1016/j.compedu.2008.12.020

Appleton, K. (2007). Elementary science teaching. In Abell, S. K., & Lederman, N. G. (Eds.), *Handbook of Research on Science Education* (pp. 493–535). Mahwah, NJ: Erlbaum.

Archive Inc. (2012). *Next generation science standards*. Retrieved 17 June 2012 from http://www.nextgenscience.org

Arends, R. I. (2000). *Learning to teach*. Boston, MA: McGraw Hill.

Aristotle. (1976). *The Nicomachean ethics* (Thomson, J. A. K., Trans.). London, UK: Penguin.

Aristotle. (2002). *The Nicomachean ethics* (Sachs, J., Trans.). New York, NY: Riverhead Books.

Ary, D., Jacobs, L. C., & Sorensen, C. (2006). *Introduction to research in education* (8th ed.). Wadsworth, UK: Cengage Learning.

Asay, L., & Orgill, M. (2010). Analysis of essential features of inquiry found in articles published in *The Science Teacher*, 1998-2007. *Journal of Science Teacher Education*, *21*(1), 57–79. doi:10.1007/s10972-009-9152-9

Aubusson, P., Griffin, J., & Kearney, M. (2012). Learning beyond the classroom: Implications for school science. In Fraser, B., Tobin, K., & McRobbie, C. (Eds.), *The Second International Handbook of Science Education* (pp. 1123–1134). Dordrecht, The Netherlands: Springer. doi:10.1007/978-1-4020-9041-7_74

Ausubel, D. P. (1963). *The psychology of meaningful verbal learning*. New York, NY: Grune and Stratton.

Ausubel, D. P., Novak, J. D., & Hanesian, H. (1978). *Educational psychology: A cognitive view* (2nd ed.). New York, NY: Holt, Rinehart, and Winston.

Bandura, A. (1986). *Social foundations of thought and action: A social cognitive theory*. Englewood Cliffs, NJ: Prentice-Hall.

Bandura, A. (1988). Self-regulation of motivation and action through goal systems. In Hamilton, V., Bower, G. H., & Frijda, N. H. (Eds.), *Cognitive Perspectives on Emotion and Motivation* (pp. 37–61). Dordrecht, The Netherlands: Kluwer Academic. doi:10.1007/978-94-009-2792-6_2

Bandura, A. (1997). *Self-efficacy: The exercise of control*. New York, NY: Freeman.

Bandura, A. (2006). Adolescent development from an agentic perspective. In Pajares, F., & Urdan, T. (Eds.), *Self-Efficacy Beliefs of Adolescents* (Vol. 5, pp. 1–43). Greenwich, CT: Information Age Publishing.

Barab, S. A., Scott, B., Siyahhan, S., Goldstone, R., Ingram-Goble, A., Zuiker, S., & Warren, S. (2009). Transformational play as a curricular scaffold: Using videogames to support science education. *Journal of Science Education, 18*(4), 305–320.

Barton, P. E. (1993). *Hispanics in science and engineering: A matter of assistance and persistence.* Princeton, NJ: Educational Testing Service.

Bazerman, C. (1988). *Shaping written knowledge : The genre and activity of the experimental article in science.* Madison, WI: University of Wisconsin Press.

Becker, K. (2007). Digital game-based learning once removed: Teaching teachers. *British Journal of Educational Technology, 38*(3), 478–488. doi:10.1111/j.1467-8535.2007.00711.x

Bee, H. (1997). *The developing child* (8th ed.). New York, NY: Longman.

Beerer, K., & Bodzin, A. (2003). *Science teacher inquiry rubric (STIR).* Retrieved April 25, 2007, from http://www.lehigh.edu/~amb4/stir/stir.pdf

Bell, B., & Cowie, B. (2001). *Formative assessment and science education.* Dordrecht, The Netherlands: Kluwer Academic Publishers.

Bell, B., & Cowie, B. (2001). The characteristics of formative assessment in science education. *Science Education, 85*, 536–553. doi:10.1002/sce.1022

Bell, P., Lewenstein, B., Shouse, A. W., & Feder, M. A. (2009). *Learning science in informal environments: People, places, and pursuits.* Washington, DC: The National Academies Press.

Bell, P., & Linn, M. C. (2000). Scientific arguments as learning artifacts: Designing for learning from the web with KIE. *International Journal of Science Education, 22*(8), 797–818. doi:10.1080/095006900412284

Bell, R. L., Lederman, N. G., & Abd-El-Khalick, F. (2000). Developing and acting upon one's conception of the nature of science: A follow-up study. *Journal of Research in Science Teaching, 37*, 563–581. doi:10.1002/1098-2736(200008)37:6<563::AID-TEA4>3.0.CO;2-N

Bell, R. L., Matkins, J. J., & Gansneder, B. M. (2011). Impacts of contextual and explicit instruction on preservice elementary teachers' understandings of the nature of science. *Journal of Research in Science Teaching, 48*, 414–436. doi:10.1002/tea.20402

Berger, K. S. (1988). *The developing person through the life span* (2nd ed.). New York, NY: Worth Publishers.

Berkman, M. B., & Plutzer, E. (2010). *Evolution, creationism, and the battle to control America's classrooms.* Cambridge, UK: Cambridge University Press. doi:10.1017/CBO9780511760914

Berland, L. K., & Reiser, B. J. (2009). Making sense of argumentation and explanation. *Science Education, 93*(1), 26–55. doi:10.1002/sce.20286

Bernstein, S. N. (2004). A limestone way of learning. *The Chronicle Review, 50*(7).

Biggs, J. (1996). Enhancing teaching through constructive alignment. *Higher Education, 32*(3), 347–364. doi:10.1007/BF00138871

Billeh, V. Y., & Hasan, O. E. (1975). Factors influencing teachers' gain in understanding the nature of science. *Journal of Research in Science Teaching, 12*, 209–219. doi:10.1002/tea.3660120303

Bixler, R. D., Carlisle, C. L., Hammitt, W. E., & Floyd, M. F. (1994). Observed fears and discomforts among urban students on school field trips to wildland areas. *The Journal of Environmental Education, 26*, 24–33. doi:10.1080/00958964.1994.9941430

Bixler, R. D., & Morris, B. (2000). Factors differentiating water-based wildland recreationists from nonparticipants: Implications for recreation activity instruction. *Journal of Park and Recreation Administration, 18*(2), 54–72.

Black, P., Harrison, C., Lee, C., Marshall, B., & Wiliam, D. (2004). Working inside the black box: Assessment for learning in the classroom. *Phi Delta Kappan, 86*(1), 9–21.

Black, P., & Wiliam, D. (1998). Assessment and classroom learning. *Assessment in Education, 5*(1), 7–74. doi:10.1080/0969595980050102

Black, P., & Wiliam, D. (1998). Inside the black box: Raising standards through classroom assessment. *Phi Delta Kappan, 80*(2).

Blake, C., & Scanlon, E. (2007). Reconsidering simulations in science education at a distance: Features of effective use. *Journal of Computer Assisted Learning, 23*(6), 491–502. doi:10.1111/j.1365-2729.2007.00239.x

Bleicher, R. E., & Lindgren, J. (2005). Success in learning science and preservice science teaching self-efficacy. *Journal of Science Teacher Education, 16,* 205–225. doi:10.1007/s10972-005-4861-1

Blikstein, P., & Wilensky, U. (2009). An atom is known by the company it keeps: Constructing multi-agent models in engineering education. *International Journal of Computers for Mathematical Learning, 14*(2), 81–119. doi:10.1007/s10758-009-9148-8

Blumenfeld, P. C., & Meece, J. L. (1988). Task factors, teacher behavior, and students' involvement and use of learning strategies in science. *The Elementary School Journal, 88,* 235–250. doi:10.1086/461536

Borko, H., & Putnam, R. T. (1996). Learning to teach. In Berliner, D. C., & Calfee, R. C. (Eds.), *Handbook of Educational Psychology* (pp. 673–708). New York, NY: McMillan.

Böttcher, F., & Meisert, A. (2011). Argumentation in science education: A model-based framework. *Science & Education, 20,* 103–140. doi:10.1007/s11191-010-9304-5

BouJaoude. S. (2010). *Competencies and educational structures needed to prepare secondary students for the 21st century*. Paper presented at a Symposium on Secondary Education Organized by the Ministry of Education in the United Arab Emirates in Cooperation with the Arab Bureau of Education for the Gulf States. Dubai, UAE.

BouJaoude, S., Salloum, S., & Abd-El Khalick, F. (2004). Relationships between selective cognitive variables and students' ability to solve chemistry problems. *International Journal of Science Education, 26,* 63–84. doi:10.1080/0950069032000070315

Boyd, M., & Rubin, D. (2006). How contingent questioning promotes extended student talk: A function of display questions. *Journal of Literacy Research, 38*(2), 141–159. doi:10.1207/s15548430jlr3802_2

Braaten, M., & Windschitl, M. (2011). Working toward a stronger conceptualization of scientific explanation for science education. *Science Education, 95,* 639–669. doi:10.1002/sce.20449

Bradley, D. B., & Kelly, A. M. (2011). Promoting inclusiveness in acoustical physics. *Academic Exchange Quarterly, 15*(4), 88–93.

Bransford, J. D., Brown, A. L., & Cocking, R. R. (2000). *How people learn: Brain, mind, experience, and school*. Washington, DC: National Academies Press.

Bredekamp, S., & Copple, C. (Eds.). (1997). *Developmentally appropriate practice in early childhood programs*. Washington, DC: National Association for the Education of Young Children.

Breier, M., & Ralphs, A. (2009). In search of phronesis: Recognizing practical wisdom in the recognition (assessment) of prior learning. *British Journal of Sociology of Education, 30,* 479–493. doi:10.1080/01425690902954646

Brem, S. K., & Rips, L. J. (2000). Explanation and evidence in informal argument. *Cognitive Science, 24*(4), 573–604. doi:10.1207/s15516709cog2404_2

Brewe, E. (2008). Modeling theory applied: Modeling instruction in introductory physics. *American Journal of Physics, 76*(12), 1155–1160. doi:10.1119/1.2983148

Bricker, L. A., & Bell, P. (2008). Conceptualizations of argumentation from science studies and the learning sciences and their implications for the practices of science education. *Science Education, 92*(3), 473–498. doi:10.1002/sce.20278

Brickhouse, N. W. (2001). Embodying science: A feminist perspective on learning. *Journal of Research in Science Teaching, 38,* 282–295. doi:10.1002/1098-2736(200103)38:3<282::AID-TEA1006>3.0.CO;2-0

Brickhouse, N. W., Dahger, Z. R., Letts, W. J., & Shipman, H. L. (2000). Diversity of students' views about evidence, theory, and the interface between science and religion in an astronomy course. *Journal of Research in Science Teaching, 37,* 340–362. doi:10.1002/(SICI)1098-2736(200004)37:4<340::AID-TEA4>3.0.CO;2-D

Brook, A. J. (1993). The Reverend William Buckland, the first paleoecologist. *Biologist (Columbus, Ohio), 40,* 149–152.

Brookhart, S., & Freeman, D. (1992). Characteristics of entering teaching candidates. *Review of Educational Research, 62*(1), 37–60.

Brown, A. (1987). Metacognition, executive, control, self-regulation, and other more mysterious mechanisms. In Weinert, F. E., & Kluwe, R. H. (Eds.), *Metacognition, Motivation, and Understanding* (pp. 65–116). Hillsdale, NJ: Lawrence Erlbaum Associates.

Brown, A. L., & Reeve, R. A. (1987). Bandwidths of competence: The role of supportive contexts in learning and development. In Liben, L. S. (Ed.), *Development and Learning: Conflict or Congruence?* (pp. 173–223). Hillsdale, NJ: Lawrence Erlbaum Associates.

Browne, M. W., & Cudeck, R. (1993). Alternative ways of assessing model fit. In Bollen, K. A., & Long, J. S. (Eds.), *Testing Structural Equation Models* (pp. 136–162). Beverly Hills, CA: Sage.

Buckley, C. J., & Boulter, C. J. (2000). Investigating the role of representations and expressed models in building mental models. In Gilbert, J. K., & Boulter, C. J. (Eds.), *Developing Models in Science Education* (pp. 119–135). Dordrecht, The Netherlands: Kluwer Academic Publishers. doi:10.1007/978-94-010-0876-1_6

Bunch, G. C., Shaw, J. M., & Geaney, E. R. (2010). Documenting the language demands of mainstream content-area assessment for English learners: participant structures, communicative modes, and genre in science performance assessments. *Language and Education*, 24(3), 185–214. doi:10.1080/09500780903518986

Burbules, N. C., & Bruce, B. C. (2001). Theory and research on teaching as dialogue. In Richardson, V. (Ed.), *Handbook of Research on Teaching* (4th ed., pp. 1102–1121). Washington, DC: American Educational Research Association.

Burke, R., & Hall, J. (2005). *Krakatoa*. [DVD]. PBS Home Video.

Burry-Stock, J., & Oxford, R. (1994). Expert science teaching edcuational evaluation model (ESTEEM): Measuring excellence in science teaching for professional development. *Journal of Personnel Evaluation in Education*, 8(3), 267–297. doi:10.1007/BF00973725

Bybee, R. (1997). *Achieving scientific literacy: From purposes to practices*. Portsmouth, NH: Heinemann Educational Books.

Bybee, R. W., Taylor, J. A., Gardner, A., Scotter, P. V., Powell, J. C., & Westbrook, A. (2006). *The BSCS 5E instructional model: Origins, effectiveness, and applications*. Colorado Springs, CO: BSCS.

Campbell, T., & Neilson, D. (2009). Student ideas and inquiries: Investigating friction in the physics classroom. *Science Activities*, 46(1), 13–16. doi:10.3200/SATS.46.1.13-16

Campbell, T., & Neilson, D. (2012). Modeling electricity: Model-based inquiry with demonstrations and investigations. *The Physics Teacher*, 50(6). doi:10.1119/1.4745686

Campbell, T., Zhang, D., & Neilson, D. (2010). Model based inquiry in the high school physics classroom: An exploratory study of implementation and outcomes. *Journal of Science Education and Technology*, 20(3), 258–269. doi:10.1007/s10956-010-9251-6

Capps, D. K., Crawford, B. A., & Constas, M. A. (2012). A review of empirical literature on inquiry professional development: Alignment with best practices and a critique of the findings. *Journal of Science Teacher Education*, 23(3), 291–318. doi:10.1007/s10972-012-9275-2

Carey, S., Evans, R., Honda, M., Jay, E., & Unger, C. (1989). An experiment is when you try it and see if it works: A study of grade 7 students' understanding of the construction of scientific knowledge. *International Journal of Science Education*, 11, 514–529. doi:10.1080/0950069890110504

Carey, S., & Spelke, E. (1994). Domain-specific knowledge and conceptual change. In Hirschfeld, L. A., & Gelman, S. A. (Eds.), *Mapping the Mind: Domain Specificity in Cognition and Culture* (pp. 169–200). Cambridge, UK: Cambridge University Press. doi:10.1017/CBO9780511752902.008

Carnevale, A. P., & Strohl, J. (2010). How increasing college access in increasing inequality, and what to do about it. In Kahlenberg, R. D. (Ed.), *Rewarding Strivers* (pp. 71–190). New York, NY: The Century Foundation.

Carr, D. (2005). Personal and interpersonal relationships in education and teaching: A virtue ethical perspective. *British Journal of Educational Studies*, 53, 255–271. doi:10.1111/j.1467-8527.2005.00294.x

Carr, D. (2007). Character in teaching. *British Journal of Educational Studies, 55*(4), 369–389. doi:10.1111/j.1467-8527.2007.00386.x

Carr, W. (1995). *For education*. Berkshire, UK: Open University Press.

Case, R., & McKeough, A. (1989). Schooling and the development of central conceptual structures: An example from the domain of children's narrative. *International Journal of Educational Research, 13*, 835–855. doi:10.1016/0883-0355(89)90068-2

Cavagnetto, A., Hand, B. M., & Norton-Meier, L. (2010). The nature of elementary student science discourse in the context of the science writing heuristic approach. *International Journal of Science Education, 32*(4), 427–449. doi:10.1080/09500690802627277

Cavagnetto, A. R. (2010). Argument to foster scientific literacy: A review of argument interventions in K-12 science contexts. *Review of Educational Research, 80*(3), 336–371. doi:10.3102/0034654310376953

Chambers, S. K., & Andre, T. (1997). Gender, prior knowledge, interest and experience in electricity and conceptual change text manipulations in learning about direct current. *Journal of Research in Science Teaching, 34*(2), 107–123. doi:10.1002/(SICI)1098-2736(199702)34:2<107::AID-TEA2>3.0.CO;2-X

Charmaz, K. (2002). Qualitative interviewing and grounded theory analysis. In Gubrium, J., & Holstein, J. A. (Eds.), *Handbook of Interview Research: Context and Method* (pp. 675–694). Thousand Oaks, CA: Sage Publications.

Chawla, L. (1988). Children's concern for the natural environment. *Children's Environments Quarterly, 5*, 13–20.

Chawla, L., & Hart, R. (1988). Roots of environmental concern. In Lawrence, D., Habe, R., Hacker, A., & Sherrod, D. (Eds.), *People's Needs/Planet Management: Paths to Coexistence* (pp. 15–18). Pomona, CA: Environmental Design Research Association.

Cheek, N. Jr, & Burch, W. Jr. (1976). *The social organization of leisure in human society*. New York, NY: Harper and Row.

Chemers, M. (2006). *Understanding undergraduate and high school science support programs: A theory-driven, interdisciplinary, multi-method approach*. Paper presented at the Annual Meeting of the American Evaluation Association. Portland, OR.

Chemers, M. M. (2003). *Assessing scientific inquiry and leadership skills: Research plan*. Santa Cruz, CA: University of California Santa Cruz.

Chen, C., Sable, J., Mitchell, L., & Liu, F. (2011). *Documentation to the NCES common core of data public elementary/secondary school universe survey: School year 2009–10 (NCES 2011-348)*. Washington, DC: National Center for Education Statistics. Retrieved May 16, 2011 from http://nces.ed.gov/pubsearch/pubs.info.asp?pubid=2011348

Chi, M. T. H. (1997). Creativity: Shifting across ontological categories flexibly. In Ward, T. B., Smith, S. M., & Vaid, J. (Eds.), *Creative Thought: An Investigation of Conceptual Structures and Processes* (pp. 209–234). Washington, DC: American Psychological Association. doi:10.1037/10227-009

Chi, M. T. H. (1997). Quantifying qualitative analyses of verbal data: A practical guide. *Journal of the Learning Sciences, 6*, 271–315. doi:10.1207/s15327809jls0603_1

Chiappetta, E. L., & Koballa, T. R. J. (2006). *Science instruction in the middle and secondary schools: Developing fundamental knowledge and skills for teaching* (6th ed.). Upper Saddle River, NJ: Pearson Perrill Prentice Hall.

Chin, C. (2007). Teacher questioning in science classrooms: Approaches that stimulate productive thinking. *Journal of Research in Science Teaching, 44*(6), 815–843. doi:10.1002/tea.20171

Chin, C., & Brown, D. (2000). Learning in science: A comparison of deep and surface approaches. *Journal of Research in Science Teaching, 37*, 109–138. doi:10.1002/(SICI)1098-2736(200002)37:2<109::AID-TEA3>3.0.CO;2-7

Chinn, C. A., & Brewer, W. F. (1993). The role of anomalous data in knowledge acquisition: A theoretical framework and implications for science instruction. *Review of Educational Research, 63*, 1–49.

Chinn, C. A., & Brewer, W. F. (1998). Theories of knowledge acquisition. In Fraser, B. J., & Tobin, K. (Eds.), *International Handbook of Science Education* (pp. 97–113). London, UK: Kluwer Academic Publishers.

Chinn, C. A., & Malhotra, B. A. (2002). Epistemologically authentic inquiry in schools: A theoretical framework for evaluating inquiry tasks. *Science Education, 86*(2), 175–218. doi:10.1002/sce.10001

Clark, C. (2005). The structure of educational research. *British Educational Research Journal, 31*, 289–308. doi:10.1080/01411920500082128

Clark, D., Nelson, B., Sengupta, P., & D'Angelo, C. (2009). *Rethinking science learning through digital games and simulations: Genres, examples, and evidence.* Paper presented at Learning Science: Computer Games, Simulations, and Education Workshop Sponsored by the National Academy of Sciences. Washington, DC.

Clark, D., & Sampson, V. (2006). *Characteristics of students' argumentation practices when supported by personally-seeded discussions.* Paper presented at the Annual Meeting of the National Association for Research in Science Teaching. San Francisco, CA.

Clark, D. B., & Sampson, V. (2008). Assessing dialogic argumentation in online environments to relate structure, grounds, and conceptual quality. *Journal of Research in Science Teaching, 45*(3), 293–321. doi:10.1002/tea.20216

Clarke, J. (2009). *Exploring the complexity of inquiry learning in an open-ended problem space.* (Unpublished Doctoral Dissertation). Harvard Graduate School of Education. Boston, MA.

Clary, R. M., Gresham, D., Bases, F., Hamlin, E., Bergeron, N., & Petry, C. ... Fischer, E. (2005). Sediment and water analysis adjacent to an active scrap yard and archived superfund site, Lafayette Parish, Louisiana. *Gulf Coast Association of Geological Societies Transactions, 55*, 89-100.

Clary, R. M., & Wandersee, J. H. (2005). Through the looking glass: The history of aquarium views and their potential to improve learning in science classrooms. *Science and Education, 14*, 579–596. doi:10.1007/s11191-004-7691-1

Clary, R. M., & Wandersee, J. H. (2006). A writing template for probing students' geological sense of place. *Science Education Review, 5*(2), 51–59.

Clary, R. M., & Wandersee, J. H. (2007). A mixed methods analysis of the effects of an integrative geobiological study of petrified wood in introductory college geology classrooms. *Journal of Research in Science Teaching, 44*(8), 1011–1035. doi:10.1002/tea.20178

Clary, R. M., & Wandersee, J. H. (2008a). Marquee fossils: Using local specimens to integrate geology, biology, and environmental science. *Science Teacher (Normal, Ill.), 75*(1), 44–50.

Clary, R. M., & Wandersee, J. H. (2008b). Earth science teachers' perceptions of an autonomous fieldwork assignment in a nationwide online paleontology course. *Journal of Geoscience Education, 56*, 149–155.

Clary, R. M., & Wandersee, J. H. (2009a). Amber: Use "tree tears turned to stone" to teach biology, ecology... and more! *Science Scope, 33*(3), 22–29.

Clary, R. M., & Wandersee, J. H. (2009b). Incorporating informal learning environments and local fossil specimens in earth science classrooms: A recipe for success. *Science Education Review, 8*. Retrieved from http://www.scienceeducationreview.com/open_access/index.html

Clary, R. M., & Wandersee, J. H. (2010a). Scientific caricatures in the earth science classroom: An alternative assessment for meaningful science learning. *Science and Education, 19*(1), 21–38. doi:10.1007/s11191-008-9178-y

Clary, R. M., & Wandersee, J. H. (2010b). Connect-the-spheres with the coal cycle. *Science Scope, 34*(2), 20–29.

Clary, R. M., & Wandersee, J. H. (2010c). The "green" root beer laboratory. *Science Teacher (Normal, Ill.), 77*(2), 25–28.

Clary, R. M., & Wandersee, J. H. (2011a). Our human-plant connection. *Science Scope, 34*(8), 32–37.

Clary, R. M., & Wandersee, J. H. (2011b). DinoViz: The history and nature of science through the progression of dinosaur visualization. *Science Scope, 34*(6), 14–21.

Clary, R. M., & Wandersee, J. H. (2011c). A coprolite mystery: Who DUNG it? *Science Scope, 34*(7), 32–42.

Clary, R. M., & Wandersee, J. H. (2011d). A "coprolitic vision" for earth science education. *School Science and Mathematics*, *111*(6), 262–273. doi:10.1111/j.1949-8594.2011.00087.x

Clary, R. M., & Wandersee, J. H. (2011e). To see a scientific world in a grain of sand.... *Science Teacher (Normal, Ill.)*, *78*(5), 29–33.

Clary, R. M., & Wandersee, J. H. (2011f). Adopt-a-dino: Creative scientific visualization. *Science Teacher (Normal, Ill.)*, *78*(6), 36–41.

Clary, R. M., & Wandersee, J. H. (2011g). 1883 news report—Krakatoa erupts! A biology-geology integration inquiry. *Science Teacher (Normal, Ill.)*, *78*(9), 42–47.

Clary, R. M., & Wandersee, J. H. (2011h). Geobiological opportunities to learn at US fossil parks. In Feig, A., & Stokes, A. (Eds.), *Qualitative Inquiry in Geoscience Education Research* (Vol. 474, pp. 113–134). GSA Special Papers. doi:10.1130/2011.2474(09)

Clary, R. M., & Wandersee, J. H. (2012). Rock on! Using gravel to promote critical thinking in the classroom. *Science Teacher (Normal, Ill.)*, *79*(3), 42–46.

Clary, R. M., Wandersee, J. H., & Carpinelli, A. (2008). The great dinosaur feud: Science against all odds. *Science Scope*, *32*(2), 34–40.

Clement, J. J. (2008). *Creative model construction in scientists and students: The role of imagery, analogy, and mental simulation*. Dordrecht, The Netherlands: Springer. doi:10.1007/978-1-4020-6712-9

Cobb, E. (1993). *The ecology of imagination in childhood* (2nd ed.). Dallas, TX: Spring Publications.

Cobb, P., Wood, T., & Yackel, E. (1990). Classrooms as learning environments for teachers and researchers. In Davis, R. B., Maher, C. A., & Noddings, N. (Eds.), *Constructivist Views of the Teaching and Learning of Mathematics* (pp. 125–146). Reston, VA: NCTM. doi:10.2307/749917

Cobern, W. (1996). Worldview theory and conceptual change in science education. *Science Education*, *80*(5), 579–610. doi:10.1002/(SICI)1098-237X(199609)80:5<579::AID-SCE5>3.0.CO;2-8

Cohen, J. A. (1960). A coefficient of agreement for nominal scales. *Educational and Psychological Measurement*, *20*, 37–46. doi:10.1177/001316446002000104

College Board. (2011). *AP biology curriculum framework*. New York, NY: The College Board.

Collier, T., & Salloum, S. (2011). *The use of discrepant events and higher order/scaffolding questions for deeper science learning*. Paper presented at the Annual Ethnography in Education Research Forum. Philadelphia, PA.

Committee on Education, Council of the City of New York. (2004). *Lost in space: Science education in New York City public schools*. New York, NY: Committee on Education, Council of the City of New York.

Corno, L. (1993). The best-laid plans: Modern conceptions of volition and educational research. *Educational Researcher*, *22*(2), 14–22.

Corno, L., & Mandinach, E. B. (1983). The role of cognitive engagement in classroom learning and motivation. *Educational Psychologist*, *18*, 88–108. doi:10.1080/00461528309529266

Costa, V. B. (1995). When science is another world: Relationships between worlds of family, friends, school, and science. *Science Education*, *79*(3), 313–333. doi:10.1002/sce.3730790306

Costello, P. J. M., & Mitchell, S. (1995). Introduction: Argument: Voices, text and contexts. In Costello, P. J. M., & Mitchell, S. (Eds.), *Competing and Consensual Voices: The Theory and Practice of Argument*. Clevedon, UK: Multilingual Matters Ltd.

Cox-Petersen, A. M., Marsh, D. D., Kisiel, J., & Melber, L. M. (2003). Investigation of guided school tours, student learning, and science reform: Recommendations at a museum of natural history. *Journal of Research in Science Teaching*, *40*(2), 200–218. doi:10.1002/tea.10072

Cranton, P. (1997). Higher education: A global community. *New Directions for Teaching and Learning, Winter*, *72*, 5-9.

Cranton, P. (2001). Interpretive and critical evaluation. *New Directions for Teaching and Learning*, *88*, 11–18. doi:10.1002/tl.33

Cranton, P. (2002). Teaching for transformation. In Ross-Gordon, O. M. (Ed.), *Contemporary Viewpoints on Teaching Adults Effectively: New Directions for Adult and Continuing Education*. San Francisco, CA: Jossey-Bass.

Crawford, B. A. (2007). Learning to teach science inquiry in the rough and tumble of practice. *Journal of Research in Science Teaching, 44*, 613–642. doi:10.1002/tea.20157

Crawford, B. A., & Cullin, M. J. (2004). Supporting prospective teachers' conceptions of modelling in science. *International Journal of Science Education, 26*(11), 1379–1401. doi:10.1080/09500690410001673775

Crawford, C. (1984). *The art of computer game design: Reflections of a master game designer*. New York, NY: Osborne/McGraw-Hill.

Crawford, T. (2005). What counts as knowing: Constructing a communicative repertoire for student demonstration of knowledge in science. *Journal of Research in Science Teaching, 42*(2), 139–165. doi:10.1002/tea.20047

Crawley, C. E. (2007). Localized debates of agricultural biotechnology in community newspapers: A quantitative content analysis of media frames and sources. *Science Communication, 28*, 314–346. doi:10.1177/1075547006298253

Cresswell, T. (2004). *Place: A short introduction*. Malden, MA: Blackwell Publications.

Crichton, M. (1990). *Jurassic park*. New York, NY: Knopf.

Cronbach, L. J., & Snow, R. E. (1977). *Aptitudes and instructional methods: A handbook for research on interactions*. New York, NY: Irvington.

Cronin, J., Kingsbury, G. G., Dahlin, M., Adkins, D., & Bowe, B. (2007). *Alternate methodologies for estimating state standards on a widely-used computerized adaptive test*. Paper presented at the National Council on Measurement in Education. New York, NY.

Crumb, G. H. (1965). Understanding of science in high school physics. *Journal of Research in Science Teaching, 3*, 246–250. doi:10.1002/tea.3660030312

Cuban, L. (1999). The technology puzzle. *Education Week, 18*(43), 58.

Dagher, Z. R., & BouJaoude, S. (1997). Scientific views and religious beliefs of college students: The case of biological evolution. *Journal of Research in Science Teaching, 34*(5), 429–445. doi:10.1002/(SICI)1098-2736(199705)34:5<429::AID-TEA2>3.0.CO;2-S

Danish, J., Peppler, K., Phelps, D., & Washington, D. (2011). Life in the hive: Supporting inquiry into complexity within the zone of proximal development. *Journal of Science Education and Technology, 20*(5), 454–467. doi:10.1007/s10956-011-9313-4

Darling-Hammond, L. (2008). A future worthy of teaching for America. *Phi Delta Kappan, 89*, 730–735.

Darling-Hammond, L., & McLaughlin, M. W. (1995). Policies that support professional development in an era of reform. *Phi Delta Kappan, 92*(6), 81–92.

de Freitas, S. (2006). *Learning in immersive worlds: A review of game-based learning*. Bristol, UK: Joint Information Systems Committee (JISC) E-Learning Programme. Retrieved March 11, 2008, from http://www.jisc.ac.uk/media/documents/programmes/elearninginnovation/gamingreport_v3.pdf

DeBacker, T. K., & Crowson, H. M. (2006). Influences on cognitive engagement: Epistemological beliefs and need for closure. *The British Psychological Society, 76*, 535–551.

DeBoer, G. (1991). *A history of ideas in science education*. New York, NY: Teachers College Press.

DeBoer, G. (2005). Historical perspectives on inquiry teaching in schools. In Flick, L. B., & Lederman, N. G. (Eds.), *Scientific Inquiry and Nature of Science: Implications for Teaching, Learning, and Teacher Education* (pp. 17–36). Dordrecht, The Netherlands: Springer.

Deci, E. L. (1975). *Intrinsic motivation*. New York, NY: Plenum Press. doi:10.1007/978-1-4613-4446-9

Deci, E. L., & Ryan, R. M. (1991). A motivational approach to self: Integration in personality. In R. Diensbier (Ed.), *Nebraska Symposium on Motivation: Perspectives on Motivation,* (pp. 237-288). Lincoln, NE: University of Nebraska Press.

Decker, D. J., & Purdy, K. G. (1986). Becoming a hunter: Identifying stages of hunting involvement for improving hunter education programs. *Wildlife Society Bulletin, 14*, 474–479.

Decker, L., & Rim-Kaufman, S. (2008). Personality characteristics and teacher beliefs among pre-service teachers. *Teacher Education Quarterly*, *35*(2), 45–62.

Dede, C. (2009). Immersive interfaces for engagement and learning. *Science*, *323*(5910), 66–69. doi:10.1126/science.1167311

DeLany, B. (1991). Allocation, choice, and stratification within high schools: How the sorting machine copes. *American Journal of Education*, *99*(2), 181–207. doi:10.1086/443978

Deniz, H., Cetin, F., & Yılmaz, I. (2011). Examining the relationships among Turkish preservice biology teachers' acceptance of evolution, religiosity and teaching preference for evolution. *Journal of Science Education and Technology*, *31*(4), 2.1-2.8

Deniz, H., Donnelly, L., & Yilmaz, I. (2008). Exploring the factors related to acceptance of evolutionary theory among Turkish preservice biology teachers: Toward a more informative conceptual ecology for biological evolution. *Journal of Research in Science Teaching*, *45*(4), 420–443. doi:10.1002/tea.20223

Deniz, H., Shrader, P. G., & Keilty, J. (2012). *Impact of evolution instruction on understanding and acceptance of evolutionary theory and the nature of relationships among understanding, acceptance, and religiosity*. Paper presented at the Annual Meeting of National Association for Research in Science Teaching. Indianapolis, IN.

Devisch, O. (2008). Should planners start playing video games? Arguments from SimCity and Second Life. *Planning Practice & Theory*, *9*(2), 209–228. doi:10.1080/14649350802042231

Dewey, J. (1933). *How we think: A restatement of the relation of reflective thinking to the educative process*. Boston, MA: D. C. Heath.

Dewey, J. (1938). *Experience and education*. London, UK: Collier-Macmillian.

DeWitt, J., & Storksdieck, M. (2008). A short review of school field trips: Key findings from the past and implications for the future. *Visitor Studies*, *11*(2), 181–197. doi:10.1080/10645570802355562

Dey, I. (1999). *Grounding grounded theory: Guidelines for qualitative inquiry*. San Diego, CA: Academic Press.

Dickes, A., & Sengupta, P. (2011). Learning natural selection in 4th grade with multi-agent-based computational models. In Sengupta, P., & Hall, R. (Eds.), *Models, Modeling, and Naïve Intuitive Knowledge in Science Learning*. Berkeley, CA: Jean Piaget Society. doi:10.1007/s11165-012-9293-2

Dieterle, E. (2010). Games for science education. In Hirumi, A. (Ed.), *Playing Games in School: Video Games and Simulations for Primary and Secondary Education* (pp. 89–112). Eugene, OR: International Society for Technology in Education.

Dieterle, E., & Clarke, J. (2007). Multi-user virtual environments for teaching and learning. In Pagani, M. (Ed.), *Encyclopedia of Multimedia Technology and Networking* (2nd ed., pp. 1033–1044). Hershey, PA: IGI Global. doi:10.4018/978-1-60566-014-1.ch139

Dillon, J. T. (1981). To question and not to question during discussion: Non-questioning techniques. *Journal of Teacher Education*, *32*(15), 15–20. doi:10.1177/002248718103200605

diSessa, A., & Abelson, H. (1986). Boxer: A reconstructible computational medium. *Communications of the ACM*, *29*(9), 859–868. doi:10.1145/6592.6595

Dole, J. A., & Sinatra, G. M. (1998). Reconceptualizing change in the cognitive construction of knowledge. *Educational Psychologist*, *33*, 109–128.

Donovan, M. S., & Bransford, J. D. (2005). *How students learn: History, mathematics, and science in the classroom*. Washington, DC: National Academies Press.

Dorner, D. (1989). *The logic of failure: Why things go wrong and what we can do to make go right*. New York, NY: Metropolitan Books.

Doyle, W. (1977). Paradigms for research on teacher effectiveness. In Shulman, L. S. (Ed.), *Review of Research in Education* (Vol. 5, pp. 163–179). Washington, DC: American Educational Research Association.

Drake, J., Worle, I., & Mehrtens, C. (1997). An introductory-level field-based course in geology and biology. *Journal of Geoscience Education*, *45*, 234–237.

Drew, C. (2011, November 4). Why science majors change their minds (it's just so darn hard). *New York Times*. Retrieved from http://www.nytimes.com/2011/11/06/education/edlife/why-science-majors-change-their-mind-its-just-so-darnhard.html?pagewanted=1&_r=1&emc=eta1

Driver, R. (1981). Pupils' alternative frameworks in science. *European Journal of Science Education*, *3*(1), 93–101. doi:10.1080/0140528810030109

Driver, R., Asoko, H., Leach, J., Mortimer, E., & Scott, P. (1994). Constructing scientific knowledge in the classroom. *Educational Researcher*, *23*, 5–12.

Driver, R., Guesne, E., & Tiberghien, A. (Eds.). (1985). *Children's ideas in science*. Philadelphia, PA: Open University Press.

Driver, R., Leach, J., Millar, R., & Scott, P. (1996). *Young people's images of science*. Philadelphia, PA: Open University Press.

Driver, R., Newton, P., & Osborne, J. (2000). Establishing the norms of scientific argumentation in classrooms. *Science Education*, *84*(3), 287–313. doi:10.1002/(SICI)1098-237X(200005)84:3<287::AID-SCE1>3.0.CO;2-A

Driver, R., Squires, A., Rushworth, P., & Wood-Robinson, V. (1994). *Making sense of secondary science: Research into children's ideas*. London, UK: Taylor & Francis Ltd.

Duffee, L., & Aikenhead, G. (1992). Curriculum change, student evaluation, and teacher practical knowledge. *Science Education*, *76*, 493–506. doi:10.1002/sce.3730760504

Duffin, C. J. (2006). William Buckland (1786-1856). *Geology Today*, *22*(3), 104–108. doi:10.1111/j.1365-2451.2006.00562.x

Dukas, G. (2009). *Characterizing student navigation in educational multiuser virtual environments: A case study using data from the river city project*. (Unpublished Doctoral Dissertation). Harvard Graduate School of Education. Boston, MA.

Dupeyrat, C., & Marine, C. (2005). Implicit theories of intelligence, goal orientation, cognitive engagement, and achievement: A test of Dweck's model with returning to school adults. *Contemporary Educational Psychology*, *30*, 43–59. doi:10.1016/j.cedpsych.2004.01.007

Duschl, R. (1990). *Restructuring science education: The importance of theories and their development*. New York, NY: Teachers College Press.

Duschl, R. (2000). *Making the nature of science explicit. Improving science education: The contribution of research*. Philadelphia, PA: Open University Press.

Duschl, R. (2008). Science education in three-part harmony: Balancing conceptual, epistemic, and social learning goals. *Review of Research in Education*, *32*, 268–291. doi:10.3102/0091732X07309371

Duschl, R. A. (1988). Abandoning the scientistic legacy of science education. *Science Education*, *72*, 51–62. doi:10.1002/sce.3730720105

Duschl, R. A. (1994). Research on the history and philosophy of science. In Gabel, D. L. (Ed.), *Handbook of Research on Science Teaching and Learning* (pp. 443–465). New York, NY: Macmillian.

Duschl, R. A., & Osborne, J. (2002). Supporting and promoting argumentation discourse in science education. *Studies in Science Education*, *38*, 39–72. doi:10.1080/03057260208560187

Duschl, R., Schweingruber, H., & Shouse, A. (Eds.). (2007). *Taking science to school: Learning and teaching science in grades K-8*. Washington, DC: National Academy Press.

Eamon, M. K. (2004). Socio-demographic, school, neighborhood, parenting influences on the academic achievement of Latino youth adolescents. *Journal of Youth and Adolescence*, *34*(2), 163–174. doi:10.1007/s10964-005-3214-x

Earth Science Literacy Initiative. (2010). Earth science literacy principles: Big ideas and supporting concepts of earth science. *National Science Foundation*. Retrieved from http://www.earthscienceliteracy.org/es_literacy_6may10_.pdf

Eichinger, D. C., Anderson, C. W., Palinscar, A., & David, Y. M. (1991). *An illustration of the roles of content knowledge, scientific argument, and social norms in collaborative problem solving*. Paper presented at the American Educational Research Association. Chicago, IL.

Eisenkraft, A. (2010). Millikan lecture 2009: Physics for all: From special needs to olympiads. *American Journal of Physics*, 78(4), 328–337. doi:10.1119/1.3293130

Eisner, E. W. (2002). From episteme to phronesis to artistry in the study and improvement of teaching. *Teaching and Teacher Education*, 18, 375–385. doi:10.1016/S0742-051X(02)00004-5

Electronic Arts. (2011). *SimCity™ 4 deluxe edition*. Retrieved on 4 December, 2011 from http://www.ea.com/simcity-4-deluxe

Elkins, J. T., & Elkins, N. M. L. (2007). Teaching geology in the field: Significant geoscience concept gains in entirely field-based introductory geology courses. *Journal of Geoscience Education*, 55(2), 126–132.

Elliot, E., & Dweck, C. (1988). Goals: An approach to motivation and achievement. *Journal of Personality and Social Psychology*, 54, 5–12. doi:10.1037/0022-3514.54.1.5

Ertzberger, J. (2009). An exploration of factors affecting teachers' use of video games as instructional tools. In *Proceedings of the Society for Information Technology & Teacher Education International Conference 2009*, (vol. 1, pp. 1825-1831). IEEE.

Eve, R. A., & Dunn, D. (1990). Psychic powers, astrology & creationism in the classroom? *The American Biology Teacher*, 52(1), 10–20. doi:10.2307/4449018

Fadigan, K. A., & Hammrich, P. L. (2004). A longitudinal study of the educational and career trajectories of female participants of an urban informal science education program. *Journal of Research in Science Education*, 41(8), 835–860.

Falk, J. (2001). *Free choice science education: How we learn science outside of school*. New York, NY: Teachers College Press.

Falk, J., & Dierking, L. (2000). *Learning from museums: Visitor experiences and the making of meaning*. Walnut Creek, CA: Alta Mira Press.

Falk, J., & Dierking, L. (2002). *Lessons without limit: How free-choice learning is transforming education*. Walnut Creek, CA: Alta Mira Press.

Federation of American Scientists. (2006). *Summit on educational games - Harnessing the power of video games for learning*. Washington, DC: Federation of American Scientists.

Feldman, A. (2002). Multiple perspectives for the study of teaching: Knowledge, reason, understanding, and being. *Journal of Research in Science Teaching*, 39, 1032–1055. doi:10.1002/tea.10051

Feltovich, P. J., Spiro, R. J., & Coulson, R. L. (1993). Learning, teaching, and testing for complex conceptual understanding. In Frederiksen, N., & Bejar, I. (Eds.), *Test Theory for a New Generation of Tests* (pp. 181–217). Hillsdale, NJ: LEA.

Felzien, L., & Cooper, J. (2005). Modeling the research process: Alternative approaches to teaching undergraduates. *Journal of College Science Teaching*, 34, 42–46.

Fenstermacher, G. D. (1994). The knower and the known in teacher knowledge research. In Darling-Hammond, L. (Ed.), *Review of Research in Education* (Vol. 20, pp. 3–56). Washington, DC: American Educational Research Association.

Flavell, J. H. (1979). Metacognition and cognitive monitoring: A new area of cognitive-developmental inquiry. *The American Psychologist*, 34, 906–911. doi:10.1037/0003-066X.34.10.906

Flavell, J. H. (1999). Cognitive development: Children's knowledge about the mind. *Annual Review of Psychology*, 50, 21–45. doi:10.1146/annurev.psych.50.1.21

Flavell, J. H., Miller, P. H., & Miller, S. A. (1993). *Cognitive development* (3rd ed.). Upper Saddle River, NJ: Prentice Hall.

Flowerday, T., & Schraw, G. (2003). Effect of choice on cognitive and affective engagement. *The Journal of Educational Research*, 96, 207–215. doi:10.1080/00220670309598810

Flyvbjerg, B. (2001). *Making social sciences matter*. Cambridge, UK: Cambridge University Press.

Ford, M. (2008). Disciplinary authority and accountability in scientific practice and learning. *Science Education*, 92(3), 404–423. doi:10.1002/sce.20263

Ford, M. J., & Forman, E. A. (2006). Redefining disciplinary learning in classroom contexts. *Review of Research in Education, 30*, 1–32. doi:10.3102/0091732X030001001

Forman, E., Barnhart, B., Deafenbaugh, L., & Ewing, M. (2010). *Fostering constructive criticism in a high school biology classroom: Understanding the social dynamics of argumentation*. Paper presented at the Annual Meeting of the National Association for Research in Science Teaching. Philadelphia, PA.

Fosnot, C. T. (1989). *Enquiring teachers, enquiring learners: A constructivist approach to teaching*. New York, NY: Teachers College Press.

Fosnot, C. T. (1996). Constructivism: A psychological theory of learning. In Fosnot, C. T. (Ed.), *Constructivism: Theory, Perspectives, and Practice* (pp. 8–33). New York, NY: Teachers College Press.

Fosnot, C. T. (1996). Teachers construct constructivism: The center for constructivist teaching/teacher preparation project. In Fosnot, C. T. (Ed.), *Constructivism: Theory, Perspectives, and Practices* (pp. 205–216). New York, NY: Teachers College Press.

Fredricks, J. A., Blumenfeld, P. C., & Paris, A. H. (2004). School engagement: Potential of the concept, state of the evidence. *Review of Educational Research, 74*, 59–109. doi:10.3102/00346543074001059

Freeman, C., & Smith, D. (1997). *Active and engaged? Lessons from an interdisciplinary and collaborative college mathematics and science course for preservice teachers*. Paper presented at the Annual Meeting of the American Educational Research Association. Chicago, IL.

Gadamer, H.-G. (1989). *Truth and method* (2nd ed.). (Weinsheimer, J., & Marshall, D. G., Trans.). New York, NY: Crossroad.

Gallagher, J. J. (1991). Prospective and practicing secondary school science teachers' knowledge and beliefs about the philosophy of science. *Science Education, 75*, 121–134. doi:10.1002/sce.3730750111

Gallagher, J. M., & Reid, D. K. (2002). *The learning theory of Piaget & Inhelder* (2nd ed.). New York, NY: Authors Choice Press.

Gallagher, S. (1992). *Hermeneutics and education*. New York, NY: SUNY.

Gallup (2012). *Evolution, creationism, intelligent design*. Retrieved from http://www.gallup.com/poll/21814/evolution-creationism-intelligent-design.aspx

Garbarino, J. (1989). An ecological perspective on the role of play in child development. In Bloch, M. N., & Pellegrini, A. D. (Eds.), *The Ecological Context of Children's Play* (pp. 16–34). Norwood, NJ: Ablex Publishing Corporation.

Garcia, T., & Pintrich, P. R. (1994). Regulating motivation and cognition in the classroom: The role of self-schemas and self-regulatory strategies. In Schunk, D. H., & Zimmerman, B. J. (Eds.), *Self-Regulation of Learning and Performance: Issues and Educational Application* (pp. 127–153). Hillsdale, NJ: Erlbaum.

Garet, M. S., Porter, A. C., Desimone, L., Birman, B. F., & Yoon, K. S. (2001). What makes professional development effective? Results from a national sample of teachers. *American Educational Research Journal, 38*(4), 915–945. doi:10.3102/00028312038004915

Gee, J. P. (2004). *Situated language and learning: A critique of traditional schooling*. London, UK: Routledge.

Gee, J. P. (2007). *What video games have to teach us about learning and literacy* (2nd ed.). New York, NY: Palgrave Macmillan. doi:10.1145/950566.950595

Gelman, R., & Brenneman, K. (2004). Science learning pathways for young children. *Early Childhood Research Quarterly, 19*, 150–158. doi:10.1016/j.ecresq.2004.01.009

Ghatala, E. S. (1986). Strategy monitoring training enables young learners to select effective strategies. *Educational Psychologist, 21*, 434–454.

Giere, R. (1988). Laws, theories, and generalizations. In *The Limits of Deductivism* (pp. 37–46). Berkeley, CA: University of California Press.

Giere, R. N. (1999). *Science without laws*. Chicago, IL: The University of Chicago Press.

Giere, R. N. (1999). Using models to represent reality. In Magnani, L., Nersessian, N. J., & Thagard, P. (Eds.), *Model-Based Reasoning in Scientific Discovery* (pp. 41–57). New York, NY: Kluwer Academic/Plenum. doi:10.1007/978-1-4615-4813-3_3

Giere, R. N., Bickle, J., & Mauldin, R. F. (Eds.). (2006). *Understanding scientific reasoning* (5th ed.). Toronto, Canada: Thomson Wadsworth.

Gilbert, J. K., & Boulter, C. J. (Eds.). (2000). *Developing models in science education*. Dordrecht, The Netherlands: Kluwer Academic Publishers. doi:10.1007/978-94-010-0876-1

Gilbert, J. K., Boulter, C., & Rutherford, M. (1998). Models in explanations, part 1: Horses for courses? *International Journal of Science Education*, *20*(1), 83–97. doi:10.1080/0950069980200106

Gilbert, S. W., & Ireton, S. W. (2003). *Understanding models in earth and space science*. Arlington, VA: NSTA Press.

Gipps, C. V. (1994). *Beyond testing: Towards a theory of educational assessment*. London, UK: Routledge.

Glaser, B. G., & Strauss, A. L. (1967). *The discovery of grounded theory: Strategies for qualitative research*. New York, NY: Aldine. doi:10.1097/00006199-196807000-00014

Gobert, J. D. (2005). The effects of different learning tasks on model-building in plate tectonics: Diagramming versus explaining. *Journal of Geoscience Education*, *53*(4), 444–455.

Gobert, J. D., & Clement, J. J. (1999). Effect of student-generated diagram versus student-generated summaries on conceptual understanding of causal and dynamic knowledge in plate tectonics. *Journal of Research in Science Teaching*, *26*(1), 39–53. doi:10.1002/(SICI)1098-2736(199901)36:1<39::AID-TEA4>3.0.CO;2-I

Gobert, J. D., & Pallant, A. (2004). Fostering students' epistemologies of models via authentic model-based tasks. *Journal of Science Education and Technology*, *13*(1), 7–22. doi:10.1023/B:JOST.0000019635.70068.6f

Goldstone, R. L., & Sakamoto, Y. (2003). The transfer of abstract principles governing complex adaptive systems. *Cognitive Psychology*, *46*, 414–466. doi:10.1016/S0010-0285(02)00519-4

Gollub & Spital. (2002). Advanced physics in high schools. *Physics Today*, *55*(5), 48–53. doi:10.1063/1.1485584

Gowin, D. B. (1981). *Educating*. Ithaca, NY: Cornell University Press.

Graham, S., & Golan, S. (1991). Motivational influences on cognition: Task involvement, ego involvement, and depth of processing. *Journal of Educational Psychology*, *83*, 187–194. doi:10.1037/0022-0663.83.2.187

Greene, B. A., & Miller, R. B. (1996). Influences on achievement: Goals, perceived ability, and cognitive engagement. *Contemporary Educational Psychology*, *21*, 181–192. doi:10.1006/ceps.1996.0015

Greene, B. A., Miller, R. B., Crowson, H. M., Duke, B. L., & Akey, K. L. (2004). Predicting high school students' cognitive engagement and achievement: Contributions of classroom perceptions and motivation. *Contemporary Educational Psychology*, *29*, 462–482. doi:10.1016/j.cedpsych.2004.01.006

Greenwood, C. R., Horton, B. T., & Utley, C. A. (2002). Academic engagement: Current perspectives on research and practice. *School Psychology Review*, *31*, 328–349.

Griffin, J., & Symington, D. (1998). Moving from task-oriented to learning-oriented strategies on school excursions to museums. *Science Education*, *81*(6), 763–779. doi:10.1002/(SICI)1098-237X(199711)81:6<763::AID-SCE11>3.0.CO;2-O

Grose, E. C., & Simpson, D. (1982). Attitudes of introductory college biology students toward evolution. *Journal of Research in Science Teaching*, *19*(1), 15–23. doi:10.1002/tea.3660190103

Gross, A. G. (1990). *The rhetoric of science*. Cambridge, MA: Harvard University Press.

Grotzer, T. A. (1993). *Children's understanding of complex causal relationships in natural systems*. (Unpublished Doctoral Dissertation). Harvard University. Cambridge, MA.

Grotzer, T. A. Dede, C., Metcalfe, S., & Clarke, J. (2009). *Addressing the challenges in understanding ecosystems: Why getting kids outside may not be enough*. Paper presented at the National Association of Research in Science Teaching (NARST) Conference. Orange Grove, CA.

Grotzer, T. A., Duhaylongsod, L., & Tutwiler, M. S. (2011). *Developing explicit understanding of probabilistic causation: Patterns and variation in young children's reasoning*. Paper presented at the American Educational Research Association (AERA) Conference. New Orleans, LA.

Grotzer, T. A. (2004, October). Putting science within reach: Addressing patterns of thinking that limit science learning. *Principal Leadership*.

Grotzer, T. A. (2012). *Learning causality in a complex world: Understandings of consequence*. Lanham, MD: Rowman & Littlefield.

Grotzer, T. A., & Basca, B. B. (2003). Helping students to grasp the underlying causal structures when learning about ecosystems: How does it impact understanding? *Journal of Biological Education*, *38*(1), 16–29. doi:10.1080/00219266.2003.9655891

Grotzer, T. A., & Lincoln, R. (2007). Educating for "intelligent environmental action" in an age of global warming. In Moser, S. C., & Dilling, L. (Eds.), *Creating a Climate for Change: Communicating Climate Change and Facilitating Social Change* (pp. 266–280). Cambridge, UK: Cambridge University Press. doi:10.1017/CBO9780511535871.020

Grotzer, T. A., Tutwiler, M. S., Dede, C., Kamarainen, A., & Metcalf, S. (2011). *Helping students learn more expert framing of complex causal dynamics in ecosystems using EcoMUVE*. Paper presented at the National Association of Research in Science Teaching (NARST) Conference. Orlando, FL.

Gunstone, R. F., & Mitchell, I. J. (1998). Metacognition and conceptual change. In Mintzes, J. J., Wandersee, J. H., & Novak, J. D. (Eds.), *Teaching Science for Understanding: A Human Constructivist View* (pp. 133–163). San Diego, CA: Academic Press.

Habermas, J. (1971). *Knowledge and human interests*. Boston, MA: Beacon.

Haim, E. (2003). Inquiry-events as a tool for changing science teaching efficacy beliefs of kindergarten and elementary school teachers. *Journal of Science Education and Technology*, *12*(4), 495–501. doi:10.1023/B:JOST.0000006309.16842.c8

Haines, S., & Blake, R. Jr. (2005). Field and natural science. *Journal of College Science Teaching*, *34*(7), 28–31.

Hall, R. C. W., Day, T., & Hall, R. C. W. (2011). A plea for caution: violent video games, the Supreme Court, and the role of science. *Mayo Clinic Proceedings*, *86*(4), 315–321. doi:10.4065/mcp.2010.0762

Halloun, I. A. (2004). *Modeling theory in science education*. Dordrecht, The Netherlands: Kluwer Academic Publishers.

Halverson, R. (2005). What can K-12 school leaders learn from video games and gaming. *Innovate: Journal of online. Education*, *1*(6).

Hammerman, D. R., Hammerman, W. M., & Hammerman, E. L. (1994). *Teaching in the outdoors*. Danville, IL: Interstate Publishers, Inc.

Hand, B., Wallace, C. S., & Yang, E. M. (2004). Using the science writing heuristic to enhance learning outcomes from laboratory activities in seventh grade science: Quantitative and qualitative aspects. *International Journal of Science Education*, *26*, 131–149. doi:10.1080/0950069032000070252

Haney, J. J., & McArthur, J. (2002). Four case studies of prospective science teachers' beliefs concerning constructivist teaching practices. *Science Education*, *86*, 783–802. doi:10.1002/sce.10038

Hanuscin, D., Akerson, V., & Phillipson-Mower, T. (2006). Integrating nature of science instruction into a physical science content course for preservice elementary teachers: NOS views of teaching assistants. *Science Education*, *90*, 912–935. doi:10.1002/sce.20149

Harackiewicz, J. M., Barron, K. E., Pintrich, P. R., & Elliot, A. J. (2002). Revision of achievement goal theory: Necessary and illuminating. *Journal of Educational Psychology*, *94*, 638–645. doi:10.1037/0022-0663.94.3.638

Harel, I., & Papert, S. (1991). Software design as a learning environment. In Harel, I., & Papert, S. (Eds.), *Constructionism*. Norwood, NJ: Ablex.

Hauge, M. R., & Gentile, D. A. (2003). Video game addiction among adolescents: Associations with academic performance and aggression. *Child Development*, *40*(306), 1–3.

Hehn, J., & Neuschatz, M. (2006). Physics for all? A million and counting! *Physics Today*, *59*, 37–43. doi:10.1063/1.2186280

Heidegger, M. (1962). *Being and time* (Macquarrie, J., & Robinson, E., Trans.). New York, NY: Harper and Row.

Helme, S., & Clarke, D. (2001). Identifying cognitive engagement in the mathematics classroom. *Mathematics Education Research Journal, 13*, 133–153. doi:10.1007/BF03217103

Hemler, D., & Repine, T. (2006). Teachers doing science: An authentic geology research experience for teachers. *Journal of Geoscience Education, 54*, 93–102.

Henderson, R. W. (1986). Self-regulated learning: Implications for the design of instructional media. *Contemporary Educational Psychology, 11*, 405–427. doi:10.1016/0361-476X(86)90032-9

Hesse, M. (1966). *Models and analogies in science*. Notre Dame, IN: University of Notre Dame Press.

Hestenes, D. (1992). Modelling games in the Newtonian world. *American Journal of Physics, 60*, 732–748. doi:10.1119/1.17080

Hestenes, D. (1993). *Modelling is the name of the game*. Paper presented at the National Science Foundation Modelling Conference. Dedham, MA.

Hidi, S. (1990). Interest and its contribution as a mental resource for learning. *Review of Educational Research, 60*, 549–571.

Hines, P. J., Jasny, B. R., & Mervis, J. (2009). Adding a T to the three R's. *Science, 323*(5910), 53. doi:10.1126/science.323.5910.53a

Hirschfeld, L. A., & Gelman, S. A. (1994). Toward a topography of mind: An introduction to domain specificity. In Hirschfeld, L. A., & Gelman, S. A. (Eds.), *Mapping the Mind: Domain Specificity in Cognition and Culture* (pp. 3–35). Cambridge, UK: Cambridge University Press. doi:10.1017/CBO9780511752902.002

Hmelo-Silver, C. E., & Azevedo, R. (2006). Understanding complex systems: Some core challenges. *Journal of the Learning Sciences, 15*, 53–61. doi:10.1207/s15327809jls1501_7

Hmelo-Silver, C. E., Marathe, S., & Liu, L. (2007). Fish swim, rocks sit, and lungs breathe: Expert-novice understanding of complex systems. *Journal of the Learning Sciences, 16*, 307–331. doi:10.1080/10508400701413401

Hodson, D. (1993). Philosophic stance of secondary school science teachers, curriculum experiences, and children's understanding of science: Some preliminary findings. *Interchange, 24*(1-2), 41–52. doi:10.1007/BF01447339

Hogan, K. (2000). Exploring a process view of students' knowledge about the nature of science. *Science Education, 84*, 51–70. doi:10.1002/(SICI)1098-237X(200001)84:1<51::AID-SCE5>3.0.CO;2-H

Hogan, K., & Maglienti, M. (2001). Comparing the epistemological underpinnings of students' and scientists' reasoning about conclusions. *Journal of Research in Science Teaching, 38*(6), 663–687. doi:10.1002/tea.1025

Hogan, K., Nastasi, B. K., & Pressley, M. (2000). Discourse patterns and collaborative scientific reasoning in peer and teacher-guided discussions. *Cognition and Instruction, 17*, 379–432. doi:10.1207/S1532690XCI1704_2

Honey, M. A., & Hilton, M. (Eds.). (2010). *Learning science through computer games and simulations*. Washington, DC: National Academy Press.

Horizon Research. (2002). *2000 national survey of science and mathematics education: Status of elementary school science teaching*. Chapel Hill, NC: Horizon Research Inc.

Horizon Research. (2002). *Inside the classroom interview protocol*. Retrieved from http://www.horizon-research.com/instruments/clas/cop.php

Howard, A., & Bray, D. (1988). *Managerial lives in transition*. New York: Guilford Press.

Hsi, S. (2007). Conceptualizing learning from the everyday activities of digital kids. *International Journal of Science Education, 29*(12), 1509–1529. doi:10.1080/09500690701494076

Hu, L., & Bentler, P. M. (1999). Cutoff criteria in fix indexes in covariance structure analysis: Conventional criteria versus new alternatives. *Structural Equation Modeling, 6*(1), 1–55. doi:10.1080/10705519909540118

Ipsos, M. O. R. I. (2006). *BBC news*. Retrieved from http://news.bbc.co.uk/2/hi/science/nature/4648598.stm

Isaacs, W. (1999). *Dialogue and the art of thinking together*. New York, NY: Currency/Doubleday.

Isaacs, W. H. (1993). Taking flight: Dialogue, collective thinking, and organizational learning. *Organizational Dynamics*, *22*(2), 24–39. doi:10.1016/0090-2616(93)90051-2

Iseke-Barnes, J. M. (1996). Issues of educational uses of the Internet: power and criticism in communications and searching. *Journal of Educational Computing Research*, *15*(1), 1–23. doi:10.2190/FLYP-YNQC-9T55-MKB5

Issenberg, S. B., Mcgaghie, W. C., Petrusa, E. R., Gordon, D. L., & Scalese, R. (2005). Features and uses of high-fidelity medical simulations that lead to effective learning: A BEME systematic review. *Medical Teacher*, *27*(1), 10–28. doi:10.1080/01421590500046924

Issler, K. (1983). A conception of excellence in teaching. *Education*, *103*, 338–344.

Jacobson, M. J. (2001). Problem-solving, cognition, and complex systems: Differences between experts and novices. *Complexity*, *6*(3), 41–49. doi:10.1002/cplx.1027

James, C. (2007). *Playing the game: Comparing teacher gamers to non-gamers*. (Unpublished Doctoral Dissertation). The University of Alabama. Birmingham, AL. Retrieved from http://proquest.umi.com/pqdlink?RQT=568&VInst=PROD&VName=PQD&VType=PQD&Fmt=7&did=1379569531&TS=1302994639&fromjs=1

James, J. J., Bixler, R. D., & Vadala, C. (2010). From play in nature, to recreation then vocation: A developmental model for natural history-oriented environmental professionals. *Children, Youth and Environments*, *20*(1), 231–256.

Jiménez-Aleixandre, M., & Erduran, S. (2008). Argumentation in science education: An overview. In Erduran, S., & Jiménez-Aleixandre, M. (Eds.), *Argumentation in Science Education: Perspectives from Classroom-Based Research* (pp. 3–27). Dordrecht, The Netherlands: Springer.

Jimenez-Aleixandre, M., Rodriguez, M., & Duschl, R. A. (2000). Doing the lesson or doing science: Argument in high school genetics. *Science Education*, *84*(6), 757–792. doi:10.1002/1098-237X(200011)84:6<757::AID-SCE5>3.0.CO;2-F

Jimoyiannis, A., & Vassilis, K. (2001). Computer simulations in physics teaching and learning: A case study on students' understanding of trajectory motion. *Computers & Education*, *36*(2), 183–204. doi:10.1016/S0360-1315(00)00059-2

Johnson, R. L., Penny, J. A., & Gordon, B. (2009). *Assessing performance: Designing, scoring, and validating performance tasks*. New York, NY: The Guilford Press.

Johnston, A. (2008). Demythologizing or dehumanizing? A response to Settlage and the ideals of open inquiry. *Journal of Science Teacher Education*, *19*, 11–13. doi:10.1007/s10972-007-9079-y

Johnston, A. T., & Southerland, S. A. (2002). *Conceptual ecologies and their influence on nature of science conceptions: More dazed and confused than ever*. Paper presented at the Annual Meeting of the National Association for Research in Science Teaching. New Orleans, LA.

Jordan, R. C., Ruibal-Villasenor, M., Hmelo-Siler, C. E., & Etkina, E. (2011). Laboratory materials: Affordances or constraints? *Journal of Research in Science Teaching*, *48*(9), 1010–1025. doi:10.1002/tea.20418

Jungwirth, E. (1970). An evaluation of the attained development of the intellectual skills needed for understanding of the nature of scientific inquiry by BSCS pupils in Israel. *Journal of Research in Science Teaching*, *7*, 141–151. doi:10.1002/tea.3660070210

Justi, R., & Gilbert, J. K. (1999). History and philosophy of science through models: The case of chemical kinetics. *Science & Education*, *8*, 287–307. doi:10.1023/A:1008645714002

Justi, R., & Gilbert, J. K. (2002a). Modelling, teachers' views on the nature of modelling, and implications for the education of modellers. *International Journal of Science Education*, *24*(4), 369–387. doi:10.1080/09500690110110142

Justi, R., & Gilbert, J. K. (2002b). Science teachers' knowledge about and attitudes towards the use of models and modelling in learning science. *International Journal of Science Education*, *24*(12), 1273–1292. doi:10.1080/09500690210163198

Justi, R., & Gilbert, J. K. (2003). Teachers' views on the nature of models. *International Journal of Science Education*, *25*(11), 1369–1386. doi:10.1080/0950069032000070324

Kafai, Y. B., Franke, M. L., Ching, C. C., & Shih, J. C. (1998). Game design as an interactive learning environment for fostering students' and teachers' mathematical inquiry. *International Journal of Computers for Mathematical Learning*, *3*(2), 149–184. doi:10.1023/A:1009777905226

Kahn, S. (2010). New pedagogies on teaching science with computer simulations. *Journal of Science Education and Technology*, *20*, 215–232. doi:10.1007/s10956-010-9247-2

Kanfer, R., & Ackerman, P. L. (1989). Motivation and cognitive abilities: An integrative aptitude treatment interaction approach to skill acquisition. *The Journal of Applied Psychology*, *74*, 657–690. doi:10.1037/0021-9010.74.4.657

Karplus, R. (1977). Science teaching and the development of reasoning. *Journal of Research in Science Teaching*, *14*, 169–175. doi:10.1002/tea.3660140212

Kebritchi, M., Hirumi, A., Kappers, W., & Henry, R. (2008). Analysis of the supporting websites for the use of instructional games in K-12 settings. *British Journal of Educational Technology*, *40*(4), 733–754. doi:10.1111/j.1467-8535.2008.00854.x

Kelly, A. M. (2010a). Differentiating the underrepresented: Physics opportunities for Bronx high school students in a university setting. In H. Oluseyi (Ed.), *2009 American Institute of Physics Conference Proceedings Series: Vol. 1280: Joint Annual Conference of the National Society of Black Physicists and the National Society of Hispanic Physicists*, (pp. 176-181). Melville, NY: American Institute of Physics.

Kelly, A. M. (2010b). Transformative informal physics in the Bronx. *Academic Exchange Quarterly*, *14*(1), 57–62.

Kelly, A. M. (2011). Teaching Newton's laws with the iPod Touch in conceptual physics. *The Physics Teacher*, *49*(4), 202–205. doi:10.1119/1.3566026

Kelly, A. M. (2012). Introducing physical and chemical properties and changes. *Science Teacher (Normal, Ill.)*, *79*.

Kelly, A. M., & Kennedy-Shaffer, R. (2011). Teaching Newton's laws to urban middle school students: Strategies for conceptual understanding. *Journal of Curriculum and Instruction*, *5*(1), 54–67.

Kelly, A. M., & Sheppard, K. (2008). Newton in the Big Apple: Access to physics in New York City. *The Physics Teacher*, *46*(5), 280–283. doi:10.1119/1.2909745

Kelly, A. M., & Sheppard, K. (2009). Secondary physics availability in an urban setting: The relationship to academic achievement and course offerings. *American Journal of Physics*, *77*(10), 902–906. doi:10.1119/1.3191690

Kelly, A. M., & Sheppard, K. (2010). The relationship between the urban small schools movement and access to physics education. *Science Educator*, *19*(1), 14–25.

Kelly, G., Chen, C., & Prothero, W. (2000). The epistemological framing of a discipline: Writing science in university oceanography. *Journal of Research in Science Teaching*, *37*(7), 691–718. doi:10.1002/1098-2736(200009)37:7<691::AID-TEA5>3.0.CO;2-G

Kelly, G., & Duschl, R. (2002). *Toward a research agenda for epistemological studies in science education*. Paper presented at the Annual Meeting of the National Association for Research in Science Teaching. New Orleans, LA.

Kelly, G., & Green, J. (1998). The social nature of knowing: Toward a sociocultural perspective on conceptual change and knowledge construction. In Guzzetti, B., & Hynd, C. (Eds.), *Perspectives on Conceptual Change* (pp. 145–182). Mahwah, NJ: Erlbaum.

Kelly, G. J. (2005). Inquiry, activity, and epistemic practice. *Rutgers University*. Retrieved from http://www.ruf.rice.edu/rgrandy/NSFConSched.html

Kelly, G. J. (2007). Discourse in science classrooms. In Abell, S. K., & Lederman, N. G. (Eds.), *Handbook of Research on Science Education*. Mahwah, NJ: Lawrence Erlbaum Associates.

Kemmis, S. (2009). Action research as a practice-based practice. *Educational Action Research*, *17*, 463–474. doi:10.1080/09650790903093284

Kemmis, S. (2010). What is to be done? The place of action research. *Educational Action Research*, *18*, 417–427. doi:10.1080/09650792.2010.524745

Kemmis, S., & McTaggart, R. (2000). Participatory action research. In Denzin, N., & Lincoln, Y. (Eds.), *Handbook of Qualitative Research* (2nd ed., pp. 567–605). Thousand Oaks, CA: Sage.

Kennedy, M. M. (2006). Knowledge and vision in teaching. *Journal of Teacher Education*, *57*, 205–211. doi:10.1177/0022487105285639

Kessels, J., & Korthagen, F. (1996). The relationship between theory and practice: Back to the classics. *Educational Researcher*, *25*, 17–22.

Ketelhut, D. J. (2007). The impact of student self-efficacy on scientific inquiry skills: An exploratory investigation in river city, a multi-user virtual environment. *Journal of Science Education and Technology*, *16*(1), 99–111. doi:10.1007/s10956-006-9038-y

Keys, C., Hand, B., Prain, V., & Collins, S. (1999). Using the science writing heuristic as a tool for learning from laboratory investigations in secondary school. *Journal of Research in Science Teaching*, *36*(10), 1065–1084. doi:10.1002/(SICI)1098-2736(199912)36:10<1065::AID-TEA2>3.0.CO;2-I

Khan, S. (2007). Model-based inquiries in chemistry. *Science Education*, *91*, 877–905. doi:10.1002/sce.20226

Khan, S. (2011). What's missing in model-based teaching. *Journal of Science Teacher Education*, *22*, 535–560. doi:10.1007/s10972-011-9248-x

Khishfe, R. (2008). The development of seventh graders' views of nature of science. *Journal of Research in Science Teaching*, *45*, 470–496. doi:10.1002/tea.20230

Khishfe, R., & Abd-El-Khalick, F. (2002). Influence of explicit and reflective versus implicit inquiry-oriented instruction on sixth graders' views of nature of science. *Journal of Research in Science Teaching*, *39*, 551–578. doi:10.1002/tea.10036

Kirriemuir, J., & McFarlane, A. (2004). *Literature review in games and learning*. Bristol, UK: Futurelab.

Kirschner, P. A., Sweller, J., & Clark, R. E. (2006). Why minimal guidance during instruction does not work: An analysis of the failure of constructivist, discovery, problem-based, experiential, and inquiry-based teaching. *Educational Psychologist*, *41*(2), 75–86. doi:10.1207/s15326985ep4102_1

Kitsantas, A., Zimmerman, B. J., & Cleary, T. (2000). The role of observation and emulation in the development of athletic self-regulation. *Journal of Educational Psychology*, *92*(4), 811–817. doi:10.1037/0022-0663.92.4.811

Klahr, D., Dunbar, K., & Fay, A. L. (1990). Designing good experiments to test bad hypotheses. In *Computational Models of Scientific Discovery and Theory Formation* (pp. 355–401). San Mateo, CA: Morgan Kaufman.

Klahr, D., & Nigam, M. (2004). The equivalence of learning paths in early science instruction: Effects of direct instruction and discovery learning. *Psychological Science*, *15*, 661–667. doi:10.1111/j.0956-7976.2004.00737.x

Klassen, S. (2006). Contextual assessment in science education: Background, issues, and policy. *Science Education*, *90*(5), 820. doi:10.1002/sce.20150

Kline, R. B. (2005). *Principles and practice of structural equation modeling* (2nd ed.). New York, NY: Guilford Press.

Klopfer, E., Yoon, S., & Um, T. (2005). Teaching complex dynamic systems to young students with StarLogo. *Journal of Computers in Mathematics and Science Teaching*, *24*(2), 157–178.

Klopfer, L., & Cooley, W. (1961). *Test on understanding science, form W*. Princeton, NJ: Educational Testing Services.

Knowles, T., & Brown, D. F. (2000). *What every middle school teacher should know*. Portsmouth, NH: Heinemann.

Koch, J. (2010). *Science stories: Science methods for elementary and middle school teachers* (4th ed.). Belmont, CA: Wadsworth Cengage Learning.

Koehler, M. J., Mishra, P., & Yahya, K. (2007). Tracing the development of teacher knowledge in a design seminar: Integrating content, pedagogy and technology. *Computers & Education*, *49*, 740–762. doi:10.1016/j.compedu.2005.11.012

Koray, O. (2011). The effectiveness of problem-based learning supported with computer simulations on academic performance about buoyancy. *Energy Education Science and Technology Part B-Social and Educational Studies*, *3*(3), 293–304.

Kortahgen, F., & Kessels, J. (1999). Linking theory and practice: Changing the pedagogy of teacher education. *Educational Researcher*, *28*, 4–17.

Korthagen, F., Loughran, J., & Russell, T. (2006). Developing fundamental principles for teacher education programs and practices. *Teaching and Teacher Education*, *22*, 1020–1041. doi:10.1016/j.tate.2006.04.022

Krathwohl, D. R. (2002). A revision of Bloom's taxonomy: An overview. *Theory into Practice*, *41*(4), 212–218. doi:10.1207/s15430421tip4104_2

Kuech, R. K., & Lunetta, C. N. (2002). Using digital technologies in the science classroom to promote conceptual understanding. *Journal of Computers in Mathematics and Science Teaching*, *21*(2), 103–126.

Kuerbis, P. J., & Mooney, L. B. (2008). Using assessment design as a model of professional development. In Coffey, J., Douglas, R., & Stearns, C. (Eds.), *Assessing Science Learning: Perspectives from Research and Practice* (pp. 409–426). Arlington, VA: National Science Teachers Association Press.

Kuhn, D. (1989). Children and adults as intuitive scientists. *Psychological Review*, *96*(4), 674–689. doi:10.1037/0033-295X.96.4.674

Kuhn, D. (1991). *The skills of argument*. Cambridge, UK: Cambridge University Press. doi:10.1017/CBO9780511571350

Kuhn, D. (1993). Science as argument: Implications for teaching and learning scientific thinking. *Science Education*, *77*(3), 319–337. doi:10.1002/sce.3730770306

Kuhn, D. (2000). Metacognitive development. *Current Directions in Psychological Science*, *9*, 178–181. doi:10.1111/1467-8721.00088

Kuhn, D., Shaw, V., & Felton, M. (1997). Effects of dyadic interaction on argumentative reasoning. *Cognition and Instruction*, *15*(3), 287–315. doi:10.1207/s1532690xci1503_1

Kuhn, L., & Reiser, B. (2005). *Students constructing and defending evidence-based scientific explanations*. Paper presented at the Annual Meeting of the National Association for Research in Science Teaching. Dallas, TX.

Kuhn, L., & Reiser, B. (2006). *Structuring activities to foster argumentative discourse*. Retrieved from http://hi-ce.org/iqwst/Papers/KuhnReiserAERA2006.pdf

Lampert, M. (1990). When the problem is not the question and the solution is not the answer: Mathematical knowing and teaching. *American Educational Research Journal*, *27*(1), 29–63.

Landis, J. R., & Koch, G. G. (1977). The measurement of observer agreement for categorical data. *Biometrics*, *33*, 159–174. doi:10.2307/2529310

Lange, M. (2002). *An introduction to the philosophy of physics: Locality, fields, energy, and mass*. Oxford, UK: Blackwell Publishing.

Langer, E. J. (1997). *The power of mindful learning*. Reading, MA: Addison-Wesley.

Lareau, A., & Horvat, E. M. (1999). Moments of social inclusion and exclusion: Race, class, and cultural capital in family-school relationships. *Sociology of Education*, *72*(1), 37–53. doi:10.2307/2673185

Latour, B. (1999). *Pandora's hope: Essays on the reality of science studies*. Boston, MA: Harvard University Press.

Lave, J. (1988). *Cognition in practice*. Cambridge, UK: Cambridge University Press. doi:10.1017/CBO9780511609268

Lave, J., & Wenger, E. (1991). *Situated learning: Legitimate peripheral participation*. Cambridge, UK: Cambridge University Press. doi:10.1017/CBO9780511815355

Lawrenz, F., Huffman, D., & Appeldoorn, K. (2005). Enhancing the instructional environment. *Journal of College Science Teaching*, *35*(7), 40–44.

Lawson, A. (2003). The nature and development of hypothetico-predictive argumentation with implications for science teaching. *International Journal of Science Education*, *25*(11), 1387–1408. doi:10.1080/0950069032000052117

Lawson, A. E. (1982). The nature of advanced reasoning and science instruction. *Journal of Research in Science Teaching*, *19*, 743–760. doi:10.1002/tea.3660190904

Leach, J. T., Hind, A. J., & Ryder, J. (2003). Designing and evaluating short teaching interventions about the epistemology of science in high school classrooms. *Science Education*, *87*(6), 831–848. doi:10.1002/sce.10072

Lederman, N. G. (1992). Students' and teachers' conceptions of the nature of science: A review of the research. *Journal of Research in Science Teaching*, *29*, 331–359. doi:10.1002/tea.3660290404

Lederman, N. G. (2007). The nature of science: Past, present, and future. In Abell, S. K., & Lederman, N. G. (Eds.), *Handbook of Research on Science Education*. London, UK: Lawrence Erlbaum & Associates, Publishers.

Lederman, N. G., Abd-El-Khalick, F., Bell, R. L., & Schwartz, R. (2002). Views of nature of science questionnaire (VNOS): Toward valid and meaningful assessment of learners' conceptions of nature of science. *Journal of Research in Science Teaching*, *39*, 497–521. doi:10.1002/tea.10034

Lee, O. (1995). Subject matter knowledge, classroom management, and instructional practices in middle school science classrooms. *Journal of Research in Science Teaching*, *32*(4), 423–440. doi:10.1002/tea.3660320409

Lee, O., & Anderson, C. W. (1993). Task engagement and conceptual change in middle school science classrooms. *American Educational Research Journal*, *30*, 585–610.

Lehrer, R., & Schauble, L. (2006). Cultivating model-based reasoning in science education. In *Cambridge Handbook of the Learning Sciences* (pp. 371–388). Cambridge, UK: Cambridge University Press.

Lehrer, R., & Schauble, L. (2010). What kind of explanation is a model? In Stein, M. K. (Ed.), *Instructional Explanations in the Disciplines* (pp. 9–22). New York, NY: Springer. doi:10.1007/978-1-4419-0594-9_2

Lehrer, R., Schauble, L., & Lucas, D. (2008). Supporting development of the epistemology of inquiry. *Cognitive Development*, *23*(4), 512–529. doi:10.1016/j.cogdev.2008.09.001

Leitão, S. (2000). The potential of argument in knowledge building. *Human Development*, *43*(6), 332–360. doi:10.1159/000022695

Lemke, J. L. (1990). *Talking science: Language, learning, and values*. Norwood, NJ: Ablex.

Lemke, J. L. (2001). Articulating communities: Sociocultural perspectives on science education. *Journal of Research in Science Teaching*, *38*, 296–316. doi:10.1002/1098-2736(200103)38:3<296::AID-TEA1007>3.0.CO;2-R

Lenhart, A., Jones, S., Macgill, A. R., & Pew Internet and American Life Project. (2008). *Adults and video games*. Retrieved February 3, 2012, from http://www.pewinternet.org/~/media//Files/Reports/2008/PIP_Adult_gaming_memo.pdf.pdf

Lent, R. W., Brown, S., & Larkin, K. (2012). Comparison of three theoretically derived variables in predicting career and academic behavior: Self-efficacy, interest congruence, and consequence thinking. *Journal of Counseling Psychology*, *34*(3), 293–298. doi:10.1037/0022-0167.34.3.293

Lerner, L. S. (2000). *Good science, bad science: Teaching evolution in the states*. Washington, DC: Thomas B. Fordham Foundation.

Levy, S. T., & Wilensky, U. (2008). Inventing a "mid-level" to make ends meet: Reasoning through the levels of complexity. *Cognition and Instruction*, *26*, 1–47. doi:10.1080/07370000701798479

Lim, M., & Calabrese Barton, A. (2006). Science learning and a sense of place in an urban middle school. *Cultural Studies of Science Education*, *1*(1), 107–142. doi:10.1007/s11422-005-9002-9

Lin, M. C., Tutwiler, M. S., & Chang, C. Y. (2011). Exploring the relationship between virtual learning environment preference, use, and learning outcomes in 10th grade earth science students. *Learning, Media and Technology*, *36*(4), 399–417. doi:10.1080/17439884.2011.629660

Lin, M. C., Tutwiler, M. S., & Chang, C. Y. (2012). Gender bias in virtual learning environments: An exploratory study. *British Journal of Educational Technology*, *43*(2), 59–63. doi:10.1111/j.1467-8535.2011.01265.x

Linn, M. C., & Eylon, B. S. (2006). Science education: Integrating views of learning and instruction. In *Handbook of Educational Psychology* (pp. 511–544). New York, NY: Macmillan.

Linn, M. C., Eylon, B., & Davis, E. A. (2004). The knowledge integration perspective on learning. In *Internet Environments for Science Education* (pp. 29–46). Mahwah, NJ: Lawrence Erlbaum Associates.

Linnenbrink, E. A., & Pintrich, P. R. (2003). The role of self-efficacy beliefs in student engagement and learning in the classroom. *Reading & Writing Quarterly*, *19*, 119–137. doi:10.1080/10573560308223

Liu, S., & Lederman, N. G. (2002). Taiwanese gifted students' views of nature of science. *School Science and Mathematics*, *102*, 114–122. doi:10.1111/j.1949-8594.2002.tb17905.x

Lizotte, D. J., Harris, C. J., McNeill, K. L., Marx, R. W., & Krajcik, J. (2003). *Usable assessments aligned with curriculum materials: Measuring explanation as scientific way of knowing*. Paper presented at the Annual Meeting of the American Educational Research Association. Chicago, IL.

Lizotte, D. J., McNeill, K. L., & Krajcik, J. (2004). Teacher practices that support students' construction of scientific explanations in middle school classrooms. In *Proceedings of the 6th International Conference of the Learning Sciences*, (pp. 310-317). Mahwah, NJ: Lawrence Erlbaum Associates, Inc.

Llewellyn, D. (2007). *Inquiry within: Implementing inquiry-based science standards in grades 3-8* (2nd ed.). Thousand Oaks, CA: Corwin Press.

Locke, E. A., & Latham, G. P. (1990). *A theory of goal setting and task performance*. Englewood Cliffs, NJ: Prentice Hall.

Locke, E. A., & Latham, G. P. (2002). Building a practically useful theory of goal setting and task motivation: A 35-year odyssey. *The American Psychologist*, *57*, 705–717. doi:10.1037/0003-066X.57.9.705

Lord, T., & Orkwiszewski, T. (2006). Moving from didactic to inquiry-based instruction. *The American Biology Teacher*, *68*, 342–345. doi:10.1662/0002-7685(2006)68[342:DTIIIA]2.0.CO;2

Lotter, C., Harwood, W. S., & Bonner, J. J. (2007). The influence of core teaching conceptions on teachers' use of inquiry teaching practices. *Journal of Research in Science Teaching*, *44*, 1318–1347. doi:10.1002/tea.20191

Louca, L. T., Zacharia, Z. T., & Constantinou, C. P. (2011). In quest of productive modeling-based learning discourse in elementary school science. *Journal of Research in Science Teaching*, *48*(8), 919–951. doi:10.1002/tea.20435

Loucks-Horsley, S., Stiles, K. E., Mundry, S., Love, N., & Hewson, P. W. (2010). *Designing professional development for teachers of science and mathematics* (3rd ed.). Thousand Oaks, CA: Corwin Press.

Loughran, J. (2002). Effective reflective practice: In search of meaning in learning about teaching. *Journal of Teacher Education*, *53*(1), 33–43. doi:10.1177/0022487102053001004

Louv, R. (2005). *Last child in the woods: Saving our children from nature-deficit disorder*. Chapel Hill, NC: Algonquin Books.

Luft, J., Bell, R. L., & Gess-Newsome, J. (2008). *Science as inquiry in the secondary setting*. Arlington, VA: National Science Teachers Association.

Lynch, M., & Woolgar, S. (Eds.). (1990). *Representation in scientific practice*. Cambridge, MA: MIT Press.

Lynch, S. (2001). Science for all is not equal to one size fits all: Linguistic and cultural diversity and science education reform. *Journal of Research in Science Teaching*, *38*, 622–627. doi:10.1002/tea.1021

Lyon, E. G., Bunch, G. C., & Shaw, J. M. (2012). Navigating the language demands of an inquiry-based science performance assessment: Classroom challenges and opportunities for English learners. *Science Education*, *96*(4). doi:10.1002/sce.21008

Maia, P. F., & Justi, R. (2009). Learning of chemical equilibrium through modelling-based teaching. *International Journal of Science Education*, *31*(5), 603–630. doi:10.1080/09500690802538045

Mandinach, E. B., & Corno, L. (1985). Cognitive engagement variations among students of different ability level and sex in a computer problem solving game. *Sex Roles*, *13*, 241–251. doi:10.1007/BF00287914

Marshall, J. C. (2009). *The creation, validation, and reliability associated with the EQUIP (electronic quality of inquiry protocol): A measure of inquiry-based instruction*. Paper presented at the National Association of Researchers of Science Teaching Conference. Washington, DC.

Marshall, J. C., Horton, B., & Edmondson, E. (2007). *4E x 2 instructional model*. Retrieved from http://www.clemson.edu/iim

Marshall, J. C., Horton, B., Igo, B. L., & Switzer, D. M. (2009). K-12 science and mathematics teachers' beliefs about and use of inquiry in the classroom. *International Journal of Science and Mathematics Education, 7*(3), 575–596. doi:10.1007/s10763-007-9122-7

Marshall, J. C., Horton, B., Smart, J., & Llewellyn, D. (2009). *EQUIP: Electronic quality of inquiry protocol.* Retrieved from http://iim-web.clemson.edu/wp-content/uploads/2009/02/equip-2009.pdf

Marshall, J. C., Horton, B., & White, C. (2009). EQUIPping teachers: A protocol to guide and improve inquiry-based instruction. *Science Teacher (Normal, Ill.), 76*(4), 46–53.

Marshall, J. C., & Horton, R. M. (2009). *Developing, assessing, and sustaining inquiry-based instruction: A guide for math and science teachers and leaders.* Saarbruecken, Germany: VDM Publishing House Ltd.

Marshall, J. C., Smart, J., & Horton, R. M. (2010). The design and validation of EQUIP: An instrument to assess inquiry-based instruction. *International Journal of Science and Mathematics Education, 8*(2), 299–321. doi:10.1007/s10763-009-9174-y

Marshall, J. C., Smart, J., Lotter, C., & Sirbu, C. (2011). Comparative analysis of two inquiry observational protocols: Striving to better understand the quality of teacher facilitated inquiry-based instruction. *School Science and Mathematics, 111*(6), 306–315. doi:10.1111/j.1949-8594.2011.00091.x

Martin, A. M., & Hand, B. (2009). Factors affecting the implementation of argument in the elementary science classroom: A longitudinal case study. *Research in Science Education, 39*(1), 17–38. doi:10.1007/s11165-007-9072-7

Martin-Hansen, L. M. (2008). First-year college students' conflict with religion and science. *Science & Education, 17*(4), 317–357. doi:10.1007/s11191-006-9039-5

Martiniello, M. (2008). Language and the performance of English-language learners in math word problems. *Harvard Educational Review, 78*(2), 333–368.

Martins, I., Mortimer, E., Osborne, J., Tsatsarelis, C., & Jiménez-Aleixandre, M. P. (2001). Rhetoric and science education. In Behrendt, H., Dahncke, H., Duit, R., Gräber, W., Komorek, M., Kross, A., & Reiska, P. (Eds.), *Research in Science Education - Past, Present, and Future* (pp. 189–198). Dordrecht, The Netherlands: Kluwer Academic Publishers.

Marzano, R. J., Pickering, D. J., & Pollock, J. E. (2001). *Classroom instruction that works: Research-based strategies for increasing student achievement.* Alexandria, VA: ASCD.

Massey, D., & Jess, P. (Eds.). (1995). *A place in the world? Places, culture and globalization.* Oxford, UK: Oxford University Press.

Mastrilli, T. (2005). Environmental education in Pennsylvania's elementary teacher education programs: A statewide report. *The Journal of Environmental Education, 36*(3), 22–30. doi:10.3200/JOEE.36.3.22-30

Matthews, M. R. (1994). *Science teaching: The role of history and philosophy in science.* New York, NY: Routledge.

Mavrommatis, Y. (1997). Understanding assessment in the classroom: Phases of the assessment process—The assessment episode. *Assessment in Education: Principles. Policy & Practice, 4*(3), 381–400.

Mayer, R. (2004). Should there be a three-strikes rule against pure discovery learning? The case for guided methods of instruction. *The American Psychologist, 38*, 79–83.

Mayo, M. J. (2009). Video games: A route to large-scale STEM education? *Science, 323*(5910), 79–82. doi:10.1126/science.1166900

McComas, W. (2006). Science teaching beyond the classroom: The role and nature of informal learning environments. *Science Teacher (Normal, Ill.), 73*(1), 26–30.

McComas, W. F. (2005). *Seeking NOS standards: What content consensus exists in popular books on the nature of science.* Paper presented at the Meeting of National Association for Research in Science Teaching. Dallas, TX.

McComas, W. F. (2008). Seeking historical examples to illustrate key aspects of the nature of science. *Science & Education, 17*(2/3), 249–263. doi:10.1007/s11191-007-9081-y

McComas, W. F., Clough, M. P., & Almazroa, H. (1998). The role and character of the nature of science in science education. *Science & Education*, *7*, 511–532. doi:10.1023/A:1008642510402

McComas, W. F., Lee, C. K., & Sweeney, S. (2009). *The comprehensiveness and completeness of nature of science content in the U.S. state science standards*. Paper presented at the National Association for Research in Science Teaching International Conference. Garden Grove, CA.

McConnell, D. A., Steer, D. A. N., & Owens, K. D. (2003). Assessment and active learning strategies for introductory geology courses. *Journal of Geoscience Education*, *51*(2), 205–216.

McDonald, C. V. (2010). The influence of explicit nature of science and argumentation instruction on preservice primary teachers' views of nature of science. *Journal of Research in Science Teaching*, *47*, 1137–1164. doi:10.1002/tea.20377

McGonigal, J. (2011). *Reality is broken: Why games make us better and how they can change the world*. New York, NY: Penguin Press.

McKellar, R., Chatterton, B., Wolfe, A., & Currie, P. (2011). A diverse assemblage of late cretaceous dinosaur and bird feathers from Canadian amber. *Science*, *333*(6049), 1619–1622. doi:10.1126/science.1203344

McLaughlin, J. S. (2005). Classrooms without walls. *Journal of College Science Teaching*, *35*(4), 5–6.

McMillan, J. H. (2007). *Formative classroom assessment: Theory into practice*. New York, NY: Columbia University.

McNeill, K. L. (2009). Teachers' use of curriculum to support students in writing scientific arguments to explain phenomena. *Science Education*, *93*(2), 233–268. doi:10.1002/sce.20294

McNeill, K. L., & Krajcik, J. (2007). Middle school students' use of appropriate and inappropriate evidence in writing scientific explanations. In *Thinking with Data: The Proceedings of 33rd Carnegie Symposium on Cognition*. Mahwah, NJ: Lawrence Erlbaum Associates, Inc.

McNeill, K. L., Lizotte, D. J., Krajcik, J., & Marx, R. W. (2006). Supporting students' construction of scientific explanations by fading scaffolds in instructional materials. *Journal of the Learning Sciences*, *15*(2), 153–191. doi:10.1207/s15327809jls1502_1

Medley, D. M. (1979). Effectiveness of teachers. In Peterson, P. L., & Walberg, H. J. (Eds.), *Research on Teaching: Concepts, Findings, and Implications*. Berkley, CA: Mc-Cutchan.

Meece, J. L., Blumenfeld, P. C., & Hoyle, R. H. (1988). Students' goal orientations and cognitive engagement in classroom activities. *Journal of Educational Psychology*, *80*, 514–523. doi:10.1037/0022-0663.80.4.514

Mehan, H. (1979). *Learning lessons: Social organization in the classroom*. Cambridge, MA: Harvard University Press.

Meichtry, Y. J. (1992). Influencing student understanding of the nature of science: Data from a case curriculum development. *Journal of Research in Science Teaching*, *29*, 389–407. doi:10.1002/tea.3660290407

Meredith, J., Fortner, R., & Mullins, G. (1997). A model of affect in nonformal education. *Journal of Research in Science Teaching*, *34*, 805–818. doi:10.1002/(SICI)1098-2736(199710)34:8<805::AID-TEA4>3.0.CO;2-Z

Metallidou, P., & Vlachou, A. (2007). Motivational beliefs, cognitive engagement, and achievement in language and mathematics in elementary school children. *International Journal of Psychology*, *42*, 2–15. doi:10.1080/00207590500411179

Metcalf, S., Kamarainen, A., Tutwiler, M. S., Grotzer, T., & Dede, C. (2011). Ecosystem science learning via multi user virtual environments. *International Journal of Gaming and Computer-Mediated Simulations*, *3*(1), 86–90. doi:10.4018/jgcms.2011010107

Meyer, H. (2004). Novice and expert teachers' conceptions of learners' prior knowledge. *Science Education*, *88*, 970–983. doi:10.1002/sce.20006

Mezirow, J. (1990). *Fostering critical reflection in adulthood: A guide to transformative and emancipatory learning*. San Francisco, CA: Jossey-Bass.

Mezirow, J. (1997). Transformative learning: Theory to practice. *New Directions for Adult and Continuing Learning*, *74*, 5–12. doi:10.1002/ace.7401

Michael, J., & Modell, H. I. (2003). *Active learning in secondary and college science classrooms: A working model for helping the learner to learn*. Mahwah, NJ: LEA Inc.

Millar, R., & Osborne, J. (1998). *Beyond 2000: Science education for the future*. London, UK: King's College.

Miller, J. D., Scott, E. C., & Okamoto, S. (2006). Public acceptance of evolution. *Science*, *313*, 765–766. doi:10.1126/science.1126746

Miller, R. B., Greene, B. A., Montalvo, G. P., Ravindran, B., & Nichols, J. D. (1996). Engagement in academic work: The role of learning goals, future consequences, pleasing others, and perceived ability. *Contemporary Educational Psychology*, *21*, 388–422. doi:10.1006/ceps.1996.0028

Millstone, J., & Levy, A. (2012). *What do teachers really think about using video games in the classroom?* Paper presented at the 9th Annual Games for Change Summit. New York, NY.

Mintzes, J. J., Wandersee, J. H., & Novak, J. D. (Eds.). (1998). *Teaching science for understanding: A human constructivist view*. San Diego, CA: Academic Press.

Mintzes, J. J., Wandersee, J. H., & Novak, J. D. (Eds.). (2000). *Assessing science for understanding: A human constructivist view*. San Diego, CA: Academic Press.

Moje, M. (1995). Talking about science: An interpretation of the effects of teacher talk in a high school science classroom. *Journal of Research in Science Teaching*, *32*, 349–371. doi:10.1002/tea.3660320405

Moore, R., & Wong, H. (1997). *Natural learning: Rediscovering nature's way of teaching*. Berkeley, CA: MIG Communications.

Morge, L. (2005). Teacher-pupil interaction: A study of hidden beliefs in conclusion phases. *International Journal of Science Education*, *27*(8), 935–956. doi:10.1080/09500690500068600

Mork, S. M. (2005). Argumentation in science lessons: Focusing on the teacher's role. *NorDiNa*, *1*(1), 17–30.

Morrison, J. A., Raab, F., & Ingram, D. (2009). Factors influencing elementary and secondary teachers' views on the nature of science. *Journal of Research in Science Teaching*, *46*, 384–403. doi:10.1002/tea.20252

Morrison, M., & Morgan, M. S. (1999). Models as mediating instruments. In Morgan, M. S., & Morrison, M. (Eds.), *Models as Mediators: Perspectives on Natural and Social Science* (pp. 10–37). Cambridge, UK: Cambridge University Press. doi:10.1017/CBO9780511660108.003

Mortimer, E. F., & Scott, P. H. (2003). *Meaning making in secondary science classrooms*. Maidenhead, UK: Open University Press.

Moseley, C., Reinke, K., & Bookout, V. (2002). The effect of teaching outdoor environmental education on pre-service teachers' attitudes toward self-efficacy and outcome expectancy. *The Journal of Environmental Education*, *34*(1), 9–15. doi:10.1080/00958960209603476

Moseley, C., & Utley, J. (2008). An exploratory study of preservice teachers' beliefs about the environment. *The Journal of Environmental Education*, *39*(4), 15–30. doi:10.3200/JOEE.39.4.15-30

Moss, D. M., Abrams, E. D., & Kull, J. R. (1998). *Describing students' conceptions of the nature of science over an entire school year*. Paper presented at the Annual Meeting of the National Association for Research in Science Teaching. San Diego, CA.

Moursund, D., & Bielefeldt, T. (1999). *Will new teachers be prepared to teach in a digital age? A national survey on information technology in teacher education*. Santa Monica, CA: Milken Exchange on Education Technology. Retrieved from http://www.eric.ed.gov/ERICWebPortal/contentdelivery/servlet/ERICServlet?accno=ED428072

Mowrer, D. E. (1996). A content analysis of student/instructor communication via computer conferencing. *Higher Education*, *32*, 217–241. doi:10.1007/BF00138397

Mulholland, J., & Wallace, J. (2008). Computer, craft, complexity, change: Exploration into science teacher knowledge. *Studies in Science Education*, *44*, 41–62. doi:10.1080/03057260701828135

Mumtaz, S. (2000). Factors affecting teachers' use of information and communications technology: A review of the literature. *Journal of Information Technology for Teacher Education*, *9*(3), 319. doi:10.1080/14759390000200096

Murray, J. P., Biggins, B., Donnerstein, E., Menninger, R. W., Rich, M., & Strasburger, V. (2011). A plea for concern regarding violent video games. *Mayo Clinic Proceedings*, *86*(8), 818–820. doi:10.4065/mcp.2011.0321

National Academies of Sciences. (2007). *Rising above the gathering storm: Energizing and employing America for a brighter economic future*. Washington, DC: The National Academies Press.

National Academies of Sciences. (2010a). *Expanding underrepresented minority participation: America's science and technology talent at the crossroads*. Washington, DC: National Academies Press.

National Academies of Sciences. (2010b). *Rising above the gathering storm, revisited: Rapidly approaching category 5*. Washington, DC: National Academies Press.

National Academy of Science. (1998). *Teaching about evolution and the nature of science*. Washington, DC: National Academy Press.

National Academy of Science. (2008). *Science, evolution, and creationism*. Washington, DC: National Academy Press.

National Association of Biology Teachers. (2011). *NABT's statement on teaching evolution*. Retrieved from http://www.nabt.org/websites/institution/index.php?p=92

National Association of Educational Progress. (2009). *The nation's report card*. Retrieved January 30, 2012, from http://www.nationsreportcard.gov/science_2009/

National Board for Professional Teaching Standards. (2000). *A distinction that matters: Why national teacher certification makes a difference*. Greensboro, NC: Center for Educational Research and Evaluation.

National Center for Education Statistics. (1998). *Pursuing excellence: A study of U.S. twelfth grade mathematics and science achievement in international context*. Washington, DC: NCES.

National Center for Education Statistics. (2004). *Highlights from TIMMS*. Washington, DC: Office of Educational research and Improvement. Retrieved from http://nces.ed.gov/pubs99/1999081.pdf

National Center for Education Statistics. (2009). *Digest of education statistics: Percentage of public and private high school graduates taking selected mathematics and science courses in high school, by sex and race/ethnicity: Selected years, 1982 through 2005*. Washington, DC: U. S. Department of Education Institute of Education Sciences. Retrieved November 9, 2010, from http://nces.ed.gov/programs/digest/d09/tables/dt09_151.asp

National Center for Education Statistics. (2009). *Science framework for the 2009 national assessments of educational progress*. Washington, DC: National Center for Education Statistics.

National Commission on Mathematics and Science Teaching for the 21st Century [NCMST]. (2000). *Before It's Too Late*. Washington, DC: Education Publications Center.

National Committee on Science Education Standards and Assessment. (1996). *National science education standards*. Retrieved from http://www.nap.edu/openbook.php?record_id=4962&page=R1

National Council for Accreditation of Teacher Education. (1997). *Technology and the new professional teacher: Preparing for the 21st century classroom*. Retrieved from http://www.eric.ed.gov/ERICWebPortal/contentdelivery/servlet/ERICServlet?accno=ED412201

National Council for Accreditation of Teacher Education. (2008). *Professional standards for the accreditation of teacher preparation institutions*. Retrieved from http://www.ncate.org/Portals/0/documents/Standards/NCATE%20Standards%202008.pdf

National Research Council. (1996). *National science education standards*. Washington, DC: National Academic Press.

National Research Council. (2000). *Inquiry and the national science education standards: A guide for teaching and learning*. Washington, DC: National Academies Press.

National Research Council. (2005). *America's lab report: Investigations in high school science*. Washington, DC: National Academy Press.

National Research Council. (2007). *Taking science to school: Learning and teaching in grades K-8*. Washington, DC: National Academy Press.

National Research Council. (2010). *A framework for science education. Preliminary Public Draft*. Washington, DC: Committee on Conceptual Framework for New Science Education Standards.

National Research Council. (2010). *Exploring the intersection of science education and 21st century skills: A workshop summary*. Washington, DC: National Academies Press

National Research Council. (2011). *Conceptual framework for new science education standards*. Washington, DC: National Academy of Sciences Board on Science Education.

National Research Council. (2012). *A framework for K-12 science education: Practices, crosscutting concepts and core idea*. Washington, DC: The National Academy Press.

National Science Board. (2006). *Science and engineering indicators 2006*. Arlington, VA: National Science Foundation.

National Science Foundation, Division of Science Resources Statistics. (2007). *Women, minorities, and persons with disabilities in science and engineering: 2007*. Arlington, VA: National Science Foundation. Retrieved from http://www.nsf.gov/statistics/wmpd

National Science Teachers Association. (1998). *Standards for science teacher preparation*. Retrieved from http://www.nsta.org/main/pdfs/nsta98standards.pdf

National Science Teachers Association. (2003). *NSTA position statement: The teaching of evolution*. Retrieved from http://www.nsta.org/pdfs/positionstatement_evolution.pdf

Naylor, S., Keogh, B., & Downing, B. (2007). Argumentation and primary science. *Research in Science Education*, *37*, 17–39. doi:10.1007/s11165-005-9002-5

Neilson, D., Campbell, T., & Allred, B. (2010). Model-based inquiry: A buoyant force module for high school physics classes. *Science Teacher (Normal, Ill.)*, *77*(8), 38–43.

Nelson, M. M., & Davis, E. A. (2012). Preservice elementary teachers' evaluations of elementary students' scientific models: An aspect of pedagogical content knowledge for scientific modeling. *International Journal of Science Education*. doi:10.1080/09500693.2011.594103

Nersessian, N. J. (1999). Model-based reasoning in conceptual change. In Magnani, L., Nersessian, N. J., & Thagard, P. (Eds.), *Model-Based Reasoning in Scientific Discovery* (pp. 5–22). New York, NY: Kluwer Academic/Plenum Publishers. doi:10.1007/978-1-4615-4813-3_1

Nersessian, N. J. (2008). *Creating scientific concepts*. Cambridge, MA: The MIT Press.

Nersessian, N. J., & Patton, C. (2009). Model-based reasoning in interdisciplinary engineering. In *Handbook of the Philosophy of Technology and Engineering Sciences* (pp. 687–718). Amsterdam, The Netherlands: North Holland. doi:10.1016/B978-0-444-51667-1.50031-8

New York City Coalition for Education Justice. (2007). *New York City's middle grade schools: Platforms for success or pathways to failure?* Retrieved January 12, 2012, from http://www.nyccej.org/117/new-york-citys-middle-grade-schools-platforms-for-success-or-pathways-to-failure

New York City Department of Education. (2006). *Annual school reports cards 2004-2005*. Retrieved January 12, 2012, from http://schools.nyc.gov/Accountability/data/AnnualSchoolReports/default.htm

New York State Education Department. (2009). *Physics/physical setting core standards*. Retrieved November 30, 2009, from http://www.emsc.nysed.gov/ciai/mst/pub/phycoresci.pdf

New York Times. (2006, August 18). *In elite new york high schools, a dip in blacks and hispanics*. Retrieved January 30, 2012, from http://www.nytimes.com/2006/08/18/education/18schools.html?pagewanted=all

Newton, P., Driver, R., & Osborne, J. (1999). The place of argumentation in the pedagogy of school science. *International Journal of Science Education*, *21*(5), 553–576. doi:10.1080/095006999290570

Ng, F., Zeng, H., & Plass, J. (2009). *Research on educational impact of games: A literature review*. Report No. 02/2009. New York, NY: Institute for Games for Learning, NYU/CUNY.

Noel, J. (1999). On the varieties of phronesis. *Educational Philosophy and Theory*, *31*, 273–289. doi:10.1111/j.1469-5812.1999.tb00466.x

Nolen, S. B. (1988). Reasons for studying: Motivational orientations and study strategies. *Cognition and Instruction*, *5*, 269–287. doi:10.1207/s1532690xci0504_2

Norris, S. P., & Phillips, L. M. (2003). How literacy in its fundamental sense is central to scientific literacy. *Science Education*, *87*(2), 224–240. doi:10.1002/sce.10066

Northwest Evaluation Association. (2003). *Technical manual*. Lake Oswego, OR: Northwest Evaluation Association.

Northwest Evaluation Association. (2004). *Reliability and validity estimates: NWEA achievement level tests and measure of academic progress*. Retrieved from http://www.nwea.org

Northwest Evaluation Association. (2005). *NWEA reliability and validity estimates: Achievement level tests and measures of academic progress*. Lake Oswego, OR: Northwest Evaluation Association.

Norton-Meier, L., Hand, B., Hockenberry, L., & Wise, K. (2008). *Questions, claims and evidence: The important place of argument in children's science writing*. Portsmouth, NH: Heinemann.

Novak, J. D. (1963). What should we teach in biology? *News & Views*, *7*(2).

Novak, J. D. (1990). Concept mapping: A useful tool for science education. *Journal of Research in Science Teaching*, *27*(10), 937–949. doi:10.1002/tea.3660271003

Novak, J. D. (1998). *Learning, creating, and using knowledge: Concept maps as facilitative tools in schools and corporations*. Mahwah, NJ: Lawrence Erlbaum Associates Publishers.

Novak, J. D. (1998). The pursuit of a dream: Education can be improved. In Mintzes, J. J., Wandersee, J. H., & Novak, J. D. (Eds.), *Teaching Science for Understanding: A Human Constructivist View* (pp. 3–29). San Diego, CA: Academic Press.

Novak, J. D., & Gowin, D. (1984). *Learning how to learn*. Cambridge, UK: Cambridge University Press. doi:10.1017/CBO9781139173469

Novak, J. D., & Musonda, D. (1991). A twelve-year longitudinal study of science concept learning. *American Educational Research Journal*, *28*, 117–153.

Nystrand, M., Gamoran, A., Kachur, R., & Prendergast, C. (1997). *Opening dialogue: Understanding the dynamics of language and learning in the English classroom*. New York, NY: Teachers College Press.

Ohio Department of Education. (2010). *Science education standards revision and model curriculum development*. Retrieved from http://www.ode.state.oh.us/GD/Templates/Pages/ODE/ODEDetail.aspx?page=3&TopicRelationID=1705&ContentID=76585

O'Loughlin, M. (1992). Rethinking science education: Beyond Piagetian constructivism toward a sociocultural model of teaching and learning. *Journal of Research in Science Teaching*, *29*, 791–820. doi:10.1002/tea.3660290805

OECD. (2005). *Definition and selection of key competencies*. Retrieved from www.oecd.org/dataoecd/47/61/35070367.pdf

OECD. (2009). *Pisa 2009 assessment framework – Key competencies in reading, mathematics, and science*. Washington, DC: OECD.

Office of Technology Assessment. (1995). *Teachers and technology: Making the connection*. Report No. OTA-EHR-616. Washington, DC: U.S. Government Printing Office. Retrieved from http://www.eric.ed.gov/ERICWebPortal/contentdelivery/servlet/ERICServlet?accno=ED386155

Ogunniyi, M. B. (1983). Relative effects of a history/philosophy of science course on student teachers' performance on two models of science. *Research in Science & Technological Education*, *1*, 193–199. doi:10.1080/0263514830010207

Oh, P. S., & Oh, S. J. (2011). What teachers of science need to know about models: An overview. *International Journal of Science Education*, *33*(8), 1109–1130. doi:10.1080/09500693.2010.502191

Okuda, Y., Bryson, E. O., DeMaria, S. Jr, Jacobson, L., Quinones, J., Shen, B., & Levine, A. I. (2009). The utility of simulation in medical education: What is the evidence? *The Mount Sinai Journal of Medicine, New York*, *76*, 330–343. doi:10.1002/msj.20127

O'Leary, J. T., Behrens-Tepper, J., McGuire, F. A., & Dottavio, F. D. (1987). Age of first hunting experience: Results from a nationwide recreation survey. *Leisure Sciences, 9*, 225–233. doi:10.1080/01490408709512164

Olson, J. K., & Clough, M. P. (2001). *Secondary science teachers' implementation practices following a course emphasizing contextualized and decontextualized nature of science instruction*. Paper presented at the 6th International History, Philosophy, and Science Teaching Conference. Denver, CO.

Oregon State University. (2011). *Volcano sounds*. Retrieved from http://bit.ly/pWMB7m

Orion, N., & Hofstein, A. (1994). Factors that influence learning during a scientific field trip in a natural environment. *Journal of Research in Science Teaching, 31*, 1097–1119. doi:10.1002/tea.3660311005

Orr, D. (1994). *Earth in mind: On education, environment, and the human prospect*. Washington, DC: National Academy Press.

Osborne, J. (2007). Towards a more social pedagogy in science education: the role of argumentation. *Revista Brasileira de Pesquisa em Educação em Ciências, 7*(1). Retrieved July 12, 2010, from http://www.fae.ufmg.br/abrapec/revista/index.html

Osborne, J. F. (2001). Promoting argument in the science classroom: A rhetorical perspective. *Canadian Journal of Science. Mathematics and Technology Education, 1*, 271–290. doi:10.1080/14926150109556470

Osborne, J. F., Erduran, S., & Simon, S. (2004b). *Ideas, evidence and argument in science: In-service training pack, resource pack and video*. London, UK: Nuffield Foundation.

Osborne, J. F., & Patterson, A. (2011). Scientific argument and explanation: A necessary distinction? *Science Education, 95*, 627–638. doi:10.1002/sce.20438

Osborne, J., Collins, S., Ratcliffe, M., Millar, R., & Duschl, R. (2003). What "ideas-about-science" should be taught in school? A Delphi study of the expert community. *Journal of Research in Science Teaching, 40*, 692–720. doi:10.1002/tea.10105

Osborne, J., & Dillon, J. (2008). *Science education in Europe: Critical reflections*. London, UK: Nuffield Foundation.

Osborne, J., Erduran, S., & Simon, S. (2004). Enhancing the quality of argumentation in school science. *Journal of Research in Science Teaching, 41*(10), 994–1020. doi:10.1002/tea.20035

Osif, B. A. (1997). Evolution and religious beliefs: A survey of Pennsylvania high school teachers. *The American Biology Teacher, 59*(9), 552–556. doi:10.2307/4450382

Pajares, M. F. (1992). Teachers' beliefs and education research: Cleaning up a messy construct. *Review of Educational Research, 62*, 307–332.

Palmer, J. (1993). Development of concern for the environment and formative experiences of educators. *The Journal of Environmental Education, 24*, 26–30. doi:10.1080/00958964.1993.9943500

Papert, S., & Harel, I. (1991). Situating constructionism. In *Constructionism*. New York, NY: Ablex.

Paris, S. G., Byrnes, J. P., & Paris, A. H. (2001). Constructing theories, identities, and actions of self-regulated learners. In Zimmerman, B. J., & Schunk, D. H. (Eds.), *Self-Regulated Learning and Academic Achievement: Theoretical Perspectives* (2nd ed., pp. 253–287). Mahwah, NJ: Earlbaum.

Paris, S. G., Cross, D. R., & Lipson, M. Y. (1984). Informed strategies for learning: A program to improve children's reading awareness and comprehension. *Journal of Educational Psychology, 76*, 1239–1252. doi:10.1037/0022-0663.76.6.1239

Parkinson, J. (2004). *Improving secondary science teaching*. London, UK: Routledge Falmer. doi:10.4324/9780203464328

Partnership for 21st Century Skills. (2009). *Framework for 21st century learning*. Retrieved from http://www.p21.org/index.php?option=com_content&task=view&id=254&Itemid=119

Passmore, C., & Stewart, J. (2002). A modeling approach to teaching evolutionary biology in high schools. *Journal of Research in Science Teaching, 39*(3), 185–204. doi:10.1002/tea.10020

Passmore, C. M., & Svoboda, J. (2012). Exploring opportunities for argumentation in modeling classrooms. *International Journal of Science Education*. Retrieved from http://www.academia.edu/233497/Exploring_Opportunities_for_Argumentation_in_Modelling_Classrooms

Paulson, A., Perkins, K., & Adams, W. (2009). *How does the type of guidance student use of an interactive simulation?* Retrieved from http://phet.colorado.edu/publications/Paulson_etal_2009/Paulson_etal_2009.pdf

Pellegrino, J. W., Chudowsky, N., & Glaser, R. (2001). *Knowing what students know: The science and design of educational assessment*. Washington, DC: National Academy Press.

Pelletier, C. (2008). Gaming in context: How young people construct their gendered identities in playing and making games. In Kafai, Y. B., Heeter, C., Denner, J., & Sun, J. Y. (Eds.), *Beyond Barbie and Mortal Kombat: New Perspectives on Gender and Gaming*. Cambridge, MA: The MIT Press.

Pemberton, S. G., & Frey, R. W. (1991). William Buckland and his coprolitic vision. *Ichnos*, *1*(4), 317–325. doi:10.1080/10420949109386367

Pendlebury, S. (1995). Reason and story in wise practice. In McEwan, H., & Egan, K. (Eds.), *Narrative in Teaching, Learning and Research* (pp. 50–65). New York, NY: Teachers College Press.

Penner, D. E., Lehrer, R., & Schauble, L. (1998). From physical models to biomechanics: A design-based modeling approach. *Journal of the Learning Sciences*, *7*(3&4), 429–449.

Perkes, V. (1975). Relationships between a teacher's background and sensed adequacy to teach elementary science. *Journal of Research in Science Teaching*, *12*(1), 85–88. doi:10.1002/tea.3660120112

Perkins, D. N., & Salomon, G. (1988). Teaching for transfer. *Educational Leadership*, *46*, 22–32.

Perkins, D., & Grotzer, T. A. (2005). Dimensions of causal understanding: The role of complex causal models in students' understanding of science. *Studies in Science Education*, *41*, 117–165. doi:10.1080/03057260508560216

Peters Burton, E. E. (2010). *Learning about the human aspect of the scientific enterprise: Gender differences in conceptions of scientific knowledge*. Advancing Women in Leadership Journal, 30(12). Retrieved from http://advancingwomen.com/awl/awl_wordpress/

Peters, E. E. (2006). Connecting inquiry and the nature of science. *The Science Education Review*, *5*(2), 37–44.

Peters, E. E. (2012). Developing content knowledge in students through explicit teaching of the nature of science: Influences of goal setting and self-monitoring. *Science & Education*, *21*(6), 881–898. doi:10.1007/s11191-009-9219-1

Peters, E. E., & Kitsantas, A. (2010a). Self-regulation of student epistemic thinking in science: The role of metacognitive prompts. *Educational Psychology*, *30*(1), 27–52. doi:10.1080/01443410903353294

Peters, E. E., & Kitsantas, A. (2010b). The effect of nature of science metacognitive prompts on science students' content and nature of science knowledge, metacognition, and self-regulatory efficacy. *Journal of School Science and Math*, *110*, 382–396. doi:10.1111/j.1949-8594.2010.00050.x

Peterson, N. (1982). *Developmental variables affecting environmental sensitivity in professional environmental educators*. (Unpublished Master's Thesis). Southern Illinois University. Carbondale, IL.

Pew Internet and American Life Project. (2011). *Teen internet usage over time*. Retrieved from http://www.pewinternet.org/Static-Pages/Trend-Data-for-Teens/~/media/Infographics/Trend Data/Teens/September 2009/Teen InternetUsageOverTime–Sep2009.zip

Phenice, L., & Griffore, R. (2003). Young children and the natural world. *Contemporary Issues in Early Childhood*, *4*(2), 167–171. doi:10.2304/ciec.2003.4.2.6

Piaget, J. (1954). *The construction of reality in the child*. New York, NY: Basic Books. doi:10.1037/11168-000

Piaget, J. (1977). *The development of thought: Equilibration of cognitive structures* (Rosin, A., Trans.). New York, NY: The Viking Press.

Pintrich, P. R., & De Groot, V. (1990). Motivation and self-regulated learning components of classroom academic performance. *Journal of Educational Psychology, 82*, 33–40. doi:10.1037/0022-0663.82.1.33

Pintrich, P. R., Marx, R. W., & Boyle, R. A. (1993). Beyond cold conceptual change: The role of motivational beliefs and classroom contextual factors in the process of conceptual change. *Review of Educational Research, 63*(2), 167–199.

Pintrich, P. R., & Schrauben, B. (1992). Students' motivational beliefs and their cognitive engagement in classroom academic tasks. In Schunk, D. H., & Meece, J. L. (Eds.), *Student Perceptions in the Classroom* (pp. 149–183). Hillsdale, NJ: Erlbaum.

Plevyak, L., Bendixen-Noe, M., Henderson, J., Roth, R., & Wilke, R. (2001). Level of teacher preparation and implementation of EE: Mandated and non-mandated EE teacher preparation states. *The Journal of Environmental Education, 32*(2), 28–37. doi:10.1080/00958960109599135

Pluta, W. J., Chinn, C. A., & Duncan, R. G. (2011). Learners' epistemic criteria for good scientific models. *Journal of Research in Science Teaching, 48*(5), 486–511. doi:10.1002/tea.20415

Plutzer, E., & Berkman, M. B. (2008). Evolution, creationism, and the teaching of human origins in schools. *Public Opinion Quarterly, 72*(3), 540–553. doi:10.1093/poq/nfn034

Posner, G. J., Strike, K. A., Hewson, P. W., & Gertzog, W. A. (1982). Accommodation of a scientific conception: Toward a theory of conceptual change. *Science Education, 66*(2), 211–227. doi:10.1002/sce.3730660207

Powers, A. (2004). Teacher preparation for environmental education: Faculty perspectives on the infusion of environmental education into pre-service methods courses. *The Journal of Environmental Education, 35*(3), 3–11.

Pratt, M. L. (1987). Linguistic utopias. In Fabb, N., Attridge, D., Durant, A., & MacCabe, C. (Eds.), *The Linguistics of Writing: Arguments between Language and Literature* (pp. 48–66). New York, NY: Methuen.

Prensky, M. (2001). Digital natives, digital immigrants part 1. *Horizon, 9*(5), 1–6. doi:10.1108/10748120110424816

Prensky, M. (2005). Engage me or enrage me: What today's learners demand. *EDUCAUSE Review, 40*(5), 60.

Quellmalz, E. S., & Pellegrino, J. W. (2009). Technology and testing. *Science, 323*(5910), 75. doi:10.1126/science.1168046

Raia, F. (2008). Causality in complex dynamic systems: A challenge in earth systems science education. *Journal of Geoscience Education, 56*(1), 81–94.

Rapp, D. N., & Sengupta, P. (2012). Models and modeling in science learning. In *Encyclopedia of the Sciences of Learning*. New York, NY: Springer.

Relph, E. (1976). *Place and placelessness*. London, UK: Pion.

Rennie, L., & Johnston, D. (2004). The nature of learning and its implications for research in learning from museums. *Science Education, 88*, S4–S16. doi:10.1002/sce.20017

Rennie, L. J., Feher, E., Dierking, L. D., & Falk, J. H. (2003). Toward an agenda for advancing research on science learning in out-of-school settings. *Journal of Research in Science Teaching, 40*(2), 112–120. doi:10.1002/tea.10067

Resnick, M. (1994). *Turtles, termites, and traffic jams: Explorations in massively parallel microworlds*. Cambridge, MA: MIT Press.

Richardson, V. (1996). The role of attitudes and beliefs in learning to teach. In Sikula, J. (Ed.), *Handbook of Research on Teacher Education* (2nd ed., pp. 102–119). New York, NY: Macmillan.

Riegl-Crumb, C., & Grodsky, E. (2010). Racial-ethnic differences at the intersection of math course-taking and achievement. *Sociology of Education, 83*(3), 248–270. doi:10.1177/0038040710375689

Riffe, D., Lacy, S., & Fico, F. (2005). *Analyzing media messages: Using quantitative content analysis in research* (2nd ed.). New York, NY: Routledge.

Riggs, I. M., & Enochs, L. G. (1990). Toward the development of an elementary teacher's science teaching efficacy belief instrument. *Science Education, 74*(6), 625–637. doi:10.1002/sce.3730740605

Rivkin, M. (1995). *The great outdoors: Restoring children's right to play outside*. Washington, DC: National Association for the Education of Young Children.

Rohrkemper, M. (1989). Self-regulated learning and academic achievement: A Vygotskian view. In Zimmerman, B. J., & Schunk, D. H. (Eds.), *Self-Regulated Learning and Academic Achievement: Theory, Research and Practice* (pp. 143–167). New York, NY: Springer. doi:10.1007/978-1-4612-3618-4_6

Rosenshine, B., & Stevens, R. (1986). Teaching functions. In Wittrock, M. C. (Ed.), *Handbook of Research on Teaching* (3rd ed., pp. 376–391). New York, NY: Macmillan.

Rowe, M. B. (1974). A humanistic intent: The program of preservice elementary education at the University of Florida. *Science Education*, *58*, 369–376. doi:10.1002/sce.3730580311

Rubin, D. B. (2005). Causal inferences using potential outcomes. *Journal of the American Statistical Association*, *100*(469), 322–331. doi:10.1198/016214504000001880

Ruiz-Primo, M. A., & Furtak, E. M. (2007). Exploring teachers' informal formative assessment practices and students' understanding in the context of scientific inquiry. *Journal of Research in Science Teaching*, *44*(1). doi:10.1002/tea.20163

Russell, M., Bebell, D., O'Dwyer, L., & O'Connor, K. (2003). Examining teacher technology use: Implications for preservice and inservice teacher preparation. *Journal of Teacher Education*, *54*(4), 297–310. doi:10.1177/0022487103255985

Rutledge, M. L., & Warden, M. A. (2000). Evolutionary theory, the nature of science & high school biology teachers: Critical relationships. *The American Biology Teacher*, *62*(1), 23–31. doi:10.1662/0002-7685(2000)062[0023:ETTNOS]2.0.CO;2

Rutten, N., van Joolingen, W. R., & van der Veen, J. T. (2012). The learning effects of computer simulations in science education. *Computers & Education*, *58*(1), 136–153. doi:10.1016/j.compedu.2011.07.017

Ryan, A. G., & Aikenhead, G. S. (1992). Students' preconceptions about the epistemology of science. *Science Education*, *76*, 559–580. doi:10.1002/sce.3730760602

Ryan, R. M., Connell, J. P., & Deci, E. L. (1984). A motivational analysis of self-determination and self-regulation in education. In Ames, C., & Ames, R. (Eds.), *Research on Motivation in Education* (Vol. 2, pp. 13–52). New York, NY: Academic Press.

Ryder, J., Leach, J., & Driver, R. (1999). Undergraduate science students' images of science. *Journal of Research in Science Teaching*, *36*, 201–219. doi:10.1002/(SICI)1098-2736(199902)36:2<201::AID-TEA6>3.0.CO;2-H

Sacks, H., Schegloff, E. A., & Jefferson, G. (1974). A simplest systematics for the organization of turn-taking for conversation. *Linguistic Society of America*, *50*, 696–735. doi:10.2307/412243

Sadler, P. M., & Tai, R. H. (2001). Success in introductory college physics: The role of high school preparation. *Science Education*, *85*(2), 111–136. doi:10.1002/1098-237X(200103)85:2<111::AID-SCE20>3.0.CO;2-O

Sadler, T. D. (2004). Informal reasoning regarding socioscientific issues: A critical review of research. *Journal of Research in Science Teaching*, *41*(5), 513–536. doi:10.1002/tea.20009

Saleh, I. M., & Khine, M. S. (Eds.). (2009). *Fostering scientific of mind: Pedagogical knowledge and best practices in science education*. Rotterdam, The Netherlands: Sense Publishers.

Salloum, S. (2006). *Teaching as practice: Blending intellectual and the moral in pursuit of science teachers' practical knowledge*. (Unpublished Doctoral Dissertation). University of Illinois at Urbana-Champaign. Champaign, IL.

Salloum, S. (2009). *Pedagogical growth and practitioner inquiry: ESL urban teachers' perceptions of their growth while conducting inquiries into their practice*. Paper presented at the Annual Meeting of the American Educational Research Association. San Diego, CA.

Salloum, S., & Abd-El-Khalick, F. (2010). Practical knowledge in teaching: Case studies from physical science classrooms. *Journal of Research in Science Teaching*, *47*, 929–951.

Salloum, S., Jennings, M., Arrabito, N., Schmidt, M., McCall, C., & Frederick, T. ... Benn-Scantlebury, A. (2010). *Novice urban teachers engaging in practitioner inquiry: Lessons learned, rewards, and challenges.* Paper presented at Annual Ethnography in Education Research Forum. Philadelphia, PA.

Samarapungavan, A., Westby, E. L., & Bodner, G. M. (2006). Contextual epistemic development in science: A comparison of chemistry students and research chemists. *Science Education, 90*, 468–495. doi:10.1002/sce.20111

Samarapungavan, A., & Wiers, R. (1997). Children's thoughts on the origin of species: A study of explanatory coherence. *Cognitive Science, 21*, 147–177. doi:10.1207/s15516709cog2102_2

Sampson, V. (2004). The science management observation protocol. *Science Teacher (Normal, Ill.), 71*(10), 30–33.

Sampson, V., & Clark, D. (2008). Assessment of the ways students generate arguments in science education: Current perspectives and recommendations for future directions. *Science Education, 92*(3), 447–472. doi:10.1002/sce.20276

Sampson, V., & Gerbino, F. (2010). Two instructional models that teachers can use to promote and support scientific argumentation in the biology classroom. *The American Biology Teacher, 72*(7), 427–431. doi:10.1525/abt.2010.72.7.7

Sampson, V., & Gleim, L. (2009). Argument-driven inquiry to promote the understanding of important concepts & practices in biology. *The American Biology Teacher, 71*(8), 465–472.

Sampson, V., & Grooms, J. (2009). Promoting and supporting scientific argumentation in the classroom: The evaluate alternatives instructional model. *Science Scope, 32*(10), 67–73.

Sampson, V., & Grooms, J. (2010). Generate an argument: An instructional model. *Science Teacher (Normal, Ill.), 77*(5), 33–37.

Sampson, V., Grooms, J., & Walker, J. (2009). Argument-driven inquiry: A way to promote learning during laboratory activities. *Science Teacher (Normal, Ill.), 76*(7), 42–47.

Sampson, V., Grooms, J., & Walker, J. (2011). Argument-driven inquiry as a way to help students learn how to participate in scientific argumentation and craft written arguments: An exploratory study. *Science Education, 95*(2), 217–257. doi:10.1002/sce.20421

Sandoval, W. A., & Millwood, K. A. (2005). The quality of students' use of evidence in written scientific explanations. *Cognition and Instruction, 23*(1), 23–55. doi:10.1207/s1532690xci2301_2

Sandoval, W. A., & Morrison, K. (2003). High school students' ideas about theories and theory change after a biological inquiry unit. *Journal of Research in Science Teaching, 40*(4), 369–392. doi:10.1002/tea.10081

Sandoval, W. A., & Reiser, B. J. (2004). Explanation-driven inquiry: Integrating conceptual and epistemic scaffolds for scientific inquiry. *Science Education, 88*, 345–372. doi:10.1002/sce.10130

Sanford, R., Ulicsak, M., Facer, K., & Rudd, T. (2006). *Teaching with games: Using commercial off-the-shelf games in formal education.* Bristol, UK: Futurelab.

Savinainen, A., & Scott, P. (2002). The force concept inventory: A tool for monitoring student learning. *Physics Education, 37*(1), 45–52. doi:10.1088/0031-9120/37/1/306

Sawada, D., Piburn, M., Falconer, K., Turley, J., Benford, R., & Bloom, I. (2000). *Reformed teaching observation protocol (RTOP).* Technical Report No. IN00-01. Phoenix, AZ: Arizona State University

Scardamalia, M., & Bereiter, C. (1991). Higher levels of agency for children in knowledge-building: A challenge for the design of new knowledge media. *Journal of the Learning Sciences, 1*(1), 37–68. doi:10.1207/s15327809jls0101_3

Scharmann, L. C., Smith, M. U., James, M. C., & Jensen, M. (2005). Explicit reflective nature of science instruction: Evolution, intelligent design, and umbrellaology. *Journal of Science Teacher Education, 16*, 27–41. doi:10.1007/s10972-005-6990-y

Schauble, L., Glaser, R., Duschl, R., Schulze, S., & John, J. (1995). Students' understanding of the objectives and procedures of experimentation in the science classroom. *Journal of the Learning Sciences, 4*(2), 131–166. doi:10.1207/s15327809jls0402_1

Schauble, L., Klopfer, L. E., & Raghavan, K. (1991). Students' transition from an engineering model to a science model of experimentation. *Journal of Research in Science Teaching, 28*, 859–882. doi:10.1002/tea.3660280910

Schein, E. H. (1993). On dialogue, culture, and organizational learning. *Reflections: The SoL Journal, 4*(4), 27–38. doi:10.1162/152417303322004184

Schiefele, U. (1991). Interest, learning, and motivation. *Educational Psychologist, 26*, 299–323.

Schmidt, W. H., McNight, C. C., & Raizen, S. A. (2002). *A splintered vision: An investigation of U.S. science and mathematics education*. Retrieved from http://imc.lisd.k12.mi.us/MSC1/Timms.html

Schneider, W. (2008). The development of metacognitive knowledge in children and adolescents: Major trends and implication for education. *Mind, Brain, and Education, 2*, 114–121. doi:10.1111/j.1751-228X.2008.00041.x

Schrader, P. G., Zheng, D., & Young, M. (2006). Teachers' perceptions of video games: MMOGs and the future of preservice teacher education. *Journal of Online Education, 2*(3).

Schunk, D. H. (1982). Verbal self-regulation as a facilitator of children's achievement and self-efficacy. *Human Learning, 1*, 265-277.

Schunk, D. H. (1985). Self-efficacy and classroom learning. *Psychology in the Schools, 22*, 208–223. doi:10.1002/1520-6807(198504)22:2<208::AID-PITS2310220215>3.0.CO;2-7

Schunk, D. H. (1987). Peer models and children's behavioral change. *Review of Educational Research, 57*, 149–174.

Schunk, D. H., & Hanson, A. R. (1988). Influence of peer-model attributes on children's beliefs and learning. *Journal of Educational Psychology, 77*, 313–322. doi:10.1037/0022-0663.77.3.313

Schunk, D. H., & Zimmerman, B. J. (Eds.). (1998). *Self-regulated learning: From teaching to self-reflective practice*. New York, NY: Guilford Press.

Schwab, J. (1962). The teaching of sicnece as inquiry. In Schwab, J., & Brandwein, P. (Eds.), *The Teaching of Science* (pp. 1–104). Cambridge, MA: Harvard University Press.

Schwandt, T. A. (1996). Farewell to criteriology. *Qualitative Inquiry, 2*, 58–72. doi:10.1177/107780049600200109

Schwandt, T. A. (2000). Three epistemological stances for qualitative inquiry. In Denzin, N. K., & Lincoln, Y. S. (Eds.), *Handbook of Qualitative Inquiry* (2nd ed., pp. 189–213). Thousand Oaks, CA: Sage.

Schwandt, T. A. (2005). On modeling our understanding of the practice fields. *Pedagogy, Culture & Society, 13*, 313–332. doi:10.1080/14681360500200231

Schwartz, R. S., Lederman, N. G., & Thompson, T. (2001). *Grade nine students' views of nature of science and scientific inquiry: The effects of an inquiry-enthusiast's approach to teaching science as inquiry*. Paper Presented at the Annual Meeting of the National Association of Research in Science Teaching (NARST). St. Louis, MO.

Schwartz, B., & Sharpe, K. (2010). *Practical wisdom*. New York, NY: Riverhead Books.

Schwartz, R. S., Lederman, N. G., & Crawford, B. A. (2004). Developing views of nature of science in an authentic context: An explicit approach to bridging the gap between nature of science and scientific inquiry. *Science Education, 88*, 610–645. doi:10.1002/sce.10128

Schwarz, C. V., & Gwekwerere, Y. N. (2007). Using a guided inquiry and modeling instructional framework (EIMA) to support preservice K-8 science teaching. *Science Education, 91*, 158–186. doi:10.1002/sce.20177

Schweizer, D. M., & Kelly, G. J. (2005). An investigation of student engagement in a global warming debate. *Journal of Geoscience Education, 53*(1), 75–84.

Scott, D., & Willits, F. K. (1989). Adolescent and adult leisure patterns: A 37-year follow up study. *Leisure Sciences, 11*, 323–335. doi:10.1080/01490408909512230

Scott, P. (1998). Teacher talk and meaning making in science classrooms: A Vygotskian analysis and review. *Studies in Science Education, 32*, 45–80. doi:10.1080/03057269808560127

Scott, P., & Ametller, J. (2007). Teaching science in a meaningful way: Striking a balance between 'opening up' and 'closing down' classroom talk. *The School Science Review, 88*(324), 77–84.

Scott, P. H., Mortimer, E. F., & Aguiar, O. G. (2006). The tension between authoritative and dialogic discourse: A fundamental characteristic of meaning making interactions in high school science lessons. *Science Education, 90*, 605–631. doi:10.1002/sce.20131

Sengupta, P. (2011). Design principles for a visual programming language to integrate agent-based modeling in K-12 science. In *Proceedings of the Eighth International Conference of Complex Systems (ICCS 2011)*, (pp. 1636 – 1637). ICCS.

Sengupta, P., & Wilensky, U. (2010). Multi-agent-based modeling and learning electricity: Design and epistemological issues. In Khine, M. S., & Saleh, I. M. (Eds.), *Dynamic Modeling: Cognitive Tool for Scientific Enquiry*. New York, NY: Springer.

Sengupta, P., & Wright, M. (2010). *ViMAP*. Nashville, TN: Vanderbilt University.

Shaffer, D. W. (2006). *How computer games help children learn* (1st ed.). New York, NY: Palgrave Macmillan. doi:10.1057/9780230601994

Shaffer, D. W., Squire, K. R., Halverson, R., & Gee, J. P. (2005). Video games and the future of learning. *Phi Delta Kappan, 87*(2), 104.

Shaftel, J., Belton-Kocher, E., Glasnapp, D., & Poggio, J. (2006). The impact of language characteristics in mathematics test items on the performance of English language learners and students with disabilities. *Educational Assessment, 11*(2), 105. doi:10.1207/s15326977ea1102_2

Shankar, G., & Skoog, G. (1993). Emphasis given evolution and creationism by Texas high school biology teachers. *Science Education, 77*(2), 221–233. doi:10.1002/sce.3730770209

Shapiro, B. L. (1996). A case study of change in elementary student teacher thinking during an independent investigation in science: Learning about the face of science that does not yet know. *Science Education, 80*, 535–560. doi:10.1002/(SICI)1098-237X(199609)80:5<535::AID-SCE3>3.0.CO;2-C

Shavelson, R. J., Baxter, G. P., & Pine, J. (1991). Performance assessment in science. *Applied Measurement in Education, 4*(4), 347–362. doi:10.1207/s15324818ame0404_7

Shavelson, R. J., Young, D. B., Ayala, C. C., Brandon, P. R., Furtak, E. M., & Ruiz-Primo, M. A. (2008). On the impact of curriculum-embedded formative assessment on learning: A collaboration between curriculum and assessment developers. *Applied Measurement in Education, 21*(4), 295–314. doi:10.1080/08957340802347647

Shaw, J. (2005). Getting things right at the classroom level. In Herman, J. L. H. (Ed.), *Uses and Misuses of Data for Educational Accountability and Improvement: The One Hundred Fourth Yearbook of the National Academy Society for the Study of Education, Part 2* (pp. 340–357). Chicago, IL: National Society for the Study of Education. doi:10.1111/j.1744-7984.2005.00036.x

Shaw, J. M., Bunch, G. C., & Geaney, E. R. (2010). Analyzing language demands facing English learners on science performance assessments: The Sald framework. *Journal of Research in Science Teaching, 47*(8), 909–928.

Sheingold, K., & Hadley, M. (1990). *Accomplished teachers: Integrating computers into classroom practice*. New York, NY: Center for Technology in Education, Bank Street College of Education. Retrieved from http://www.eric.ed.gov/ERICWebPortal/contentdelivery/servlet/ERICServlet?accno=ED322900

Shen, J., & Confrey, J. (2007). From conceptual change to transformative modeling: A case study of an elementary teacher in learning astronomy. *Science Education, 91*(6), 948–966. doi:10.1002/sce.20224

Shepard, L. A. (2000). The role of assessment in a learning culture. *Educational Researcher, 29*(7), 4–14.

Sheppard, K., & Robbins, D. M. (2003). Physics was once first and was once for all. *The Physics Teacher, 41*, 420–424. doi:10.1119/1.1616483

Sherin, B., diSessa, A. A., & Hammer, D. M. (1993). Dynaturtle revisited: Learning physics through collaborative design of a computer model. *Interactive Learning Environments, 3*(2), 91–118. doi:10.1080/1049482930030201

Shin, N., Jonassen, D. H., & McGee, S. (2003). Predictors of well-structured and ill-structured problem solving in an astronomy simulation. *Journal of Research in Science Teaching, 40*(1), 6–33. doi:10.1002/tea.10058

Shulman, L. (2007). Practical wisdom in the service of professional practice. *Educational Researcher, 36*, 560–563. doi:10.3102/0013189X07313150

Shulman, L. S. (1986). Paradigms and research programs in the study of teaching: A contemporary perspective. In Wittrock, M. C. (Ed.), *Handbook on Research on Teaching* (3rd ed., pp. 3–36). New York, NY: Macmillan.

Shulman, L. S. (1987). Knowledge and teaching: Foundations of the new reform. *Harvard Educational Review*, *57*(1), 1–23.

Shulman, L., & Keisler, E. (Eds.). (1966). *Learning by discovery: A critical appraisal*. Chicago, IL: Rand McNally.

Siegel, H. (1989). The rationality of science, critical thinking, and science education. *Synthese*, *80*(1), 9–42. doi:10.1007/BF00869946

Siegel, M. A. (2007). Striving for equitable classroom assessments for linguistic minorities: Strategies for and effects of revising life science items. *Journal of Research in Science Teaching*, *44*(6). doi:10.1002/tea.20176

Siegler, R. S. (1989). Mechanisms of cognitive development. *Annual Reviews*, *40*, 353–379. doi:10.1146/annurev.ps.40.020189.002033

Siegler, R. S. (1994). Cognitive variability: A key to understanding cognitive development. *Current Directions in Psychological Science*, *3*, 1–5. doi:10.1111/1467-8721.ep10769817

Siegler, R. S., & Booth, J. L. (2004). Development of numerical estimation in young children. *Child Development*, *75*, 428–444. doi:10.1111/j.1467-8624.2004.00684.x

Siegler, R. S., & Chen, Z. (2002). Development of rules and strategies: Balancing the old and the new. *Journal of Experimental Child Psychology*, *81*, 446–457. doi:10.1006/jecp.2002.2666

Simmons, P. E., Emory, A., Carter, T., Coker, T., Finnegan, B., & Crockett, D. (1999). Beginning teachers: Beliefs and classroom actions. *Journal of Research in Science Teaching*, *36*, 930–954. doi:10.1002/(SICI)1098-2736(199910)36:8<930::AID-TEA3>3.0.CO;2-N

Simon, S., Erduran, S., & Osborne, J. (2006). Learning to teach argumentation: Research and development in the science classroom. *International Journal of Science Education*, *28*(2-3), 235–260. doi:10.1080/09500690500336957

Simpson, E., & Stansberry, S. (2009). Video games and teacher development: Bridging the gap in the classroom. In Miller, C. T. (Ed.), *Games: Purpose and Potential in Education* (pp. 163–185). New York, NY: Springer. doi:10.1007/978-0-387-09775-6_7

Simpson, J. C. (2001). Segregated by subject – Racial differences in the factors influencing academic major between European Americans, Asian Americans, and African, Hispanic, and Native Americans. *The Journal of Higher Education*, *72*(1), 63–100. doi:10.2307/2649134

Sinatra, G. M., Southerland, S. A., McConaughy, F., & Demastes, J. W. (2003). Intentions and beliefs in students' understanding and acceptance of biological evolution. *Journal of Research in Science Teaching*, *40*(5), 510–528. doi:10.1002/tea.10087

Singer, S. R., Hilton, M. L., & Schweingruber, H. A. (Eds.). (2005). *America's lab report: Investigations in high school science*. Washington, DC: National Academies Press.

Singh, K., Granville, M., & Dika, S. (2002). Mathematics and science achievement: Effects of motivation, interest, and academic engagement. *The Journal of Educational Research*, *95*, 323–332. doi:10.1080/00220670209596607

Skoric, M. M., Teo, L. L. C., & Neo, R. L. (2009). Children and video games: Addiction, engagement, and scholastic achievement. *Cyberpsychology & Behavior*, *12*(5), 567–572. doi:10.1089/cpb.2009.0079

Smit, J. J. A., & Finegold, M. (1995). Models in physics: Perceptions held by final-year prospective physical science teachers studying at South African universities. *International Journal of Science Education*, *17*(5), 621–634. doi:10.1080/0950069950170506

Smith, C. L., Maclin, D., Houghton, C., & Hennessey, M. G. (2000). Sixth-grade students' epistemologies of science: The impact of school science experiences on epistemological development. *Cognition and Instruction*, *18*, 349–422. doi:10.1207/S1532690XCI1803_3

Smith, J. P., diSessa, A. A., & Roschelle, J. (1993). Misconceptions reconceived: A constructivist analysis of knowledge in transition. *Journal of the Learning Sciences*, *3*(2), 115–163. doi:10.1207/s15327809jls0302_1

Smith, L., & Southerland, S. A. (2007). Reforming practice or modifying reforms? Elementary teachers' response to the tools of reform. *Journal of Research in Science Teaching, 43,* 396–423. doi:10.1002/tea.20165

Smith, M. U., & Scharmann, L. C. (1999). Defining versus describing the nature of science: A pragmatic analysis for classroom teachers and science educators. *Science Education, 83,* 493–509. doi:10.1002/(SICI)1098-237X(199907)83:4<493::AID-SCE6>3.0.CO;2-U

Snyder, T. D., & Dillow, S. A. (2011). *Digest of education statistics: 2010.* Washington, DC: U.S. Department of Education, Institute of Education Sciences, National Center for Education Statistics. Retrieved from http://nces.ed.gov/pubs2011/2011015.pdf

Sofranko, A. J., & Nolan, M. F. (1972). Early life experiences and adult sports participation. *Journal of Leisure Research, 4,* 6–18.

Solano-Flores, G., & Nelson-Barber, S. (2001). On the cultural validity of science assessments. *Journal of Research in Science Teaching, 38*(5), 553–573. doi:10.1002/tea.1018

Solomon, K. (1987). Social influences on the construction of pupils' understanding of science. *Studies in Science Education, 14,* 63–82. doi:10.1080/03057268708559939

Southerland, S. A., Gess-Newsome, J., & Johnson, A. (2003). Portraying science in the classroom: The manifestation of scientists' beliefs in classroom practice. *Journal of Research in Science Teaching, 40,* 669–691. doi:10.1002/tea.10104

Southerland, S., & Gess-Newsome, J. (1999). Preservice teachers' views of science teaching as shaped by images of teaching. *Science Education, 83*(2), 131–151. doi:10.1002/(SICI)1098-237X(199903)83:2<131::AID-SCE3>3.0.CO;2-X

Southerland, S., Sinatra, G. M., & Mathews, M. (2001). Beliefs, knowledge, and science education. *Educational Psychology Review, 13,* 325–351. doi:10.1023/A:1011913813847

Soy, E. (1967). Attitudes of prospective elementary teachers toward science as a field of specialty. *School Science and Mathematics, 67,* 507–520. doi:10.1111/j.1949-8594.1967.tb15233.x

Sprague, D. (2004). Technology and teacher education: Are we talking to ourselves? *Contemporary Issues in Technology & Teacher Education, 3*(4). Retrieved from http://www.citejournal.org/vol3/iss4/editorial/article1.cfm

Squire, K. (2003). Videogames in education. *International Journal of Intelligent Simulations and Gaming, 2*(1), 49–62.

Squire, K. (2005). Changing the game: What happens when video games enter the classroom. *Innovate: Journal of Online Education, 1*(6), 25–49.

Staver, J. R. (1999). When public understanding of science thwarts standards-based science education. *Electronic Journal of Science Education, 3*(4). Retrieved from http://ejse.southwestern.edu/article/viewArticle/7613/5380

Steffen, A. (Ed.). (2007). *Principle 1: The backstory.* Retrieved from http://www.worthchanging.com/archives/004534.html

Steinkuehler, C., & Chmiel, M. (2006). Fostering scientific habits of mind in the context of online play. In *Proceedings of the 7th International Conference on Learning Sciences,* (pp. 723-729). New York, NY: International Society of the Learning Sciences.

Stenhouse, D. (1985). *Active philosophy in education and science.* London, UK: George Allen & Unwin.

Stiggins, R. (2005). From formative assessment to assessment FOR learning: A path to success in standards-based schools. *Phi Delta Kappan, 87*(4), 324–328.

Stigler, J. W., & Hiebert, J. (1999). *The teaching gap: Best ideas from the world's teachers for improving education in the classroom.* New York, NY: The Free Press.

Stoddart, T., Pinal, A., Latzke, M., & Canaday, D. (2002). Integrating inquiry science and language development for English language learners. *Journal of Research in Science Teaching, 39*(8), 664–687. doi:10.1002/tea.10040

Strickland, T. (2009). *Reforming Ohio's education system for the 21st century: Governor Ted Strickland's education reform and funding plan.* Retrieved from http://www.conversationoneducation.org/information/

Strike, K., & Posner, G. J. (1992). A revisionist theory of conceptual change. In Duschl, R. A., & Hamilton, R. J. (Eds.), *Philosophy of Science, Cognitive Psychology, and Educational Theory and Practice* (pp. 147–176). New York, NY: State University of New York.

Supreme Education Council. (2002). *Science standards*. Retrieved February 16, 2012, from http://www.english.education.gov.qa/section/sec/education_institute/cso/science_standards

Sweller, J. (2004). Instructional design consequences of an analogy between evolution by natural selection and human cognitive architecture. *Instructional Science*, *32*, 9–31. doi:10.1023/B:TRUC.0000021808.72598.4d

Tamir, P. (1972). Understanding the process of science by students exposed to different science curricula in Israel. *Journal of Research in Science Teaching*, *9*, 239–245. doi:10.1002/tea.3660090309

Tanner, T. (1980). Significant life experiences: A new research area in environmental education. *The Journal of Environmental Education*, *11*(4), 20–24. doi:10.1080/00958964.1980.9941386

Tao, P. (2003). Eliciting and developing junior secondary students' understanding of the nature of science through a peer collaboration instruction in science stories. *International Journal of Science Education*, *25*, 147–171. doi:10.1080/09500690210126748

Taylor, B. M., Pearson, P. D., Peterson, D. S., & Rodriguez, M. C. (2003). Reading growth in high-poverty classrooms: The influence of teacher practices that encourage cognitive engagement in literacy learning. *The Elementary School Journal*, *104*, 3–28. doi:10.1086/499740

Taylor, I., Barker, M., & Jones, A. (2003). Promoting mental model building in astronomy education. *International Journal of Science Education*, *25*(10), 1205–1225. doi:10.1080/0950069022000017270a

Tenner, E. (1996). *Why things bite back*. New York, NY: Vintage Books.

Thomas, J. W., & Rohwer, W. D. Jr. (1986). Academic studying: The role of learning strategies. *Educational Psychologist*, *21*, 19–41.

Tobin, K., & McRobbie, C. J. (1997). Beliefs about the nature of science and the enacted science curriculum. *Science & Education*, *6*, 355–371. doi:10.1023/A:1008600132359

Tobin, K., & Roth, W.-M. (2005). Implementing coteaching and cogenerative dialoguing in urban science education. *School Science and Mathematics*, *105*, 313–322. doi:10.1111/j.1949-8594.2005.tb18132.x

Toulmin, S. (1958). *The uses of argument*. Cambridge, UK: Cambridge University Press.

Treadwell, M. (2011). Whatever happened? In Wan, G., & Gut, D. (Eds.), *Bringing Schools into the 21st Century* (pp. 7–40). New York, NY: Springer. doi:10.1007/978-94-007-0268-4_2

Trent, J. (1965). The attainment of the concept "understanding science" using contrasting physics courses. *Journal of Research in Science Teaching*, *3*, 224–229. doi:10.1002/tea.3660030309

Trigwell, K., Prosser, M., & Waterhouse, F. (1999). Relation between teachers' approaches to teaching and students' approaches to learning. *Higher Education*, *37*, 57–70. doi:10.1023/A:1003548313194

Tsai, C.-C. (2002). Nested epistemologies: Science teachers' beliefs of teaching, learning, and science. *International Journal of Science Education*, *24*, 771–783. doi:10.1080/09500690110049132

Turner, T. (2000). The science curriculum: What is it for? In Sears, J., & Sorenson, P. (Eds.), *Issues in Science Teaching* (pp. 4–15). London, UK: Routledge Falmer. doi:10.1145/3166.3168

Tyson, W., Lee, R., Borman, K. M., & Hanson, M. A. (2007). Science, technology, engineering, and mathematics (STEM) pathways: High school science and math coursework for postsecondary degree attainment. *Journal of Education for Students Placed at Risk*, *12*, 243–270. doi:10.1080/10824660701601266

U.S. Census Bureau. (2012). *Statistical abstract of the United States, 2012*. Washington, DC: Government Printing Office. Retrieved May 18, 2012, from http://www.census.gov/compendia/statab/cats/education.html

U.S. Department of Education. (2000). *Teacher's tools for the 21st century: A report on teacher' use of technology*. Washington, DC: National Center for Education Statistics.

U.S. Government Accountability Office. (2005). *Higher education: Federal science, technology, engineering, and mathematics programs and related trends*. Retrieved October 11, 2011, from http://www.gao.gov/new.items/d06114.pdf

United States Census Bureau Website. (2011). *Factfinder*. Retrieved 4 December, 2011 from http://factfinder.census.gov/servlet/SAFFFacts?_event=Search&geo_id=&_geoContext=&_street=&_county=danville&_cityTown=danville&_state=04000US17&_zip=&_lang=en&_sse=on&pctxt=fph&pgsl=010&show_2003_tab=&redirect=Y

United States Department of Labor. (2010). *Occupational employment statistics*. Retrieved from http://www.bls.gov/oes/current/largest_occs.htm

Urdan, T. C., Midgley, C., & Anderman, E. M. (1998). The role of classroom goal structure in students' use of self-handicapping strategies. *American Educational Research Journal, 35*, 101–122.

Urhahne, D., Schanze, S., Bell, T., Mansfield, A., & Homes, J. (2010). Role of the teacher in computer-supported collaborative inquiry learning. *International Journal of Science Education, 32*(2), 221–243. doi:10.1080/09500690802516967

van Driel, J. H., Beijaard, D., & Verloop, N. (2001). Professional development and reform in science education: The role of teachers' practical knowledge. *Journal of Research in Science Teaching, 38*, 137–158. doi:10.1002/1098-2736(200102)38:2<137::AID-TEA1001>3.0.CO;2-U

van Driel, J. H., & Verloop, N. (1999). Teachers' knowledge of models and modelling in science. *International Journal of Science Education, 21*(11), 1141–1153. doi:10.1080/095006999290110

van Driel, J. H., & Verloop, N. (2002). Experienced teachers' knowledge of teaching and learning of models and modelling in science education. *International Journal of Science Education, 24*(2), 1255–1272. doi:10.1080/09500690210126711

van Joolingen, W. (2004). *Roles of modeling in inquiry learning*. Paper presented at the IEEE International Conference on Advanced Learning Technologies. Joensuu, Finland.

van Manen, M. (1999). The practice of practice. In M. Lange, J. Olson, & W. BŸnder, (Eds.), *Changing Schools/Changing Practices: Perspectives on Educational Reform and Teacher Professionalism*. Luvain, Belgium: Garant.

van Manen, M. (1995). On the epistemology of reflective practice. *Teachers and Teaching: Theory and Practice, 1*, 33–50. doi:10.1080/1354060950010104

van Zee, E. H., Iwasyk, M., Kurose, A., Simpson, D., & Wild, J. (2001). Student and teacher questioning during conversations about science. *Journal of Research in Science Teaching, 38*(2), 159–190. doi:10.1002/1098-2736(200102)38:2<159::AID-TEA1002>3.0.CO;2-J

Vanfossen, B. E., Jones, J. D., & Spade, J. Z. (1987). Curriculum tracking and status maintenance. *Sociology of Education, 60*(2), 104–122. doi:10.2307/2112586

Varelas, M., & Pineda, E. (1999). Intermingling and bumpiness: Exploring meaning making in the discourse of a science classroom. *Research in Science Education, 29*(1), 25–49. doi:10.1007/BF02461179

Vellom, R. P., & Anderson, C. W. (1999). Reasoning about data in middle school science. *Journal of Research in Science Teaching, 36*(2), 179–199. doi:10.1002/(SICI)1098-2736(199902)36:2<179::AID-TEA5>3.0.CO;2-T

Verhey, S. D. (2005). The effect of engaging prior learning on student attitudes toward creationism and evolution. *Bioscience, 55*(11), 991–1003. doi:10.1641/0006-3568(2005)055[0996:TEOEPL]2.0.CO;2

Vianna, E., & Stetsenko, A. (2006). Embracing history through transforming it. *Theory & Psychology, 16*, 16–81.

Victor, E. (1962). Why are our elementary school teachers reluctant to teach science? *Science Education, 46*(2), 185–192. doi:10.1002/sce.3730460231

von Aufschnaiter, C., Erduran, S., Osborne, J., & Simon, S. (2008). Arguing to learn and learning to argue: Case studies of how students' argumentation related to their scientific knowledge. *International Journal of Science Education, 45*(1), 101–131.

Vygotsky, L. (1978). *Mind in society: The development of higher psychological processes*. Cambridge, MA: Harvard University Press.

Vygotsky, L. S. (1986). *Thought and language*. Cambridge, MA: MIT Press.

Walker, B., & Salt, D. (2006). *Resilience thinking: Sustaining ecosystems and people in a changing world*. Washington, DC: Island Press.

Walker, C. O., Greene, B. A., & Mansell, R. A. (2006). Identification with academics, intrinsic/extrinsic motivation, and self-efficacy as predictors of cognitive engagement. *Learning and Individual Differences*, *16*, 1–12. doi:10.1016/j.lindif.2005.06.004

Wandersee, J. H., & Clary, R. M. (2006). Fieldwork: New directions and examples in informal science education research. In Mintzes, J., & Leonard, W. (Eds.), *NSTA Handbook of College Science Teaching: Theory, Research, & Practice* (pp. 167–176). Arlington, VA: NSTA Press.

Wandersee, J. H., & Clary, R. M. (2006a). On seeing flowers: Are you missing anything? *The Human Flower Project*. Retrieved from http://www.humanflowerproject.com/index.php/weblog/on_seeing_flowers_are_you_missing_anything

Wandersee, J. H., & Clary, R. M. (2012). Envisioning a rainbow bridge: Eight research studies that reveal optimal opportunities to learn biology and geology at informal science education sites. In P. Kurtz & F. Ren (Eds.), *Proceedings of the 11th World Congress for Center for Inquiry-Transnational: Scientific Inquiry and Human Development*. Amherst, NY: Prometheus Books.

Wandersee, J. H., Clary, R. M., & Guzman, S. M. (2006). How-to-do-it: A writing template for probing students' botanical sense of place. *The American Biology Teacher*, *68*(7), 419–422. doi:10.1662/0002-7685(2006)68[419:AWTFPS]2.0.CO;2

Wandersee, J. H., Mintzes, J. J., & Novak, J. D. (1994). Research on alternative conceptions in science. In Gabel, D. (Ed.), *Handbook of Research on Science Teaching and Learning* (pp. 177–210). New York, NY: Macmillan.

Wandersee, J. H., & Schussler, E. (1999). Preventing plant blindness. *The American Biology Teacher*, *61*, 84–86. doi:10.2307/4450624

Wan, G., & Gut, D. (Eds.). (2011). *Bringing schools into the 21st century*. New York, NY: Springer. doi:10.1007/978-94-007-0268-4

Wang, M. C., & Peverly, S. T. (1986). The self-instructive process in classroom learning contexts. *Contemporary Educational Psychology*, *11*, 370–404. doi:10.1016/0361-476X(86)90031-7

Watson, J. D., & Crick, F. H. C. (1953). A structure for deoxyribose nucleic acid. *Nature*, *171*, 737–738. doi:10.1038/171737a0

Welch, W. W., & Pella, M. O. (1968). The development of an instrument for inventorying knowledge of the processes of science. *Journal of Research in Science Teaching*, *5*, 64. doi:10.1002/tea.3660050115

Wells, G., & Mejía Arauz, R. (2006). Dialogue in the classroom. *Journal of the Learning Sciences*, *15*, 379–428. doi:10.1207/s15327809jls1503_3

Werth, A. (2012). Avoiding the pitfall of progress and associated perils of evolutionary education. *Evolution: Education & Outreach*, *5*(2), 249–265. doi:10.1007/s12052-012-0417-y

Wertsch, J. V. (1979). From social interaction to higher psychological processes: A clarification and application of Vygotsky's theory. *Human Development*, *22*(1), 1–22. doi:10.1159/000272425

Wertsch, J. V. (1984). The zone of proximal development from a comparative and organismic point of view. In Rogoff, B., & Wertsch, J. V. (Eds.), *New Directions for Child Development*. San Francisco, CA: Jossey-Bass.

White, B. Y., & Frederiksen, J. R. (1998). Inquiry, modeling, and metacognition: Making science accessible to all students. *Cognition and Instruction*, *16*(1), 3–118. doi:10.1207/s1532690xci1601_2

White, B. Y., & Frederiksen, J. R. (2005). A theoretical framework and approach for fostering metacognitive development. *Educational Psychologist*, *40*(4), 211–223. doi:10.1207/s15326985ep4004_3

White, P. A. (1997). Naive ecology: Causal judgments about a simple ecosystem. *The British Journal of Psychology*, *88*, 219–233. doi:10.1111/j.2044-8295.1997.tb02631.x

Whitten, B. L., Foster, S. R., Duncombe, M. L., Allen, P. E., Heron, P., & McCullough, L. (2004). Like a family: What works to create friendly and respectful student-faculty interactions. *Journal of Women and Minorities in Science and Engineering*, *10*(3), 229–242. doi:10.1615/JWomenMinorScienEng.v10.i3.30

Wieman, C. E., Adams, W. K., & Perkins, K. K. (2008). PhET: Simulations that enhance learning. *Science*, *322*, 682–683. doi:10.1126/science.1161948

Wigfield, A., Tonks, S., & Eccles, J. S. (2004). Expectancy-value theory in cross-cultural perspective. In McInerney, D. M., & Van Etten, S. (Eds.), *Big Theories Revisited* (Vol. 4, pp. 165–198). Greenwich, CT: Information Age.

Wiggins, G. (1989). A true test: Toward more authentic and equitable assessment. *Phi Delta Kappan*, *70*(9), 703–713.

Wiggins, G., & McTighe, J. (1998). *Understanding by design*. Alexandria, VA: ASCD.

Wildy, H., & Wallace, J. (1995). Understanding teaching or teaching for understanding: Alternative frameworks for science classrooms. *Journal of Research in Science Teaching*, *32*, 143–156. doi:10.1002/tea.3660320205

Wilensky, U., & Resnick, M. (1999). Thinking in levels: A dynamic systems perspective to making sense of the world. *Journal of Science Education and Technology*, *8*(1). doi:10.1023/A:1009421303064

Wilensky, U., & Reisman, K. (2006). Thinking like a wolf, a sheep, or a firefly: Learning biology through constructing and testing computational theories—An embodied modeling approach. *Cognition and Instruction*, *24*(2), 171–209. doi:10.1207/s1532690xci2402_1

Wiliam, D., & Leahy, S. (2007). A theoretical foundation for formative assessment. In McMillan, J. H. (Ed.), *Formative Classroom Assessment* (pp. 29–42). New York, NY: Teachers College Press.

Wilson, L. (2009). *Best practices for using games & simulations in the classroom guidelines for k–12 educators*. New York, NY: SIIA Education Division.

Wilson, R. (1996). *Starting early: Environmental education during the early childhood years*. Columbus, OH: ERIC Clearinghouse for Science, Mathematics, and Environmental Education.

Windschitl, M. (1999). A vision educators can put into practice: Portraying the constructivist classroom as a cultural system. *School Science and Mathematics*, *99*, 189–196. doi:10.1111/j.1949-8594.1999.tb17473.x

Windschitl, M. (2003). Inquiry projects in science teaching teacher education, what can investigative experiences reveal about teacher thinking and eventual classroom practice? *Science Education*, *87*(1), 112–144. doi:10.1002/sce.10044

Windschilt, M. (2004). *What types of knowledge do teachers use to engage learners in "doing science"? Rethinking the continuum of preparation and professional development for secondary science educators*. Paper presented at the Meeting High School Science Laboratories: Role and Vision. Washington, DC: National Academy of Sciences.

Windschitl, M. (2012). *Ambitious teaching as the "new normal" in American science classrooms: How will we prepare the next generation of professional educators?* Paper presented at Penn State University. State College, PA.

Windschitl, M., & Thompson, J. (2006). Transcending simple forms of school science investigation: The impact of preservice instruction on teachers' understanding of model-based inquiry. *American Educational Research Journal*, *43*(4), 783–835. doi:10.3102/00028312043004783

Windschitl, M., Thompson, J., & Braaten, M. (2008a). How novice science teachers appropriate epistemic discourses around model-based inquiry for use in classrooms. *Cognition and Instruction*, *26*, 310–378. doi:10.1080/07370000802177193

Windschitl, M., Thompson, J., & Braaten, M. (2008b). Beyond the scientific method: Model-based inquiry as a new paradigm of preference for school science investigations. *Science Education*, *92*, 941–967. doi:10.1002/sce.20259

Woltemade, C., & Staniski-Martin, D. (2002). A student-centered field project comparing NEXRAD and rain gauge precipitation values in mountainous terrain. *Journal of Geoscience Education*, *50*, 296–302.

Woolfolk, A. E., & Galloway, C. M. (1985). Nonverbal communication and the study of teaching. *Theory into Practice*, *24*, 77–85. doi:10.1080/00405848509543150

Woolnough, B. (1994). Why students choose physics, or reject it. *Physics Education*, *29*(6), 368–374. doi:10.1088/0031-9120/29/6/006

Yerrick, R. (2000). Lower track science students' argumentation and open inquiry instruction. *Journal of Research in Science Teaching*, *37*(8), 807–838. doi:10.1002/1098-2736(200010)37:8<807::AID-TEA4>3.0.CO;2-7

Yore, L. D., Bisanz, G. L., & Hand, B. M. (2003). Examining the literacy component of science literacy: 25 years of language arts and science research. *International Journal of Science Education*, *25*(6), 689–725. doi:10.1080/09500690305018

Zacharia, Z. C. (2007). Comparing and combining real and virtual experimentation: An effort to enhance students' conceptual understanding of electric circuits. *Journal of Computer Assisted Learning*, *23*(2), 120–132. doi:10.1111/j.1365-2729.2006.00215.x

Zagzebski, L. T. (1996). *Virtues of the mind: An inquiry into the nature of virtue and the ethical foundations of knowledge*. Cambridge, UK: Cambridge University Press. doi:10.1017/CBO9781139174763

Zeidler, D. (1997). The central role of fallacious thinking in science education. *Science Education*, *81*, 483–496. doi:10.1002/(SICI)1098-237X(199707)81:4<483::AID-SCE7>3.0.CO;2-8

Zimmerman, B. J. (1986). Development of self-regulated learning: Which are the key subprocesses? *Contemporary Educational Psychology*, *11*, 307–313. doi:10.1016/0361-476X(86)90027-5

Zimmerman, B. J. (1990). Self-regulating academic learning and achievement: The emergence of a social cognitive perspective. *Educational Psychology Review*, *2*, 173–201. doi:10.1007/BF01322178

Zimmerman, B. J. (2000). Attaining self-regulation: A social-cognitive perspective. In Boekaerts, M., Pintrich, P., & Zeidner, M. (Eds.), *Handbook of Self-Regulation* (pp. 13–39). San Diego, CA: Academic Press. doi:10.1016/B978-012109890-2/50031-7

Zimmerman, B. J. (2008). Goal setting: A key proactive source of academic self-regulation. In Schunk, D. H., & Zimmerman, B. J. (Eds.), *Motivation and Self-Regulated Learning: Theory, Research, and Applications* (pp. 267–295). Hillsdale, NJ: Lawrence Erlbaum Associates.

Zimmerman, B. J., & Bandura, A. (1994). Impact of self-regulatory influences on writing course attainment. *American Educational Research Journal*, *31*, 845–862.

Zimmerman, B. J., Bandura, A., & Martinez-Pons, M. (1992). Self-motivation for academic attainment: The role of self-efficacy beliefs and personal goal setting. *American Educational Research Journal*, *29*, 663–676.

Zimmerman, B. J., & Kitsantas, A. (1997). Developmental phases in self-regulation: Shifting from process to outcome goals. *Journal of Educational Psychology*, *89*, 1–10. doi:10.1037/0022-0663.89.1.29

Zimmerman, B. J., & Kitsantas, A. (2002). Acquiring writing revision and self-regulatory skill through observation and emulation. *Journal of Educational Psychology*, *94*(4), 660–668. doi:10.1037/0022-0663.94.4.660

Zimmerman, B. J., & Kitsantas, A. (2007). A writer's discipline: The development of self-regulatory skill. In Hidi, S., & Boskolo, P. (Eds.), *Motivation to Write*. New York, NY: Kluwer Publishers.

Zimmerman, B. J., & Martinez-Pons, M. (1992). Perceptions of efficacy and strategy use in the self-regulation of learning. In Schunk, D. H., & Meece, J. (Eds.), *Student Perceptions in the Classroom: Causes and Consequences* (pp. 185–207). Hillsdale, NJ: Erlbaum.

Zint, M. (2002). Comparing three attitude-behavior theories for predicting science teachers' intentions. *Journal of Research in Science Teaching*, *39*(9), 819–844. doi:10.1002/tea.10047

Zohar, A., & Nemet, F. (2002). Fostering students' knowledge and argumentation skills through dilemmas in human genetics. *Journal of Research in Science Teaching*, *39*(1), 35–62. doi:10.1002/tea.10008

About the Contributors

Myint Swe Khine is a Professor in the field of Learning Sciences and Technology and Head of Graduate Programs and Research at the University of Bahrain. He received his Master degrees from the University of Southern California, Los Angeles, USA, and the University of Surrey, Guildford, UK, and his Doctor of Education from Curtin University of Technology, Australia. He worked in the Learning Sciences Technology Academic Group at Nanyang Technology University, Singapore, for several years. He publishes widely in academic journals and has edited some books. Recent publications include *Learning to Play: Exploring the Future of Education with Video Games* (Peter Lang, USA), *Advances in Nature of Science Research: Concepts and Methodologies* (Springer, 2012), and *Perspectives on Scientific Argumentation: Theory, Practice, and Research* (Springer, 2012).

Issa M. Saleh is the Head of Education Studies Division and Associate Professor at the University of Bahrain. He received Bachelor of Science, Master, and Doctorate in Education from the University of North Florida, Jacksonville, USA. Dr. Issa's credentials include the distinctive professional certification required to teach Science and administer schools (K-12) in the State of Florida, USA. He has taught at A. Phillip Randolph Academy of Technology (APR) for several years and also taught at Emirates College for Advanced Education, University of North Florida, University of Atlanta, and Florida Community College in Jacksonville Florida, USA. Dr. Issa was the science department head and principal investigator in understanding of science teaching and learning at APR for in-service science teachers for several years and later the Dean of Students Services and Curriculum Instruction at APR. His most recent books are *Fostering Scientific Habits of Mind: Pedagogical Knowledge and Best Practices in Science Education* (Sense Publishers, 2009), *Transformative Leadership and Educational Excellence: Learning Organizations in the Information Age* (Sense Publishers, 2009), *New Science of Learning: Computers, Cognition and Collaboration in Education* (Springer, 2010), and *Teaching Teachers: Approaches in Improving Quality of Education* (Nova Science Publishers, 2011).

* * *

Matthew J. Benus, Ph.D., is an Assistant Professor of Education at Indiana University Northwest where he teaches science methods. His current research interests focus on dialogic interactions in a science classroom with students and teacher. Additionally, he researches how science teachers establish and refine whole-class dialogue around knowledge construction, consensus-making, and critique, while using argument-based inquiry.

About the Contributors

Saouma BouJaoude graduated from the University of Cincinnati, Cincinnati, Ohio, USA, in 1988 with a Doctorate in Curriculum and Instruction with emphasis on science education. After serving as an Assistant Professor of Science Education at the Department of Science Teaching, Syracuse University, Syracuse, New York, USA, he joined the faculty of the American University of Beirut in 1993. Dr. BouJaoude served as Director of the Science and Math Education Center, Chair of the Department of Education, and is presently director of SMEC and the Center of Teaching and Learning. Dr. BouJaoude has published in international journals such as the *Journal of Research in Science Teaching*, *Science Education*, and the *International Journal of Science Education*, has written chapters in edited books in English and Arabic, and has been an active presenter at local, regional, and international education and science education conferences. Dr. BouJaoude serves on the editorial boards of a number of science education journals.

Todd Campbell is an Associate Professor at Utah State University. His research focuses on factors influencing current reform in science education. This is supported by investigating science teacher professional development, scientific inquiry/modeling instructional practices, and science/technology integration. Dr. Campbell is the principal investigator for a National Science Foundation project focused on integrating technology into science instruction and a state-level Mathematics Science Partnership professional development project partnering science teachers and scientists in curriculum development. He has published in *International Journal of Science Education*, *Journal of Science Teacher Education*, and *Journal of Science Education and Technology*, among others.

Douglas B. Clark is an Associate Professor of Science Education at Vanderbilt University. Clark completed his Doctoral and Postdoctoral work at UC Berkeley and his Master's at Stanford. His research analyzes students' science learning processes in technology-enhanced environments, simulations, and digital games with a particular focus on conceptual change, representations, and argumentation in these environments. Clark's current work focuses specifically on digital games to support students' understanding of core science concepts by integrating and overlaying popular game dynamics with (a) formal science concepts and representations and (b) supports for engaging students in critique and argumentation about the underlying science concepts.

Renee Clary is co-founder of EarthScholars Research Group, which seeks to improve the public's scientific literacy through the optimization of interdisciplinary geological-biological instruction. The EarthScholars group conducts research in multiple educational settings, including informal, traditional, and online environments. Renee Clary is an Assistant Professor of Geology in the Department of Geosciences, College of Arts and Sciences, at Mississippi State University. She also serves as the Director of the Dunn-Seiler Museum on Mississippi State University's campus. She was elected fellow of the Geological Society of London in 2006, and received the Mississippi Science Teachers Association Outstanding College Science Teacher Award in 2011. She has authored 35 research articles in journals, 15 chapters within compendia, and more than 50 publications in electronic media and proceedings.

Danielle Dani holds a B.S. in Biology and a M.S. in Biology. She received her Ed.D. in Curriculum and Instruction from the University of Cincinnati. Dr. Dani teaches graduate and undergraduate courses in science education and teacher education. Her major research interests include the beliefs and practices of teachers engaged in teaching culturally responsive, place-based science through the methods and perspectives of inquiry and problem solving.

Hasan Deniz is an Assistant Professor of Science Education at University of Nevada Las Vegas. He teaches undergraduate, Masters, and Doctoral level courses in the Science Education Program at University of Nevada Las Vegas. His research agenda includes epistemological beliefs in science and evolution education.

Eric Frauman is an Associate Professor at ASU and also teaches in Health, Leisure, and Exercise Science. His teaching specialties include outdoor recreation resource management, nature-based tourism, and risk management. His interests are in non-formal education and expertise in evaluation and research methods.

Lisa Gross is an Assistant Professor in Curriculum and Instruction at Appalachian State University in Boone, North Carolina. Her research interests include the identity development of elementary preservice educators and candidates' comfort in teaching science content. At present, she is exploring how the early socialization experiences of elementary teaching candidates influence their attitudes toward teaching science-related concepts in the outdoors.

Tina Grotzer is an Associate Professor at the Harvard Graduate School of Education, a Principal Investigator at Harvard Project Zero, and a faculty member at the Center for Health and the Global Environment at Harvard Medical School. She directs the Understandings of Consequence Project, funded by the National Science Foundation. She is a cognitive scientist whose work considers how people reason about causal complexity. She studies the kinds of default assumptions that people typically make when learning new complex scientific information and how to best frame scientific research for public understanding. She was recently awarded an NSF Career award.

Brian Hand, Ph.D., is Professor of Science Education at the University of Iowa. His research focuses on two major areas. The first is on how we can use language as a learning tool to improve students' understanding of science. The second area of research is the development of scientific argument through the use of the Science Writing Heuristic (SWH). This research is aimed at helping students learn about and use science argument to construct science knowledge. To support these efforts, in the last five years he has received funding from NSF and both Federal and State Departments of Education.

Rola Khishfe is an Assistant Professor at the American University of Beirut (AUB). She received the Ph.D. in Science Education from Illinois Institute of Technology; M.A. in Science Education, Certification in Teaching Secondary Science, and B.S. in Biology from AUB. Prior to joining AUB as faculty, she was an Assistant Professor at Loyola University, Chicago, and has also had teaching experience at the elementary, middle, and high school levels. She has served on the editorial and review boards of

About the Contributors

several journals. She has published in prestigious science education journals such as Journal *of Research in Science Teaching* and *International Journal of Science Education*. Dr. Khishfe's research interests focus on the teaching and learning about nature of science, argumentation, and decision making in relation to socioscientific issues.

J. Joy James teaches in the Health, Leisure, and Exercise Department at Appalachian State University. She has spent 10+ years teaching in both the non-formal and outdoor education fields before coming to academia. Her research interests include outdoor education, environmental interpretation, and management of outdoor recreation.

Syh-Jong Jang is a Professor at the Graduate School of Education and Center for Teacher Education, Chung Yuan Christian University in Taiwan. He received his PhD in Science Education from the University of Texas at Austin. His expertise is PCK and TPACK, Educational Technology in Teacher Education and Innovative Science Teaching. He have published many articles in distinguished Journals and also served as a reviewer in *Computers & Education*, *Educational Researcher*, *Higher Education*, and *International Journal of Science Education*.

Angela M. Kelly is Assistant Professor of Physics and Astronomy at Stony Brook University, and serves on the Doctoral faculty of the Center for Science and Mathematics Education (CESAME). She earned her Ph.D. in Science Education from Columbia University. Her research interests include inequities in secondary physics access, science teacher recruitment and retention, and pedagogical content knowledge in the physical sciences. She has recently published articles in *American Journal of Physics, School Science & Mathematics, The New Educator Journal, Journal of Curriculum & Instruction, The Physics Teacher*, and *Science Educator*. In 2010, she received the Provost's Faculty Recognition Award for Excellence in Scholarship and Research from Lehman College, City University of New York.

Edward Lyon is an Assistant Professor at Mary Lou Fulton Teachers College at the Arizona State University. A former high school science teacher, Dr. Lyon has been actively involved in classroom-based research projects that explore K-12 science teaching and assessing for linguistically diverse students. His own research aims to better understand how science teachers assess student learning and become prepared to assess in linguistically diverse classrooms. Mr. Lyon has been awarded a University of California All Campus Consortium on Research for Diversity (UC/ACCORD) dissertation fellowship to support his dissertation research that explores the changes in preservice science teachers' assessment expertise.

Jeff C. Marshall's work and expertise has led to national recognition in teaching and research. Jeff serves as Director for the Inquiry in Motion Institute and Co-Director for the Center of Excellence for Inquiry in Mathematics and Science at Clemson University. Further, Jeff earned the Presidential Award of Excellence for Mathematics and Science Teaching; he is Nationally Board Certified in AYA Science, he continues to write and present work on inquiry teaching and learning in science education, and he consults regularly with universities and school districts across the nation.

Mario Martinez-Garza is a pre-Doctoral fellow of Learning Sciences and Learning Environment Design at Vanderbilt University. His main areas of interest are investigating the potential of play as a vehicle for learning through cognitive perspectives, and applications of theory-based design principles to support learning through game environments of all kinds. He holds a Master's degree in Education and has served as a middle-school math and science teacher, and as a game designer for several commercial and educational games companies.

Drew Nielson is a Physics Teacher at Logan High School, Logan, Utah. He has been teaching AP Physics, General Physics, and Chemistry for 15 years. While teaching science is his primary focus, he has collaborated with Dr. Todd Campbell to develop new techniques and methods of teaching physics. He is currently developing ways that Model-Based Inquiry can be used effectively in physics courses. He has published in *The Science Teacher*, *Journal of Science Education and Technology*, and *Science Activities*. In addition, he was a presenter at the Association for Science Teacher Education (ASTE) 2011 International Conference.

Lori Norton-Meier, PhD., is an Associate Professor of Literacy Education at the University of Louisville. Her current research projects include The Science Literacy Project, a series of studies on inquiry-based science and literacy, and an ethnographic lifespan study of play and literacy.

Phil Seok Oh is an Associate Professor at Gyeongin National University of Education, Korea. His research focus is on scientific reasoning and discourse in the contexts of abductive inquiry of earth science and modeling. He, as a science teacher educator, has collaborated extensively with pre- and in-service science teachers, and produced science curriculum materials and relevant research reports. He was a visiting research professor at Utah State University, supported by the National Research Foundation of Korea. He has published in *Science Education*, *International Journal of Science Education*, and *Journal of Korean Association for Science Education*, among others.

Erin Peters-Burton is an Assistant Professor of Educational Psychology and Science Education at George Mason University. Her research agenda focuses on how self-regulated learning processes help students who feel excluded in science classes become more aware of the scientific enterprise and how scientific knowledge is generated. Her fifteen years of experience as a secondary science teacher informs her research on enhancing student scientific epistemologies. She has published extensively on self-regulated learning processes, student-centered learning, and the nature of science. Dr. Peters-Burton has won national awards including Albert Einstein Distinguished Educator Fellow and Virginia University Science Educator of the Year.

Sara Salloum is an Assistant Professor of Science and Adolescent Urban Education at Long Island University – Brooklyn. She completed a Masters in Science Education at the American University of Beirut and Ph.D. at the University of Illinois at Urbana-Champaign. Prior to her Ph.D., Dr. Salloum was a science teacher in Lebanon. She currently teaches secondary school curriculum and methods and teacher-conducted inquiry courses. Dr. Salloum's research is on teacher knowledge and its development towards equitable inquiry-based teaching. Specifically, she studies the role of collaborative practitioner inquiry in supporting new teachers' pedagogic growth along different domains: academic and practical-moral knowledge, and the craft of teaching.

About the Contributors

Pratim Sengupta is a learning scientist and a physicist whose interests are located along two strands of learning sciences research: design and cognition. At Vanderbilt, he directs the Mind, Matter, and Media Lab. He is particularly interested in studying how new forms of computational representational systems (e.g., multi-agent-based modeling, tangible programming, etc.) can lead to rethinking how knowledge is represented in various scientific domains (with a particular emphasis on physics), and the implications for learnability of those domains.

Meng-Fang Tsai received her Bachelor in Food Science from Tunghai University, Taiwan, Master in Secondary Education from Indiana University, South Bend, USA, and Ph.D. in Educational Psychology from Purdue University, West Lafayette, USA. Her research interests focus on students' cognitive engagement, self-regulated learning, and the influence of these two theoretical concepts on students' acquisition and application of scientific literacy as well as instructional practices involved in the process. She currently expands her research to science teachers' development of PCK and TPACK. She is also interested in the employment of quantitative content analysis in the research of science education.

M. Shane Tutwiler holds a Master of Education degree with distinction from the Harvard Graduate School of Education, where he is currently a Doctoral candidate in Human Development and Education. He is an experienced science teacher with a background in nuclear engineering. His research interests focus on the use of technology and cognitive science to enhance science education.

James Wandersee is co-founder of the EarthScholars Research Group, which focuses on designing innovative visual approaches that integrate and improve botanical and geological instruction, with the aim of increasing public understanding of science. He is the W. H. LeBlanc Alumni Association Professor (Biology Education) in the Department of Educational Theory, Policy, and Practice in Louisiana State University's College of Education. He was elected a lifetime fellow of the Linnean Society of London and of the American Association for the Advancement of Science—Biological Sciences section. He received the 2007 Charles Edwin Bessy Medal from the Botanical Society of America "for inspiring students and the public to explore and appreciate the wonders of botany." His books, articles, and professional presentations span more than 40 years and 15 countries; they have been translated into seven languages.

Morgan B. Yarker is a PhD candidate in the Science Education program at the University of Iowa. She has a B.S. in Meteorology and a M.S. in Atmospheric Science, which has aided her work in using computer-based weather forecast models. Her current research in science education looks at dialogic interactions among students as they work towards understanding science models.

Index

A

academic engagement 65, 82-83
active learning 167-178, 181, 200, 284
Agent-Based Model (or ABM) 93
argumentation 22, 85-94, 96, 98-105, 107, 121-122, 125, 225-228, 238-245, 256, 263, 285
Argument-Driven Inquiry (ADI) approach 89-91, 99-100
Aristotelian concept 27-28, 31
Assessing Scientific Inquiry and Leadership Skills (AScILS) Project 254-255, 257-258, 260-261
assessment coherence 247, 252, 255
assessment equity 247, 252-253, 258
assessment use 247, 252, 257-258

B

belief 8, 12, 14, 19-20, 24-25, 28-30, 35-37, 42, 48-50, 54, 56-58, 61-62, 81-83, 110, 148-153, 156-158, 187, 200, 205, 208, 210-211, 219, 221-222, 267-268, 304-306
best practice 163, 290-291, 303
big ideas 87, 175, 180, 225, 229, 236
biodiversity 52-53, 177
Biological Sciences Curriculum Study (BSCS) 4, 21, 303
Bronx Institute Program 185, 189, 191, 194-195, 198, 200-201

C

causal interactions 127
classroom activity 82, 148, 150, 156-157, 168, 227, 239
classroom assessment 247-248, 252-253, 255, 259-263
classroom community 33, 226, 228-229, 231, 238-239
classroom interaction 7, 73, 228-232, 238, 296, 298
classroom practice 28, 41, 50, 112, 120, 148, 151-153, 155, 157, 159, 163, 223, 226-228, 247-248, 251, 253, 259-261, 264, 278, 292, 295, 302
cognitive engagement 19, 64-83
computer technology 150-151, 154, 251
conceptual domain 53
conceptual ecology 52, 61
constructivism 21, 27, 29-30, 48-50, 54, 88, 92, 105, 166-168, 170, 175, 177, 181-182, 184, 190-191, 193, 199, 206, 241-242, 291-293, 300, 302-303
content area 42, 73, 169, 175, 210-211, 249-250
conversational pattern 229, 231, 236
core academic content 269, 272-273, 276, 278-279, 281, 283
cultural border crossing 59-60
curriculum 3-4, 20, 22, 25, 30, 33, 47, 53, 55, 57-59, 64, 87, 92, 94, 96-98, 100, 103, 106, 111, 117, 121, 129, 131, 139, 141, 153, 155-156, 158-160, 167, 170, 174, 176, 189, 193-196, 202, 204, 206, 211, 218, 226, 229, 243, 248-249, 251, 255-256, 262-263, 269-286, 290, 294-297, 299-300, 302
cyclic modeling 106-107, 110-112, 116, 119-120

D

digital games 147-153, 155-157, 160
digital native 152, 157, 162, 270, 286
domain 5-6, 14, 16-17, 36, 41, 52-54, 56-58, 60, 66, 73, 80-82, 92, 100, 129-130, 149, 177, 199, 209, 226, 249, 258, 296, 301

E

EarthScholars Research Group 165-168, 175, 177
educational games 148-150, 152-157, 161
educational research 19-20, 24-25, 28-29, 41, 46-49, 61-62, 81-83, 101, 103, 126, 144, 149, 155, 167, 182, 203, 220-221, 240-242, 302, 304-305

Index

education standards 49, 53, 55-56, 62, 103, 106, 169, 172, 182, 238, 243, 249, 262, 286, 292, 305
Electronic Quality of Inquiry Protocol (EQUIP) 243, 290, 292-302, 304-305
emancipatory knowledge 267-268
emotional engagement 65
environmental education 205, 212-213, 216-223
Environmental Socialization (ES) 205, 218-219
epistemic domain 53-54
evaluative modeling 106-107, 110-111, 119, 122
evidence-based idea 228-233, 236
evolutionary theory 24, 52-63, 169, 171, 173, 175, 177, 249
experimental modeling 106-107, 110-111, 114-116, 119
Expert Science Teaching Educational Evaluation Model (ESTEEM) 292, 303
exploratory modeling 106-107, 110, 119-120
expressive modeling 106-107, 110-114, 119, 121-122

F

field-based learning 205, 212-213, 216-219
Force Concept Inventory (FCI) 197
formal assessment 252-253
framework for the analysis of education programs (FAEP) 266, 268-269

G

game-based learning 147-156, 158-161
geobiology 165, 167, 169-171, 175-180

H

High-Level Implementation (HLI) 238-240
History and Nature of Science (HNOS) 272-273
human constructivism 167, 170, 177
hypotheses 78-79, 87, 102, 110, 114-116, 128, 137, 157, 195, 261, 277

I

independent learner 1
informal learning 166, 175-176, 178-179, 181, 184-185, 195, 209
information processing theory 67
inquiry-based instruction 181, 251, 290-302, 304-305

inquiry learning 110, 120, 125, 143, 165, 168, 170, 174, 300
instructional discourse 64-65, 72-74, 76, 78, 80
integrated science learning 165, 170
intelligent design 24, 54, 59, 61, 110
interactive 12, 80, 89, 104-105, 127, 130-131, 133, 136, 139-143, 145, 159, 161, 239
interdisciplinary science 165, 168-170, 175-176, 178
internal consistency 295

K

K-12 Science Education 53, 62, 106-107, 109, 124, 167, 227-228, 243, 247, 249, 262, 291, 305
knowledge construction 64-65, 67, 149, 242

L

learning environment 8, 10, 13, 66-68, 72, 74, 76, 80, 92, 94, 96, 99-100, 102, 104-105, 145, 152, 161, 165, 175-176, 179, 181, 185, 195, 218, 267, 303
learning sequence 106
learning tool 149-150
legal domain 53, 58
lifeworld 206, 208
Low-Level Implementation (LLI) 238-239

M

Medium-Level Implementation (MLI) 238-239
metacognition 7, 17, 23, 65-70, 72, 80-83, 167, 181, 250, 259, 263, 306
middle school 34, 38, 44, 70-71, 82, 103, 105, 165-166, 168-173, 184-185, 187-190, 200-203, 221, 265, 270, 276, 278, 304
Model-Based Inquiry (MBI) 106, 109, 111-112, 114-117, 119-120
Model Curriculum 271-275, 284, 286
modeling 12, 41, 49, 85-89, 91-101, 103-117, 119-125, 181, 201, 250, 263, 296, 304, 306

N

Nature of Science (NOS) 1-25, 28, 33, 47-48, 52, 54-56, 60-62, 90, 101, 122, 168-169, 171, 177, 180, 269, 272, 278
non-rural teacher 205, 213-218

O

Organization for Economic Co-operation and Development (OECD) 250, 262, 266-269, 286

P

Partnership for 21st Century Skills (P21) 266-268, 286

pedagogy 29, 31, 36, 43, 48-49, 54, 82, 103, 106, 110, 121, 127, 156, 165, 211, 243

Personal and Social Perspectives (PSP) 272-273

phronesis 27, 29, 31-35, 42-43, 46-47, 49

physical science 4, 20, 34, 38, 49, 125, 166, 169, 172, 184-192, 197, 199-201, 249, 251, 269-270, 272, 274, 283, 297

Physical Science Study Curriculum (PSSC) 4

physical setting 189, 192, 204, 275-276

Piaget 29, 65-67, 81-82, 101, 206, 209, 222

place-identity 207

practical hermeneutics 27, 29

practical-moral knowledge 27-29, 31, 35-38, 41-43, 46

practical wisdom 27, 29, 31-36, 38, 41-43, 46-47, 50

problem-based-learning 129

process-product paradigm 150

Q

quantitative content analysis 64, 68, 72, 74-77, 79-81, 83

question complexity 232

R

reflective approach 1-2, 4, 6, 9, 18

Reform Teaching Observation Protocol (RTOP) 224-225, 230, 244, 292-293, 306

religion 19, 52, 54, 62

religious domain 53, 57

role of teachers 50, 147, 156

rural teacher 205, 213-214, 216-218

S

scaffolded concepts 195

science classroom 3, 38, 49-50, 72, 82, 93, 104-112, 114, 116, 120-122, 126, 130, 166-167, 169-174, 176-179, 181, 187, 203, 224-227, 229, 240, 243-244, 247-248, 250, 252-254, 259-261, 290-294, 301, 303-305

science education 1-2, 8, 13-14, 18-25, 27-31, 34-36, 41, 46-56, 60-63, 75, 83, 85-86, 100-107, 109, 111, 120-128, 130, 142-143, 145-146, 160-161, 167, 169, 172, 177, 179-182, 184-185, 188, 191, 201-204, 219-223, 227-228, 238, 240-245, 247-255, 260-262, 265-266, 269, 278, 285-286, 291-293, 295, 302-306

science education community 54, 120, 247, 285

science instruction 3, 16, 18, 20-22, 24, 54, 57, 61, 101, 107-109, 111, 120, 166-167, 169, 178, 185, 189, 252, 275, 291-292, 303

science lab 17, 89, 188

science learning 13, 45-47, 64-68, 70-72, 75-81, 101, 103, 107-108, 130, 144-145, 157, 160, 165-166, 170, 176, 178-179, 195-196, 200, 204, 221, 251-253, 262, 291

science literacy 60, 175, 180, 225, 227, 245, 249, 261, 285

Science Management Observation Protocol (SMOP) 292

Science Teacher Inquiry Rubric (STIR) 292, 302

Science Writing Heuristic (SWH) approach 224, 227, 229, 231, 235

scientific achievement 188

scientific community 7, 18, 56, 58, 86, 88, 99, 170-171, 186, 201, 240

scientific inquiry 15-17, 21, 24, 41, 45, 62, 85-86, 99-100, 104, 107, 109, 111, 125, 145, 161, 165, 167, 178, 180, 182, 189, 225, 244, 249, 251, 254, 261, 263, 269, 272, 276-277, 282-283, 293

scientific investigation 89, 166, 171, 178, 272, 274

scientific knowledge 2-4, 9, 13, 17, 19, 23, 31-32, 41, 54, 71, 78, 91, 101, 113, 168, 171, 241, 244, 248, 250, 278, 280-281

scientific literacy 2, 18, 52, 64, 86, 91, 177-178, 240-241, 248-250, 253, 255, 260-262, 266, 275, 285

scientific phenomena 15, 87, 89, 110, 112, 119, 131

scientific practice 99, 103, 107, 109-110, 116, 119, 122, 199, 241, 249-252, 255, 257

scientific reasoning 102, 123, 140, 193, 225, 242, 247-248, 254-260

scientific theory 6, 55, 86-87, 121

scientific thinking 1-3, 16, 102, 242

Secondary Science Teaching Analysis Matrix (STAM) 292

self-regulated learning 1-2, 6-9, 13-14, 16, 18, 20, 23, 25, 83, 104

SimCity 132-140, 142-143, 146, 149

simulation 92-95, 100, 105, 110, 123, 127-133, 135-146, 149, 156-157, 160-161, 163, 254-259

socio-cultural domain 53, 56

Index

sociocultural perspective 27, 29-30, 48, 242
standards 3, 8, 11, 14, 18, 22, 49, 53, 55-56, 62, 86-87, 100, 103, 106, 109, 153, 155, 162, 168-170, 172, 178, 182, 189-193, 200-201, 204, 238, 243, 249, 256, 259, 261-262, 265-266, 268, 270-277, 279-286, 288, 291-292, 294, 301, 303-305
STEM (Science, Technology, Engineering, and Mathematics) 162, 184-189, 200-201, 204
student achievement 39, 42, 197, 290-293, 300-302, 305
student-centered learning 168

T

teacher knowledge 27-31, 34-36, 42, 46, 48-49, 82
teacher performance 290, 293, 300-301
teacher preference 205
Test on Understanding Science (TOUS) 4
theoretical knowledge 28, 31, 36, 43, 51

Tool for Analyzing Science Standards and Curricula (TASSC) 265-266, 268-272, 274, 276, 278-285, 288
twenty-first century skills 160, 265, 284

V

Virtual Environments (VEs) 131
virtual simulation 128-130, 133, 139
virtual world 130-131, 140, 142
Vygotsky 29, 67, 83, 206, 209, 222, 228, 244-245, 296, 306

W

whole-class 224-225, 228-229, 231, 235, 238-239

Y

young learner 20, 29, 64, 66-68, 72, 74-75, 79-80, 147

CPSIA information can be obtained at www.ICGtesting.com
Printed in the USA
LVOW02*0035150214

373846LV00010B/38/P

9 781466 628090